Full text and further information: www.ms-journal.de

Macromolecular Symposia
is published 12 times a year

Annual subscription rate 2013

Europe	Euro	2,529
Switzerland	Sfr	4,003
All other areas	US$	3,330
		electronic only

All Wiley-VCH prices are exclusive of VAT.
Prices are subject to change.
Online ISSN: 1521 – 3900

Copyright Permission:
Fax: +49 (0) 62 01/6 06-332,
E-mail: rights@wiley-vch.de

Postage and handling charges included. All Wiley-VCH prices are exclusive of VAT. Prices are subject to change.
Contact: www.wileycustomerhelp.com

Cancellation of subscriptions: The publishers must be notified not later than three months before the end of the calendar year.
Order through your bookseller or directly at the publisher:
www.wileycustomerhelp.com

Typesetting: Thomson Digital (India) Ltd., India

Macromolecular Symposia Vol. 333

Polymer Reaction Engineering – 11th International Workshop

Selected Contributions from:
Polymer Reaction Engineering – 11th
International Workshop
Hamburg, Germany,
May 21 – 24, 2013

Symposium Editors:
Hans-Ulrich Moritz, Universität Hamburg,
Germany
Werner Pauer, Universität Hamburg,
Germany

WILEY-VCH

Macromolecular Symposia: Vol. 333

Articles published on the web will appear through:

wileyonlinelibrary.com

Cover: The 11th International Workshop "Polymer Reaction Engineering" was held in Hamburg, Germany, in May 21–24, 2013. The cover is selected from the article by Lueth and co-authors.

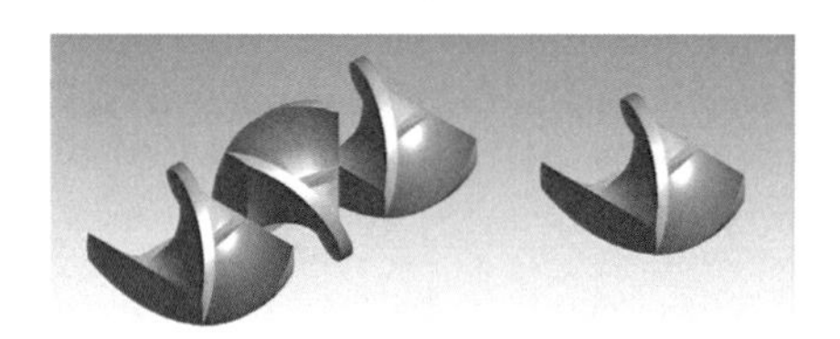

Polymer Reaction Engineering – 11th International Workshop

Hamburg, Germany

This volume represents the proceedings of the 11th International Workshop on "Polymer Reaction Engineering" held at the University of Hamburg, Germany, in cooperation with DECHEMA (Gesellschaft für Chemische Technik und Biotechnologie e.V., Frankfurt a. M.), May 21–24, 2013. Since three decades the workshop series and the corresponding proceedings successfully accompany the chemical community promoting the update, exchange and discussion of new findings in the field of polymer reaction engineering between experts from academia and industry alike. The 11th International Workshop on "Polymer Reaction Engineering" was a mile stone in the long tradition of this conference. Never before such a huge number of contributions were presented. The workshop was switched from autumn to spring and a was extended in time.

122 contributions were presented in the form of keynote lectures, short lectures or posters by numerous experts from 28 countries of all parts of the world. With 191 participants the 10th International Workshop was again one of the largest Meetings of this kind worldwide. As long as this workshop exists the organizers fund young scientists to take part in the conference. This year a quarter of the participants were young PhD students, and thanks the generous support of WILEY-VCH Publisher, awards for the two most outstanding posters could be granted. Our congratulation to Callista Preusser and Fabian Lüth.

After all, one third of the participants from industry document special relevance attached to the conference. A significant part of the contributions are compiled in the present proceedings. Excellent papers of emerging new concepts and promising developments, technologies from neighboring fields of chemical engineering and industrial solutions in process and product design are widely discussed from a superior perspective. New catalysts and catalytic polymerization processes, controlled radical polymerization, high-throughput and micro technologies, new reactor and process design and intensified processes are included. Polymer thermodynamics, process analytics, modeling and process control enable development and application of these technologies. Distinctive spotlights were put on contributions of new environmentally benign polymerization processes in consideration of economical needs. Furthermore nanotechnologies perform new and extended characteristics to polymer materials.

We thank all the contributors for providing their manuscripts and thankfully acknowledge WILEY-VCH Publisher for publishing this volume, thus making the proceedings available not only to the conference attendees.

We are indebted to DECHEMA for organizing this workshop as an outstanding international conference and especially encouraging young scientists to join the scientific community.

Finally we like to thank the members of the Scientific Advisory Committee for helping to establish the scientific program of high quality and for careful revision of the contributions.

H.-U. Moritz
W. Pauer

Detailed Investigations into Radical Polymerization Kinetics by Highly Time-Resolved SP–PLP–EPR

*Hendrik Kattner, Michael Buback**

Summary: Instantaneous initiation of radical polymerization by applying an intense laser pulse on a monomer-photoinitiator solution in conjunction with time-resolved detection of radical concentration provides access to a detailed picture of polymerization kinetics. Examples are given of measurement of chain-length-dependent termination, of intramolecular chain transfer as well as of reversible-deactivation radical polymerization, namely of RAFT and ATRP. The enormous potential of the method relates to the fact that the kinetics of the major acting species, i.e., of the radicals, may be directly monitored with high time resolution.

Keywords: ESR/EPR; kinetics (polym.); laser-induced polymerizations; radical polymerization; termination rate coefficient

Introduction

Control of polymerization processes, e.g., of monomer-to-polymer conversion, polymer microstructure and heat production, on a technical scale requires precise knowledge of the kinetics and mechanism of the major reaction steps. Thus for radical polymerization, information about initiation, propagation, termination and transfer kinetics is required. Despite the eminent relevance of radical polymerization for producing huge amounts of polymer, the underlying detailed kinetics is far from being fully understood. The situation has enormously improved upon the advent of pulsed-laser polymerization (PLP) in conjunction with size-exclusion chromatography (PLP–SEC) which method allows for reliably measuring propagation rate coefficients, k_p. Single-pulsed PLP (SP–PLP) in conjunction with near-infrared spectroscopic detection (SP–PLP–NIR) of monomer conversion induced by a single laser pulse, allows for measuring the ratio k_p/k_t which, together with k_p from PLP–SEC, affords for determination of the termination rate coefficient, k_t, actually of the mean value, $\langle k_t \rangle$, averaged over the chain lengths of terminating radicals.[1–3] Both techniques take advantage of the instantaneous production of an intense burst of primary radicals by photo-dissociation of a suitable initiator. They are however indirect in that overall quantities are analyzed, such as molar mass distribution in PLP–SEC and monomer conversion in SP–PLP–NIR. It appeared rewarding to directly monitor radicals, which are the relevant reactive components of radical polymerization. This strategy gave rise to the development of SP–PLP–EPR from 2004 on with major contributions having been provided by Barth.[4] The beauty and the power of the method consist of the combination of well-defined instantaneous initiation of radical polymerization by a laser pulse with sensitive detection of the time evolution of radical concentration to values as low as 10^{-8} mol · L^{-1}. Partly more than one type of radical species may be monitored, which allows for measurement of transformation rates of radical species.[5] Pulsed-laser techniques have in common that the size of growing radicals scales with time t after applying the laser pulse, unless transfer reactions come into play. Therefore, SP–PLP–EPR is perfectly suited for studies into chain-length dependent (CLD) kinetics.[5,6]

Institute for Physical Chemistry, University of Göttingen, Tammannstr. 6, D-37077 Göttingen, Germany
E-mail: mbuback@gwdg.de

Chain-length dependence is particularly important with termination reactions, which are diffusion-controlled processes that may be largely affected by the size of the reacting species. The SP–PLP–EPR technique is unrivalled in detailed measurement of termination kinetics, including systems where, by intramolecular chain-transfer, e.g., by the backbiting reaction, the kinetics of secondary and tertiary radicals may be analyzed for acrylate-type monomers, as both types of radicals exhibit clearly different EPR spectra.[7,8] Reversible-deactivation (controlled) radical polymerizations, such as RAFT[9,10] and ATRP[11] are also systems with more than one type of radicals, which may be studied via SP–PLP–EPR. The present article describes experimental aspects of SP–PLP–EPR and illustrates the potential of the new method with examples from both conventional and reversible-deactivation polymerization and also addresses backbiting[12] and composite-model behavior of CLD termination.[7,13–18]

Experimental Part

Sample Preparation

The monomers under investigation: dibutyl itaconate, vinyl acetate, butyl acrylate, and dodecyl methacrylate were purified by passing through a glass column filled with inhibitor remover. Dissolved oxygen, in solvents and monomers, is removed by several freeze-pump-thaw cycles. Photoinitiators are used as received and are added under an argon atmosphere in a glove box. Samples are filled into EPR tubes sealed with *Parafilm*®. For quantitative measurements, the volumes of all samples have to be the same. In general, large sample volumes are preferred, but in case of highly polar samples, smaller volumes have to be used in order to reduce dielectric loss. Photoinitiator concentration should be minimized to avoid concentration gradients of macroradicals due to the UV absorbance of the photoinitiator. On the other hand, high radical concentrations are required for good signal-to-noise quality. Optimum sample volume and photoinitiator concentrations, which are typically in the millimolar range, are obtained from screening experiments.

Photoinitiators and UV-Initiation

Samples are placed in the cavity (Figure 1) and irradiated by an excimer laser (COMPex 102; XeF, 351 nm; Lambda Physik) either at repetition rates from 5 to 25 Hz under pseudo-stationary conditions or by

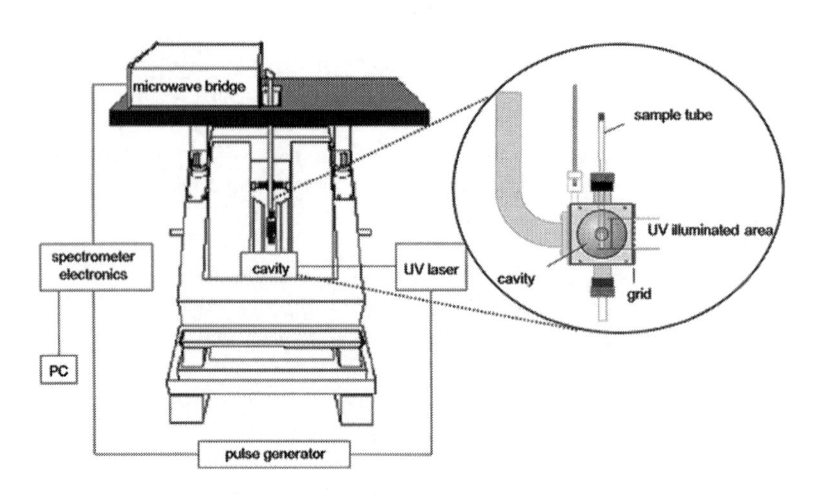

Figure 1.
Scheme of the SP–PLP–EPR setup consisting of an UV laser and an EPR spectrometer with PC. An enlarged view of the cavity is shown on r.h.s. of the figure, which was taken from Ref.[23] Copyright Wiley-VCH Verlag GmbH & Co. KGaA. Reproduced with permission.

single pulses, where the EPR spectrometer and the excimer laser are triggered by a pulse generator (Scientific Instruments 9314). Continuous irradiation may be applied by using a mercury UV lamp (500 W, LAX 1450, Mueller Electronik).

Within the single-pulse experiments, the delay between two successive pulses is selected such that radical concentration from the preceding pulse has completely decayed. These experiments should preferably be carried out using photoinitiators, which decompose into two radical fragments, that both immediately add to monomer.[19] This requirement is perfectly met by α-methyl-4(methylmercapto)-morpholinopropiophenone (MMMP), which was our preferred photoinitiator for acrylate, methacrylate and itaconate polymerizations.[20,21] Photoinitiation in aqueous solution may be carried out with 2-hydroxy-2-methylpropiophenone (*Darocour*®). The absorption of both initiators at the laser wavelength 351 nm is sufficiently high, whereas most monomers exhibit very low absorption at 351 nm and are stable towards radiation of this wavelength.

EPR Setup

EPR spectra and $c_R(t)$-curves were recorded on a Bruker Elexsys E 500 series CW EPR spectrometer (Figure 1) consisting of a microwave bridge containing a microwave source and a detector, a cavity (ER 4122SHQE-LC, Version V1.1, Bruker), which helps to amplify weak signals, a console (spectrometer electronics) for electronic data processing and two tunable magnets.[22] The console is connected to a PC. Temperature was controlled by an EVR 4131 VT unit (Bruker). Temperatures from −195 to +350 °C may be reached by purging the cavity with nitrogen. SP–PLP–EPR measurements were typically performed in the range −40 to +70 °C depending on the melting and boiling points of monomer or of the monomer/solvent system and on the thermal stability of the photoinitiator. In general, low temperatures are preferred due to the higher population of the lower spin levels and the slower rate of radical termination.

For samples with small dielectric loss, standard EPR tubes (ER 221TUB/2, 3, 4 CFQ, Bruker) with inner diameters of 2, 3 or 4 mm have been used. Highly polar samples require special flat cells (WG-808-S-Q, Suprasil TE102 Aqueous Cell, VT, 100 μL sample volume, Rototec-Spintec). Adding polymer to the system prior to PLP may enhance signal-to-noise ratio.[12] The continuous wave (CW) EPR spectrometer varies the magnetic field B_x at constant microwave frequency (9.85 GHz (X-band) for the empty cavity). The first derivative of the absorption spectrum is recorded and subsequently the double integral of the dispersion spectrum $\iint I(B_x)$ is calculated, which is proportional to the number of spin carriers.[22] $I(B_x)$ stands for the dispersion intensity at constant magnetic field B_x, which quantity is recorded during the single-pulse experiment as a function of time with a resolution down to microseconds.

SP–PLP–EPR Experiment

A typical single-pulse experiment starts by recording an EPR spectrum under periodic laser pulsing, i.e., under pseudo-stationary PLP conditions to determine the magnetic field position(s), at which the EPR intensity should be traced. A short sweep time is selected to prevent significant monomer-to-polymer conversion.[23] The basic principle of an SP–PLP–EPR experiment is illustrated in Figure 2 by an EPR spectrum of dibutyl itaconate (DBI) at 0 °C. The decay of the intensity at the maximum position, $I(B_x)$, is plotted as a function of time t after firing the laser pulse at $t = 0$. The decay has also been recorded at other field positions to check for the accuracy of the measurement.

Quantitative SP–PLP–EPR experiments require the entire sample to be irradiated. Experimental parameters such as modulation amplitude, receiver gain and attenuation are optimized at sufficiently high signal-to-noise ratio.[15,22] Further improvement of signal quality may be reached by accumulating several $I(B_x)$ vs. t curves during one measurement or by co-adding EPR traces

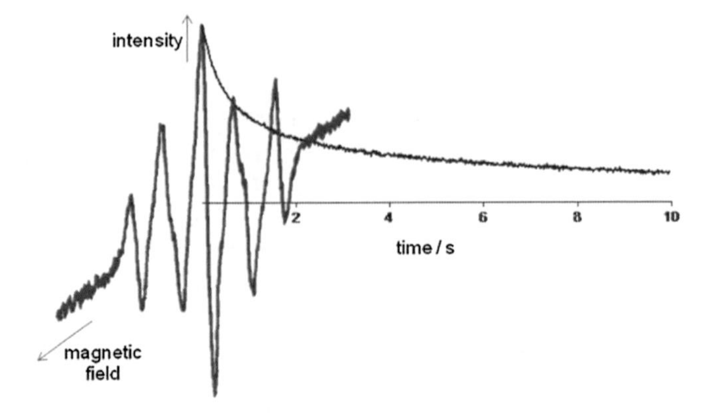

Figure 2.
EPR spectrum of DBI at 0 °C measured under pseudo-stationary conditions using a laser repetition rate of 25 Hz (l.h.s.); decay of the intensity $I(B_x)$ after single-pulse initiation (r.h.s.). The figure is from Ref.[23] Copyright Wiley-VCH Verlag GmbH & Co. KGaA. Reproduced with permission.

from a series of independent experiments.[22,23] Monomer-to-polymer conversion is additionally checked by gravimetry or by UV–vis spectroscopy during and after each SP–PLP–EPR experiment.

Potential Artifacts of SP–PLP–EPR Experiments

An experimental artifact may result from a microwave signal induced by laser pulsing and from overlap of this signal with the EPR contour of radical species. In what follows, only this type of artifact will be addressed. Details about other, less important artifacts are detailed in Ref.[19]

The "laser artifact" may result from the intense laser pulse interfering with the microwave detection unit and inducing a signal response of the microwave bridge. An example is given in Figure 3. Characteristics of the "laser artifact" are a sharp signal with negative and positive wings. Due to the statistical nature of this artifact, its impact is minimized by co-adding data from several EPR experiments. Moreover, the "laser artifact" may be largely reduced by selecting low laser energy and low time resolution.

Calibration of the EPR Setup

Calibration of SP–PLP–EPR experiments aims at correlating the measured time evolution of an EPR signal to the time-dependent concentration of the associated radical species. It is known that the double integral of an EPR signal, $\iint I(B_x)$, is proportional to absolute radical concentration, c_R. The calibration procedure has been detailed in Ref.[15,19,23] and thus will only briefly be addressed here. Three calibration steps need to be carried out: (1) EPR spectra of a stable radical (reference) species are recorded on solutions of known concentration and the resulting double integrals are correlated with the concentration of the reference compound. Typically, 2,2,6,6-tetramethyl-1-

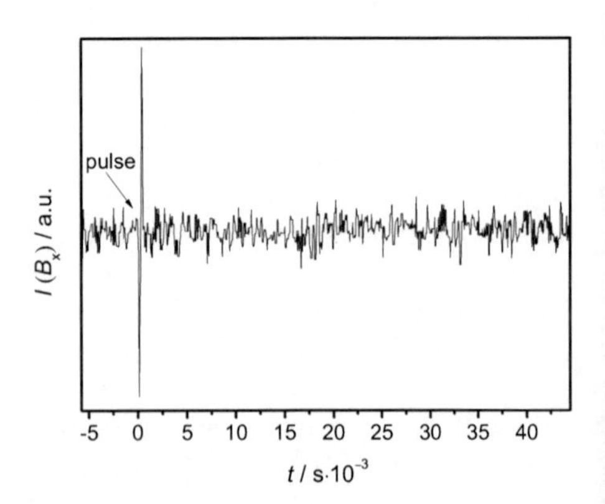

Figure 3.
Typical "laser artifact" signal shortly after applying the laser pulse at $t = 0$. The EPR spectrum has been recorded with a time resolution of 10 μs at −65 °C on pure vinyl acetate (VAc) without photoinitiator being present, i.e., no primary radicals are produced by photoinititor decomposition.

piperidinyloxyl (TEMPO) is used as the calibration standard for organic systems and 4-hydroxy-2,2,6,6-tetramethylpiperidine-1-oxyl (TEMPOL) for aqueous systems. (2) The double integrals of the EPR spectra measured on the polymerization system under investigation at pseudo-stationary PLP conditions, $\iint I(B_x)$, are correlated with the dispersion intensity at the magnetic field position $I(B_x)$ used for highly time-resolved detection after firing the laser pulse at $t = 0$ during an SP–PLP–EPR experiment. (3) Finally, the signal intensity of the detection unit used for the highly instationary SP–PLP–EPR experiments has to be calibrated against the EPR detection unit used within the EPR measurements under quasi-stationary conditions. The correlations between c_R, $\iint I(B_x)$ and $I(B_x)$ are quantified by the system-specific coefficients h_1, h_2 and h_3 (see Scheme 1) with the indices referring to the numbering of steps given above.

Calibration step (1) has to be performed under identical conditions, e.g., of solvent type, temperature, concentration, modulation amplitude, receiver gain and sample volume, as used in the single-pulse experiment. Calibration coefficients h_1 and h_2 are deduced from the slope of linear fits of c_R(TEMPO(L)) plotted against $\iint I(B_x)$ (Figure 4) and of $\iint I(B_x)$ plotted against $I(B_x)$, respectively.

The quantity $c_R(t)$, which constitutes the final SP –PLP–EPR signal, is obtained via Eq. (1):

$$c_R(t) = h_1 \cdot h_2 \cdot h_3 \cdot I(B_X, t) \quad (1)$$

Selected Results from SP–PLP–EPR

Investigations into Chain-Length Dependent Termination

Pulse-laser irradiation of a suitable photoinitiator allows for almost instantaneous

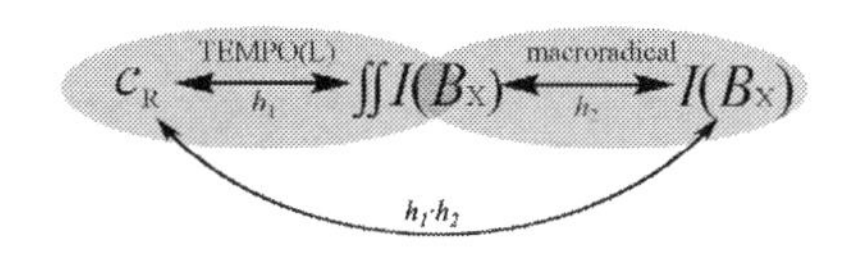

Scheme 1.
Calibration for SP–PLP–EPR experiments (see text; calibration step 3 is not included in the scheme).

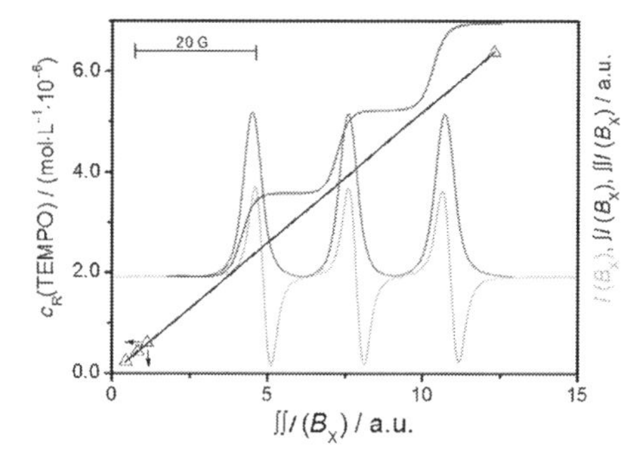

Figure 4.
Determination of the calibration coefficient h_1 from the slope to the straight line which correlates the double integral of the EPR spectra taken at −47 °C with the associated TEMPO concentration. As an example, the TEMPO spectrum for a concentration of $6.08 \cdot 10^{-7}$ mol · L^{-1} is shown together with the associated data obtained by single and double integration.

initiation of chain growth. Kinetic chain length i scales with time t after applying the laser pulse (Eq. (2)) resulting in a close-to-monodisperse (Poissonian) chain-length distribution of growing radicals unless chain-transfer processes come into play. As termination occurs between two radicals of more or less identical size,[6] the chain-length dependence (CLD) of k_t may be investigated via the correlation between time (after pulsing) and chain length.

$$i = k_p \cdot c_M \cdot t \quad (2)$$

where k_p is the propagation rate coefficient and c_M is monomer concentration. The validity of Eq. (2) depends on the absence of transfer reactions and on initiator fragment–to–monomer addition being fast compared to the rate of a single propagation step. Co-addition of c_R vs. t traces from separate experiments allows for investigation of the CLD of termination at low degrees of monomer-to-polymer conversion. Hence, the CLD of k_t may be separated from viscosity effects on termination, which is a major advantage of SP–PLP–EPR over the RAFT-CLD-T technique.[6]

It is by now generally accepted that the CLD of k_t exhibits "composite-model" behavior (Eq. (3)).[24–27]

$$k_t^{i,i} = k_t^{1,1} \cdot i^{-\alpha_s} \qquad i \le i_c$$
$$k_t^{i,i} = k_t^{1,1} \cdot i_c^{-\alpha_s+\alpha_l} \cdot i^{-\alpha_l} = k_t^0 \cdot i^{-\alpha_l} \qquad i > i_c \tag{3}$$

The rate coefficient $k_t^{i,i}$ represents termination of two radicals, both of chain length i. $k_t^{1,1}$ refers to termination of two radicals of chain length unity and k_t^0 refers to the hypothetical situation of termination of two entangled radicals both of $i=1$. The CLD of $k_t^{i,i}$ is given by the power-law exponents α_s and α_l. In the short-chain regime, i.e., at chain lengths below a crossover value, i_c, the exponential decay of $k_t^{i,i}$ is more pronounced than in the long-chain regime above i_c, as $\alpha_s > \alpha_l$.[6,7,14,27–30] It is assumed that termination of short chains is mostly determined by center-of-mass diffusion, whereas segmental diffusion is dominant with long (entangled) chains. The observed differences between α_s and α_l are consistent with this picture.[31,32] In order to determine the composite-model parameters α_s, α_l, i_c and $k_t^{1,1}$, the rate law for termination (Eq. (4)) has to be integrated (Eq. (5)) and the chain-length averaged termination coefficient, $<k_t>$, to be replaced by Eq. (3) to yield Eq. (6). The time required for one propagation step, t_p, is given by $(k_p \cdot c_M)^{-1}$.

$$\frac{dc_R}{dt} = -2 \cdot \langle k_t \rangle \cdot c_R^2 \tag{4}$$

$$\frac{c_R^0}{c_R(t)} = 1 + 2\langle k_t \rangle \cdot c_R^0 \cdot t \tag{5}$$

$$\frac{c_R^0}{c_R(t)} - 1 = \frac{2 \cdot k_t^{1,1} \cdot c_R^0 \cdot t_p^{\alpha_s}}{1-\alpha_s} \cdot t^{1-\alpha_s} \qquad 1 << i \le i_c$$
$$\frac{c_R^0}{c_R(t)} - 1 = \frac{2 \cdot k_t^0 \cdot c_R^0 \cdot t_p^{\alpha_l}}{1-\alpha_l} \cdot t^{1-\alpha_l} \qquad i >> i_c \tag{6}$$

The experimental data are linearized by plotting $\log(c_R^0/c_R(t) - 1)$ vs. $\log t$. Both α values are obtained from the slope to the straight-line fits at small and large chain lengths, respectively, and i_c is found from the point of intersection of the two straight lines (see Figure 5). From the y-intercept of the fit to the short-chain regime, $k_t^{1,1}$ is determined with c_R^0 being obtained by separate calibration of the EPR setup. As can be seen from Figure 5, linear regression for $i < i_c$ does not fit the experimental data points for very short chains, i.e., below $i < 3$. The reason behind this failure is that Eq. (3) is not valid under these conditions. The more adequate description of chain length for t approaching zero is given by Eq. (7).[33]

$$i = k_p \cdot c_M \cdot t + 1 \tag{7}$$

Implementing this equation into the integrated rate law, in conjunction with Eq. (5), yields:

$$\frac{c_R^0}{c_R(t)} - 1 = \frac{2 \cdot k_t^{1,1} \cdot c_R^0 \cdot \left(\left(k_p \cdot c_M \cdot t + 1 \right)^{1-\alpha_s} - 1 \right)}{k_p \cdot c_M \cdot (1-\alpha_s)} \qquad 1 \le i < i_c \tag{8}$$

$$\frac{c_R^0}{c_R(t)} - 1 = \frac{2 \cdot k_t^{1,1} \cdot c_R^0 \cdot \left((i_c)^{1-\alpha_s} - 1 \right)}{k_p \cdot c_M \cdot (1-\alpha_s)} - \frac{2 \cdot k_t^{1,1} \cdot c_R^0 \cdot (i_c)^{1-\alpha_s}}{k_p \cdot c_M \cdot (1-\alpha_l)} + \frac{2 \cdot k_t^0 \cdot c_R^0 \left(k_p \cdot c_M \cdot t + 1 \right)^{1-\alpha_l}}{k_p \cdot c_M \cdot (1-\alpha_l)} \qquad i_c \le i \tag{9}$$

Determination of composite-model parameters via (Eq. (8) and (9)) is performed by an iterative procedure. Chain-

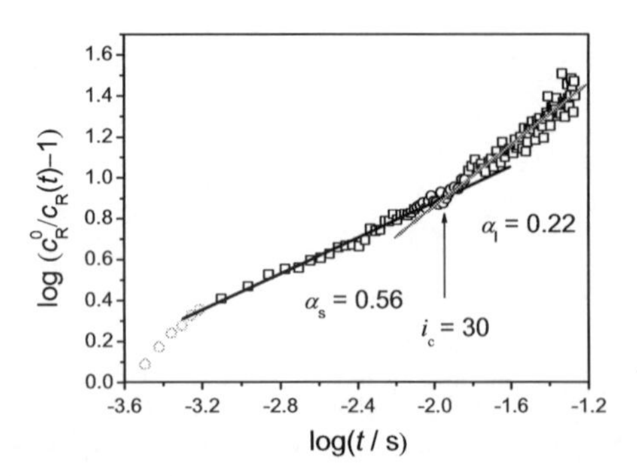

Figure 5.
Linear fitting of experimental data for n-butyl acrylate (1.52 M in toluene) at −40 °C via Eq. (11). An α_s value is obtained from the slope to the straight line at small t, corresponding to the chain-length range: $3 < i < 25$), whereas α_l is obtained from the range slightly above $i_c = 30$, i.e., from $i > 40$. The grey points (○) at very low t have not been included into the fitting for α_s.

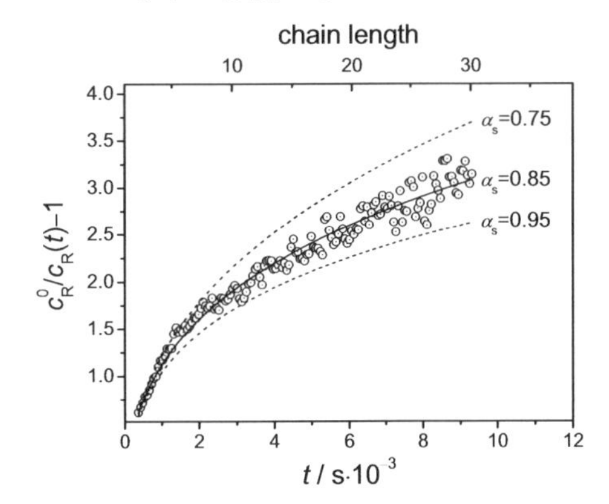

Figure 6.
Iterative fit of experimental data for BA radical polymerization (1.52 M in toluene) at −40 °C in the short-chain regime by using Eq. (8). Data from Ref.[4] is fitted up to $i_c = 30$. At the low polymerization temperature of −40 °C, backbiting of secondary BA radicals, which scrambles the i vs. t proportionality, is negligible.[15,19,23]

length-dependent propagation is not taken into account. The IUPAC-recommended chain-length-independent k_p values from PLP–SEC were adopted for analysis. The data for the small-chain and long-chain regions are fitted separately. The crossover chain length is identified by applying the double-log procedure described above.

Because of the low radical concentration at longer times after single-pulse initiation, signal-to-noise quality is usually not sufficient to allow for reasonably applying Eq. (9) toward measuring α_l. Therefore, α_l values are deduced from Eq. (11), which equation is well applicable at higher chain lengths. It should be noted that the determination of α values requires no EPR calibration. The primary pulse-induced radical concentration, however, needs to be known for estimates of $k_t^{1,1}$. Composite-model parameters as deduced from SP–PLP–EPR experiments are collated in Table 1.

The ease by which SP–PLP–EPR experiments may be carried out depends on the size of k_p and k_t of the monomer under investigation. High concentrations of slowly growing and slowly terminating radicals are easily detected. This is why the technique has first been applied to dibutyl itaconate (DBI). The bulk termination coefficient $k_t^{1,1}$ of this less common monomer at 0 °C is as low as about $10^5\,\mathrm{L\cdot mol^{-1}\cdot s^{-1}}$.[23,34] SP–PLP–EPR studies were expanded to dodecyl methacrylate (DMA) and benzyl methacrylate (BzMA), which monomers exhibit termination rate coefficients $k_t^{1,1}$ that are by about two orders of magnitude above $k_t^{1,1}$ of DBI.[25] By now, even acrylates with small alkyl ester moiety as well as *n*-butyl methacrylate (*n*-BMA) and *tert*-butyl methacrylate (*tert*-BMA) with $k_t^{1,1}$ values in the range $10^8\,\mathrm{L\cdot mol^{-1}\cdot s^{-1}}$ have been investigated.[15,35] SP–PLP–EPR experiments are challenging for monomers, which exhibit low k_p and high k_t and thus are associated with a rapid decay of radical concentration already at small radical chain lengths, which poses problems towards elucidating the CLD of termination rate. Methyl methacrylate (MMA) is such a difficult monomer.[2,14] To some extent, the problems could be overcome by carrying out experiments on per-deuterated MMA in bulk. The so-obtained rate coefficient from SP–PLP–EPR was: $k_t^{1,1} = 4.4\cdot10^8\,\mathrm{L\cdot mol^{-1}\cdot s^{-1}}$ at 0 °C.[14] Vinyl acetate (VAc) and styrene (Sty) are further

Table 1.
Composite-model parameters deduced from SP–PLP–EPR experiments for bulk and partly for solution polymerization of a range of monomers. The uncertainties are estimated to be ±20% for $k_t^{1,1}$, ±0.15 for α_s, ±0.06 for α_l, ±15 for i_c and ±2 kJ · mol^{-1} for $E_A(k_t^{1,1})$. The data is from Ref.[23]

	$k_t^{1,1}$ (0 °C)/L · mol · s^{-1}	$E_A(k_t^{1,1})$/kJ · mol^{-1}	i_c	α_s	α_l
n-BMA	$1.3\cdot10^8$	10	50	0.65	0.20
tert-BMA	$9.1\cdot10^7$	12	70	0.56	0.20
MMA	$5.4\cdot10^8$	9	–	0.63	–
MAA/aqueous phase	$8\cdot10^7$	–	–	0.65	–
DBI	$1.3\cdot10^5$	28	100	0.5	0.16
BA/toluene	$3.2\cdot10^8$	8.4	30	0.85	0.16

examples of such challenging monomers. Even for these monomers, detailed SP–PLP–EPR investigations are underway.[2,36–38]

As predicted by theory, α_s is well above α_l, irrespective of the type of monomer. With the exception of slowly terminating monomers, such as DBI, where steric hindrance plays a major role, clear evidence for diffusion control of termination has been found: The measured $k_t^{1,1}$ values are in close agreement with associated numbers predicted by the Smoluchowski approach. Moreover, the activation energy of $k_t^{1,1}$ is close to the one of fluidity.

Investigations into Kinetics Associated with Intramolecular Chain Transfer

In case of intramolecular chain-transfer, e.g., by 1,5 H-shift (backbiting) reactions, which are particularly important with acrylate-type monomers, the linear correlation of i and t breaks down and two types of radicals, i.e., secondary chain-end radicals and tertiary midchain radicals have to be considered. The detailed kinetic analysis requires numeric integration of the differential equations for both types of radicals, e.g., via the program package PREDICI. The SP–PLP–EPR experiments are particularly informative in case of the two types of radicals being easily distinguished by their hyperfine splitting pattern. Midchain radicals (MCRs), e.g., in BA polymerization are produced by the backbiting reaction of secondary propagating radicals (SPRs) (Figure 7). The relative amount of both types of radicals, e.g., the fraction of MCR radicals, x_{MCR}, is deduced from the double integrals of the individual species (Figure 8). The individual spectra are obtained from the experimental (mixed) spectrum by a fitting procedure which is based on the simulated spectra of each species. These spectra are integrated twice (Eq. (10) and Eq. (11)) in order to determine the associated radical concentrations.

$$\frac{c_{SPR}}{c_{MCR}} = \frac{\iint I(B_x)_{SPR}}{\iint I(B_x)_{MCR}} \tag{10}$$

$$x_{MCR} = \frac{\iint I(B_x)_{MCR}}{\iint I(B_x)_{MCR} + \iint I(B_x)_{SPR}} \tag{11}$$

The $c_R(t)$-curves measured at positions, which are characteristic for SPRs and MCRs, respectively, are fitted by PREDICI

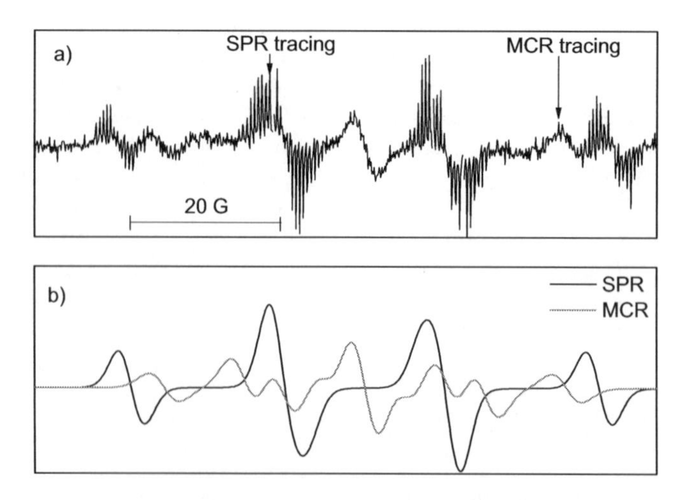

Figure 7.
a) Experimental EPR spectrum taken during a BA polymerization (1.52 M in toluene) at 0 °C with laser radiation (351 nm) being applied at a repetition rate of 20 Hz. The oscillating structure is characteristic of SPRs, which rapidly terminate after each laser pulse. MCRs, on the other hand, are slowly terminating and thus do not reflect laser pulsing. (b) From the simulated SPR and MCR spectra, the optimum magnetic field positions for time-resolved SP–PLP–EPR measurement of individual SPR and MCR concentrations may be identified. The arrows indicate the field positions at which the time-resolved measurements were carried out.

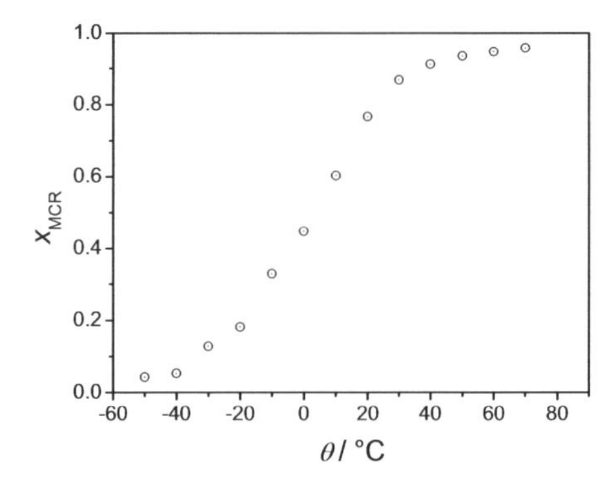

Figure 8.
Fraction of MCRs, x_{MCR}, during BA polymerizations (1.52 M in toluene) at different temperatures. For the calibration-free determination of x_{MCR} see text. The data is from Ref.[19,23]

(Figure 9). Additional reaction steps that need to be considered in the PREDICI fitting are: cross-termination, i.e., termination between an SPR and an MCR, with rate coefficient $k_t^{s,t}$, termination between two tertiary MCRs, with rate coefficient $k_t^{t,t}$, and propagation of an MCR associated with rate coefficient, k_p^t. This latter reaction reverses the backbiting step (with rate coefficient k_{bb}) in the sense that, by propagation, an MCR transforms into an SPR. All rate coefficients including the CLD of termination may be obtained from PREDICI refinement. The detailed reaction scheme may be found in Ref.[13]

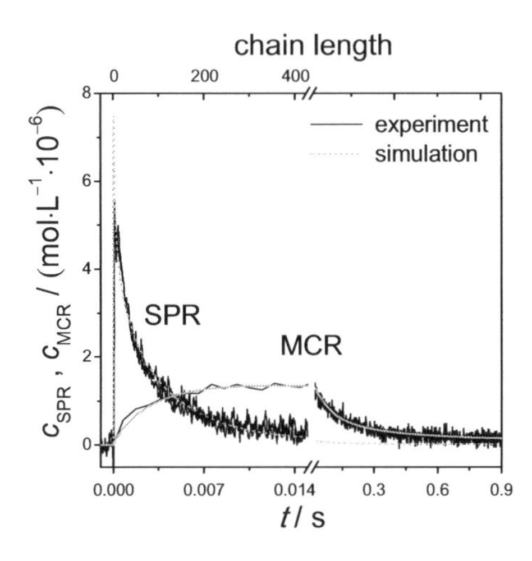

Figure 9.
Experimental and simulated (fitted) curves of SPR and MCR concentration for BA polymerization (1.52 M in toluene) at 30 °C. The traces demonstrate the initial rapid decay of SPRs under partial formation of MCRs through backbiting and the subsequent slow decay of MCRs. Note the different time scales. The figure is from Ref.[23] Copyright Wiley-VCH Verlag GmbH & Co. KGaA. Reproduced with permission.

Recently, the SP–PLP–EPR technique has been applied to SPR and MCR kinetics in aqueous solution. The transfer and termination kinetics of non-ionized and fully-ionized acrylic acid were successfully investigated and a strong dependence on monomer concentration and degree of ionization has been observed for $k_t^{s,s}(1,1)$, k_{bb}, k_p^t and $<k_t^{s,t}>$.[12]

Investigations into RAFT-Mediated Polymerization

Time-resolved measurement of concentrations of two types of radicals opens the avenue for detailed SP–PLP–EPR analysis of reversible-deactivation radical polymerizations, such as RAFT and ATRP, whenever the characteristic EPR bands are not seriously overlapping. Figure 10 illustrates that this requirement is fulfilled in the case of dithiobenzoate-mediated RAFT polymerization of BA. The EPR band of the BA radical, BA$^•$, is well separated from the one of the intermediate radical species, INT$^•$.[9] SP–PLP–EPR experiments may thus be carried out on both BA$^•$and INT$^•$ species.

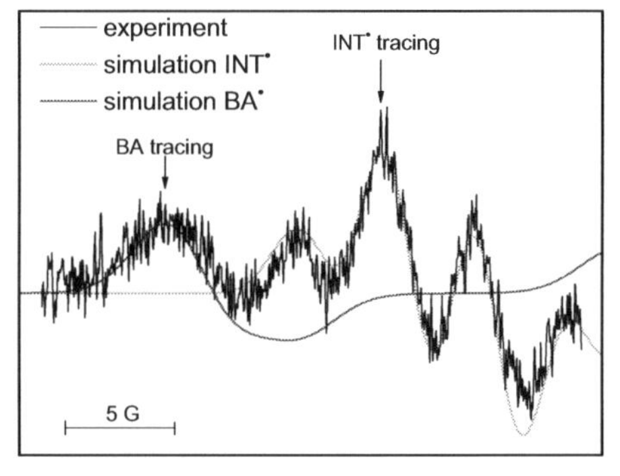

Figure 10.
Experimental and simulated spectra measured during dithiobenzoate-mediated RAFT polymerization of BA (1.5 M in toluene) at −40 °C under pseudo-stationary conditions with MMMP acting as the photoinitiator.

Rate coefficients for addition, k_{ad}, and fragmentation, k_{frag}, were determined under main- equilibrium conditions at different temperatures. Analysis of the experimental $c_R(t)$ traces (Figure 11) indicates fast fragmentation of the intermediate radical and cross-termination of propagating and resonance-stabilized intermediate radicals, with rate coefficient k_t^{cross}, which says that cross-termination is responsible for retardation in dithiobenzoate-mediated polymerizations rather than slow fragmentation of the INT$^{\bullet}$ species.[9] This mechanistic picture is fully supported by the formation of so-called "missing step" products with the systems cyano-*iso*-propyl radical – cyano-*iso*-propyl dithiobenzoate and phenylethyl radical – phenylethyl dithiobenzoate.[39,40,41] The occurrence of these products explains why no significant amounts of three-arm-star material are found despite the proven cross-termination process.

According to Eq. (12), the RAFT equilibrium constant, K_{eq}, is determined from the ratio of k_{ad} over k_{frag} with these coefficients being obtained via SP–PLP–EPR measurement (Figure 11). Alternatively, K_{eq} may be deduced from the EPR-spectroscopically measured ratio of P$^{\bullet}$/INT$^{\bullet}$concentrations (Eq. (13)), which is directly accessible from the ratio of associated double integrals measured under pseudo-stationary conditions, thus requiring no calibration for individual $c_P^{\bullet}$ and $c_{INT}^{\bullet}$.[9,42]

$$K_{eq} = \frac{k_{add}}{k_{frag}} = \frac{c(\mathrm{INT}^{\bullet})}{c(\mathrm{P}^{\bullet}) \cdot c(\mathrm{RAFT})} \tag{12}$$

$$K_{eq} = \frac{c(\mathrm{INT}^{\bullet})}{c(\mathrm{P}^{\bullet})} \cdot \frac{1}{c(\mathrm{RAFT})} = \frac{\iint B_x(\mathrm{INT}^{\bullet})}{\iint B_x(\mathrm{P}^{\bullet})} \cdot \frac{1}{c(\mathrm{RAFT})} \tag{13}$$

As can be seen from the two lower entries in Table 3, K_{eq} from the two approaches is in satisfactory agreement, which further suggests that the kinetic model underlying PREDICI simulation of the individual SP–PLP–EPR traces is adequate. For more details see Ref.[9]

Investigations into ATRP Kinetics

So far, ATRP rate coefficients have been determined by combination of the rate coefficient for activation, k_{act}, with the ATRP equilibrium constant, $K_{ATRP} = k_{act}/k_{deact}$. SP–PLP–EPR measurements provide direct access to the deactivation rate coefficient, k_{deact}. The new method has been applied towards ATRP of dodecyl methacrylate (DMA).[11] The underlying ATRP mechanism is illustrated in Scheme 3. The principle of measuring k_{deact} consists of producing an intense burst of P$^{\bullet}$ radicals by applying a laser single pulse in the presence of Cu^{II} halogen-ligand complex species. By transfer of a halogen atom from the Cu^{II}

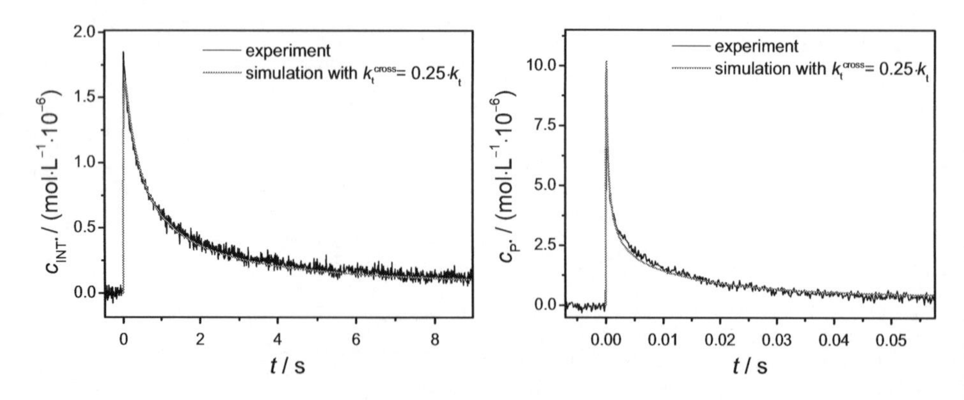

Figure 11.
Experimental and simulated $c_R(t)$ traces of propagating (P$^{\bullet}$) = (BA$^{\bullet}$) and INT$^{\bullet}$ radicals in dithiobenzoate-mediated BA polymerization (1.5 M in toluene) at −40 °C. The best fit is obtained for $k_t^{cross} = 0.25 \cdot k_t$. The different time scales for radical decay should be noted. For more details see Ref.[9]

Table 2.
Rate coefficients of termination, propagation and chain transfer in BA polymerization (1.52 M in toluene) at temperatures from 0 to 60 °C. The coefficient for termination of two SPR $k_t^{s,s}(1,1)$ corresponds to $k_t^{1,1}$, given above. The uncertainties of the pre-exponential and of k are estimated to be ±20% and are ±2 kJ · mol^{-1} for the activation energy. The data is from Ref.[13]

	pre-exponential factor/L · mol^{-1} · s^{-1} or s^{-1}	activation energy/kJ · mol^{-1}	k at 50 °C/L · mol^{-1} · s^{-1} or s^{-1}
$k_t^{s,s}(1,1)$	$1.3 \cdot 10^{10}$	8.4	$5.7 \cdot 10^{8}$
$k_t^{s,t}(1,1)$	$4.2 \cdot 10^{9}$	6.6	$3.6 \cdot 10^{8}$
k_{bb}	$1.6 \cdot 10^{8}$	34.7	$3.9 \cdot 10^{2}$
k_p^t	$9.2 \cdot 10^{5}$	28.3	25

Table 3.
Rate coefficients for addition, k_{ad} and fragmentation, k_{frag}, as obtained via SP–PLP–EPR investigation of dithiobenzoate-mediated BA polymerization (1.5 M in toluene) at −40 °C assuming cross-termination to occur with a rate coefficient of $k_t^{cross} = 0.25 \cdot k_t$. For comparison, K_{eq} as determined from pseudo-stationary EPR experiments, see Eq. (13), is also given in Table 3. The uncertainties of the rate coefficients and equilibrium constants are estimated to be ±30%.

	k at −40 °C/L · mol^{-1} · s^{-1}, s^{-1} or L · mol^{-1}
k_{add}	$1.4 \cdot 10^{6}$
k_{frag}	4.7
K_{eq} (single-pulsed)	$2.9 \cdot 10^{5}$
K_{eq} (pseudo-stat.)	$2.3 \cdot 10^{5}$

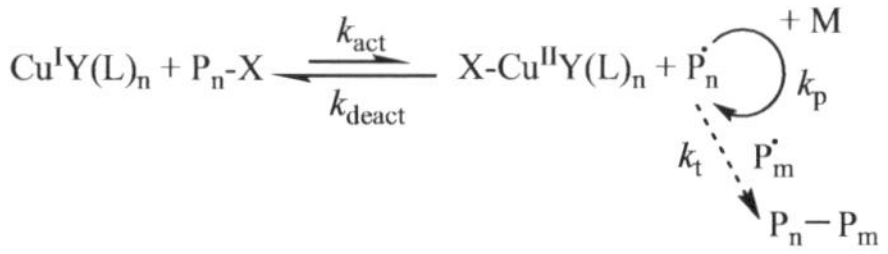

$$\mathrm{Cu^{I}Y(L)_n + P_n\text{-}X \underset{k_{deact}}{\overset{k_{act}}{\rightleftharpoons}} X\text{-}Cu^{II}Y(L)_n + P_n^{\bullet}} \quad (+\mathrm{M},\ k_p); \quad \mathrm{P_n^{\bullet} + P_m^{\bullet} \xrightarrow{k_t} P_n\text{–}P_m}$$

Scheme 3.
Mechanism of Cu-mediated ATRP with "dormant" species, P_n-X, active (propagating) radical, $P^{\bullet}$, "dead" polymer, $P_n - P_m$, and ligand, L. This scheme is superimposed on the conventional radical polymerization.

complex to $P^{\bullet}$, with rate coefficient k_{deact}, $P^{\bullet}$ concentration decays and a Cu^{I} species is formed. Due to the low equilibrium constant $K_{eq} = k_{act}/k_{deact}$ of the HMTETA (1,1,4,7,10,10-hexamethyltriethylenetetramine) ligand system and to the small amount of Cu^{I} complex produced during the experiment, activation may be ignored within the kinetic analysis. On the other hand, termination kinetics may not be ignored at the high levels of initial $c_{P^{\bullet}}$. Thus termination has to be measured within a separate SP–PLP–EPR experiment, i.e., in the absence of the Cu^{II} complex species.

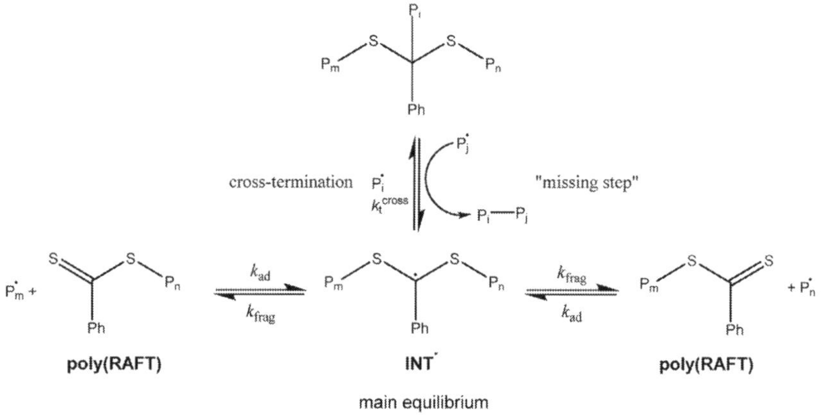

Scheme 2.
Reaction scheme for RAFT-mediated polymerizations under main-equilibrium conditions comprising cross-termination and "missing-step" reactions. $P^{\bullet}$ stands for a propagating radical ($P^{\bullet} = BA^{\bullet}$ in the present case).

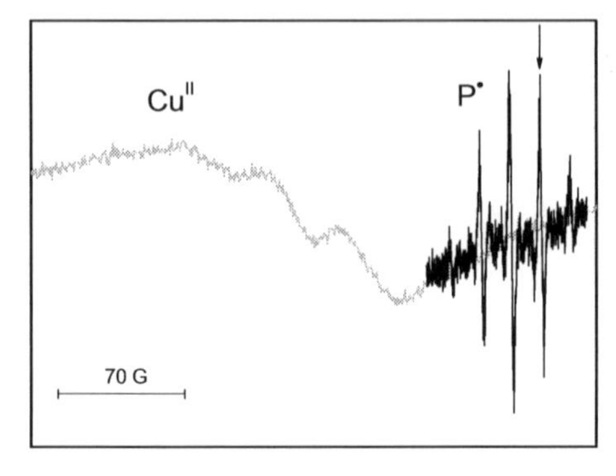

Figure 12.
Experimental EPR spectra of Cu^{II} species (l.h.s.) and propagating DMA (r.h.s.) radicals in pulsed-laser induced polymerization at 0 °C with MMMP acting as the photoinitiator. The field position used for single-pulsed experiments is indicated by the arrow. The data is from Ref.[11]

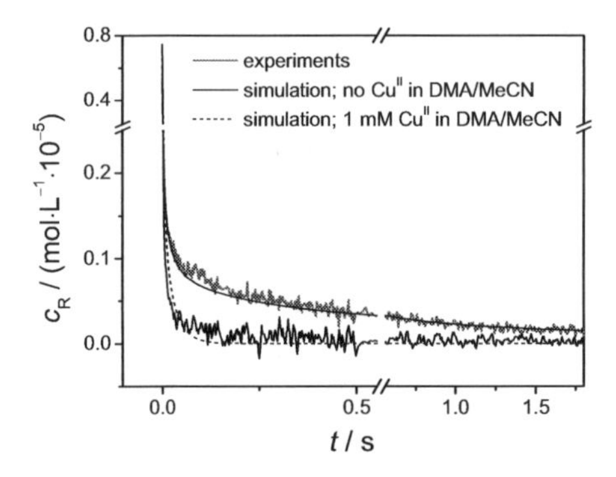

Figure 13.
Decay of DMA radical concentration after laser single-pulse initiation with (lower curve) and without (upper curve) deactivation by the Cu^{II} complex at 0 °C. Experiments were performed upon addition of 15 wt. % acetonitrile to enhance solubility of the Cu^{II} complex. Data is taken from Ref.[11]

In contrast to the Cu^{II} complex, the Cu^{I} species is diamagnetic and thus EPR-silent. The EPR-spectra of the Cu^{II} species and of the propagating DMA radical, P$^{\bullet}$, are shown in Figure 12. Illustrated in Figure 13 are the two experimental traces that need to be measured for k_{deact} determination via SP–PLP–EPR. The upper curve represents the decay of initial DMA radical concentration in solution of acetonitrile, but in the absence of the Cu^{II} complex. The lower trace has been recorded under otherwise identical experimental conditions, but with 1 mM Cu^{II}/HMTETA. The faster decay of P$^{\bullet}$ concentration is due to both conventional termination and deactivation being operative. Fitting the $c_R(t)$ traces via PREDICI, yields $k_{\text{deact}} = (4 \pm 3) \cdot 10^5$ L · mol^{-1} · s^{-1} for the particular system at 0 °C.

Conclusion

The SP–PLP–EPR technique allows for directly monitoring the time evolution of radical species and thus enables measurement of chain-length-dependent termination kinetics, of chain-transfer reactions, and of rate coefficients associated with reversible-deactivation polymerization, such as ATRP and RAFT polymerization. The fundamentals of this novel method have been worked out, but the potential of the new technique is far from being fully exploited.

Acknowledgement: The authors are grateful to *Deutsche Forschungsgemeinschaft* (BU 426/11) for generous support of this work.

[1] O. F. Olaj, I. Bitai, F. Hinkelmann, *Macromol. Chem. Phys.* **1987**, *188*, 1689.
[2] S. Beuermann, M. Buback, *Prog. Polym. Sci.* **2002**, *27*, 191.
[3] C. Barner-Kowollik, M. Buback, M. Egorov, T. Fukuda, A. Goto, O. F. Olaj, G. T. Russell, P. Vana, B. Yamada, P. B. Zetterlund, *Prog. Polym. Sci.* **2005**, *30*, 605.

[4] J. Barth, *Radical Polymerization Kinetics in Systems with Transfer Reactions Studied by Pulsed- Laser-Polymerization and Online EPR-Detection*, PhD Thesis, Göttingen, **2011**.
[5] M. Buback, M. Egorov, T. Junkers, E. Panchenko, *Macromol. Rapid Commun.* **2004**, *25*, 1004.
[6] C. Barner-Kowollik, G. T. Russell, *Prog. Polym. Sci.* **2009**, *34*, 1211.
[7] J. Barth, M. Buback, G. T. Russell, S. Smolne, *Macromol. Chem. Phys.* **2011**, *212*, 1366.
[8] J. Barth, M. Buback, *Macromolecules* **2011**, *44*, 1292.
[9] W. Meiser, J. Barth, M. Buback, H. Kattner, P. Vana, *Macromolecules* **2011**, *44*, 2474.
[10] W. Meiser, M. Buback, J. Barth, P. Vana, *Polymer* **2010**, *51*, 5977.
[11] N. Soerensen, J. Barth, M. Buback, J. Morick, H. Schroeder, K. Matyjaszewski, *Macromolecules* **2012**, *45*, 3797.
[12] J. Barth, W. Meiser, M. Buback, *Macromolecules* **2012**, *45*, 1339.
[13] J. Barth, M. Buback, P. Hesse, T. Sergeeva, *Macromolecules* **2010**, *43*, 4023.
[14] J. Barth, M. Buback, *Macromol. Rapid Commun.* **2009**, *30*, 1805.
[15] J. Barth, M. Buback, P. Hesse, T. Sergeeva, *Macromolecules* **2009**, *42*, 481.
[16] J. Barth, R. Siegmann, S. Beuermann, G. T. Russell, M. Buback, *Macromol. Chem. Phys.* **2012**, *213*, 19.
[17] J. Barth, M. Buback, G. Schmidt-Naake, I. Woecht, *Polymer* **2009**, *50*, 5708.
[18] J. Barth, M. Buback, C. Barner-Kowollik, T. Junkers, G. T. Russell, *J. Polym. Sci., Part A: Polym. Chem.* **2012**, *50*, 4740.
[19] P. Hesse, *Radical Polymerization Kinetics in Aqueous Solution and in Systems with Secondary and Tertiary Radicals Studied by Novel Pulsed- Laser Techniques*, Cuvillier Verlag, Göttingen **2008**.
[20] Z. Szablan, T. Junkers, S. P. S. Koo, T. M. Lovestead, T. P. Davis, M. H. Stenzel, C. Barner-Kowollik, *Macromolecules* **2007**, *40*, 6820.
[21] P. Vana, T. P. Davis, C. Barner-Kowollik, *Aust. J. Chem.* **2002**, *55*, 315.
[22] G. R. Eaton, S. S. Eaton, D. P. Barr, R. T. Weber, *Quantitative EPR*, Springer, **2010**.
[23] J. Barth, M. Buback, *Macromol. React. Eng.* **2010**, *4*, 288.
[24] G. B. Smith, G. T. Russell, J. P. A. Heuts, *Macromol. Theory Simul.* **2003**, *12*, 299.
[25] M. Buback, E. Müller, G. T. Russell, *J. Phys. Chem. A* **2006**, *110*, 3222.
[26] D. S. Achilias, *Macromol. Theory Simul.* **2007**, *16*, 319.
[27] G. Johnston-Hall, M. J. Monteiro, *J. Macromol. Sci., Polym. Rev.* **2008**, *46*, 3155.
[28] B. Friedman, B. O'Shaughnessy, *Macromolecules* **1993**, *26*, 5726.
[29] O. F. Olaj, G. Zifferer, *Macromolecules* **1987**, *20*, 850.
[30] A. Khokhlov, *Makromol. Rapid Commun.* **1981**, *2*, 633.
[31] S. Benson, A. North, *J. Am. Chem. Soc.* **1959**, *81*, 1339.
[32] S. W. Benson, A. M. North, *J. Am. Chem. Soc.* **1962**, *84*, 935.
[33] G. B. Smith, G. T. Russell, *Z. Phys. Chem.* **2005**, *219*, 295.
[34] M. Buback, M. Egorov, T. Junkers, E. Panchenko, *Macromol. Chem. Phys.* **2005**, *206*, 333.
[35] J. Barth, M. Buback, P. Hesse, T. Sergeeva, *Macromol. Rapid Commun.* **2009**, *30*, 1969.
[36] R. Hutchinson, J. Richards, M. Aronson, *Macromolecules* **1994**, *27*, 4530.
[37] R. Hutchinson, M. Aronson, J. Richards, *Macromolecules* **1993**, *26*, 6410.
[38] D. R. Taylor, K. Y. van Berkel, M. M. Alghamdi, G. T. Russell, *Macromol. Chem. Phys.* **2010**, *211*, 563.
[39] W. Meiser, *Investigation of the Kinetics and Mechanism of RAFT Polymerization via EPR Spectroscopy*, PhD Thesis, Göttingen, **2012**.
[40] W. Meiser, M. Buback, *Macromol. Rapid Commun.* **2012**, *33*, 1273.
[41] W. Meiser, M. Buback, O. Ries, C. Ducho, A. Sidoruk, *Macromol. Chem. Phys.* **2013**, *214*, 924.
[42] J. Barth, M. Buback, W. Meiser, P. Vana, *Macromolecules* **2010**, *43*, 51.

DOI: 10.1002/masy.201300050

Mass Transfer in Miniemulsion Polymerisation

T.G.T. Jansen,[1] *P.A. Lovell,**[2] *J. Meuldijk,*[1] *A.M. van Herk*[1,3]

Summary: The use of very hydrophobic monomers in heterogeneous polymerisations is restricted because of insufficient mass transfer as a result of limited aqueous phase diffusion. Experimental evidence for a mass transfer mechanism in miniemulsion polymerisation that is based on the direct interaction between droplets as well as between droplets and particles ('collisions') rather than monomer diffusion is presented. Mass transfer that results in the formation of a copolymer in the miniemulsion polymerisation of a mixture of lauryl methacrylate and 4-tert-butyl styrene, is almost non-existing when physical contact between individual droplets is prevented by a membrane. However, monomer transport by diffusion is observed in reference experiments with more hydrophilic monomers.

Keywords: emulsion polymerisation; lauryl methacrylate; mass transfer; membranes; miniemulsion polymerisation

Introduction

In the 40 years since it was first reported,[1] miniemulsion polymerisation has developed towards a useful and very versatile method for the production of polymer in water dispersions. The characteristic feature that distinguishes miniemulsion polymerisation from conventional emulsion polymerisation is that particle formation takes place by entry of radical species into submicron monomer droplets rather than monomer-swollen micelles or via homogeneous nucleation. The most important requirements for droplet nucleation are the absence of surfactant micelles, the reduction of droplet sizes (*i.e.* a very large specific area per unit volume of the reaction mixture) to enable efficient radical capture and the colloidal stability of the formed droplets, at least during the time-frame needed for polymerisation. The key factor in miniemulsion polymerisation is the formation and stabilisation of the monomer droplets. Formation of droplets requires shear, to be applied by, for instance, ultrasound, high-pressure homogenisers, static mixers or rotor-stator devices, *e.g.* ultra-turrax stirrers.[2] Once monomer droplets have been created, coalescence (i.e. fusion of two droplets) and Ostwald ripening (transfer of monomer from small to large droplets) reduce the number of droplets and change droplet diameters continuously. Successful retardation of this process can be established by a combination of a surfactant (against coalescence) and a so-called costabiliser. The costabiliser should possess very limited water solubility in order to create an osmotic pressure inside the monomer droplets that restricts Ostwald ripening. Historically, long-chain alkanes (*e.g.* hexadecane) have been used, but the use of monomer-soluble initiators,[3,4] polymers[5,6] or reactive comonomers[7,8] has been reported to be successful as well.

One of the main advantages of miniemulsion polymerisation is the possibility to incorporate very hydrophobic monomers in the final latex particles, since the need for monomer transport across the aqueous phase is much less a prerequisite, given that the monomer is already at the locus of

[1] Eindhoven University of Technology, Department of Chemical Engineering and Chemistry, Postbus 513 5600, MB, Eindhoven, The Netherlands
E-mail: j.meuldijk@tue.nl
[2] Materials Science Centre, The University of Manchester, Grosvenor Street, Manchester, M1 7HS, United Kingdom
[3] Institute of Chemical and Engineering Sciences, 1 Pesek Road, Jurong Island 627833, Singapore

polymerisation. It is surprising, therefore, that the available literature on very hydrophobic monomers is scarce. Miniemulsion polymerisation studies towards hydrophobic monomers comprise lauryl methacrylate,[9,10] isobornyl acrylate,[11] isooctyl acrylate[12] and 4-tert-butyl styrene[10] The use of hydrophobic monomers as costabilizer is more widespread, for instance lauryl methacrylate,[13] stearyl methacrylate[7,13] or octadecyl acrylate.[5]

Besides offering a versatile route for the production very hydrophobic submicron polymer particles that are not accessible via conventional emulsion polymerisation, miniemulsion polymerisation of very hydrophobic monomers can also be of great academic interest. Since it restricts aqueous phase effects and allows the possibility to work in very diluted systems, it can give insight into mass transfer processes between individual droplets or particles.

A common misconception in miniemulsion polymerisation is that the compartmentalised nature of the monomer droplets automatically results in the absence of (significant) monomer transport between individual droplets. *Rodriguez et al*[14] have shown mathematically as well as experimentally that mass transfer takes place between submicron miniemulsion droplets in a miniemulsion consisting of either methyl methacrylate or styrene. However, the resistance of monomer transport in the membrane used in their compartmented diffusion cell was too high to allow for any monomer transport within the time-frame of polymerisation. *Delgado et al*[15] modelled the diffusion of monomer in the miniemulsion copolymerisation of vinyl acetate and butyl acrylate and compared this to experimental results.

In this paper, we report investigations of monomer transfer in the miniemulsion (co) polymerisation of 4-tert-butyl styrene and lauryl methacrylate. The very hydrophobic nature of these monomers, resulting in water solubilities in the order of 10^{-5} mol/dm^3_w and 10^{-7} mol/dm^3_w respectively,[16] substantially enlarges the resistance against mass transfer through the aqueous phase.

Experimental Part

Lauryl methacrylate (LMA, 96%), 4-tert-butyl styrene (TBS, 95%), methyl methacrylate (MMA, >99%) and n-butyl acrylate (BA, >99%), hexadecane (HD, 99%), sodium dodecyl sulfate (SDS, >99%) lauroyl peroxide (LPO, >97%), sodium persulfate (SP, >98%) and sodium carbonate (SC, >99%)) were all obtained from (Sigma-)Aldrich. Inhibitor-removal was done by passing the monomers over a column containing inhibitor remover (Sigma-Aldrich); all other chemicals were used as received. Deionized water (MilliQ standard) has been used throughout the experiments.

Monomer conversion was followed using a Perkin-Elmer Clarus 500 gas chromatograph, equipped with an Agilent Technologies HP-FFAP column. The conversion was measured against both an internal (HD) and external (toluene) standard.

Differential Scanning Calorimetry was performed using a TA-Instruments Q100 modulated differential scanning calorimeter. Polymer samples were precipitated in acetone (or ethanol in the case of PMMA/PBA) and washed with acetone (ethanol) at least 4 times. The washing liquid was kept aside after every washcycle and evaporated to check for any residue. The cleaned polymer was dried overnight at 80 °C. Three consecutive heating and cooling cycles (10 °C/min, nitrogen atmosphere) were used for every polymer sample. Samples, typically 5–10 mg, were weighed into aluminium hermetic pans (TA-Instruments).

Particle and droplet sizes for the diffusion measurements were determined by photon correlation spectroscopy (PCS) using a Brookhaven Instruments BI-9000AT correlator with a Brookhaven BI-200SM goniometer set to a scattering angle of 90° and a 17 mW HeNe laser (632.8 nm wavelength). Samples were diluted with 3.3 g/L SDS solution to give a count rate of ~150 kcounts s^{-1} at the detector and, after temperature equilibration, were subjected to 10 successive analyses of 1 min each. The vat temperature was held at ~25 °C (controlled to ± 0.1 °C) and the exact vat temperature input when analysed the data using Brookhaven Particle Sizing Software v3.72 to obtain individual values of the particle diameter for each of the 10 measurements, the average of which was used to obtain the reported value (the standard deviation of which was typically ~2 nm). Blank measurements indicated that SDS micelles had no detectable influence on the results was. Reported diameters are z-averaged diameters (d_z), the width of the distribution is expressed by the so-called *poly*-value. For the calorimetric measurements, the d_z was determined on a Malvern Zetasizer Nano ZS. Latices were diluted with a 3.3 g/L SDS solution. The Malvern Zetasizer has an internal filter which is automatically adjusted to give a count rate of ~400 kcounts s^{-1} Samples were thermostated at 20,0 °C (± 0.1 °C). A single measurement consisted of 16 runs of 20 seconds each. The reported values are averaged over 3 measurements.

Miniemulsions were prepared according to the formulation given in Table 1. Water, SDS and (if applicable) SC were mixed together. When sodium persulfate was used, a small fraction of the water (usually 5-10 ml) was kept aside to dissolve the initiator. In a separate flask, the HD and LPO were dissolved in the monomer. Both the aqueous and monomer mixture were then transferred to a 400 ml beaker and stirred for at least 10 minutes using a magnetic stirrer bar. Subsequently, the mixture was ultrasonified for 30 minutes, using a Branson CV33 horn powered by a Sonics Vibracell VCX 750W at 65% output.

To suppress polymerisation during ultrasonication, the beaker was immersed in an ice bath. The resulting miniemulsion was checked for polymerisation by precipation of a few droplets into acetone, which showed no evidence of polymer formation.

In the case of miniemulsion copolymerisation or mixed miniemulsions, the monomer mass ratio was kept at 124.6 g LMA to 75.4 g TBS (51/49 mol/mol).

Diffusion measurements (with and without polymerisation) were carried out in an 800 ml stainless steel reactor (Figure 1), designed and built at the School of Materials in the University of Manchester. The reactor is divided into two equally-sized, symmetrical compartments using a Spectra Por 4 Regenerated Nitrocellulose membrane sheet (Spectrum Laboratories Inc.), clamped into a Teflon window frame. This membrane was soaked in deionized water for at least 30 minutes prior to reaction to remove glycerine traces and was replaced after every experiment. The measured membrane thickness was 38 μm, the reported molecular weight cut-off 12–14 kDalton. The exposed membrane area is approximately 40 cm^2.

Each compartment is stirred using a two-bladed impeller. The vessel is covered by a stainless steel lid which contains sample ports for the withdrawal of liquid samples and the supply of argon. The air tightness of the reactor was secured by a silicone material, fitted between the flanges of each compartment. Heating the reactor contents was done by immersing the reactor in a water bath, thermostated at 60 °C. The time required to bring the reactants to reaction temperature, *i.e.*10 minutes, is comparable to jacketed vessels and glass reactors.

Table 1.
Formulations used in the polymerisations (all amounts in grams)

	Monomer-phase initiated	Aqueous-phase initiated
Deionized water	200	200
Monomer	50	50
Hexadecane	2	2
Sodium Dodecyl Sulfate	1	1
Lauroylperoxide	0.795	–
Sodium Persulfate	–	0.476
Sodium Carbonate	–	0.212

Figure 1.
The diffusion reactor (left) and the open reactor showing the individual compartments separated by a membrane (right).

Each reactor compartment was charged with a miniemulsion, the amount being equivalent to the recipe used in Table 1. Both compartments were subsequently purged with argon for at least 30 minutes, after which the reactor was pressurized by closing the argon exit. Polymerisations were started by immersing the reactor in the thermostated water bath. In the case of persulfate initiation, this was followed by adding the initiator solution to each compartment. Equivolumetric samples were taken throughout the reaction from each compartment. Polymerisation of these samples was short stopped by addition of a few milligrams of hydroquinone. The reactor was kept under argon pressure during the reaction to prevent inhibition by oxygen.

A low monomer content (0.5% by volume) miniemulsion polymerisation was carried out in a 250 ml flanged round bottom glass vessel, equipped with a 4-neck glass lid, a reflux condenser and a 4-bladed impeller. 1.56 g of a standard LMA miniemulsion (see Table 1) and 0.94 g of a standard TBS miniemulsion were each diluted in 99 g of deionized water and subsequently added together, to obtain a 100-fold dilution (based on monomer/water ratio). Both miniemulsions contained LPO as the initiator.

All other miniemulsion polymerisations were carried out in a closed 1.6 L stainless steel Mettler-Toledo *RC1e* HP60 reaction calorimeter. The calorimetric profile was determined using the Mettler Toledo Icontrol 5.0 software and correlated to the final conversion measured by gas chromatography. Miniemulsions, made according to Table 1 but containing 4 times the indicated quantities, were charged to the reactor and subsequently degassed by bubbling argon through it. After this purging, the reactor was put under argon pressure and the reactor contents were heated to the reaction temperature (60 °C). In the case of initiation by SP, the necessary amount (dissolved in 20 g of water) was added using a piston pump controlled by a balance. Throughout the reaction, the reactor was stirred at 250 rpm. The reactor internals acted as baffles.

Results and Discussion

In a first set of experiments, we compared the miniemulsion homopolymerization of LMA and TBS with the miniemulsion copolymerisation of these monomers, *i.e.* where the miniemulsion was prepared from a mixture of the two monomers. In a parallel run, we mixed separate LMA and TBS miniemulsions and allowed the mixture to react. We refer to the latter experiments as 'mixed miniemulsions'. Reactions were carried out with both SP (Figure 2A) and LPO as the initiator

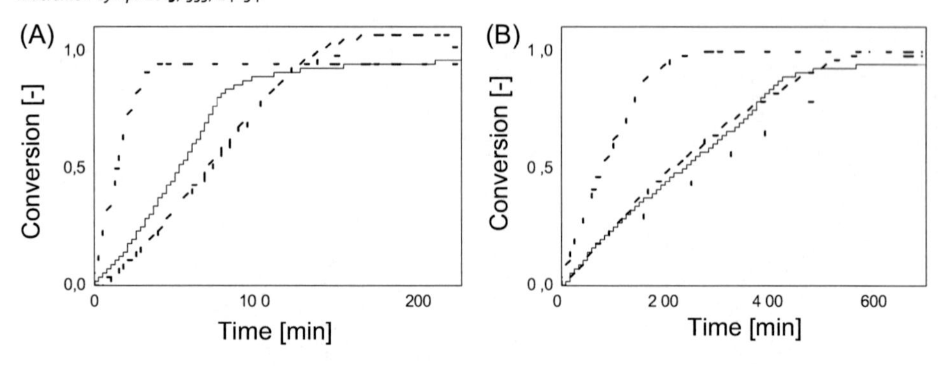

Figure 2.
Conversion-time profiles for sodium persulfate (A) and lauroyl peroxide initiated miniemulsion polymerisations. (-●●) LMA, (straight line) TBS, (- -) copolymerisation (●●) mixed miniemulsions.

(Figure 2B). For a SP initiated system, it is apparent that the conversion-time history for the mixed miniemulsions is similar to the one for a miniemulsion copolymerisation, and does not show any resemblance to a (combination of) the miniemulsion homopolymerisations. For the LPO initiated systems, we see again that the conversion-time history for the copolymerisation and the mixed miniemulsions are similar, albeit with a slight difference in the reaction rate caused by a difference in the number of particles.

Particle sizes of the final latices and glass transition temperatures (T_g) of the synthesized polymers are collected in Table 2. The measured glass transition temperatures indicate that the polymer particles created in the mixed miniemulsions polymerisation comprise copolymer particles rather than particles only containing poly-lauryl methacrylate or poly-tert-butyl styrene. Since the monomer droplets are the locus of polymerisation, the monomer composition in every droplet has to be the same and equal to the overall monomer composition. Dissolving of dried polymer in an (organic) solvent proved to be impossible, thus restricting the possibility for NMR or Gradient Polymer Elution Chromatography (GPEC) to obtain information about the copolymer composition. However, the fact that the conversion-time histories for the mixed miniemulsions and the copolymerisation do not differ at low conversion and the absence of transitions around the glass transition temperatures of the homopolymers demonstrate that the exchange of monomer between the droplets is rapid and is already accomplished before significant polymerisation occurs.

Note that these observations conflict with similar experiments published by *Ouzineb et al.* In a series of experiments,

Table 2.
Glass transition temperatures, final particle sizes and polydispersities for the sodium persulfate and lauroyl peroxide initiated calorimetric runs of the miniemulsion polymerisations of separately prepared miniemulsions of lauryl methacrylate and 4-tert butyl styrene.

		T_g [°C]	d_z [nm]	Poly [–]
Sodium persulfate initiated	*LMA (homopolymer)*	−46	127	0.01
	TBS (homopolymer)	128	109	0.03
	Copolymerisation	−1	133	0.01
	Mixed miniemulsions	3	143	0.07
Lauroyl peroxide initiated	*LMA (homopolymer)*	−46	168	0.10
	TBS (homopolymer)	119	151	0.06
	Copolymerisation	−1	155	0.09
	Mixed miniemulsion	−8	170	0.11

these authors showed by using DSC[8] and IR spectroscopy[17] that by polymerising a blend of miniemulsions containing either n-butyl methacrylate or styrene droplets, two homopolymers were obtained.

Although the water solubilities of these monomers are orders of magnitude larger than of TBS or LMA, no monomer transfer was observed. We have not been able to give a satisfying explanation for this yet, although the difference in the emulsifiers and costabilisers is worth mentioning. *Ouzineb et al* used a combination of an anionic and a neutral emulsifier (SDS and Triton X-405) together with the application of a reactive costabiliser. Our results, however, are based on hexadecane as the costabiliser and only an anionic emulsifier (SDS).

Mass transfer in (mini)emulsion systems is thought to occur mainly by diffusion. However, a second, parallel mass transfer mechanism based on particle-particle collision as a result of Brownian motion has received much less attention.[18,19] This mechanism has been postulated to be responsible for the transport of reagents with negligible water solubility such as inhibitor,[20] long-chain alkanes,[20] chain-transfer agents[21–23] or even monomers.[11,12] Since transfer of monomer by collision is a second-order process with respect to the number of droplets or particles, whereas transfer by diffusion is only a first-order process,[18,24] working in a very dilute system would affect collision-based transport much more than diffusion-based transport.

For our miniemulsions, a 100-fold dilution was the the maximum possible dilution without losing droplet stability. The very low monomer amounts do not allow the use of SP as the initiator, since the ratio initiator to monomer ratio is then too high to overcome aqueous phase inhibition and termination. A too high cation concentration may also result in compression of the charged bilayer. Hence, LPO was used as initiator for this experiment.

DSC analysis of the final polymer indicated one glass transition temperature at −1.5 °C which coincides with the glass transition of the copolymer obtained by mixing the undiluted miniemulsions and is only slightly higher than the glass transition temperature observed for the product of the miniemulsion copolymerisation. Although monitoring of the conversion by GC has not been done, since the low monomer concentrations do not allow for reliable measurement, it was apparent that the overall reaction rate was higher than the reaction rate observed for both the miniemulsion copolymerisation as well as for the mixed miniemulsions.

Because dilution of the miniemulsions did not clarify the mechanism of monomer transfer, spatial limitation of droplet-droplet interaction was chosen as a route to providing more insight into the mass transfer processes in (mini)emulsion polymerisations. Through separating two miniemulsion by a regenerated nitrocellulose membrane with molecular weight cut off in the order of 12–14 KDalton, we can distinguish between monomer transfer by collision and monomer transfer by diffusion. The small pores (pore sizes around 2 nm) form no barrier for single monomer molecules, but the pores are too small to permit the passage of monomer droplets across the membrane.

The water solubilities of MMA (0.15 M, 25 °C)[25] and BA (0.015 M, 25 °C)[26] are orders of magnitude higher than the aforementioned water solubilities of TBS and LMA. With a difference in homopolymer glass transition temperature of approximately 150 °C, MMA and BA serve as an excellent reference pair since monomer transport by diffusion is likely to occur.

In a first experiment, the permeability of the membrane for monomer was tested by equilibrating water, saturated with MMA, against deionized water at 60 °C (Figure 3). Using Fick's law of diffusion (see equation 1), the molar flux through the membrane is given by

$$J_m = -D_{m,aq} \cdot \frac{\Delta C_m}{\Delta x} \qquad (1)$$

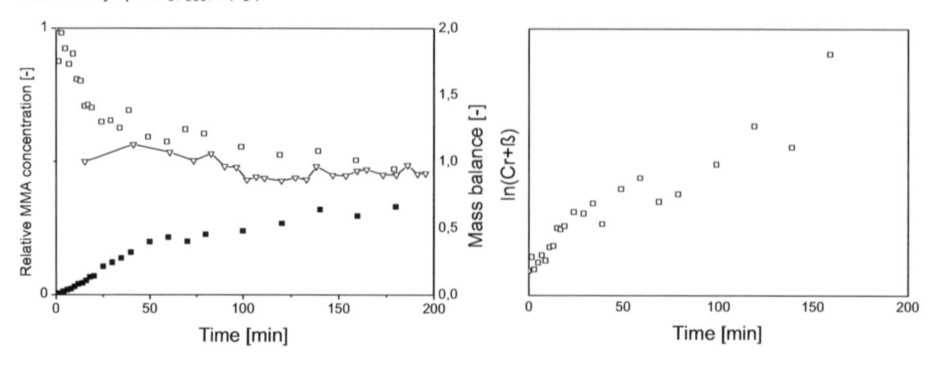

Figure 3.
MMA diffusion across the membrane in a single phase system and. (□) MMA concentration in the MMA retentate, (■) MMA concentration in the permeate, (▽) total MMA detected in the reactor (mass balance).

in which $D_{m,aq}$ represents the diffusion coefficient, ΔC_m the monomer concentration difference between the retentate and the permeate side of the membrane and Δx the thickness of the membrane. a plot of the concentration against the logarithm of time should give a straight line, at least early in the experiment. Figure 3 shows that the membrane is readily permeable to MMA. By applying Equation 1 in a non-steady state mass balance, it can be shown that a plot of $\ln(C_r + \beta)$ against time should give a straight line, as can be seen in Figure 3. The retentate concentration is symbolised by C_r, whereas β is a constant, depending on the volume of permeate, retentate and the total amount of monomer. The time necessary for complete exchange shows that there is still a significant resistance against diffusion through the membrane. Note that the time constant for diffusion through the membrane is in the order of minutes/hours and not in the order of days.[14] The initial decrease in the mass balance is due to the absorption of a small fraction of the MMA by the rubber seals that prevent the reactor from leaking.

Permeability of the membrane for LMA and TBS has been tested in an 80/20 (wt/wt) acetone/water mixture. The use of acetone for this particular measurement was required due to the very low water solubilities of these monomers. The addition of water was necessary for membrane wetting. For LMA and TBS, membrane permeability was observed, although insufficient wetting of the membrane did not permit quantitative measurement.

To quantify monomer transfer in heterogeneous systems, we allowed 2 miniemulsion to equilibrate at 60 °C. The equilibration of an MMA and a BA miniemulsion (Figure 4A) demonstrates that significant mass transfer is taking place within the time-scale for polymerisation. The diffusion rate of MMA is approximately 8 times higher than for BA, which is in agreement with the difference in water solubility.

Note that the effect observed by *Rodriguez et al*[20] where more than 50% of the most water-soluble monomer is transported initially has not been observed, our measurement had to be terminated due to wearing of the seals.

Figure 4B shows the same measurement with an LMA and a TBS miniemulsion. Despite its very low water solubility, TBS does diffuse through the membrane, but this diffusion rate is too slow to have an influence during polymerisation. As expected, the extremely low water solubility of LMA restricts diffusion and not more than 0.05% of the total LMA amount has been found in the TBS rich compartment.

By adding initiator to the miniemulsions, we can monitor the influence of (the absence of) diffusion on the polymerisation. Figure 5 shows the compartmentalised miniemulsion polymerisations of LMA

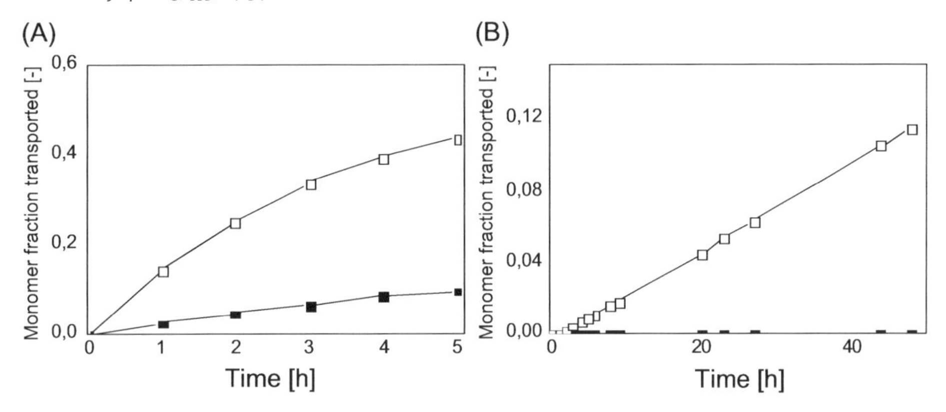

Figure 4.
Monomer diffusion in heterogeneous systems for (A) (□) MMA, (■) BA and (B) (□) TBS, (■) LMA.

and TBS. For SP as the initiator (Figure 5 left), the conversion-time histories for both LMA and TBS are similar to those in Figure 2, which is in line with the short reaction times. By extending the reaction times (Figure 5 right), as a result of using LPO instead of SP, it becomes apparent that the rates of reaction are still similar to those observed for the miniemulsion homopolymerisations, indicating that the monomers are confined in their own compartments. Although according to Figure 4 not more than 0.5–1% of the total TBS amount is transported across the membrane and the measured TBS monomer fraction (g/g) in the LMA miniemulsion is around 0.1%, the influence on the rate of polymerisation is visible. The decrease in the polymerisation rate of LMA after approximately 50% conversion can be attributed to the reduction of the apparent propagation rate coefficient due to the TBS transferred. However, for LPO-initiated polymerisations, we also observed a decrease in the measured z-average particle diameters for the LMA particles and a strong increase in latex viscosity in both the TBS and LMA latex. The origin of this viscosity change is not clear yet. Note that addition of a small amount of inorganic salts (NaCl or Na_2CO_3) brought the viscosity back to normal values for latex systems with the solids contents used. Not surprisingly, the polymers in each latex had

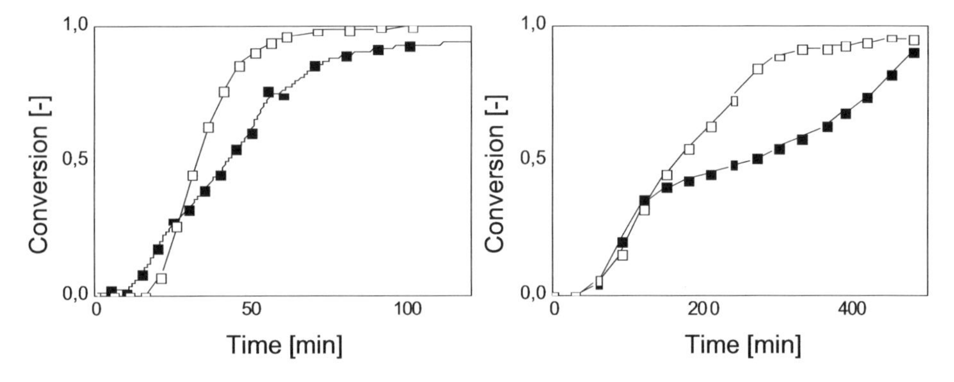

Figure 5.
Conversion-time profiles for compartmentalised miniemulsion polymerisations of (■) LMA and (□) TBS for sodium persulfate (left) and lauroyl peroxide (right) initiation.

Table 3.
Glass transition temperatures for the compartimented reactions.

	T_g [°C]	d_z [nm]	Poly [–]		T_g [°C]	d_z [nm]	Poly [–]
LMA homopolymer	−48	–	–	*BA homopolymer*	−48	–	
LMA (SP initiated)	−49	127	0.04	*BA rich polymer*	−36	128	0.12
LMA (LPO initiated)	−47	141	0.11	*MMA homopolymer*	125	–	
TBS homopolymer	152	–	–	*MMA rich polymer*	123	106	0.12
TBS (SP initiated)	153	100	0.04				
TBS (LPO initiated)	152	132	0.07				

glass transition temperatures in agreement with those measured for the polymers produced in independent miniemulsion homopolymerisations (Table 3). Note that the DSC analysis of these diffusion measurements has been done on a different, but similar DSC than for the calorimetric runs, which accounts for the deviation in the measured glass transition temperatures. Moreover, the final particle sizes were different for each compartment (see Table 3), again indicating the presence of two different latices.

The polymers from the compartmented miniemulsion polymerisation of MMA and BA (see Table 3) had glass transition temperatures different from the corresponding homopolymers, indicating that mass transfer has taken place to some extent.

Again, to allow sufficient time for mass transfer, we included LPO as the initiator. Figure 6 shows the conversion-time histories and the instantaneous monomer composition during the reaction.

The relatively high rate of diffusion of MMA ensures an MMA fraction of 5–8% (mol/mol) in the BA minemulsion droplets. The BA fraction in the MMA droplets, however, does not exceed 1 mol%. Using the reactivity ratios $r_{MMA} = 2.24$ and $r_{BA} = 0.414$,[27] the fraction F_{MMA} incorporated in the polymer chains is expected to be around 0.15. Applying Fox' rule, the measured glass transition temperature of −36 °C is close to the calculated value of −34 °C.

Visual inspection of the reactor contents after polymerisation revealed that the liquid levels were differing by approximately 23 ml, the BA rich latex having the largest volume. Nevertheless, the solid content of this BA latex was 21.2% and therefore higher than expected from the original

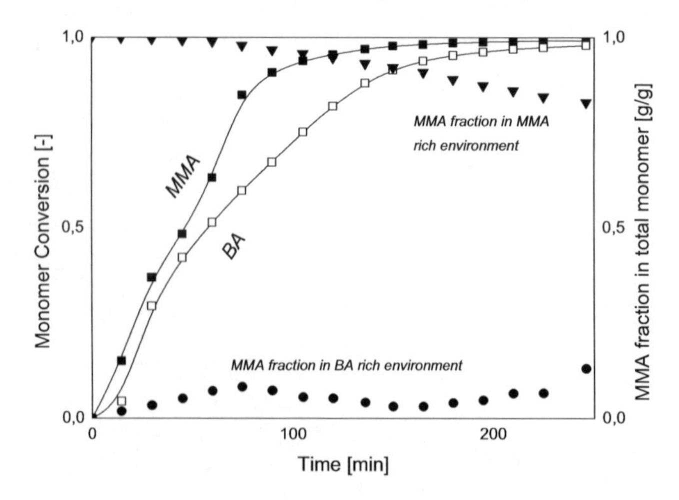

Figure 6.
Conversion-time profiles for the compartimented miniemulsion polymerisations of (■)MMA and (□) BA and the MMA monomer fraction in (●) BA rich compartment and (▼) MMA rich compartment.

formulation, whereas the MMA latex had a lower solid content (15%) compared to the formulation. This difference in liquid level cannot solely be attributed to MMA diffusion. The higher monomer weight fraction in the BA miniemulsion, as a result of the MMA diffusion, creates an osmotic pressure between the two miniemulsions, resulting in a net transfer of water across the membrane.

The presented experimental results in this paper demonstrate that, in mixed miniemulsions, monomer redistribution among the monomer droplets takes place and is completed in the early stages of polymerisation. The experimental results also prove that neither monomer water solubility, nor the amount of monomer droplets is a determining factor in monomer mass transfer in miniemulsion polymerisation. When droplet-droplet interaction is prevented, by the introduction of a monomer-permeable membrane, it becomes apparent that the exchange of monomer is almost completely restricted for very hydrophobic monomers, but still takes place for sparsely water-soluble monomers, albeit severely limited.

Conclusion

Despite the compartmentalised nature of a miniemulsion, mass transfer between individual droplets takes place. Mass transfer is a rapid process when two miniemulsions with different monomers are mixed together, the exchange of the monomer is accomplished at the very early stage of polymerisation, regardless the hydrophobicity of the monomer.

This rapid exchange observed in miniemulsions containing monomers such as lauryl methacrylate that are too hydrophobic to be used in conventional emulsion polymerisation, raises a question about whether this mass transfer is predominantly diffusion-based or collision-based. In the work described in this paper it was demonstrated that a 100-fold dilution in miniemulsion polymerisation, thereby increasing the diffusion distance and decreasing the collision probability, still resulted in rapid monomer transfer between the droplets.

When droplet-droplet interaction was restricted as a result of the use of a membrane, separating two miniemulsions, a difference in the mass transfer between sparsely water-soluble monomers such as methyl methacrylate and butyl acrylate and very hydrophobic monomers such as lauryl methacrylate and 4-tert-butyl styrene was observed. While monomer transfer still takes place for sparsely water-soluble monomers, as observed in the formation of copolymers, it is restricted compared to mixed miniemulsions and monomer transfer was almost completely absent in the case of lauryl methacrylate and 4-tert-butyl styrene.

The results presented in this paper support the idea that mass transfer by collisions makes a major contribution in (mini)emulsion polymerisation.

The authors would like to acknowledge the *Foundation Emulsion Polymerization* (SEP) Eindhoven for financially supporting this research, and the *School of Materials, University of Manchester*, where part of the research has been executed.

[1] J. Ugelstad, M. S. El-Aasser, J. W. Vanderhoff, *J. Polym. Sci. Lett. Ed.* **1973**, *11*, 503.
[2] J. M. Asua, *Prog. Polym. Sci*, **2002**, *27*, 1283.
[3] J. A. Alduncin, J. Forcada, J. M. Asua, *Macromolecules*, **1994**, *27*, 2256.
[4] J. L. Reimers, F. J. Schork, *Ind. Eng. Chem. Res.*, **1997**, *36*, 1085.
[5] J. L. Reimers, F. J. Schork, *J. Appl. Polym. Sci*, **1996**, *59*, 1833.
[6] Z. Yu, P. Ni, J. Li, X. Zhu, *Coll. Surf. A, Ph. Eng Asp*, **2004**, 242, 9.
[7] C. T. Lin, W. Y. Chiu, H. C. Lu, Y. Meliana, C. S. Chern, *J. Appl. Polym. Sci*, **2010**, *115*, 2786.
[8] K. Ouzineb, C. Graillat, T. F. McKenna, *J. Appl. Polym. Sci*, **2004**, *91*, 115.
[9] U. Yildiz, K. Landfester, M. Antonietti, *Macromol. Chem. Phys.* **2003**, *204*, 1966.
[10] K. Tauer, A. M. Imroz Ali, U. Yildiz, M. Sedlak, *Polymer*, **2005**, *46*, 1003.
[11] A. J. Back, F. J. Schork, *J. App. Polym. Sci*, **2007**, *103*, 819.

[12] A. J. Back, F. J. Schork, *J. App. Polym. Sci*, **2006**, *102*, 5649.
[13] C. S. Chern, J. C. Sheu, *Polymer*, **2001**, *42*, 2349.
[14] V. S. Rodriguez, J. Delgado, C. A. Silebi, M. S. El-Aasser, *Ind. Eng. Chem. Res.*, **1989**, *28*, 65.
[15] J. Delgado, M. S. El-Aasser, C. A. Silebi, J. W. Vanderhoff, J. Guillot, *J. Polym. Sci B*, **1988**, *26*, 1495.
[16] S. H. Yalkowsky, S. Banerjee, *Aqueous solubility – methods of estimation for organic compounds*, Marcel Dekker, New York **1992**, p.41–127.
[17] K. Ouzineb, C. Graillat, M. A. Dubé, R. Jovanovic, T. F. McKenna, *Macromol. Symp.*, **2004**, *206*, 107.
[18] J. M. Asua, V. S. Rodriguez, C. A. Silebi, M. S. El-Aasser, *Makromol. Chem., Macromol. Symp.*, **1990**, *35/36*, 59.
[19] V. S. Rodriguez, J. M. Asua, M. S. El-Aasser, C. A. Silebi, *J. Pol. Sci, B*, **1991**, *29*, 483.
[20] V. S. Rodriguez, M. S. El-Aasser, J. M. Asua, C. A. Silebi, J. Polym, A. Sci, **1989**, *27*, 3659.
[21] N. M. B. Smeets, T. G. T. Jansen, A. M. van Herk, J. Meuldijk, J. P. A. Heuts, *Polym. Chem.*, **2011**, *2*, 1830.
[22] N. M. B. Smeets, T. G. T. Jansen, T. J. J. Sciarone, J. P. A. Heuts, J. Meuldijk, A. M. van Herk, *J. Polym. Sci A*, **2010**, *48*, 1038.
[23] D. Kukulj, T. P. Davis, K. G. Suddaby, D. M. Haddleton, R. G. Gilbert, *J. Polym. Sci A*, **1997**, *35*, 859.
[24] H. F. Hernandez, K.o Tauer, *Macromol. React. Eng.*, **2009**, *3*, 375.
[25] J. W. Vanderhoff, *Chem. Eng. Sci*, **1993**, *48*, 203.
[26] J. S. Arellano, J. F. Flores, F. Zuluaga, E. Mendizabal, I. Katime, *J. Pol. Sci. A*, **2011**, *49*, 3014.
[27] R. A. Hutchinson, J. H. McMinn, D. A. Paquet, S. Beuermann, C. Jackson, *Ind. Eng. Chem. Res.* **1997**, *36*, 1103.

Microencapsulation of Fragrance and Natural Volatile Oils for Application in Cosmetics, and Household Cleaning Products

Rumeysa Tekin,[1] Nurcan Bac,[1] Huseyin Erdogmus[2]*

Summary: In this study, polyurethane-urea microcapsules containing fabric softener fragrance (Teddysoft™, by EPS) as active agents were produced by interfacial polymerization. Size, shape and morphology of the microcapsules were studied by using an optical microscope and scanning electron microscope. The effectiveness of the fragrance containing microcapsules was tested by washing hand towels in a commercial washing machine. Textile impregnation and the durability of the fragrance impregnation effect were evaluated qualitatively by perfumers, and by headspace GC analysis. Optical and scanning electron microscopic results indicated that increasing the stirring rate during the emulsion step results in microcapsules with smaller particle size, but at some point morphologies were adversely affected.

Keywords: interfacial polymerization; microencapsulation; volatile oils

Introduction

Fragrance chemicals are used in numerous products to enhance the consumer's enjoyment. They are added to consumer products such as laundry detergents, fabric softeners, soaps, personal care products, such as shampoos, body washes, deodorants etc.[1] However, fragrances exhibit high volatility, and rapidly lose their aroma if they exposed to the atmosphere.[2] In order to enhance the effectiveness of the fragrance materials, various technologies have been employed to enhance the delivery of the fragrance materials at the desired time. A widely used technology is encapsulation of the fragrance material in a protective coating.[1]

Microcapsules are small particles that contain an active agent or core material surrounded by a coating layer or shell. Their diameter may vary from 1 to 1000 micrometers. Microcapsules may have variety of structures ranging from spherical and irregular shapes with one core to multicores and even multilayer coatings. The structure of the microcapsules depend on many parameters, such as rigidity of the coating layer, the type of core material and the preparation method.[3] The encapsulation is used to protect fragrances or other active agents from oxidation caused by heat, light, humidity, and exposure to other substances over their lifetime. It has been also used to prevent the evaporation. of volatile compounds and to control the rate of release.[4] The capsule shell can be compromised by various factors such as temperature so that the contents are delivered when the capsule begins to melt. Alternatively, the capsules can be compromised by physical forces, such as crushing, or other methods that compromise the integrity of the capsule. Additionally, the capsule contents may be delivered via diffusion through the capsule wall during a desired time interval.[5]

According to Carles [6] the pyramid structure of a perfume is classified in three parts. Very volatile materials, which disappear first, comprise the top notes of a finished perfume, those of intermediate volatility and tenacity are the middle

[1] Yeditepe University Chemical Engineering Dept, Istanbul, Turkey
E-mail: rumeysatekin@gmail.com
[2] EPS Fragrances, Kemerburgaz, Istanbul, Turkey

notes, and those with the lowest volatility, tenacious products constitute the base notes.[7]

Several microencapsulation techniques have been used such as coacervation [8] and in situ polymerization [9–11] for the microencapsulation of fragrances. Interfacial polymerization techniques probably the most popular one. In interfacial polymerization process there is typically an oily core containing one of the reactants dispersed in an aqueous media containing the other reactant. These two reactants meet at an interface and react rapidly, then, the aqueous media causes a wall.[3]

In this project, polyurethane-urea microcapsules containing fabric softener fragrance were synthesized by using interfacial polymerization. Size, shape and morphology of the microcapsules were determined by an optical microscope and SEM. Microcapsules containing fragrance were applied in the fabric softener base and laundering test was performed for textile impregnation. The persistence of the fragrance impregnation was evaluated by perfumers, and headspace GCMS analysis.

Materials and Methods

Materials

Toluemne 2,4 diisocyanate (TDI) from Acros Organics, polyethylene glycol with molecular weight of 400 (PEG 400) from Merck, ethylenediamine (EDA) from Sigma-Aldrich, hydrazine (HYD) from Aldrich, dibuthyltindilaurate (DBTDL) from Merck, polyvinylalcohol with molecular weight of 88000 (PVA) from Acros Organics and Teddysoft™ (fragrance of fabric softener) from EPS were used.

Microcapsules Preparation

1L three-necked Teknosem TMK-1000 aluminium reactor with temperature controller and a mechanical stirrer was used for the microencapsulation process. The experimental setup was shown in Figure 1. Oil phase was prepared by mixing 150 mL

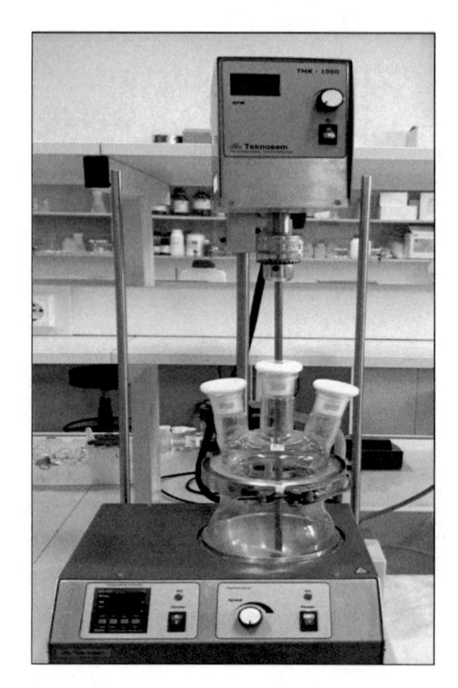

Figure 1.
Experimental set-up for microencapsulation process.

Teddysoft and 30 mL TDI. The first aqueous phase including 500 mL water and 8 gr PVA, that was used to stabilize the droplets, was mixed with oil phase with WiseTis HG-15D homogenizer at different stirring rates: 2000 rpm, 4000 rpm and 6000 rpm for 5 minutes at room temperature. The emulsion was transferred to the reactor. The second aqueous phase including 100 mL water, 46 mL PEG and 2 mL DBDTL, as a catalyst, was prepared and mixed with emulsion at 60–80 °C and 200 rpm for 1 h to form the polyurethane wall. The third aqueous phase containing 40 mL water and 8 mL EDA was added to the reactor and mixed at the same

Table 1.
Compositions for laundering test.

	Mixture	Fragrance (per cent)
1	100% Teddysoft	0.5
2	50% Teddysoft, 50% microcapsules	0.5
3	80% Teddysoft, 20% microcapsules	0.5

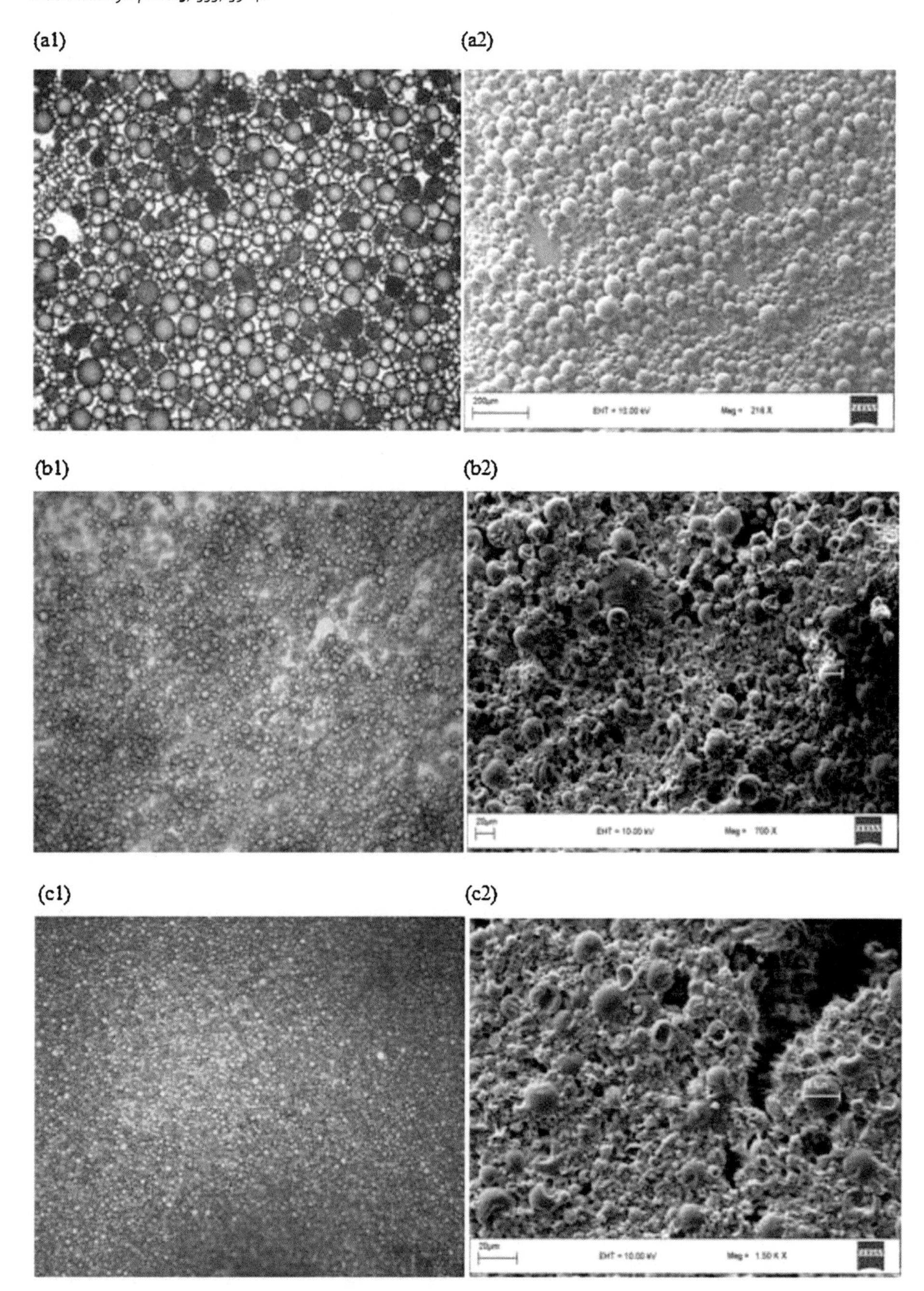

Figure 2.
Optic Microscope (a1, b1 and c1) and SEM images (a2, b2 and c2) of fragrance containing microcapsules. a1, a2: 2000 rpm; b1, b2: 4000 rpm; c1, c2: 6000 rpm.

temperature for 1 h to proceed with urea formation. The final aqueous phase containing 3 mL HYD, was added and let to react for 1h. The final product was separated by sedimentation and microcapsules were collected. Then, they were washed with water to get rid of the excess reactants.

Optical Microscopy

DMS-633 Digital Optic Mikroskope was used to analyse the size and shape of the microcapsules.

Scanning Electron Microscopy

Karl Ziess EVO 40 with Thungsten Light SEM was used to determine the size and morphology of microcapsules.

Laundering Test

Microcapsules containing fragrance (2000 rpm strring rate of emulsion) and fragrance without microcapsules were mixed with the fabric softener base. The compositions of the mixture of the fragrance and the microcapsules containing fragrance were shown in Table 1. The hand towels (20cm × 20cm) were washed in a Samsung Ag^+ Silver Nano Technology Washing Machine.

Headspace-GCMS Analysis

After laundering of the hand towels, they were left to dry. After that, the hand towels were put in separate plastic bags, and waited for a week for release of the fragrance. The gas phase was sampled and injected to the GCMS.

Results and Discussion

The optic microscope pictures showed the spherical formation of the microcapsules containing fragrance (Figure 2. a1, b1 and c1). The SEM pictures also showed the size and the morphology of the capsules (Figure 2. a2, b2 and c2). Emulsion parameters are critical for the size distribution of the microcapsules. Therefore, the microcapsules were prepared with different stirring rate of emulsion. When the stirring rate of emulsion was increased from 2000 rpm (Figure 2. a1) to 4000 rpm (Figure 2. b1) and 6000 rpm (Figure 2. c1), the size of the microcapsules got smaller as expected. The diameter of the microcapsules were approximately between 50–80 μm, 20–30 μm and 10–20 μm, when the stirring rate of the emulsion were 2000 rpm, 4000 rpm and 6000 rpm respectively. The morphology of the microcapsules prepared with the stirring rate 2000 rpm during emulsion process was smoother and more homogeneous compared to 4000 rpm and 6000 rpm as shown in Figure 2. a2, b2 and c2. Increasing the stirring rate of the emulsion resulted in cracks on microcapsules which may cause quick release of the fragrance from the microcapsules. The agglomeration also increased with increasing stirring rate during emulsion.

The laundering tests were performed by washing hand towels with the fabric softener including different amount of the fragrance containing microcapsules and without microcapsules in a commercial washing machine. The hand towels were let to dry for a week after laundering and nine perfumers were evaluated them qualitatively. According to the evaluations of perfumers shown in Table 2, the towels that were washed with the fabric softener base including more microcapsules smelt strongest.

In Figure 3, the GCMS results at the top indicate the case where fragrance with microcapsules were used in the washing step. The presence of higher

Table 2.
Evaluations of perfumers one week after laundering test.

Perfumers	Evaluation
Perfumer 1	2>3>1
Perfumer 2	2>1>3
Perfumer 3	2>3>1
Perfumer 4	2>1>3
Perfumer 5	2>1>3
Perfumer 6	2>1>3
Perfumer 7	2>1>3
Perfumer 8	2>1>3
Perfumer 9	2>1>3

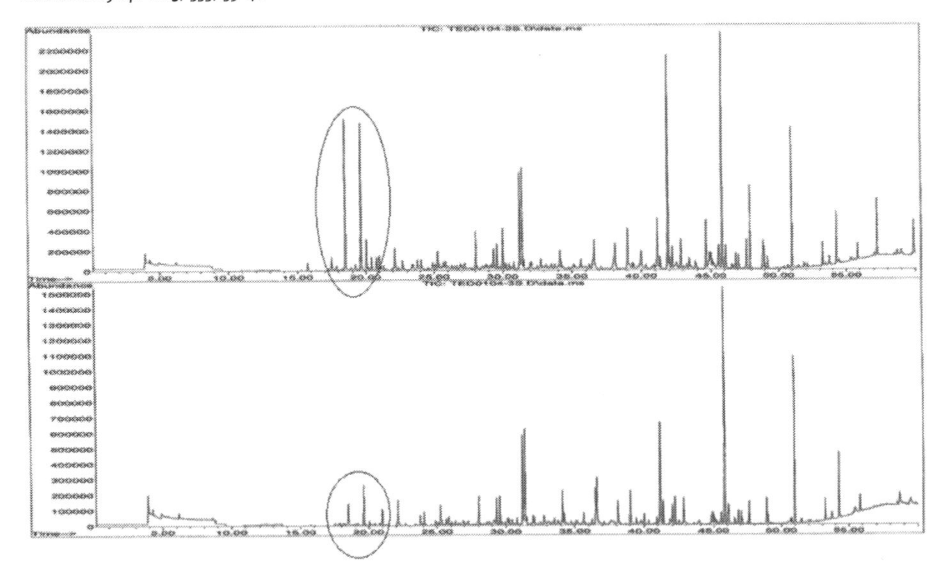

Figure 3.
Headspace-GCMS analysis. The hand towel that was washed with the softener base including 50% Teddysoft, and 50% microcapsules (above), the hand towel that was washed with the softener base with 100% Teddysoft (below).

concentration of fragrance molecules indicated as the first couple of peaks in the top section clearly proves the positive effect of microcapsules for extended release of the fragrance.

Conclusion

In this study, the fragrance of fabric softener was encapsulated in a polymer coating to extend the release rate and increase the persistence of the fragrance. The polyurethane-urea microcapsules including the fragrance were synthesized by interfacial polymerization and the emulsion parameter was investigated. The optical microscope and SEM pictures confirmed the spherical form of microcapsules. The diameters of microcapsules were ranging from 10 μm to 80 μm. It was concluded that increasing the stirring rate during emulsion caused smaller size microcapsules. However, SEM pictures also showed that the morphology of the microcapsules were adversely affected at increased stirring rates. The persistence of the fragrance was checked with both the laundering test and headspace GC analysis. The evaluations of perfumers showed that the hand towels washed with the fabric softener including more microcapsules smelt strongest after a week. The headspace-GC analysis also indicated that quantitatively more volatile compounds of the fragrance existed on the hand towel when the fragrance was encapsulated.

Acknowledgements: This project was carried out in Yeditepe University. We also acknowledge the support of our partner Erdogmus Fragrance Co. (EPS) and its perfumers.

[1] Barin et al., **2009**, *Encapsulated Fragrance Chemicals*, US 7632789 B2
[2] Smets et al., **2011**, *Microcapsule and Method of Producing Same*, US 7901772 B2
[3] S. Magdassi, E. Touitou, **1999**, *Novel Cosmetic Delivery Systems*, Newyork.
[4] S. N. Rodrigues, I. M. Martins, I. P. Fernandes, P. B. Gomes, V. G. Mata, M. F. Barreiro, A. E. Rodrigues, *Chemical Engineering Journal* **2009**, *149*, 463–472.

[5] Popplewell et al., **2009**, *Encapsulated Materials*, US 7491687 B2
[6] J. Carles, *Soap Perfumes Cosmet.* **1962**, *35*, 328.
[7] S. J. Herman, *Fragrance Applications: A Survival Guide*. Allured, Carol Stream, Il. **2002**.
[8] J. C. Soper, Y. D. Kim, M. T. Thomas, **2000**. *Method of encapsulating flavours and fragances by controlled water transport into microcapsules*, U.S. Patent 6106875
[9] J. Hwang, J. Kim, Y. Wee, J. Yun, H. Jang, S. Kim, H. Ryu, *Biotechnology and Bioprocess Engineering*, **2006**, *11*, 332–336.
[10] J. Hwang, J. Kim, Y. Wee, J. Yun, H. Jang, S. Kim, H. Ryu, *Biotechnology and Bioprocess Engineering*, **2006**, *11*, 391–395.
[11] K. Hong, S. Park, *Materials Chemistry and Physics*, **1999**, *58*, 128–131.

Macromol. Symp. **2013**, 333, 41–54 DOI: 10.1002/masy.201300045

Stimuli-Responsive Hydrogels Synthesis using Free Radical and RAFT Polymerization

Miguel A. D. Gonçalves,[1] *Virgínia D. Pinto,*[1] *Rita A. S. Costa,*[1] *Rolando C. S. Dias,**[1] *Julio C. Hernándes-Ortiz,*[2] *Mário Rui P. F. N. Costa*[2]

Summary: Temperature and pH stimuli-responsive hydrogel particles were synthesized using inverse-suspension polymerization in batch stirred reactor. Different water soluble co-monomers were present in the initial mixture (e.g. N-isopropylacrylamide and acrylic acid) as well as crosslinkers with different functionalities. Different operating conditions such as polymerization temperature, monomers dilution, neutralization and the initial ratios of co-monomers and monomers/crosslinker were also tried. Hydrogel particles were produced considering classical free-radical polymerization (FRP) and also RAFT polymerization. Commercially available RAFT agents 4-cyano-4-phenylcarbonothioylthio-pentanoic acid (CPA), 2-(dodecylthiocarbonothioylthio)-2-methylpropionic acid (DDMAT) and cyanomethyl dodecyl trithiocarbonate (CDT) were alternatively used. Sampling at different polymerization times allowed the study of the kinetics of polymerization through the analysis by SEC of the soluble phase. A tetra-detector array with simultaneous detection of refractive index, light scattering, intrinsic viscosity and ultra-violet signals was used in these studies. Usefulness of *in-line* FTIR-ATR monitoring to study the building process of such networks was also assessed. The performance of hydrogel beads was studied through drug delivery tests triggered by changes in the environmental temperature and pH. This research aims to contribute for the elucidation of the connection between the synthesis conditions, molecular architecture and properties/performance of such advanced materials.

Keywords: crosslinking; hydrogels; RAFT; stimuli-responsive polymers

Introduction

Hydrogels have been extensively studied in last decades due to their potential new applications in biotechnology and biomedicine[1,2]. Researches on this field are specially focused on the so called smart hydrogels (or stimuli responsive hydrogels) which microscopic properties are sensitive to changes triggered by the environmental conditions. Generically speaking, formation of soluble networks and gels has been experimentally and theoretically studied since the beginning of polymer science. With systems involving vinyl/multivinyl monomers, classical free radical polymerization (FRP) mechanisms were mainly considered in these studies. A new importance was given to this subject with the advent of controlled radical polymerization (CRP). New studies in this research area were driven by the possibility of improvement of networks and gels properties as result of higher structural homogeneity. In fact, in the last years, the three main CRP techniques (ATRP[3–9], NMRP[8–15], and RAFT[16–19]) were exploited aiming the production of advanced polymer networks belonging to different classes (e.g. organic or water compatible materials).

[1] LSRE-Instituto Politécnico de Bragança, Quinta de Santa Apolónia, 5300, Bragança, Portugal
E-mail: rdias@ipb.pt

[2] LSRE-Faculdade de Engenharia da, Universidade do Porto, Rua Roberto Frias s/n, 4200-465, Porto, Portugal

RAFT polymerization can be used in a broad range of operation conditions and with different monomer classes, including water compatible monomers. These advantages are explored in this work considering the inverse-suspension synthesis of different classes of stimuli-responsive hydrogels particles and using three different commercially available RAFT agents. FRP synthesis of the same materials was also performed in order to highlight the differences between the two processes (namely in the observed kinetics of polymerization). The use of different polymer characterization methods (such as SEC with tetra-detection and *in-line* FTIR-ATR monitoring) to study the formation of such networks is illustrated. The usefulness of such techniques to describe the crosslinking process is discussed. Final applications of the produced materials are also tested through drug delivery studies. Experimental results here reported aims to contribute to the sought linking between the production conditions of advanced materials and their structure/properties. Development of tools helping in the synthesis of tailored materials is the ultimate goal of this research.

Experimental

Materials

N-isopropylacrylamide (NIPA) of 99% purity, N,N-dimethylacrylamide (DMA) of 99% purity stabilized with 500 ppm monomethyl ether hydroquinone (MEHQ), 2-(dimethylamino)ethyl methacrylate (DMAEMA) of 98% purity stabilized with 700-1000 ppm MEHQ, acrylic acid (AA) of 99% purity stabilized with 180-200 ppm MEHQ, methacrylic acid (MAA) of 99% purity stabilized with 250 ppm MEHQ, N,N'-methylenebisacrylamide (MBAm) of 99% purity, ethylene glycol dimethacrylate (EGDMA) of 98% purity stabilized with 90-110 ppm MEHQ, trimethylolpropane triacrylate (TMPTA) stabilized with 100 ppm methylethylhydroquinone, 1,1,2,2 - tetraallyloxyethane (TAO), hydroquinone of 99% purity, AIBN of 98% purity, 2,2'-azobis(2-methylproprionamidine) dihydrochloride (V50) of 98% purity, ammonium persulfate (APS) of 98% purity and N,N,N',N'-tetramethylethylenediamine (TEMED) of 99% purity were purchased from Sigma Aldrich and used as received. The commercially available RAFT agents 2-(dodecylthiocarbonothioylthio)-2-methylpropionic acid (DDMAT) of 98% purity, 4-cyano-4-(phenylcarbonothioylthio)pentanoic acid (CPA) of 97% purity and cyanomethyl dodecyl trithiocarbonate (CDT) of 98% purity were also purchased from Sigma Aldrich and used as received. Dimethylformamide (DMF) of 99.5% purity (Fisher Scientific), tetrahydrofuran (THF) of 99% purity (Fisher Scientific) and cyclohexane of 99% purity (Sigma Aldrich) were also used as received. A current grade of liquid paraffin was used when needed. Caffeine of 98.5% purity, ibuprofen of 99% purity, 5-fluorouracil (5Fu) of 98.5% purity and isonicotinic acid hydrazide (isoniazid) of 99% purity were purchased from Acros Organics and used as model chemicals in the drug release tests performed. Ibuprofen was also transformed in its sodium salt in order to increase the solubility of the drug in the aqueous solutions considered.

Polymerization Runs

Hydrogels were synthesized in batch reactor using the inverse suspension process. Polymerizations were performed at 200 mL total volume scale with stable suspension formation using a volumetric ratio aqueous/organic phases = 1/5, 1% (w/w) of surfactant (span 80) in the continuous phase and agitation speed at 300 rpm. Gel production at isothermal conditions and keeping a good stirring of the reaction vessel was possible using these conditions. When applicable, reactants were previously bubbled with argon, that was also sweep in the reaction medium during the polymerizations in order to prevent inhibition by oxygen. At prescribed polymerization times, reaction samples were collected from the reactor, quenched at low temperature in a solution containing hydroquinone to stop the reactions, and afterwards prepared for injection of the soluble polymer in the SEC system[20]. Morphology of the final gel beads, after products purification, was also characterized by SEM. Micro appearance of the produced materials is illustrated in Figure 1.

Tables 1–3 describe the details of a set of experiments performed in this research. Different polymerizations were performed combining different water compatible vinyl monomers and crosslinkers. Initiation system was also changed along the experimental program. Conventional (FRP) and RAFT mechanisms were used. Besides these conditions, the following main parameters describe the set of polymerization runs:

- Initial mole fraction of co-vinyl monomer (M_1) in the binary mixture of $M_1 + M_2$ (Y_{M1}).
- Initial mole ratio between initiator and monomers (Y_I).
- Initial mole fraction of crosslinker in the total monomer mixture (Y_{CL}).
- Initial mass fraction of the monomers in the dispersed phase (Y_m).
- Initial mole ratio between RAFT agent and initiator (Y_I^{RAFT}).

Slightly different reaction conditions were used in the experimental runs 1

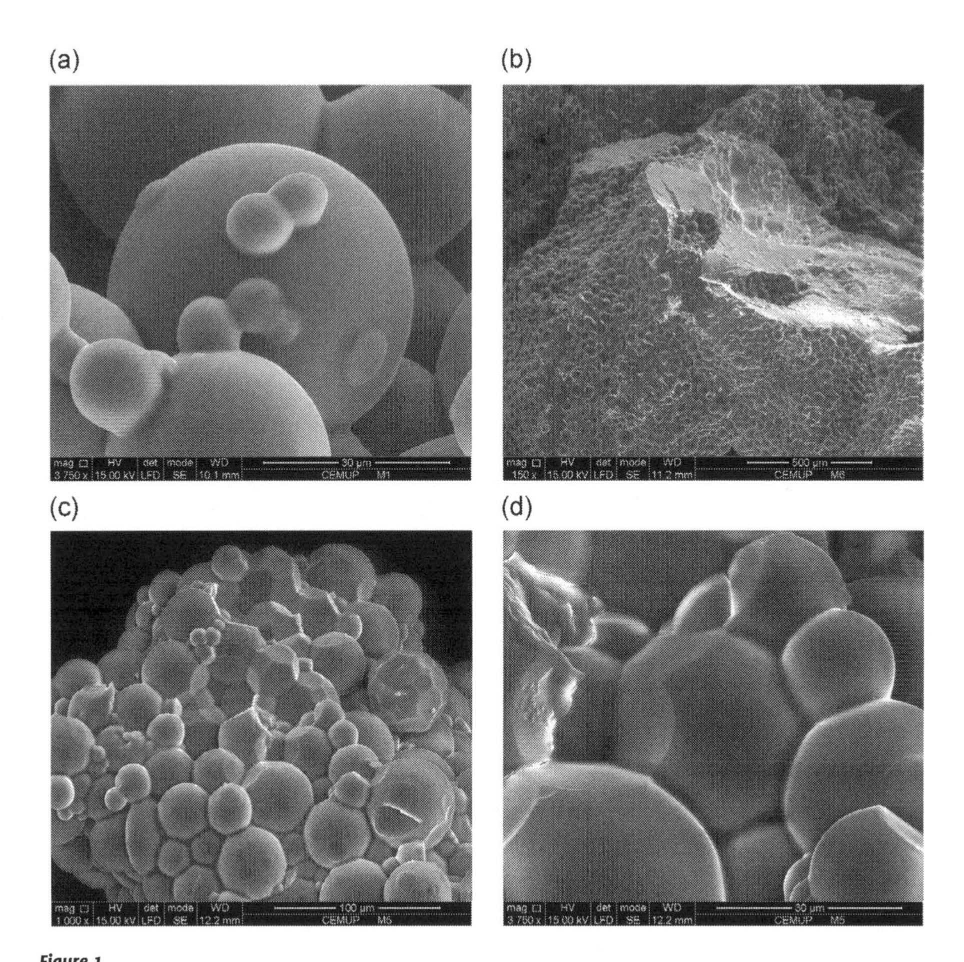

Figure 1.
SEM micrographs of some hydrogel beads synthesized in this work using the inverse-suspension technique. (a) AA/MBAm hydrogel. (b) NIPA/MBAm hydrogel. (c) and (d) NIPA/AA/MBAm hydrogel. In some cases, formation of fused material was observed due to the post-treatment of the products (precipitation/drying).

Table 1.
A set of polymerization runs performed in the inverse-suspension FRP synthesis of pH/Temperature responsive hydrogels. Water was used as solvent in the dispersed phase and cyclohexane was considered as continuum medium. Polymerizations at 20 °C.

Run	M_1	M_2	CL	I	Y_{M1}(%)	Y_{CL}(%)	Y_I(%)	Y_m(%)
1	NIPA	–	MBAm	APS	100	1	0.25	10
2	NIPA	AA	MBAm	APS	50	1	0.25	14.5
3	NIPA	MAA	MBAm	APS	67	1	0.25	13.8
4	NIPA	AA	MBAm	APS	88	1	0.26	11.1
5	NIPA	–	TAO	APS	100	1	0.25	10
6	NIPA	–	TMPTA	APS	100	1	0.25	10
7	NIPA	–	MBAm	APS	100	2	0.25	10

Table 2.
A set of polymerization runs performed in the inverse-suspension RAFT synthesis of water compatible polymers and hydrogels. DMF was used as solvent in the dispersed phase and liquid paraffin was considered as continuum medium. Polymerizations at 70 °C. DDMAT was used as RAFT agent.

Run	M_1	M_2	CL	I	Y_{M1}(%)	Y_{CL}(%)	Y_I(%)	Y_m(%)	Y_I^{RAFT}
1	NIPA	–	–	AIBN	100	0	0.24	25.6	4.18
2	AA	–	–	AIBN	100	0	0.23	17.9	4.42
3	DMA	–	–	AIBN	100	0	0.06	33.2	5.03
4	MAA	–	–	AIBN	100	0	0.05	33.6	5.00
5	AA	–	–	AIBN	100	0	0.03	40.4	8.69
6	AA	–	MBAm	AIBN	100	1	0.04	40.3	8.54
7	NIPA	–	MBAm	AIBN	100	1	0.04	30.0	8.96
8	NIPA	AA	MBAm	AIBN	90	1	0.04	31.0	9.00

and 3 described in Table 3, aiming the improvement of the *in-line* FTIR-ATR measurements. Liquid paraffin was used to minimize the IR absorption of the continuum medium and solvent (e.g. water or DMF) was not used in the polymerization phase to eliminate the usual strong influence of these compounds on the IR spectra of monomers and produced polymers. Liquid paraffin was also considered in order to promote a low thermodynamic affinity with monomers and produced gels. Precipitation of the products in the particulate form along the polymerization was thus observed and good stirring conditions, as well as, good heat dissipation could be maintained during the reactions. After purification, powder gels were obtained as final products. Effect of other synthesis conditions on the crosslinking process, such as, presence of solvent and the nature of monomers and RAFT agents, was also

Table 3.
A set of polymerization runs performed in the synthesis of water compatible polymers and hydrogels considering liquid paraffin as continuum medium. EGDMA was used as crosslinker in runs 1-3 and MBAm in runs 4-5. Polymerizations at 60 °C with exception of run 4 (50 °C). Water as solvent in runs 2 and 4 and DMF in run 5. RAFT agents used: CPA in runs 3-4 and CDT in run 5.

Run	M_1	M_2	I	Y_{M1}(%)	Y_{CL}(%)	Y_I(%)	Y_{M1}(%)	Y_I^{RAFT}
1	DMAEMA	–	AIBN	100	4.55	0.5	100	0
2	DMAEMA	–	V50	100	4.88	0.3	73	0
3	DMAEMA	–	AIBN	100	4.76	0.5	100	1.98
4	DMAEMA	MAA	V50	35	2	0.14	50	1.94
5	NIPA	AA	AIBN	88	1	0.36	22	1.99

assessed in this set of polymerizations as reported with the remaining experimental runs of Table 3.

Product Analysis by SEC with a Tetra Detector Array

The SEC apparatus used is composed of a Viscotek GPCmax VE 2001 integrated solvent and sample delivery module coupled to a tetra detector array including refractive index (RI), light scattering (LS), viscosity (IV-DP) and ultraviolet (UV) detection. Analysis were performed directly in aqueous eluents (pH of the eluent was changed according the polymer analysed) and using typically a flow-rate of 0.5 mL/min. Temperature of the analysis (in the range 30 to 50 °C) was also changed considering the different conformations of the polymers in aqueous solutions (e.g. collapsing of NIPA based materials at around 37 °C). A train of 3 SEC columns (Viscotek A2000 + Viscotek A3000 + Viscotek A6000) was considered to fractionate the polymers by size (different configurations were also used in order to not exceed the recommended maximum columns pressures). Simultaneous measurement of RI, LS and intrinsic viscosity signals yield absolute molecular weight, branching factors, hydrodynamic radius and radius of gyration of the soluble phase. Monomer conversion was also estimated through the measurement of the monomers peak areas in these chromatograms. Typical results concerning the analysis of water soluble polymers with this apparatus are illustrated in Figure 2. Simultaneous detection of three signals (RI, LS and Intrinsic Viscosity) allows the detailed characterization of the molecular architecture of the soluble phase and the observation of the influence of the production conditions (e.g. comparison FRP/RAFT synthesis) on the dynamics of crosslinking. Light scattering measurements proved to be especially important because show the possibility of occurrence of non-ideal RAFT polymerization, as below described. In this research, UV detection was used to carry out drug release studies with the synthesized hydrogels, as also below described.

In-Line FTIR-ATR Measurements

An Attenuated Total Reflection (ATR) immersion probe, coupled to a Fourier Transform Infra-Red (FTIR) spectrophotometer (which technical features were described elsewhere[7]) was used to perform the *in-line* monitoring of polymerization runs, aiming the measurement of the building process of the networks. Typical results obtained with the *in-line* FTIR-ATR monitoring of hydrogel formation are illustrated in Figure 3. These runs were designed in order to have optimum

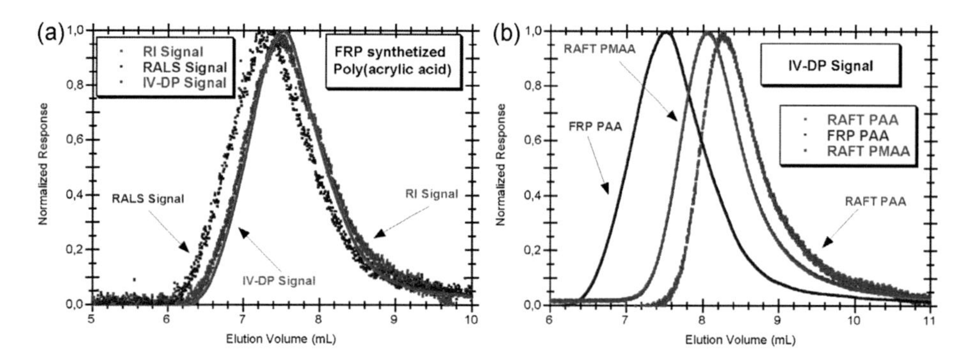

Figure 2.
(a) Refractive Index (RI), Right Angle Light Scattering (RALS) and Intrinsic Viscosity - Differential Pressure (IV-DP) signals simultaneously observed in the SEC analysis of a water soluble PAA sample. (b) IV-DP signals observed in the SEC analysis of different water soluble polymers synthesized in this research, highlighting the influence of operation conditions (e.g. FRP/RAFT) on the products molecular structure and properties.

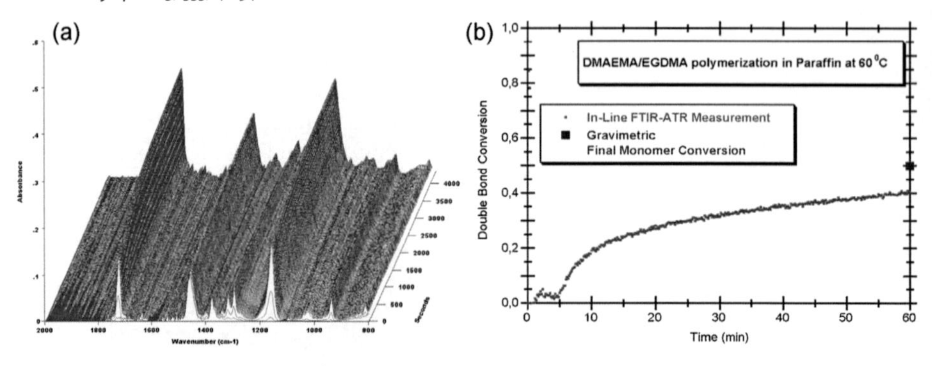

Figure 3.
(a) *In-line* FTIR-ATR spectra observed during DMAEMA/EGDMA FRP polymerization (run 1 in Table 3). Absorption peak at around 935 cm^{-1} was considered to estimate the double bonds conversion, using also the peak at around 1720 cm^{-1} as internal reference. (b) FTIR-ATR estimated dynamics of monomer conversion for DMAEMA/EGDMA polymerization considering FRP polymerization (run 1 in Table 3). Similar measurements were performed with runs 2 and 3 in Table 3 but even lower monomer conversions were observed in these experiments (almost negligible after 8 hours of polymerization in run 3).

conditions for FTIR measurements (minimizing possible interferences with monomers/polymer spectra) but a limited information concerning the crosslinking process was obtained. Coating of the probe at relative low monomer conversion (e.g. around 40% as recently reported in other studies[22]) is a possible shortcoming of these *in-line* measurements. *Off-line* FTIR analysis of previously isolated network samples collected at different polymerization times seems to lead to a better description of the crosslinking process, namely concerning pendant double bonds reactivity[23]. Crosslinker amount used in hydrogels preparation is very low (a few percent) and even with *off-line* FTIR monitoring the study of the crosslinking

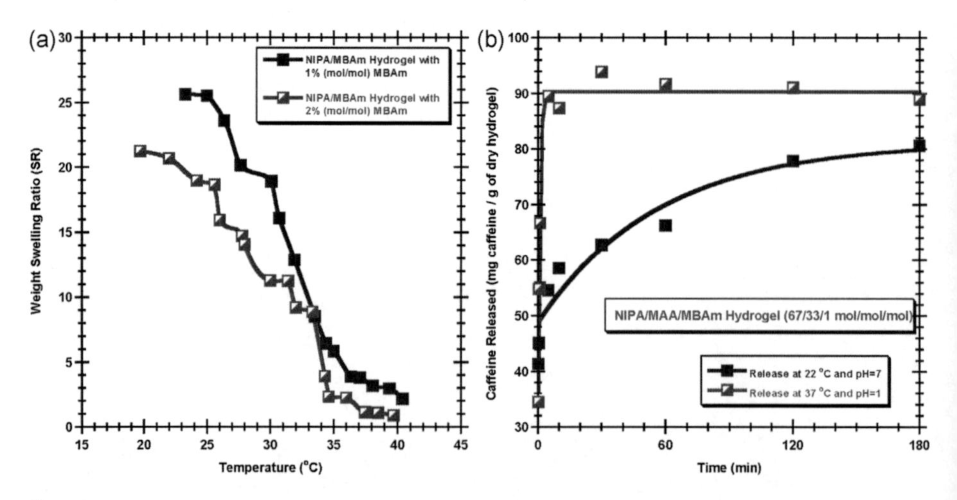

Figure 4.
(a) Measured equilibrium weight swelling ratio of NIPA/MBAm hydrogels in aqueous solutions at different temperatures illustrating networks sensitivity to changes in this parameter. (b) Dynamics of caffeine release from a pH/temperature sensitive synthesized hydrogel (NIPA/MAA/MBAm) measured by UV detection at 270 nm. Two different surrounding water solutions were considered: pH = 1/T = 37 °C (collapsed particles) and pH = 7/T = 22 °C (swollen particles). Drug loading was performed by swelling the hydrogel beads in caffeine aqueous solution during 48 hr.

process is a difficult task. ^{13}C labelling of the crosslinker (as before performed with trimethylolpropane triacrylate (TMPTA) in the framework of network formation in superabsorbent gels[24]) should be a better option in this context.

Swelling Ratio Sensitivity Measurements

Synthesized hydrogel particles, after isolation, were tested in order to assess their sensitivity to stimulations triggered by changes in the surrounding media. In particular, was measured the variation of the hydrogels weight swelling ratio, in water solutions, at different temperature and/or pH values. Stimulation of the networks by changes in these parameters was thus observed. Typical results obtained are presented in Figure 4 (a) where sensitivity of NIPA/MBAm hydrogels to temperature changes is used as illustration example.

Drug Release Testing

Performance of the produced hydrogels was also assessed considering drug delivery applications. Different model drugs (caffeine, 5-fluorouracil, isoniazid and ibuprofen) were considered in these studies that were also carried out pouring pre-incubated network particles in aqueous solutions at different conditions (changing pH/temperature). Drug release was measured by UV detection in aqueous samples collected at different elapsed times. Typical results are presented in Figure 4 (b) using the caffeine release from a NIPA/MAA/MBAm hydrogel at different pH/temperature conditions as illustration example. In spite of the complexities associated with the mathematical modelling of drug delivery[1,2], a good agreement is observed fitting the experimental data to exponential rise laws (effect of different stimulations on the drug release profiles observed is also here highlighted).

Applications of the different classes of "smart" hydrogels here studied is further enhanced in Figures 5 and 6. In Figure 5 (a) is presented the measured equilibrium weight swelling ratio of anionic (AA based) and cationic (DMAEMA based) hydrogels in aqueous solutions at different pH values. These results illustrate the networks sensitivity to changes in this parameter. Note that inverse effect of the pH on the swelling ratio of these hydrogels can be explored to trigger different macroscopic effects, as for instance the transition between shrunk to swollen networks by changing the pH from 1 to 8 (e.g. resembling the stomach/intestine pH change in human body) with AA hydrogels and the opposite with DMAEMA hydrogels. In Figure 5 (b) is showed the comparison for the change of the equilibrium weight swelling ratio with pH considering FRP and RAFT synthesized AA hydrogels. These results illustrate the high effect of the synthesis technique used on the swelling properties of the hydrogels. In fact, the primary chain length of the networks is strongly affected when FRP is replaced by RAFT. This effect can eventually be used to tune the swelling properties of the hydrogels (e.g. designing the initial ratios between monomers/RAFT agent/initiator). It is worth to note that measurements presented in Figures 5 (a) and (b) were obtained using buffer aqueous solutions at different pH values. These buffer solutions were prepared using the proper amounts of HCl, NaOH, KCl, KHP (potassium hydrogen phthalate - $C_8H_5KO_4$), KH_2PO_4 (potassium dihydrogen phosphate), $Na_2B_4O_710H_2O$ (borax) and Na_2HPO_4 (disodium hydrogen phosphate). Nevertheless, the swelling ratio of hydrogels is also strongly dependent on the ionic strength and size of the ions and counterions present in the used aqueous solutions. Accordingly, a different dependence of the hydrogels swelling ratio on pH changes can be observed if other aqueous solutions at the same pH values are considered (e.g. changing the used salts and/or using just HCl and NaOH to prepare the aqueous solutions with the desired pH values).

Measured dynamics of release of different drugs from cationic and anionic hydrogels is illustrated in Figure 6. In Figure 6 (a) and (b) is showed the dynamics of release of 5-fluorouracil from DMAEMA based (cationic) and AA based (anionic) hydrogels,

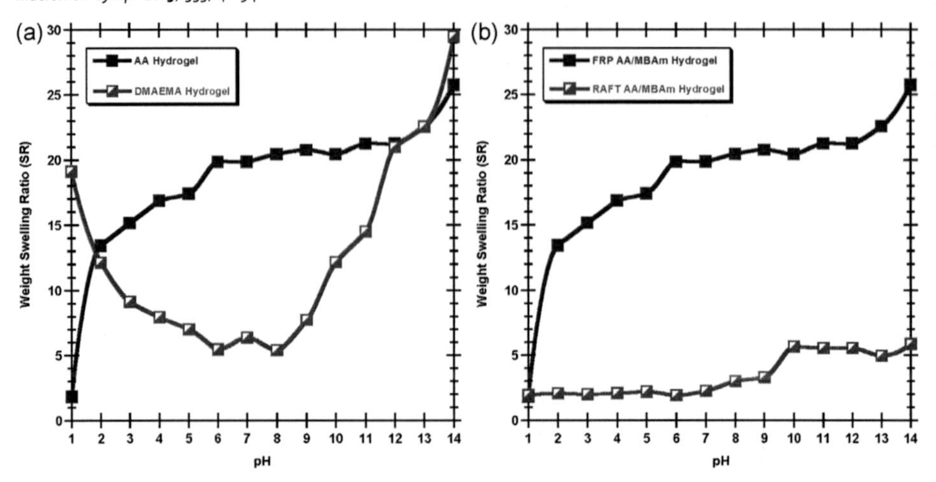

Figure 5.
(a) Measured equilibrium weight swelling ratio of anionic (AA based) and cationic (DMAEMA based) hydrogels in aqueous solutions at different pH values illustrating networks sensitivity to changes in this parameter. Inverse effect of the pH on the swelling ratio of these hydrogels can be explored to trigger different macroscopic effects (e.g. shrunk to swollen networks by changing the pH from 1 to 8 with AA hydrogels and the opposite with DMAEMA hydrogels). (b) Comparison of the change of the equilibrium weight swelling ratio with pH for FRP and RAFT synthesized AA hydrogels. High effect of the synthesis technique used on this parameter is observed. The primary chain length of the networks is strongly affected when FRP is replaced by RAFT which can eventually be used to tune the swelling properties of the hydrogels. Note that results presented in both Figures ((a) and (b)) were obtained using buffer aqueous solutions at different pH values. The swelling ratio of hydrogels is also strongly dependent on the ionic strength and size of the ions/counterions present in the solutions. A different dependence of SR on pH can be observed if other aqueous solutions at the same pH values are considered (e.g. changing the used salts).

respectively. In both cases, the release of the drug was measured in acidic (pH = 1) and alkaline (pH = 10) aqueous solutions. In spite of the differences between the two hydrogels, slightly higher steady state release of the drug was always observed with the alkaline environment. Note that some other complex effects such as hydrogel/drug interaction[1,2] (e.g. see discussions about diffusion and chemically controlled delivery systems in chapter 11 of ref[1].) should be take into account when drug release studies are performed, as for instance formation of complexes between drugs and polymer networks. These issues also have a strong effect on the amount of a specific drug that is possible to load in a hydrogel. The effect of the combination between specific drugs and hydrogels is illustrated in Figure 6 (c) where the dynamics of release of ibuprofen from DMAEMA and AA polymer networks, both placed in aqueous solution at pH = 10, is showed. Comparison of the dynamics of release of 5-fluorouracil from FRP and RAFT synthesized pH sensitive hydrogels is illustrated in Figure 6 (d). Amount of drug released is in this case expressed as the fraction of drug loaded in the hydrogel that is transferred to the aqueous solution. Note that much more lower release fractions were observed (both at pH = 1 and 10) when RAFT hydrogels were considered. These results should be a consequence of the different molecular architectures associated with FRP and RAFT networks (affecting namely their swelling ratio, as showed in Figure 5 (b)) and highlights the relation between structure and end use properties of these materials.

Results and Discussion

Reversible addition-fragmentation chain transfer polymerization is probably the

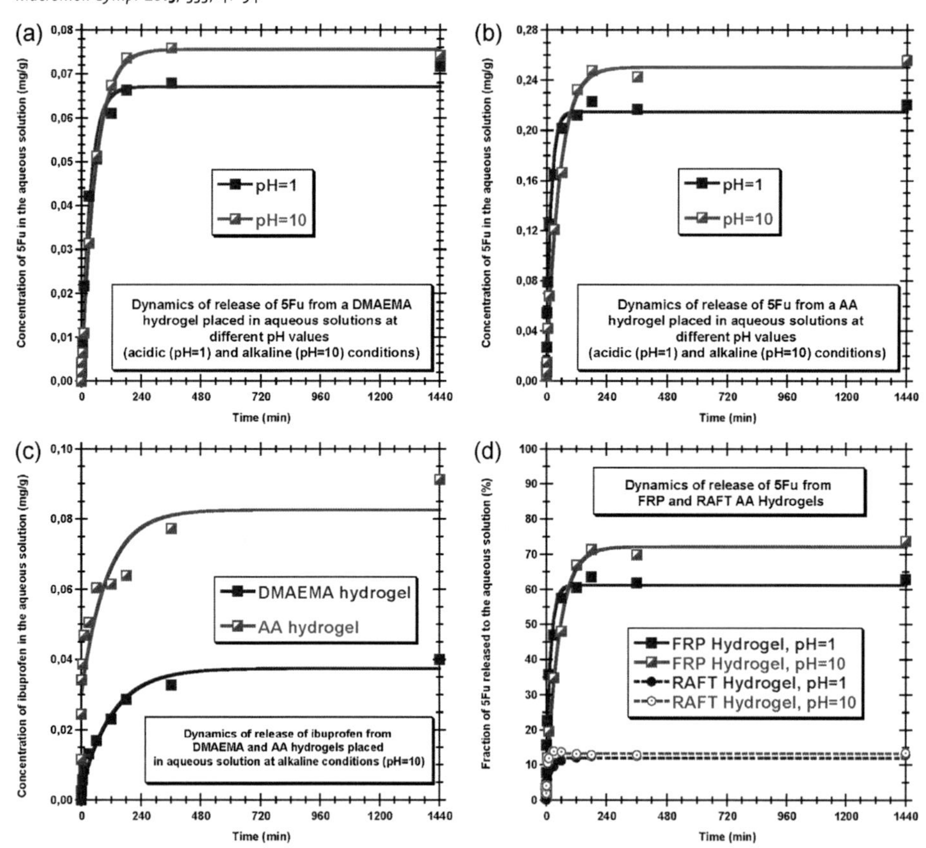

Figure 6.
(a) Dynamics of release of 5-fluorouracil from a pH sensitive hydrogel (cationic hydrogel based on DMAEMA) measured by UV detection at 270 nm. (b) Dynamics of release of 5-fluorouracil from a pH sensitive hydrogel (anionic hydrogel based on AA) measured by UV detection at 270 nm. (c) Comparison of the dynamics of release of ibuprofen from cationic (DMAEMA based) and anionic (AA based) hydrogels, both placed in aqueous solution at pH = 10 (release measured by UV detection at 223 nm). (d) Dynamics of release of 5-fluorouracil from FRP and RAFT synthesized pH sensitive hydrogels illustrating the effect of the molecular architecture of the networks on their performance. Amount of drug released is here expressed as the fraction of drug loaded in the hydrogel that is transferred to the aqueous solution (release measured by UV detection at 270 nm). In all cases presented in this Figure, drugs loadings were performed by swelling the hydrogels in 5-fluorouracil or ibuprofen aqueous solutions during 48 hr.

most versatile CRP technique allowing the polymerization of different classes of monomers. Nevertheless, the degree of control of polymerization that is attained with RAFT is strongly dependent on the reaction conditions used. Specific combination between monomer, RAFT agent, initiator and solvent used in the polymerization is a central issue to obtain tailored products with RAFT polymerization. Temperature and initial proportions monomer/RAFT agent/initiator/solvent also have a huge effect on the kinetics of formation and on the control of the molar masses of RAFT polymers. Some other issues arise when RAFT is directly performed in water, namely the low solubility of most RAFT agents in pure water (forcing the use organic co-solvents) and their potential hydrolysis (pH dependent) with loss of control on the polymerization process (see[25] and references therein). When aqueous dispersed systems are considered (e.g. the industrially important emulsion/

miniemulsion/suspension processes and their inverse counterparts) some other aspects like the transport of reactants (monomers, initiators, RAFT agents) between organic and aqueous phases become also of crucial importance. In this context, the RAFT inverse miniemulsion of acrylamide and acrylic acid were recently reported[26,27] and the effect of pH on the hydrolysis of the RAFT agent and polymerization in the continuous phase (eventually in the absence of RAFT agent) were identified as phenomena potentially involved in some loss of control observed with particular conditions. A secondary peak was observed in the RI curve (see discussion below in the context of the results here presented) which was attributed to different polymer populations formed in both phases (aqueous and organic). These aspects get an additional importance in the framework of RAFT dispersed systems that has been very recently explored to produce amphiphilic copolymers and nanoparticles/nano-objects with different morphologies (e.g. spheres, fibers, vesicles)[28–30]. Bellow are discussed some of our findings involving probably related mechanisms that are present in the RAFT inverse suspension formation of hydrogels or their linear counterparts.

Very fast reactions are generally involved in the FRP synthesis of hydrogels, as before showed with different classes of monomers[20,21]. Fast gelation is observed with a few percent of crosslinker agent and parameters such as polymerization temperature, monomer concentration and neutralization have a very strong influence on the dynamics of gelation (see results before presented in refs[20,21]). Similar behaviour was observed in the FRP polymerization runs detailed in Table 1, even considering a low polymerization temperature (T = 20 °C).

It is known that more amenable kinetics of polymerization can be achieved replacing FRP by RAFT polymerization. Design of operation conditions, namely the initial proportions between initiator/RAFT agent/monomer, can be used to manipulate reaction rates and also to design the degree of polymerization. With network formation, these parameters can be used to try the manipulation of the primary chain length (thus affecting gelation) and the minimization of intramolecular reactions (cyclizations) leading to the decrease of crosslinking efficiency.

If direct aqueous polymerization is intended, as in many cases involving hydrogels, a major problem to be faced with RAFT polymerization is the low water solubility of usual RAFT agents. This issue applies to DDMAT and therefore the aqueous RAFT polymerization is not possible. This problem can be circumvented using a different solvent, such as DMF, as described in Table 2. Under these conditions reasonable reaction rates were

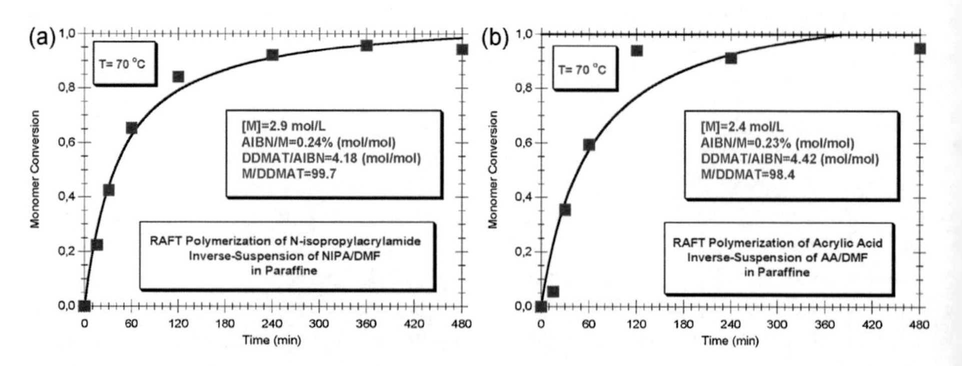

Figure 7.
Examples of observed dynamics of monomer conversion in different runs concerning the inverse-suspension RAFT synthesis of water compatible polymers and hydrogels (see Table 2). (a): NIPA RAFT polymerization with $Y_I^{RAFT} = 4.18$ (run 1). (b): AA RAFT polymerization with $Y_I^{RAFT} = 4.42$ (run 2).

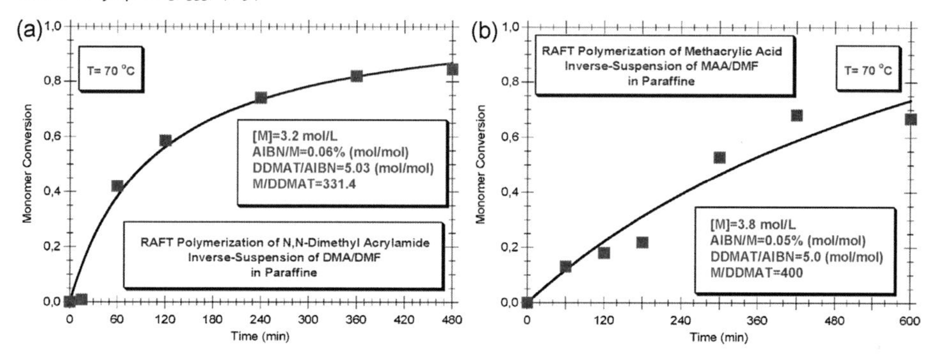

Figure 8.
Examples of observed dynamics of monomer conversion in different runs concerning the inverse-suspension RAFT synthesis of water compatible polymers and hydrogels (see Table 2). (a): DMA RAFT polymerization with $Y_I^{RAFT} = 5.03$ (run 3). (b): MAA RAFT polymerization with $Y_I^{RAFT} = 5.00$ (run 4).

observed with RAFT polymerization at T = 70 °C considering relatively high initial mole ratios RAFT agent/initiator (see Table 2). Kinetics of these RAFT polymerizations are illustrated in Figures 7-9 considering different water compatible monomers.

When RAFT network formation was sought using similar conditions (runs 6 to 8 in Table 2), lower gel fractions were observed in the products indicating a different crosslinking process when compared with FRP (theoretical and experimental kinetic studies on hydrogel formation by FRP where before reported[20,21]). Presence of such soluble phase is highlighted in Figure 10 (a) where the SEC traces of the final sample correspondent to run 6 in Table 2 are used to illustrate this issue. Note that besides the RI signal of the polymer (indicating the presence of soluble material), the LS signal is also presented and the coexistence of secondary population with low concentration but very high size can be observed. This secondary population was also observed in linear RAFT polymerizations, as below discussed, and is a possible consequence of non-ideal steps involved in the RAFT mechanism. Lower amounts of gel observed in these runs can result of the low primary chain length imposed by the RAFT process (mole ratio monomer/RAFT agent around 290). Dynamics of gel formation in RAFT

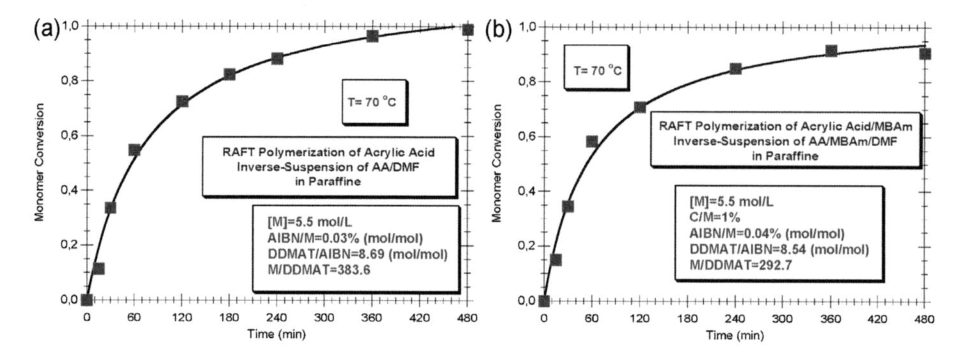

Figure 9.
Examples of observed dynamics of monomer conversion in different runs concerning the inverse-suspension RAFT synthesis of water compatible polymers and hydrogels (see Table 2). (a) AA RAFT polymerization with $Y_I^{RAFT} = 8.69$ (run 5). (b) AA/MBAm RAFT polymerization with $Y_I^{RAFT} = 8.54$ (run 6).

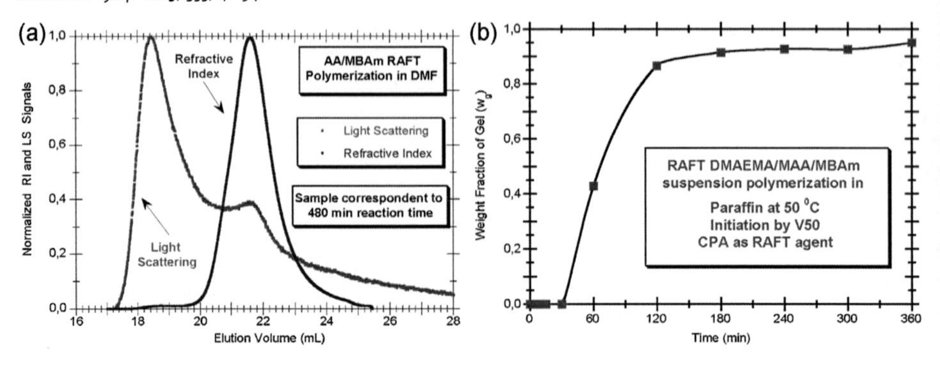

Figure 10.
(a) SEC chromatographic traces showing the presence of soluble material in AA/MBAm RAFT polymerization in DMF (run 6 in Table 2) and the formation of a secondary population with high molecular size and very low concentration. (b) Dynamics of gel formation during the RAFT synthesis of stimuli-responsive hydrogels. DMAEMA/MAA/MBAm RAFT polymerization at 50 °C using CPA agent (run 4 in Table 3) is here considered for illustration purposes.

hydrogel synthesis is also illustrated in Figure 10 (b) considering different operation conditions, namely a lower mole ratio Y_I^{RAFT} (see run 4 in Table 3), here with measurement of high final gel content. Mathematical modelling of RAFT non-linear polymerization can be especially useful to aid with the interpretation of such results and to elucidate the mechanistic differences between RAFT and FRP cross-linking processes[19,31].

Importance of multiple detection in the the SEC analysis of the RAFT products is also highlighted in Figure 11. Formation of a secondary population with very low concentration (almost negligible with RI or viscosity detection) but with very high molecular size (strong LS signal) was observed even with mono-vinyl monomer RAFT polymerization, as illustrated in Figure 11 (a). Dynamics of growth of such population is showed in Figure 11 (b). A possible justification for this phenomenon can be found in the framework of the RAFT slow fragmentation model leading to bimodal distributions formation. High concentration of intermediate radicals and the longer polymer chains associated to such species can be at the source of such non-ideal behaviour, as recently showed using

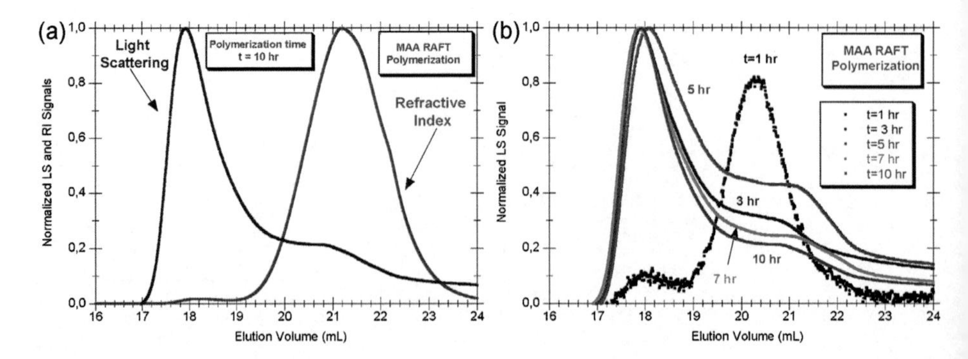

Figure 11.
(a) SEC chromatographic traces showing the formation of a secondary population with high molecular size and very low concentration in RAFT polymerization (data correspondent to run 4 in Table 2). (b) Dynamics of formation of such secondary population observed by light scattering in the same run.

mathematical modelling[32]. Nevertheless, hydrolysis of the RAFT agent and/or polymerization both in aqueous and organic phases (with formation of controlled and non-controlled polymer populations) are also mechanisms with potential effect on the loss of control in these polymerization systems, as discussed above. Additional theoretical/experimental studies should help in the interpretation of the experimental observations here presented.

Conclusion

A comprehensive experimental program concerning the synthesis, characterization and testing (e.g. for drug delivery applications) of smart hydrogels was performed. Combining different aqueous compatible monomers, pH and temperature sensitive materials were obtained. Anionic, cationic and amphoteric network gel beads were produced using the inverse-suspension technique and the kinetics of their building process was studied by SEC with a tetra detection array and also using *in-line* FTIR-ATR. Comparison between the synthesis processes of these kinds of networks using FRP and RAFT polymerization (considering three different commercially available RAFT agents) could thus be experimentally performed. Important dissimilitudes in the dynamics of gel formation were identified when FRP mechanism is replaced by the RAFT polymerization. Some peculiarities of the RAFT synthesis were studied (e.g. solubility of the used RAFT agents in water and monomers considered, effect of initial composition/kinetics on gelation) in order to find proper conditions to produce the sought hydrogels using this CRP technique.

Only a limited information on the crosslinking process was possible to obtain with *in-line* FTIR-ATR monitoring, even using optimized conditions for IR measurements (bulk monomer polymerization in a dispersed media with low IR absorbance). Possibility of probe coating during the synthesis process and the low crosslinker content associated to hydrogels production are factors affecting negatively the use of such technique in this context. Use of *off-line* FTIR analysis of isolated polymers[23,31], chemical analysis of pendant double bonds[15,23] or ^{13}C labelling of the crosslinker[24] should provide improved information concerning the crosslinking process. Conversely, the use of multiple detection (especially including light scattering) in SEC analysis proved to be very important in order to obtain a rigorous characterization of products molecular architecture. In fact, the formation of an unexpected higher size secondary polymer population was detected by light scattering when RAFT polymerization is used, even with mono-vinyl monomers. These results can eventually be explained in the framework of recent theoretical findings describing the RAFT slow fragmentation model leading to bimodal distributions formation[32]. Nevertheless, other mechanisms causing loss of control such as hydrolysis of the RAFT agent and/or simultaneous polymerization in aqueous and organic phases should also be considered in further developments on this research line.

Ongoing work in this research should hopefully also elucidate if really important gains in the sensitivity of these hydrogels can be attained replacing FRP by RAFT polymerization. Complementation of the experimental work here reported with new theoretical developments on the kinetic description of gels formation by the RAFT process[31] should also be explored to built up simulation tools aiding in the specification of synthesis conditions leading to tailored advanced materials.

Acknowledgments: Miguel Gonçalves acknowledges the financial support by FCT and FSE (Programa Operacional Potencial Humano/ POPH) through the PhD scholarship SFRH/ BD/76587/2011. Virgínia Pinto also acknowledges the financial support by FCT through researcher scholarship in the framework of the project PTDC/EQU-EQU/098150/2008 (Ministry of Science and Technology of Portugal/ Program COMPETE - QCA III/ and European

Community/FEDER). This research has also been supported through the Marie Curie Initial Training Network "Nanopoly" (Project: ITN-GA-2009-238700) and by the program SAESCTN - PIIC&DT/1/2011, Programa Operacional Regional do Norte (ON.2), contract NORTE-07-0124-FEDER-000014 (RL2_P3 Polymer Reaction Engineering).

[1] I. Galaev, B. Mattiasson, "Smart Polymers: *Applications in Biotechnology Biomedicine*", CRC Press, **2008**.
[2] A. Bajpai, S. Shukla, R. Saini, A. Tiwari, "Stimuli Responsive Drug Delivery Systems: *From Introduction to Application*", i Smithers, **2010**.
[3] Q. Yu, M. Zhou, Y. Ding, B. Jiang, S. Zhu, *Polymer.* **2007**, *48*, 7058.
[4] H. Gao, K. Min, K. Matyjaszewski, *Macromolecules.* **2007**, *40*, 7763.
[5] H. Gao, W. Li, K. Matyjaszewski, *Macromolecules.* **2008**, *41*, 2335.
[6] M. A. D. Gonçalves, R. C. S. Dias, M. R. P. F. N. Costa, *Macromol. Symp.* **2010**, *289*, 1.
[7] M. A. D. Gonçalves, V. D. Pinto, R. C. S. Dias, M. R. P. F. N. Costa, *Macromol. Symp.* **2010**, *296*, 210.
[8] M. A. D. Gonçalves, R. C. S. Dias, M. R. P. F. N. Costa, *Chem. Eng. Technol.* **2010**, 33, 1797.
[9] M. A. D. Gonçalves, I. M. R. Trigo, R. C. S. Dias, M. R. P. F. N. Costa, *Macromol. Symp.* **2010**, *291-292*, 239.
[10] N. Ide, T. Fukuda, *Macromolecules.* **1997**, 30, 4268.
[11] S. Abrol, P. A. Kambouris, M. G. Looney, D. H. Solomon, *Macromol. Rapid Commun.* **1997**, *18*, 755.
[12] P. B. Zetterlund, M. N. Alam, H. Minami, M. Okubo, *Macromol. F Rapid Commun.* **2005**, *26*, 955.
[13] E. Tuinman, N. T. McManus, M. Roa-Luna, E. Vivaldo-Lima, L. M. F. Lona, A. Penlidis, *J. Macromol. Sci. A.* **2006**, *43*, 995.
[14] J. C. Hernándes-Ortiz, E. Vivaldo-Lima, A. Penlidis, *Macromol. Theory Simul.* **2012**, *21*, 302.
[15] M. A. D. Gonçalves, V. D. Pinto, R. C. S. Dias, M. R. P. F. N. Costa, L. G. Aguiar, R. Giudici, *Macromol. React Eng.* **2013**, *7*, 155.
[16] Q. Yu, Y. Zhu, Y. Ding, S. Zhu, *Macromol. Chem. Phys.* **2008**, *209*, 551.
[17] Q. Yu, S. Xu, H. Zhang, Y. Ding, S. Zhu, *Polymer.* **2009**, *50*, 3488.
[18] G. Jaramillo-Soto, E. Vivaldo-Lima, *J. Aust. Chem.* **2012**, *65*, 1177.
[19] D. Wang, X. Li, W. J. Wang, X. Gong, B. G. Li, S. Zhu, *Macromolecules.* **2012**, 45, 28.
[20] M. A. D. Gonçalves, V. D. Pinto, R. C. S. Dias, M. R. P. F. N. Costa, *Journal of Nanostructured Polymers Nanocomposites.* **2013**, *9*, 40.
[21] M. A. D. Gonçalves, V. D. Pinto, R. C. S. Dias, M. R. P. F. N. Costa, *Macromol. Symp.* **2011**, *306-307*, 107.
[22] S. Salehpour, M. A. Dubé, *Macromol. React. Eng.* **2012**, *6*, 85.
[23] M. Hecker, "Experimentelle Untersuchungen und Monte-Carlo-Simulation netzwerkbildender Copolymerisationen". Fortschritte der Polymerisationstechnik. II (H.U. Moritz ed.) Wissenschaft Technik Verlag Berlin, ISBN 3-89685-353-8, 2000.
[24] D. J. Arriola, S. S. Cutié, D. E. Henton, C. Powell, P. B. Smith, *J. Appl. Polym. Sci.* **1997**, *63*, 439.
[25] I. Chaduc, M. Lansalot, F. D'Agosto, B. Charleux, *Macromolecules.* **2012**, 45, 1241.
[26] L. Ouyang, L. Wang, F. J. Schork, *Macromol. React. Eng.* **2011**, *5*, 163.
[27] G. Qi, C. W. Jones, F. J. Schork, *Macromol. Rapid. Commun.* **2007**, *28*, 1010.
[28] W. Zhang, B. Charleux, P. Cassagnau, *Macromolecules.* **2012**, 45, 5273.
[29] (a) W. Zhang, F. D'Agosto, O. Boyron, J. Rieger, B. Charleux, *Macromolecules.* **2012**, 45, 4075. (b) I. Chaduc, M. Girod, R. Antoine, B. Charleux, F. D'Agosto, M. Lansalot, *Macromolecules.* **2012**, *45*, 5881.
[30] W. Zhang, F. D'Agosto, P. Dugas, J. Rieger, B. Charleux, *Macromolecules.* **2013**, *54*, 2011.
[31] M. A. D. Gonçalves, V. D. Pinto, R. C. S. Dias, M. R. P. F. N. Costa, J. C. Hernándes-Ortiz, Modeling Studies on RAFT copolymerization of Styrene/Divinylbenzene in Aqueous Suspension, Presented at the 11th Workshop on Polymer Reaction Engineering, Hamburg, 2013.
[32] I. Zapata-González, E. Saldívar-Guerra, J. Ortiz-Cisneros, *Macromol. Theory Simul.* **2011**, *20*, 370.

DOI: 10.1002/masy.201300082

Study of the Segregated Behavior of Anionic Microfluidic Polymerization

Bruno Cortese,[*1] Simon Schulze,[2] Mart de Croon,[1] Volker Hessel,[1] Elias Klemm[2]*

Summary: In order to explain the reasons behind the good performance of a microfluidic reactor for anionic polymerization of a commercial polymer - narrow polydispersity index, PDI, of 1.04 (unpublished results of FP7 COPIRIDE project) - modeling of styrene was selected as test system. A comprehensive model accounting for the changes of all the relevant physical parameters was developed and used to compute the fluid dynamics inside the microreactor. The results show that despite the small characteristic dimension of the system an almost segregated behavior is obtained as a consequence of the reaction. This effect leads to the counterintuitive finding that, despite having a very large residence time distribution in the microreactor, polymers with very low PDIs are obtained in a quality as mentioned above.

Keywords: anionic polymerization; computational fluid dynamics; microreactor; novel process windows; polystyrene

Introduction

In the recent past various experimental proofs were found regarding the high potential of microfluidic systems for the synthesis of polymers with a high degree of control, both via radical and anionic polymerization techniques. This demonstrated that, despite an all at once addition of the monomer a high degree of control over the system can be maintained.[1–4]

However, less effort was devoted in developing a detailed physical explanation for these findings; most of the simulations performed almost invariably assumed a plug flow behavior in the reactor and/or perfectly mixed reaction environments. An experimental and theoretical study regarding the behavior of radical polymerizations performed in a microsystem showed the possibility of reaching values of a polydispersity index, PDI, close to the theoretical minimum, however significantly less effort was placed in theoretical investigations on other polymerization methodologies. [5–7]

We recently published a study on the behavior of anionic polymerization in microflow systems, where the variations of the physical parameters of the polymer mixture as a function of the reaction progress are taken into account. On the other hand that model is simplified considering perfect isothermicity and assuming the fluid behaves in a Newtonian manner. [8] In the present work the previous findings were expanded, considering the effect the polymer weight have on heat transfer and heat capacity and the variations of viscosity connected with the non-Newtonian behavior of polymer solutions.

The results show that overheating is present in the center of the microchannel, with a strong dependence from the linear velocity and the initial temperature settings, as well as with the obtainable reaction rates. This effect has also the consequence of reducing the extra reactor length needed to compensate for non parabolic flow patterns.

[1] Eindhoven University of Technology, Department of Chemical Engineering and Chemistry, Eindhoven, The Netherlands
E-mail: b.cortese@tue.nl

[2] University of Stuttgart, Institute of Chemical Technology, Stuttgart, Germany

Model

General Settings

To perform a complete fluid dynamic simulation of the microreactor various aspects must be taken into account. In this case the momentum, heat and mass transfer equations are the starting point, as they are summarized in equations 1-4.

$$\begin{aligned} \nabla \bullet u &= 0 \\ \rho \frac{\partial u}{\partial t} + \rho(u \bullet \nabla u) &= -\nabla p + \nabla \bullet \left(\eta \left(\nabla u + (\nabla u)^T \right) \right) \\ \rho C_P \frac{\partial T}{\partial t} + \nabla \bullet (-k_h \nabla T) &= Q - \rho C_P u \bullet \nabla T \\ \frac{\partial C_i}{\partial t} + \nabla \bullet (-D_i \nabla C_i + C_i u) &= r_i \end{aligned} \tag{1--4}$$

Where ρ is the density of the fluid, T is the temperature, u is the velocity vector, Q is the heat released/absorbed by the reaction, p is the pressure r_i is rate of formation/consumption of specie I, C_p is the heat capacity of the solution, ι is the time, D_i is the diffusion coefficient, η is the viscosity, C_i is the concentration of the i-th species and k_h is the heat transfer coefficient. Equations 1 to 4 are respectively the continuity equation, the Navier-Stokes equation, and the standard descriptions for heat and mass transfer.

All the considered phenomena are in this case heavily coupled, being them determinant for the product quality and influenced by it and by each other.

The main quality indicator followed is the PDI, which is defined in equation 5:

$$PDI = \frac{M_w}{M_n} = \frac{\sum M_i^2 N_i}{\sum M_i N_i} \Big/ \frac{\sum M_i N_i}{\sum N_i} \tag{5}$$

Where N_i represent the number of polymer molecules with a MW of M_i and M_w and M_n are respectively the weight average molecular weight and the number average molecular weight. This value is obtained from the simulations integrating the value of interest along the reactor outlet and correcting it for the interpretation of a circular geometry as exemplified in equation 6

$$\sum M_i N_i = 2\pi \int_0^y y \left(C^* \overline{MW} \right)_y \tag{6}$$

Where y denotes the radial position on the reactor and C is the total concentration of species.

The geometry of the microreactor was simplified from 3D to 2D taking advantage of the symmetry of a cylindrical reactor. All the solutions were tested to be mesh independent.

Solvent Properties

For non isothermal conditions the physical parameters of the solvent (toluene) are changing as a function of temperature. The initial data set was taken from the engineering toolbox and correlated using linear correlations (value = mT + q) for thermal conductivity, specific heat and density, the dynamic viscosity fit on a power law (value = m*T^q) in the range of interest. [9] All the parameters of the correlations are reported in Table 1 (temperatures are in K).

All the numerical values are considered according to the international system, therefore the temperature is expressed in K, k_h in W m^{-1} K^{-1}, ρ in kg m^{-3}, Cp in J kg^{-1} K^{-1} and η in Pa s. The units of measures of the m and q parameters are chosen accordingly.

Solute Properties

The weight concentration and the molecular weight of the dissolved polymers influence the physical parameters of the solution. The diffusion coefficient and the viscosity are correlated to the polymer weight and concentration by the Huggins, Han and

Table 1.
Numerical values of the numerical correlations to calculate solvent properties.

	m	Q
k_h	−1.80E−04	1.94E−01
ρ	−1.06E + 00	1.18E + 03
Cp	4.93E + 00	2.33E + 02
η	3.84E + 10	−3.17E + 00

Hayduck-Cheng equations (equations 7-9); the numerical values for toluene and polystyrene are reported in Table 2. [8,10–14]

$$\eta_{sp} = [\eta]c + \kappa([\eta]c)^2 \quad (7)$$

$$[\eta] = \alpha MW^{\beta} \quad (8)$$

$$D_{AB} = A\eta^{\beta} \quad (9)$$

$$Log_{10}(A) = \alpha\left(\frac{1}{T}\right) + \beta \quad (10)$$

It shall be noted that the intrinsic viscosity of the polymer is also affected by the temperature, according to equation 10. $[\eta]$ is in $m^3\ kg^{-1}$, D_{AB} in $m^2\ s^{-1}$, η_{sp} is adimensional, dimensions of α, β and A are chosen accordingly.[15]

The effect of the solute on the thermal conductivity, which is attributed to the heat transfer capabilities of the polymer chains, can be expressed as a heat transfer ratio where the denominator (λ_s) is the the heat transfer coefficient of the pure solvent, according to equation 11.

$$\lambda_r = \frac{\lambda}{\lambda_s} = \alpha C_w + \beta \quad (11)$$

Where C_w is the weight concentration of the solute (in $kg\ m^{-3}$) and alpha and beta are respectively 8.55e-3 $m^3\ kg^{-1}$ and 1 (adimensional). [16]

Finally the non-Newtonian behavior of the solution must be quantified for application in the simulation. This can be done using a modified Carreau model in the following form (equations 12-15).

$$\eta(MW,\gamma) = \frac{\eta(\gamma)}{\eta_0} = \left((1+\lambda_c\gamma)^2\right)^{-\frac{d}{2}}$$

$$d = f(c[\eta]) = \alpha c[\eta]^{\beta}$$

$$\lambda_c = \frac{1}{\gamma_0}$$

$$\gamma_0 = \frac{\pi^2 RT}{6}\left(\frac{c}{\eta_0 MW} + \frac{c^2}{\rho_2\eta_0 M_c}\right) \quad (12\text{–}15)$$

Table 2.
Parameters to be used in the equations 7-10.

Equation	κ	α	β
7	0.34	–	–
8	–	0.0092	0.7317
9	–	Equation (10)	−0.682
10	–	31.8	−1.91

With ρ_2 and M_c given by Yasusa et al.,[17] η_0 is the solution viscosity at 0 shear, $\alpha = 0.1$ and β =3.16e-1, c is the weight concentration of the polymer and γ the shear rate.

Results and Discussion

General Flow Pattern

In all the simulations performed a specific flow pattern was found, as is exemplified in Figure 1. According to this pattern the microreactor can be separated into two different sections, a sigmoidal one and a parabolic one.

In the sigmoidal section (which is the closest to the entrance, and which is 1 m long in this simulation) the velocity profile of the fluid undergoes significant variations as a consequence of the viscosity gradient which is present between fluid lamellae as a function of conversion. This leads to a flow profile with a sigmoidal shape (see Figure 2). In the parabolic section the fluid flow turn back into the well known laminar profile.

This effect can be explained considering that a strong viscosity gradient is present between the wall and the center of the microreactor (see Figure 2) and this, combined with the fixed total volumetric flowrate, leads to a flow profile in which a strong acceleration is imposed on the central stream. This effect reduces its intensity along the microreactor while conversion increase and finally disappear (entering the parabolic section) because of the absence of strong viscosity gradients in the final stage of the reaction.

Another direct effect of the increase in viscosity is the presence of a highly segregated flow behavior. This is due to the direct dependence of the diffusion coefficient from this physical parameter (equation 9). The segregation effect can be clearly seen if the magnitude of convective and diffusive mass transport is taken into account (Figure 3).

Along the reactor convection accounts for all the mass transport, with absolute values in the order of magnitude of 10^1 to

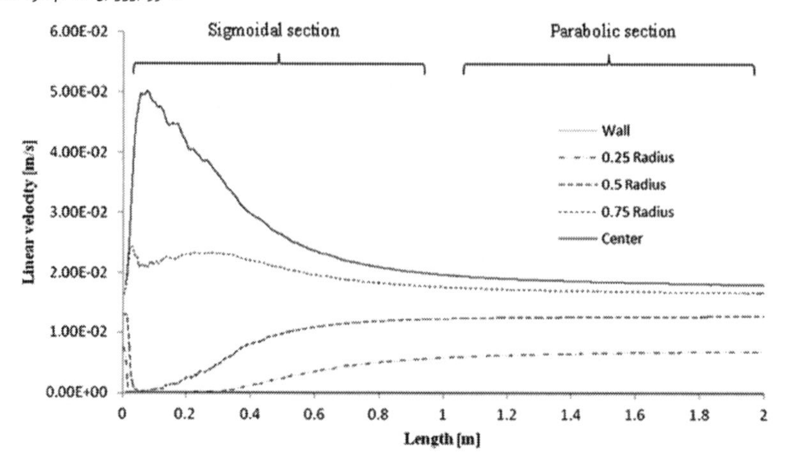

Figure 1.
Velocity profiles along the reactor at different radial positions; the sigmoidal and parabolic sections are highlighted.

10^2 [mol/m^{2}s], while diffusion accounts for values in the order of 10^-4 to 10^-7 [mol/m^{2}s], or in other words, even in the most favorable case diffusion accounts for only a tiny fraction of the total mass transfer. Similar results can be obtained considering the Fourier number of the reactor.

The main direct consequence of this behavior is the need to use longer micro-reactors to reach full conversion if compared to the length which a completely parabolic flow profile would require. The secondary effect is the absence of Taylor-Aris dispersion, or other similar phenomena, due to the almost independent nature of the different streamlines inside the reactor.

Effect of Non-Newtonian Behavior

Solutions of polymer are known to be non-Newtonian fluids; one of our objectives is to quantify the discrepancy in the simulations results if this effect is accounted for or not. To do so we run simulations for both the conditions at a given set of settings and then compared conversion, PDI, streamlines shape and velocity.

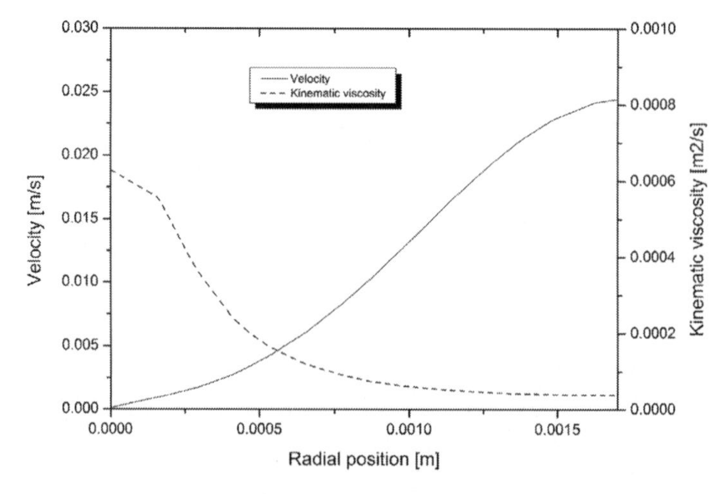

Figure 2.
Velocity and viscosity profiles along the radial positions at x = 0.5 m; the wall is at length 0.

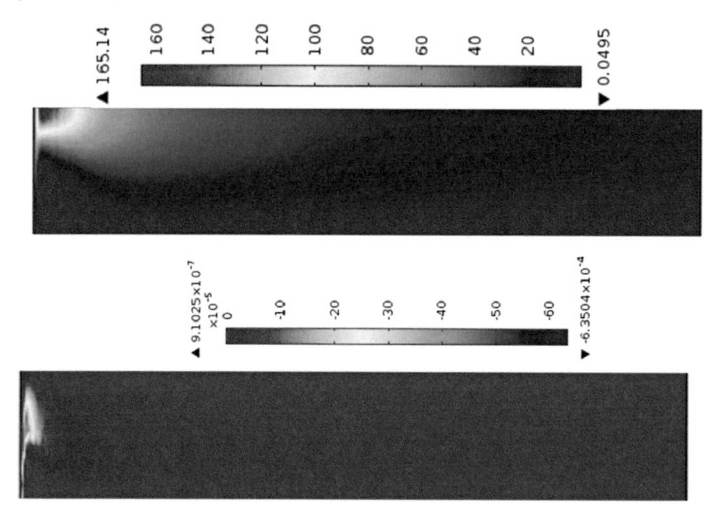

Figure 3.
Total magnitude of convection (up) and diffusion (down) in mol/m^{2}s in the reactor (x and y have different scales), flow from left to right, wall at the bottom.

In both the cases no significant differences of any kind were found, meaning that in this condition, no significant losses of precision will be obtained if the fluid is considered Newtonian. This information may be helpful especially with regard to reactor design tasks, in which will be possible to reduce the required computational effort and the level of complexity, dealing with faster and more robust systems.

Temperature Effect

In opposition to the small effect on the results which the non-Newtonian behavior have assuming the reactor isothermal can lead to larger deviations. This is due to its stronger influence on viscosity and reaction rates, which we identified as the most influential quantities in defining the system behavior. This can be shown if the second part of the sigmoidal section of the reactor is compared (Figure 4).

It is possible to note that the central streamline, when the system is not supposed isothermal, reaches higher velocities while, on the other hand, require shorter times to transition towards the parabolic section. If the temperature is plotted on the whole reactor (Figure 5) it is possible to identify a warmer spot in the sigmoidal section. This combined with the enhanced velocity of the central stream allows for an explanation.

In the initial stage, when the concentration is higher not all the produced heat is removed fast enough. This leads to an increase in the temperature at the center, which as a consequence become less viscous (according to equation 10). Being less viscous and faster the central line tends to transfer heat axially along the reactor, enhancing the differences in velocity.

After the initial section the central line enter the subsequent part of the reactor at a higher temperature then the one it would have in isothermal conditions, and the already heavy polymer layer present at the sides prevent the monomer to escape. This leads to a faster polymerization and consequently to a shorter equilibration.

Regarding the overall magnitude of the temperature gradients which can be obtained, in most cases they are contained into 2 to 3 degrees, but they can be enhanced up to 10K (see Figure 5) or more depending on the reaction rate. Increases in the processing temperature are just enhancing the aforementioned effects, making the central streams faster for a shorter period of time and leading to stronger temperature gradients.

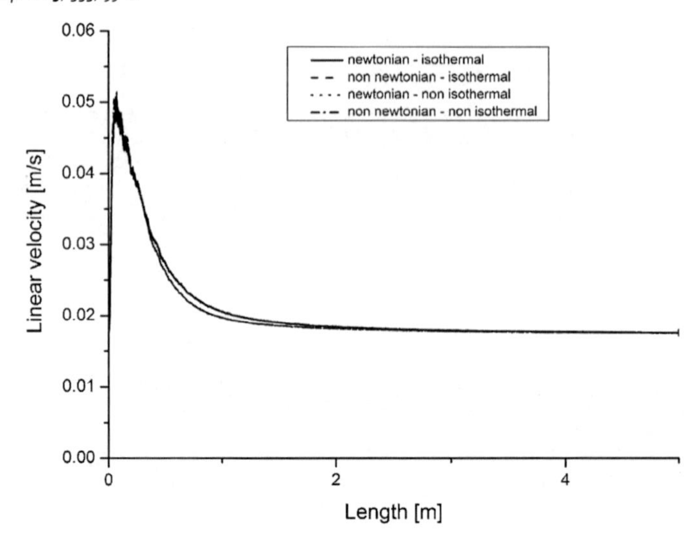

Figure 4.
Central streamline velocity with 4 different degrees of simplification.

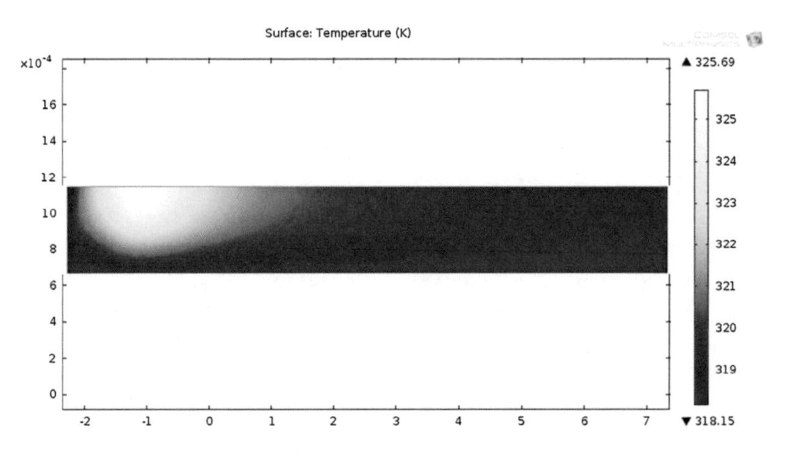

Figure 5.
Thermal distribution in the microreactor (axes are not at the same scale).

From this it can be concluded that, while in most cases considering the reactor isothermal will be a safe assumption to simplify the simulation, if particularly extreme designs are searched for this aspect cannot be neglected, even when the typical length are in the order of a millimeter or below. This also applies in case the reaction is tweaked to become really fast, which can be done with anionic polymerizations acting on the boundary conditions of the reaction (e.g. ionic strength, initiator).

As a final consideration all the degrees of simplification considered led to PDI as low as 1.01, meaning that the changes in assumptions does not influence the overall segregation behavior which leads to this result.

Conclusion

In this work the effect that different assumptions/simplifications have on the simulation of the anionic polymerization in microfluidic systems was determined.

While the non-Newtonian behavior has only a minor effect in the investigated conditions (velocity up to 1 cm/s and radii up to 1mm) it cannot be excluded that it may become relevant if high linear velocities and/or small diameters are chosen, due to the increase in the shear rate this will create. However, for the flowrates generally used in microflow systems this effect can be safely neglected in most cases.

Thermal assumptions require more care because even if the general variations in T found are in the order of 3 K or less (for reaction rates below 500 mol m^{-3} s^{-1}) they become relevant for high reaction rates (more than 2500 mol m^{-3} s^{-1} at the operation temperature). This means that for process intensification oriented designs it will not be possible to neglect this aspect.

Acknowledgments: The research leading to these results has received funding from the European Community's Seventh Framework Programme [FP7/2007-2013] under grant agreement no. CP-IP 228853-2.

[1] F. Bally, C. a. Serra, V. Hessel, G. Hadziioannou, *Macromolecular Reaction Engineering* **2010**, *4*, 543.
[2] F. Bally, C. a. Serra, V. Hessel, G. Hadziioannou, *Chemical Engineering Science* **2011**, *66*, 1449.
[3] A. Nagaki, A. Miyazaki, Y. Tomida, J. Yoshida, *Chemical Engineering Journal* **2011**, *167*, 548.
[4] A. Nagaki, Y. Tomida, J. Yoshida, *Macromolecules* **2008**, *41*, 6322.
[5] C. Rosenfeld, C. Serra, C. Brochon, G. Hadziioannou, *Lab on a chip* **2008**, *8*, 1682.
[6] C. Serra, N. Sary, G. Schlatter, G. Hadziioannou, V. Hessel, *Lab on a chip* **2005**, *5*, 966.
[7] M. M. Mandal, C. Serra, Y. Hoarau, K. D. P. Nigam, *Microfluidics and Nanofluidics* **2010**, *10*, 415.
[8] B. Cortese, *et al.*, *Macromolecular Reaction Engineering* **2012**, *6*, 507.
[9] http://www.engineeringtoolbox.com/toluene-thermal-properties-d_1763.html.
[10] H. McCormick, *Journal of Colloid Science* **1961**, 635.
[11] F. McCrackin, *Polymer* **1987**, *28*, 1847.
[12] W. Hayduk, S. C. Cheng, **1970**, **26,** 635.
[13] W. Kulicke, R. Kniewske, *Rheologica acta* **1984**, *83*, 75.
[14] Y. Fan, R. Qian, M. Shi, J. Shi, *Journal of Chemical & Engineering Data* **1995**, 40, 1053.
[15] U. Bianchi, V. Magnasco, *Journal of Polymer Science* **1959**, *XLI*, 177.
[16] R. J. Brunson, **1975**, **2,** 435.
[17] K. Yasuda, R. Armstrong, R. Cohen, *Rheologica Acta* **1981**, *178*, 163.

 DOI: 10.1002/masy.201300080

Production of Copolymers in a Tubular Reactors Through Nitroxide Mediated Controlled Free-Radical Polymerization

*Carolina L. Araujo, José Carlos Pinto**

Summary: This work presents experimental results for styrene / methyl methacrylate (MMA) copolymerizations performed in a tubular reactor equipped with lateral feed lines through the nitroxide mediated controlled free-radical mechanism. Reactions were carried out in solution with a primary feed containing a mixture of styrene, toluene, initiators and TEMPO (2,2,6,6-tetramethyl-1-piperidinoxyl) and a lateral feed containing a mixture of comonomers and TEMPO. The obtained results show that copolymers can be obtained with weight average molecular weights ranging from 30×10^3 to 60×10^3 g/gmol and polydispersities ranging from 1.3 to 1.6, with monomer conversions ranging from 20 to 70 wt%. When the average residence time is manipulated, it can be observed that the average molecular weights increase linearly with time, confirming the "living" character of the reaction system.

Keywords: copolymerization; lateral feed; methyl methacrylate; nitroxide mediated controlled polymerization; styrene; TEMPO; tubular reactor

Introduction

There is a growing interest in polymers with well-defined molecular structures and many studies have shown that resins with controlled molecular architecture can be obtained through controlled free-radical polymerizations (CFRP),[1–3] such as the nitroxide-mediated free-radical polymerizations (NMRP). The "*living character*" of CFRP's allows for production of materials with narrow molecular weight distributions and low polydispersities (IPD).

In NMRP's the "*livingness*" of the reaction mechanism is provided by a nitroxide species that is able to react with growing radicals through a reversible thermal process.[3–6] High temperatures (usually between 120 and 145 °C) and moderate nitroxide concentrations (nitroxide/initiator molar feed ratios between 1.1 and 1.5) are required for the reversible process to be of use, as nitroxides behave as inhibitors at low temperatures and high concentrations.[7,8] For this reason, reactor operation conditions must be carefully designed in order to keep the polymerization process under control, without compromising the molecular structure of the final polymer chains.

Distinct reaction vessels can be used to perform polymerization reactors. However, tubular reactors offer many competitive advantages, including the high area/volume ratio, easier temperature control and low energy demand for heating and mixing. Besides, residence times can be controlled easily through manipulation of feed flow-rates, allowing for simple tuning of the final properties of the produced resins, especially when controlled free-radical mechanisms are employed. Additionally, the use of lateral feed lines can allow for production of copolymers and in-situ polymer blends with interesting molecular properties, including

1 - Programa de Engenharia Química / COPPE - Universidade Federal do Rio de Janeiro, Cidade Universitária, CP:68502, Rio de Janeiro – 21941-972 RJ, Brazil
E-mail: carolleitearaujo@gmail.com

the manufacture of gradient copolymers and materials with bimodal molecular weight distributions.[9–12]

2,2,6,6-tetramethyl-1-piperidinoxyl (TEMPO) is the commonest nitroxide species used in NMRP processes. TEMPO presents many comparative advantages, such as the good chemical stability, commercial availability and low cost.[6] Nevertheless, TEMPO-based NMRP processes usually lead to low monomer conversions, even when reactions are performed in bulk. Besides, it is generally difficult to perform copolymerizations with TEMPO, even when one of the comonomers is styrene.

Based on the previous paragraphs, styrene / methyl methacrylate (MMA) copolymerizations are performed in a tubular reactor equipped with lateral feed lines through the nitroxide mediated controlled free-radical mechanism in order to produce block copolymers. Reactions are carried out in solution with a primary feed containing a mixture of styrene, toluene, initiators and TEMPO and a lateral feed containing a mixture of comonomers and TEMPO. It is shown that copolymers with large average molecular weights and low polydispersities can be obtained with high overall monomer conversions, even in presence of the comonomer.

Experimental Part

Experimental Unit

Figure 1 shows the experimental setup. The primary feed tank (1) and the lateral feed tank (4) were made of amber glass and had capacity of 5 L. Two pumps (2,5) (Prominent, model: GALA1000S-ST200UA002100) were used to feed the primary and the lateral mixtures. The tubular reactor (3) was made of seamless 316 stainless steel, presenting length of 12 m and nominal diameter of $^{1/4}$ in. The effluents were collected in an amber collector tank (7) with capacity of 10 L. The reactor was monitored by a process computer (6) equipped with data acquisition boards for data acquisition and control (PCI-1002H, for data input, and ADVANTECH PCI-1017, for data output). The reactor was placed inside a heating chamber built with insulating material, in order to minimize heat loss to the external surroundings.

Materials

Styrene (STY, with minimum purity of 99.9 wt%), methyl methacrylate (MMA, with minimum purity of 99.9 wt%), 2,2,6,6-tetramethyl-1-piperidinoxyl (TEMPO, with minimum purity of 97 wt%) and *tert*-butylperoxy 2-ethylhexyl carbonate (TBEC, with minimum purity of 95 wt%)

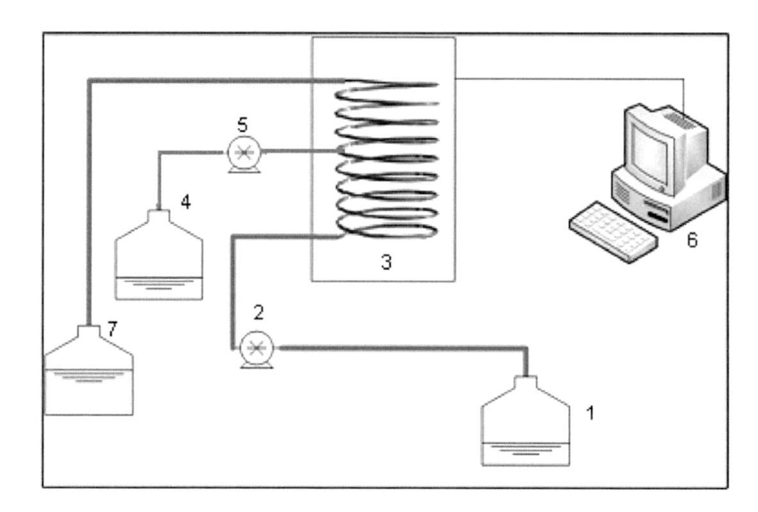

Figure 1.
Experimental setup.

were purchased from Aldrich. Toluene (TOL, with minimum purity of 99.9 wt%), benzoyl peroxide (BPO, with minimum purity of 97 wt% and containing 25 wt% of water) and hydroquinone (with minimum purity of 99 wt%) were purchased from Vetec. Tetrahydrofuran (THF, PA) was purchased from Tedia Brazil. All chemicals were used as received, without any further purification.

Polymerization Reaction

Homopolymerization reactions used styrene, toluene, BPO, TBEC and TEMPO in the primary feed stream, while the lateral feed stream was switched off. Copolymerization reactions used styrene, toluene, BPO, TBEC and TEMPO in the primary feed stream and methyl methacrylate and TEMPO in the lateral feed stream. Reactions were always performed at 135 °C. Table 1 shows some typical reaction recipes.

When the lateral feed contains MMA, a comonomer composition profile is generated in order to induce the production of block copolymers in a single reaction step. In the first part of the reactor (before the lateral feed) polystyrene blocks are necessarily formed, while in the second part of the reactor (after the lateral feed) styrene and MMA can lead to formation of a second poly(styrene-co-MMA) block. Figure 2 illustrates the formation of blocks in a tubular reactor equipped with a lateral feed, when it is assumed that the chain growth mechanism presents a *"living character"*. The primary feed stream is pumped into the reactor entrance (far left of the figure) and a homopolymer block (polystyrene) is formed until the average residence time of τ_h , when the lateral feed is introduced. At this point, a second copolymer block starts to be formed onto the end of the initial homopolymer block and grows for an additional residence time of τ_c, when the product leaves the tubular reactor.

Characterization

Monomer conversion was obtained by gravimetry.

GPC analyses were carried out in THF at 40 °C with a flow rate of 1 mL/min. 100 μL of the polymer solution (5 mg/ml) were injected into the chromatograph (Viscotek, model VE2001 GPC Solvent / Sample Module), which was equipped with a refractive index detector (Viscotek VE 3580) and 4 Phenomenex columns (pore sizes of 5×10^2 A, 10^4 A, 10^5 A and 10^6 A).

Fourier Transform InfraRed Spectroscopy (FTIR) analyses were performed in a Nicolet 6700 spectrometer (Thermo Fisher Scientifc Inc., Massachusetts, USA) equipped with a MCT detector / B Smart

Table 1.
Typical recipes for the analyzed controlled free-radical polymerizations.

	Parameter (mol/L)	Reaction R1	Reaction R2
Primary Feed	Styrene	5.18	4.32
	Toluene	5.86	4.88
	BPO	0.0032	0.0027
	TBEC	0.0026	0.0021
	TEMPO	0.0040	0.0032
Lateral Feed	Methyl methacrylate	-	9.99
	TEMPO	-	0.0034
	Conditions (min)		
Steady State 1	τ_{h1}	82	82
	τ_{c1}	-	73
	τ_{total1}	82	155
Steady State 2	τ_{h2}	328	328
	τ_{c2}	-	219
	τ_{total2}	328	547

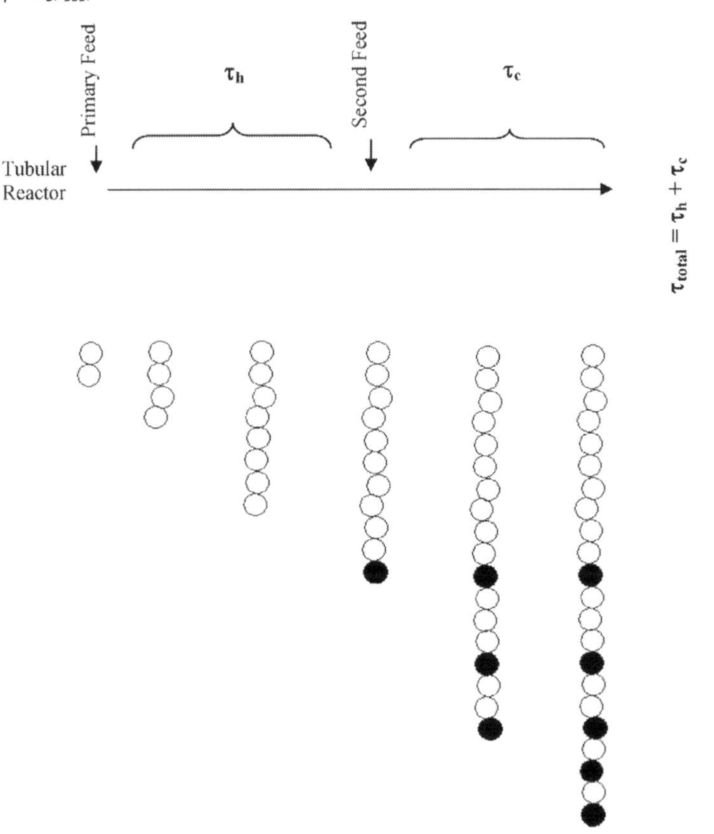

Figure 2.
Schematic representation of the block copolymer formation.

Orbit, in resolution 4 and using averages of 128 scans (64 scans in background).

Polymer composition was determined by ^{13}C NMR analyses, using a Varian Analytical Instruments spectrometer (model DX 300). The equipment was operated at 75.4 MHz with an interval time of 1s and a pulse of 90°, using probes of 10 mm. Analyses were conducted at ambient temperature when ClD was used as solvent, whereas analyses were performed at 90 °C when TCE was used as solvent.

Results and Discussion

In this work, many distinct NMRP styrene homopolymerizations and styrene/MMA copolymerizations were successfully performed in the tubular reactor, in presence and absence of the lateral feed (although not shown here due to lack of space). Figure 3 presents Mn and Mw values for the outlet products for reactions R1 and R2 in Table 1 during the transient periods between the two successive steady states. For the tubular reactor geometry, one can assume that the samples collected between the two steady states contain information about the effects of the reaction time on the final product properties (assuming that the plug flow hypothesis is valid), as it takes different time lengths for elementary volumes to leave the reactor after disturbing the feed flowrates. As one can see in Figure 3, variations of Mn and Mw values during the transients were essentially linear with time, confirming the *"living character"* of the polymerization reaction.

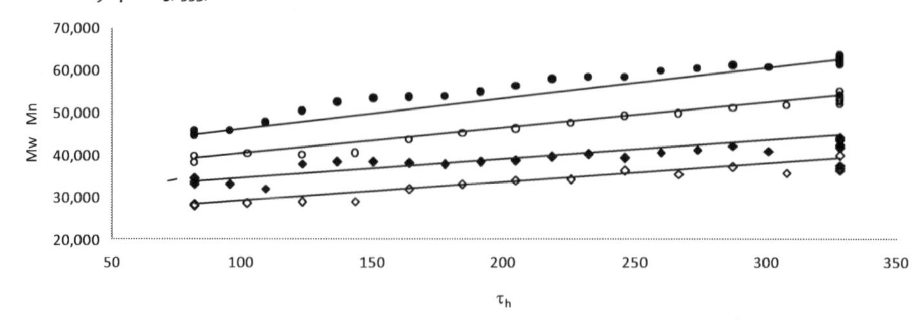

Figure 3.
Mw of the polymer product in (○) Reaction R1 and (●) Reaction R2 and Mn of the polymer product in (◇) Reaction R1 and (◆) Reaction R2 as functions of the reaction time.

Figure 4 shows the IPD values and the overall monomer conversions for reactions R1 and R2. One can see that the IPD values range from 1.3 to 1.6 (although close to 1.4 most of the time), which are much smaller that IPD values of typical styrene (1.7–2.2) and MMA (2.3–3.2) homopolymerizations, although larger than expected for "*living*" reactions (1.05–1.2). The large IPD values can be assigned to the well-known secondary mixing effects caused by the existence of radial velocity profiles in typical tubular polymerization reactors. From a practical point of view, the IPD values are not very important when one is interested in producing compatibilizers and stabilizers, so that the obtained IPD values can be regarded as appropriate. On the other hand, overall monomer conversions can be regarded as high, as they get close to 70 wt % when the residence time is increased. This can be regarded as a very interesting result, as NMRP reactions are well-known for the relatively low monomer conversions, even when styrene homopolymerizations are performed. The high monomer conversions can probably be related to the mixture of initiators, as also observed in other studies.[14,15]

Figure 5 shows the FTIR spectra of the final polymer products of reactions R1 and R2. The product of reaction R2 presents the characteristic carbonyl peaks placed at 1900 cm^{-1} and 1260 cm^{-1}, which are characteristic of esters such as MMA. This seems to confirm the presence of MMA in the copolymer and suggests that the polymer product presents the expected block architecture.

Figure 6 shows a NMR spectrum of a poly(styrene-co-MMA) sample collected at

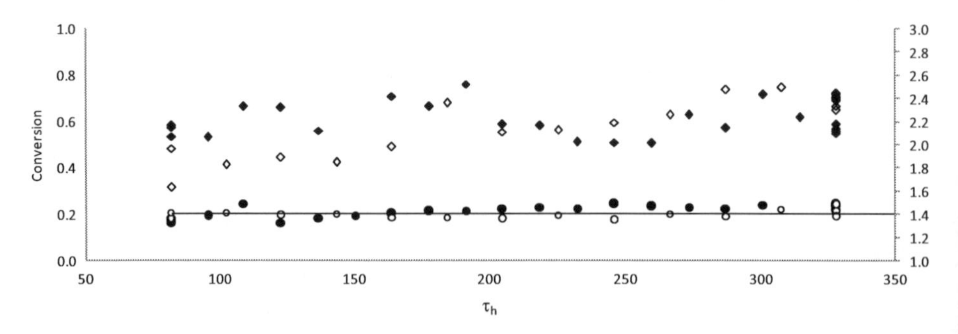

Figure 4.
IPD of the polymer product in (○) Reaction R1 and (●) Reaction R2 and overall monomer conversions in (◇) Reaction R1 and (◆) Reaction R2 as functions of the reaction time.

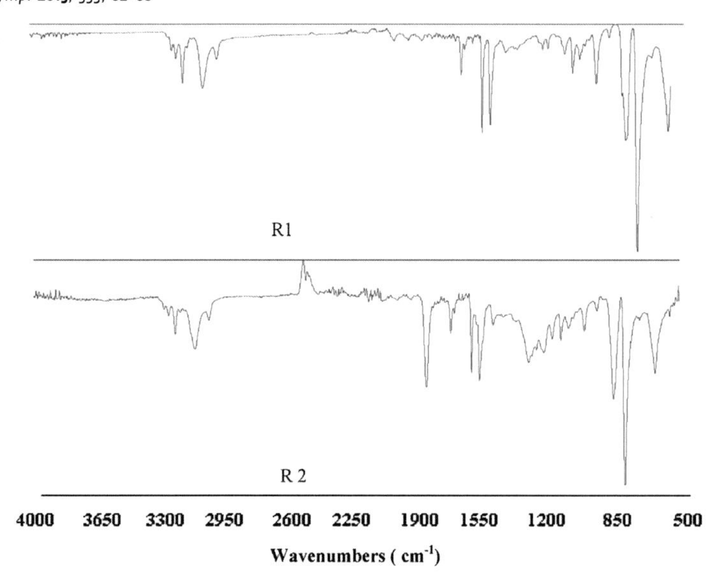

Figure 5.
FTIR spectra of the polymer products of reactions R1 and R2.

the reactor outlet stream, confirming the successful incorporation of MMA into the final polymer product. According to the NMR spectrum, the MMA composition can be as high as 30% in molar basis,[16] indicating that the proposed operation strategy can indeed lead to preparation of block copolymers in tubular reactors through controlled free-radical polymerization.

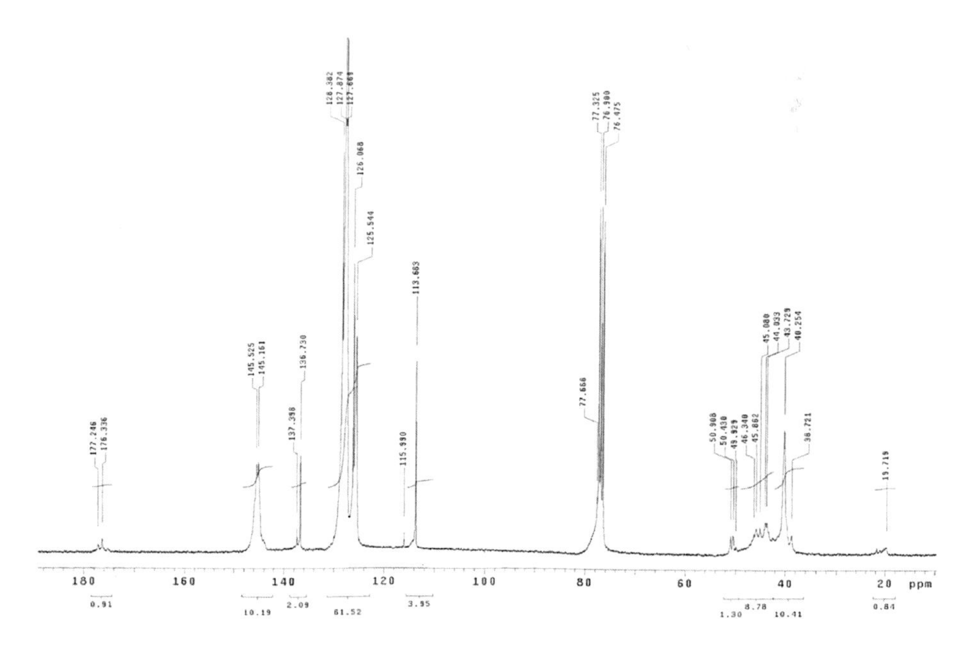

Figure 6.
NMR spectra of the copolymer poly(sty-co-MMA) products of reactions.

Conclusion

Styrene / methyl methacrylate (MMA) copolymerizations were performed in a tubular reactor equipped with lateral feed lines through the nitroxide mediated controlled free-radical mechanism. Reactions were carried out in solution with a primary feed containing a mixture of styrene, toluene, initiators and TEMPO and a lateral feed containing a mixture of comonomers and TEMPO. The obtained results showed that copolymers could be obtained with weight average molecular weights (Mw) ranging from 30×10^3 to 60×10^3 g/gmol and IPD ranging from 1.3 to 1.6, with monomer conversions ranging from 20 to 70 wt%. When the average residence time was manipulated, it could be observed that the average molecular weights increased linearly with time, confirming the "living" character of the reaction system. Finally, it was observed that MMA molecules could be incorporated into the final polymer product at large compositions, as observed through FTIR and NMR analyses.

[1] J. L. Hedrickk, T. Magbitang, E. F. Connor, T. Glauser, W. Volksen, C. J. Hawker, V. Y. Lee, R. D. Miller, *Chemistry European Journal*, **2002**, v. *8*, n. 15, p. 3308–3319
[2] C. J. Hawker, A. W. Bosman, E. Harth, *Chemical Reviews*, **2001**, v. *101*, p. 3661–3688.
[3] C. J. Hawker, *Journal of the American Chemical Society*, **1994**, v. *116*, p. 1158–1186.
[4] M. K. Lenzi, M. F. Cunningham, E. L. Lima, J. C. Pinto, *Industrial & Engineering Chemistry Research*, **2005**, v. *44*, p. 2568–2578.
[5] K. Matyjaszewski, J. Spanswick, *Materials Today, Review Feature*, **2005**, 26–33.
[6] D. Greszta, K. Matyjaszewski, *Macromolecules*, **1996**, v. *29*, p. 7661–7670.
[7] G. Moad, E. Rizzardo, D. H. Solomon, *Macromolecules*, **1982**, v. *15*, p. 909–914.
[8] M. K. Georges, R. P. N. Veregin, P. M. Kazmaier, G. K. Hamer, Macromolecules, **1993**, v. **26**, p. 2987–2988.
[9] M. Zhang, W. H. Ray, *Journal of Applied Polymer Science*, **2002**, v. *86*, p. 1047–1056.
[10] M. Zhang, W. H. Ray, *Journal of Applied Polymer Science*, **2002**, v. *86*, p. 1630–1662.
[11] M. Zhang, W. H. Ray, *Macromolecular Symposia*, **2002**, v. *182*, p. 169–180.
[12] S. Zhu, *Macromol. React. Eng.* **2010**, 4, 163–164.
[13] R. M. Silverstein, F. X. Webster, D. J. Kiemle, *Spectrometric Identification of Organic Compound*. 7, ed., New Jersey John Wiley & Sons, Inc. **2005**,
[14] C. P. R. Malere, **2011**, M. Sc. Thesis, University of Campinas – Unicamp, School of Chemical Engineering, Campinas, Brazil.
[15] C. P. R. Malere, L. M. F. Lona, **2011**, *XI Brazilian Congress of Polymers*, Campos do Jordao, SP, Brazil
[16] A. J. Brandolini, D. D. Hills, **2000**, *NMR Spectra of Polymers and Polymers Additives*, Marcel Dekker Inc., New York, USA

Properties of Smart-Scaled PTFE-Tubular Reactors for Continuous Emulsion Polymerization Reactions

Fabian Gabriel Lueth, Werner Pauer, Hans-Ulrich Moritz*

Summary: This article presents a novel design of smart scale PTFE tubular reactor systems for continuous polymerization reactions. To optimize the flow pattern, CFD calculations and RTD experiments were done. These results led to an unique tube geometry combined with a specialized arrangement of static mixers. A comparison of heat transfer properties of two different types of tubular reactors shows the feasibility of using PTFE instead of stainless steel as reactor construction material. The versatility of the operation modes is demonstrated for emulsion copolymerization experiments with the same recipe but at the different modes.

Keywords: CFD; emulsion polymerization; flow pattern; heat transfer; polytetrafluoroethylene (PTFE); residence time distribution; smart scale; tubular reactor

Introduction

Nowadays, most industrial emulsion polymerization processes are still carried out in batch- or semi-batch processes in conventional dimensioned stirred tank reactors. There are not only approaches in transferring these processes into a continuous production because of the known disadvantages of conventional reactor types like long set-up times, lower space time yield, high labor expense, or inconsistent product quality. There is also a rising interest in so called smart scale reactor systems, which are characterized by a production output up to 100 t/a.[1,2] These systems are normally designed as continuous processes to reach the described production output while having reactor volumes between lab scale and pilot plants. Using tubular reactors for continuous reaction systems has the key benefit of a large surface to volume ratio which allows a more accurate and safe control of exothermic reactions with high heat flux compared to continuous stirred tank reactors. Smart scale processes offer a high potential of process intensification, which makes cost and time optimized made-to-measure solutions for specific product qualities and quantities possible. By using this technology, the cost reduction of specific investments can be about 1/3rd which allows the development and production of high specialized products.[2,3] The intensification of polymerization processes is realized by taking advantage of the high sensitivity of the reaction system to variations of the process parameters, such as temperature or feed composition. Using smart scale reactor systems enables the control of these parameters particularly well and also small procedure changes can lead to measurable differences of the product properties. New insights from experiments carried out at novel process windows provided by the fundamental increase of the specific surface of such reactor types can be made.[3,4] In addition, there are generally only few efforts in process intensification of polymerization processes in tubular reactors and in detail in emulsion copolymerization even less.[5–7]

The challenge of operating smart scaled emulsion polymerization processes is

Institute for Technical and Macromolecular Chemistry, University of Hamburg, Bundesstrasse 45, 20146 Hamburg, Germany
E-mail: lueth@chemie.uni-hamburg.de
Fax: (+49) 040 42838 6008

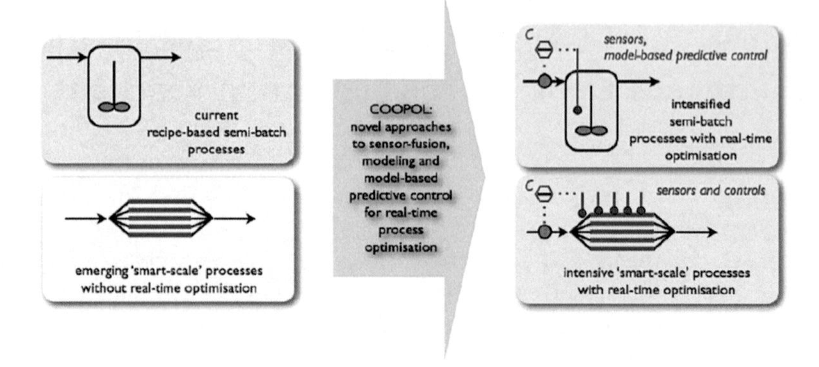

Figure 1.
Steps to real-time optimized processes on the basis of validated models derived from smart scale polymerization experiments.[3]

founded in various phenomena like mass transport limitation, phase separation, educt depletion, fouling and different solubilities of the different species in every phase of the dispersed system.[8] By choosing this combination of smart scale technology and complex reaction systems for developing the product and the process simultaneously, the chance of transferring the advanced knowledge to fundamental understanding of less complicated systems is given.

This work describes the experimental set-up of a smart scaled tubular reactor with its main properties and features. The reactor development is part of the COOPOL[1] research project funded by the European Union. The main goal of this cooperation is the real-time feed-back control by a novel sensor-fusion. This concept combines robust sensors e.g. for temperature and pressure with software based soft sensor approaches in order to enable the on-line inference of parameters into the polymerization process. This concept on the basis of advanced mathematical models, validated by smart scale experiments, is supposed to lead to a realistic optimization control for industrial applications, which is schematically described in Figure 1.[3]

One of the main interesting opportunities of running tubular reactor systems is the use of several approaches to customize the flow pattern within the tube. Therefore, a narrowed residence time distribution (RTD), an improved heat transfer due to a reduced laminar boundary layer thickness, homogenization effects and other mixing tasks can be realized. Using polytetrafluoroethylene (PTFE) tubes as reactor material is an appropriate method to solve fouling problems. Beyond that, the physical properties of PTFE enable various reactor geometries, which makes secondary flow phenomena like Dean Vortices easily accessible.[9–14] In addition, the usage of static mixers in tubular reactors is a promising approach to achieve comparable influences on the radial mixing properties. The geometry of the mixers can be chosen according to the mixing task regarding the expected pressure drop and its function as a possible emulsification device.[15–20]

The characteristic performance of a specific reactor set-up is mainly influenced by its flow pattern. Therefore, it is important to evaluate the corresponding factors that lead to a reduced backmixing, a homogenious reaction mass and to improved heat transfer properties.

This article describes a novel smart scale reactor design for which these factors and

[1] COOPOL: Control and real time optimisation of intensive polymerization processes.[3] (www.coopol.eu)

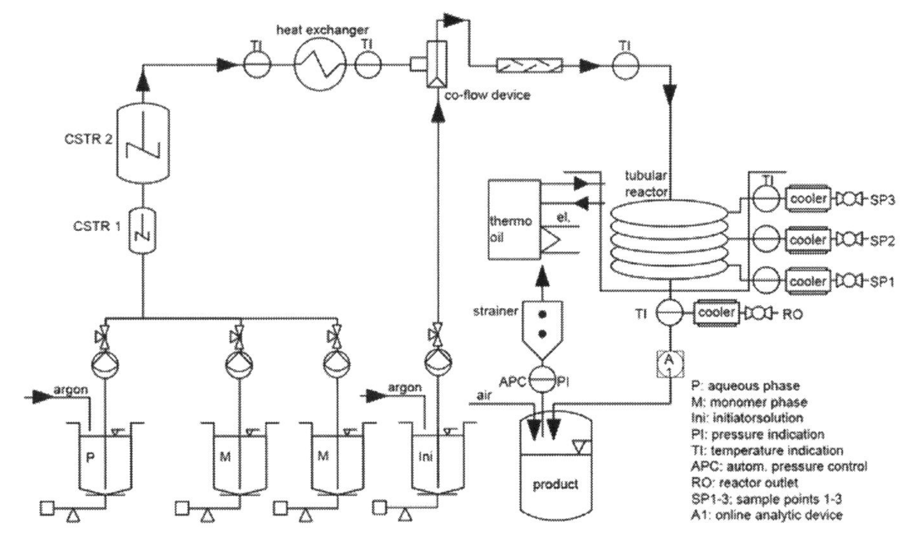

Figure 2.
Piping and instrumentation diagram of the smart scale set-up for continuous polymerization processes (designed with RI-CAD 2.0.3[2] Courtesy of HITEC ZANG GmbH.). The monomers are used as received so that argon lines on the monomer tanks are unnecessary.

operating parameters were elaborated. The resulting optimization of the reactor performance provides a maximum of possibilities for process intensification.

Reactor Set-Up for Emulsion Copolymerization

The piping and instrumentation diagram of the smart scaled experimental set-up is shown in Figure 2. The system is controlled by a LABVISION interface from HITEC ZANG. Dosing is realized by using PROMINENT pumps and scales. The mixing and emulsification process is done in two steps in two continuous stirred tank reactors of 20 mL (CSTR 1) and 200 mL (CSTR 2) volume. The initiator phase is mixed with the emulsion by a co-flow device followed by a static mixing section with an *ID* of 4 mm.

Sampling of the reaction mass is possible at the reactor end as well as at the three sample points. The sampling lines are cooled by double jacket cooling devices to avoid skin formation on the dispersions and product loss through vaporization. The pressure in the reactor is regulated by the pressure on the gas phase in the product tank. By this technique, there is no need to use piece parts with small *ID* which are sensitive to fouling processes.

Having the emulsion and the initiator phase in two separated lines, two operating modes are possible. The reaction can be carried out either under isothermal starting conditions or polytropic. By using a separately controlled plate heat exchanger, the emulsion can be heated up so that adding the initiator phase at room temperature leads to the desired operating temperature of the polymerization process. However, if the heat exchanger is switched off the reaction mass is at room temperature at the reactor inlet. It is then heated up along the reactor line through the heated thermo oil vessel. Running experiments with the same recipe but at different modes (isothermal or polytropic) can lead to nearly the same conversion but different product properties, like the molecular weight distribution (MWD).

Figure 3 shows the different conversion curves of polymerization experiments

[2] Courtesy of HITEC ZANG GmbH.

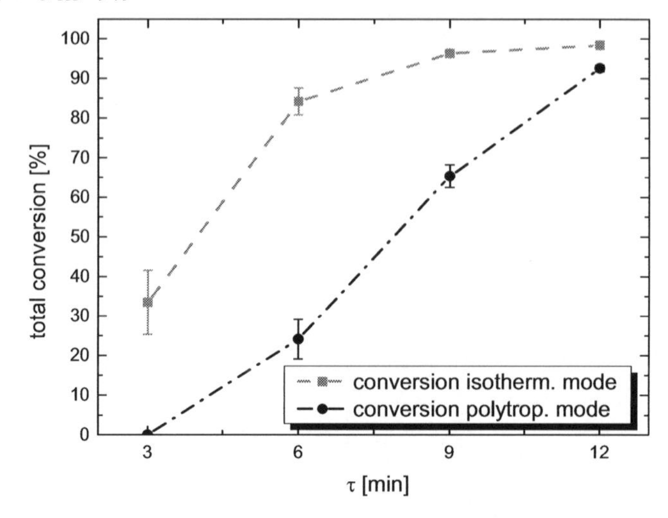

Figure 3.
Comparison of the total conversion (by microwave drying gravimetric analysis) of an emulsion copolymerization of styrene and *n*-butyl acrylate (20:80 mol%) with 15 w% monomer content under isothermal and polytropic starting conditions.

applying the same recipe but running it at different modes. In both cases, the total conversion is almost complete after a residence time of 12 min. For the polymerization with isothermal starting conditions, the conversion starts at the reactor inlet and has a higher slope compared to the polytropic operating mode, for which conversion was only detectable after a residence time of 3 min. Due to the fact that polymers are products by process, this discrepancy in conversion to time behavior and the difference in temperature at the corresponding residence times influence the product properties.

Under isothermal starting conditions, smaller molecular masses were obtained compared to the product made under polytropic conditions, as shown in Figure 4. As expected, the higher temperature at the

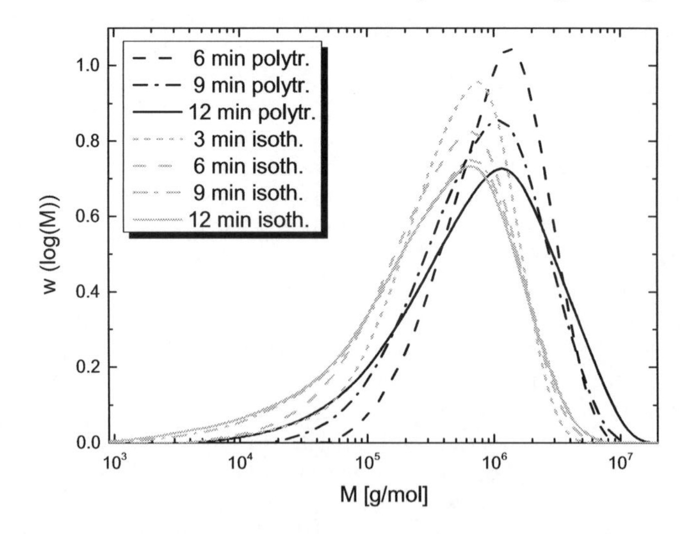

Figure 4.
Comparison of the MWD (by GPC) of the emulsion copolymerization of styrene and *n*-butyl acrylate (20:80 mol %) with 15 w% monomer content under isothermal and polytropic starting conditions.

beginning of the reaction leads to the formation of a higher amount of radicals. The increase of the polydispersity index (PDI) along the reaction is remarkable for the product made under polytropic conditions. A comparison of the products received after 12 min residence time shows that the PDI for the product received under polytropic conditions is almost twice as high as the one under isothermal.

Flow Pattern of the Tubular Reactor

The well-known challenge of fouling within continuous reactor set-ups was solved by using PTFE tubes with a length of 10 m, an inner diameter (*ID*) of 10 mm and a wall thickness (*WT*) of 1 mm (ESSKA). A helical coil with a bending diameter of 22 cm was chosen as reactor design. This design provides a maximum of compactness and the possibility to benefit from secondary flow phenomena as a synergetic effect.

Figure 5 shows the flow pattern in this tube geometry, calculated by the CFD tool ANSYS FLUENT. It is obvious that the primarily parabolic velocity profile at the reactor inlet is forced to the outside wall of the coil because of the centrifugal forces which have a stronger effect on fluid elements with higher impulses. Tangential vectors make the formation of Dean Vortices observable. These vectors, indicating the radial velocity ratio of the fluid elements, are a projection onto the contour planes, which are normal to the main flow direction. The axis of the Dean Vortices is slightly inclined due to the helical motion profile of the main flow.

A second method to induce non-laminar flow behavior, beside the utilization of the tube geometry, was implemented by using

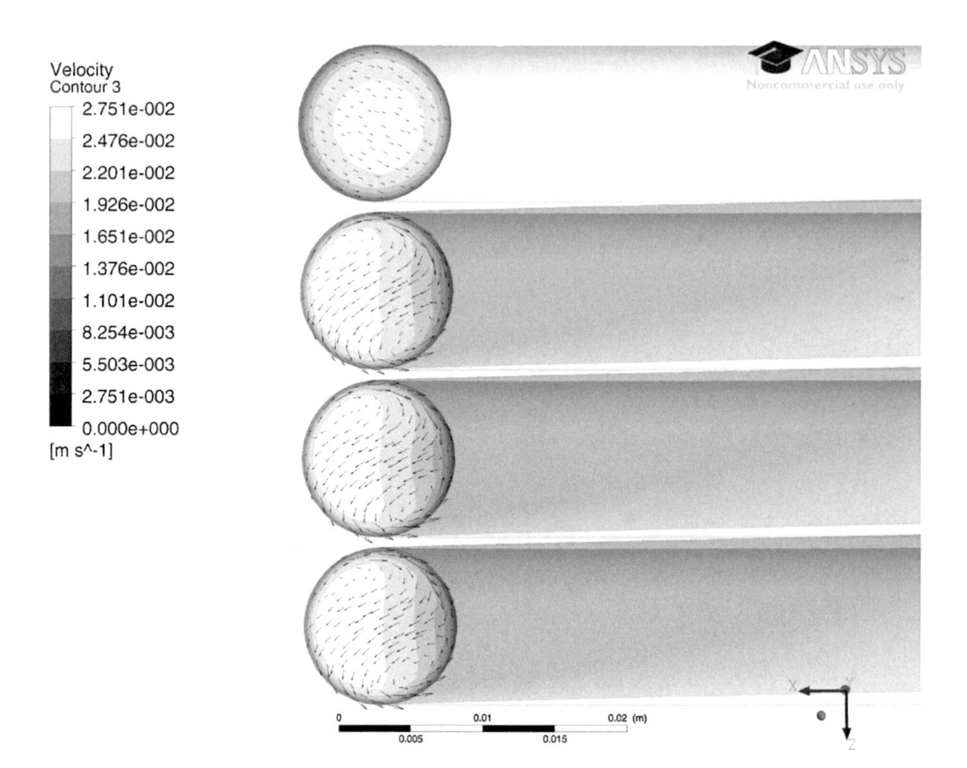

Figure 5.
Velocity contour and tangential vectors calculated by CFD (k-ε model) for a mass flow (water) through a helical coiled tube without static mixers with a mass flow rate of 2 g/s at 22 °C. The inlet is at the top. The vectors indicate the Dean Vortices.

Kenics static mixers (ESSKA) with three windings (Figure 6 bottom left). In this case, high pressure drop and volume displacement can be avoided by equipping the tube only partially with static mixer elements. This partial equipping is easily possible if PTFE is used as reactor construction material.

Since in this work the mixing different components was not of interest, partial equipping is sufficient due to the memory effect, which can be demonstrated by the vortex core region calculated by CFD with the SST model. This model is a more complicated and time-consuming calculation method but more sensitive to the

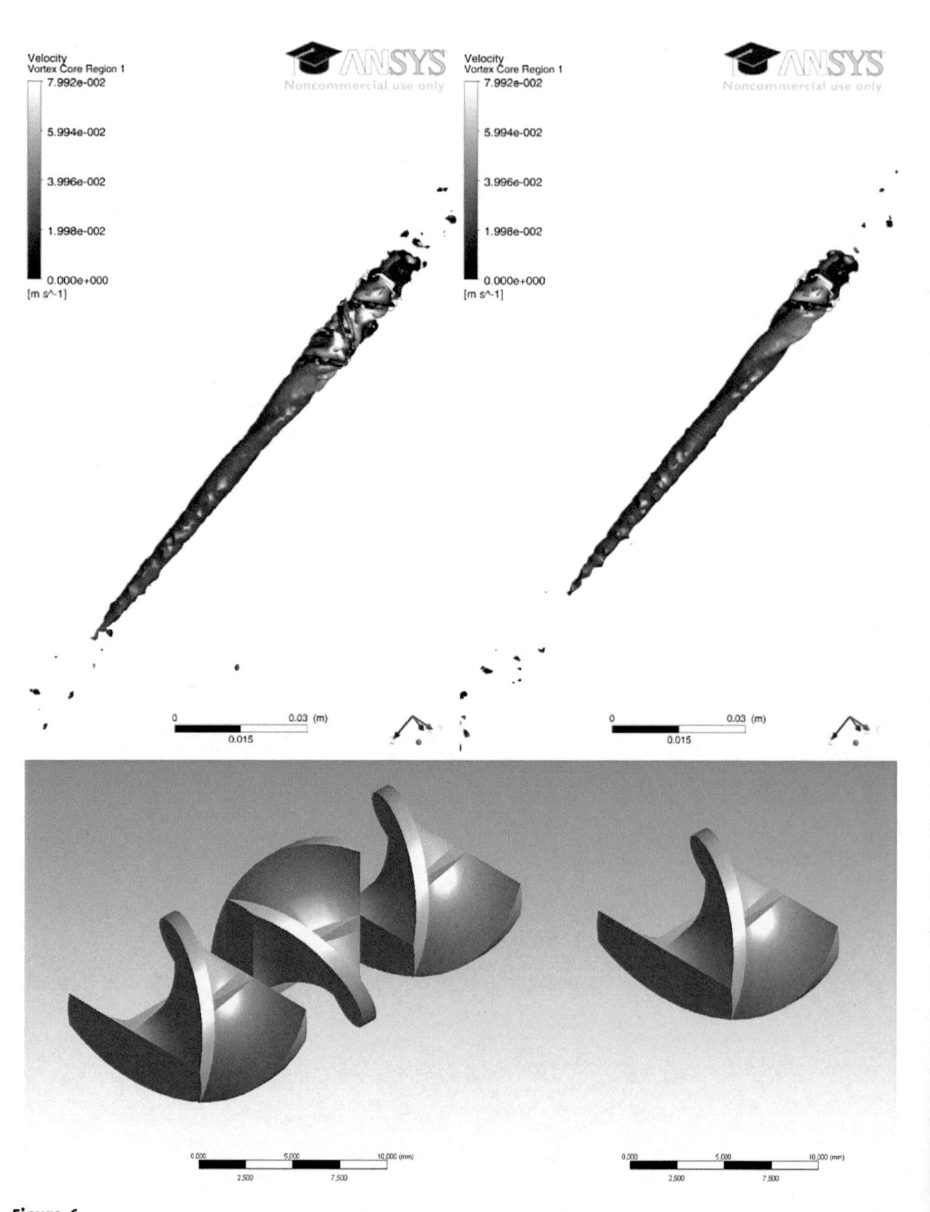

Figure 6.
Vortex core regions (top left and right) obtained by modeling a water mass flow of 1.1 g/s through a linear tube equipped with a three-coiled winding (bottom left) and a single-coiled winding static mixing element (bottom right). The number of coiled windings does not affect the length of the vortex core region.

transition from near wall areas to the interior.[21,22] The result of such a calculation for a linear tube is shown in Figure 6, in which the influence of the number of windings on the vortex length is pointed out as well.

Regarding the vortex length, the total number of windings is irrelevant because of the strong convection at every coiled winding. Therefore, further calculations with more complicated geometries can be simplified by using single-coiled mixers (Figure 6 bottom right) so that the computing time per iteration step is accelerated.

It was shown that the use of static mixers cause a forced convection which leads to a vorticity. An interaction of this vorticity with secondary flow phenomena, obtained by the helical coiled structure of the tube, was expected and investigated by CFD simulation.

Figure 7 shows the accentuation of the velocity of the fluid elements in y-direction and the core vortex region of a flow through a helical coiled tube, having a Kenics static mixer at the inlet. Figure 7 clarifies the positive interaction of these two phenomena in terms of non-laminar flow behavior. This positive interaction of these flow patterns results in a vortex length of ca. 11.25 cm, which is 3.75 cm more than the vortex length in a linear tube, shown in Figure 6. In order to take advantage of this finding, the static mixers were placed in the reactor with a distance of 12.5 cm, as shown in Figure 8. By using elements with three windings, the distance between the opposite sides of two mixers in the reactor is 10.1 cm and thus a little bit smaller than the calculated vortex length. The reactor itself was built by connecting four of 2.5 m PTFE tube pieces with stainless steel tees. The

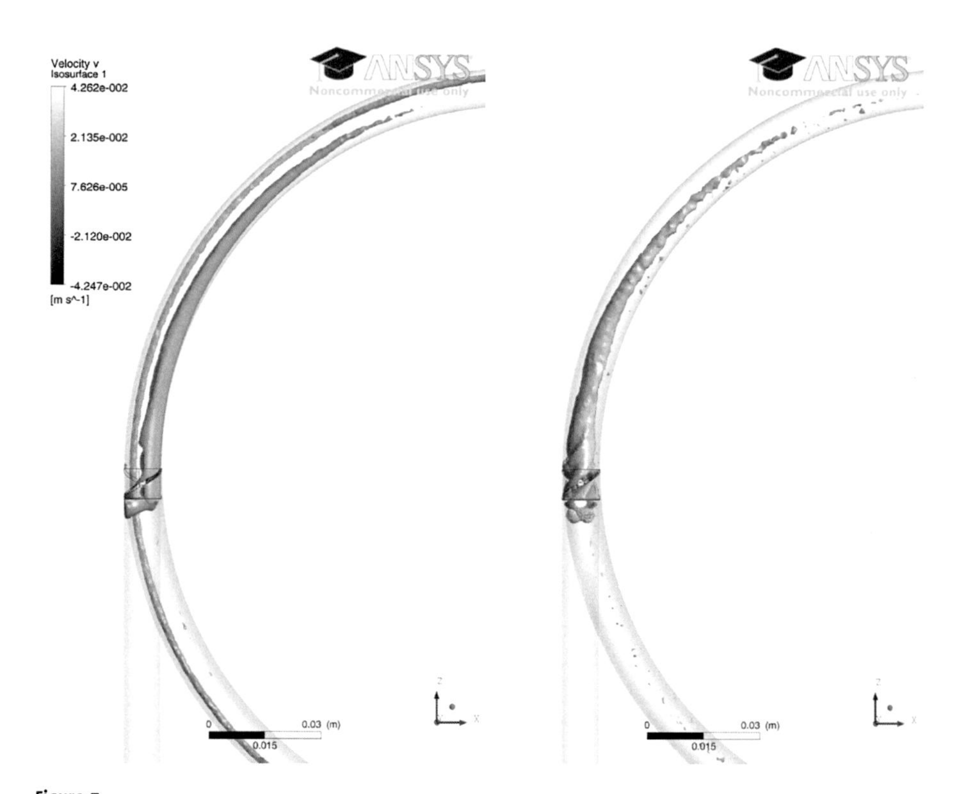

Figure 7.
CFD simulated velocity in y-direction (left) and the q-criterion of the core vortex region of a water mass flow of 1.1 g/s at 90 °C in a helical coiled tube with one single-coiled static mixer (right). The interaction of different flow phenomena is obvious at both pictures.

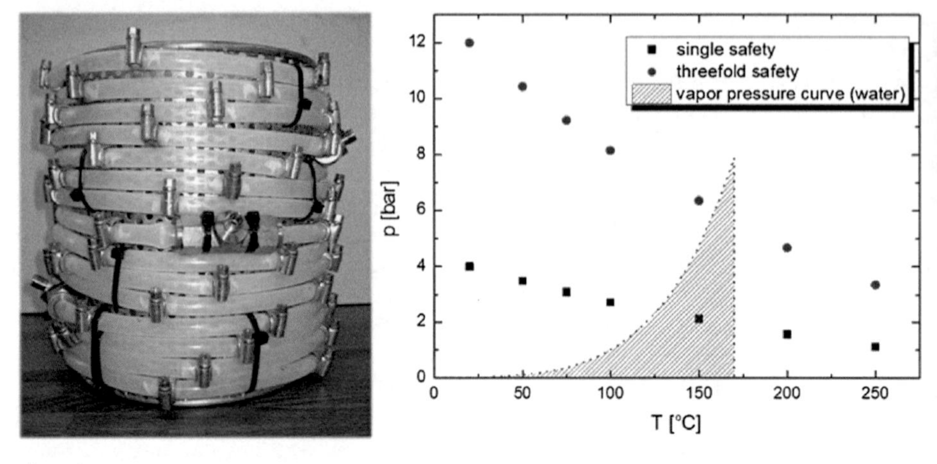

Figure 8.
Tubular reactor with PTFE as reactor material. The tube has a total length of 10 m and an *ID* of 10 mm. Static mixers were positioned in the tube every 12.5 cm. SWAGELOK tees were used to connect 2.5 m segments to enable sample taking (left). The typical operating values for pressure and temperature for emulsion polymerization reactions are mostly within the possible operating range of the PTFE tubes (right).[23]

reactor volume was determined to be 777 mL, which is ca. 8 mL less than the volume of a tube of 10 m without static mixers and tees. However, a reactor with improved flow properties without high volume displacement and high pressure drop was found. The built-in tees can be used for sample taking as well. This feature allows the observation of the reaction progress and the product qualities every 2.5 m of the reactor line for a single experiment as a snap-shot method at steady state conditions. In addition, this technique offers sample-taking for different mean residence times without changing the flow properties. This would be the case if generating different residence times was classically done by changing the mass flow rate.

The residence time distribution (RTD) of the described reactor geometry with the shown arrangement of static mixing elements was determined conductometrically by using KCl solutions in a step experiment.

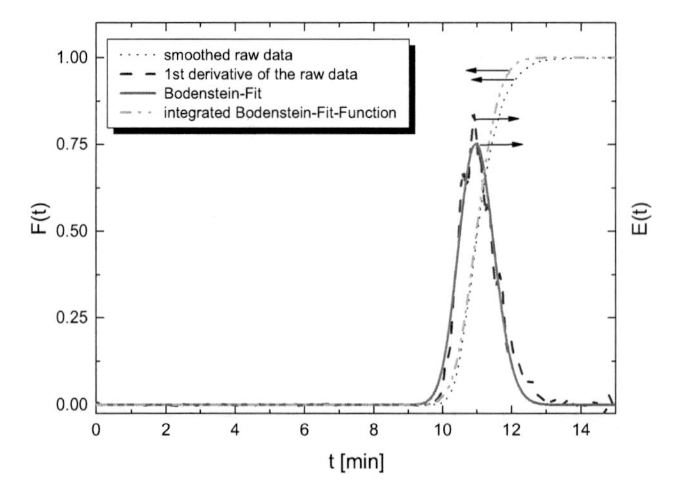

Figure 9.
The Bodenstein Number was calculated by fitting the derivative of the smoothed raw signal. The integral of the fit function was calculated for a comparison with the raw signal.[26]

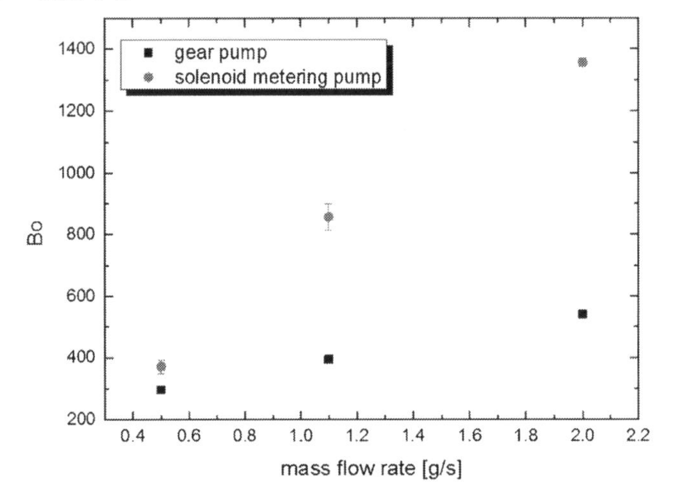

Figure 10.
Bodenstein Number for different mass flow rates (gear pump ■; pulsed pump ●) for water at 22 °C through a 10 m helical coiled tubular reactor partially equipped with static mixers.[26] Higher mass flow rates are leading to increased *Bo* especially with pulsed flow.

The real input step signals were close to ideal step signals. A Fourier Transformation of the received signal $C(t)$ and the real input signal was done, followed by a deconvolution. After an inverse Fourier Transformation, the received function was nearly in the same shape as the derivative of the raw signal. Therefore, the first derivative was assumed to be the derived RTD density function $C(t) = E(t)$ (Equation 1.). Due to this assumption, the flow behavior could be described by the Dispersion Model.[24] This model gives the deviation from ideal plug flow behavior analog to Fick's Law as an axial dispersion coefficient D_{ax}.[24,25] The characteristic flow behavior of a specific reactor under defined conditions is described by the dimensionless Bodenstein Number *Bo* (Equation 2.).

$$C = \frac{1}{2}\sqrt{\frac{Bo}{\pi\theta}}\, e^{\frac{-Bo(1-\theta)^2}{4\theta}} \tag{1}$$

$$Bo = \frac{uL}{D_{ax}} \tag{2}$$

$$\theta = \frac{t}{\tau} \tag{3}$$

Where u is the flow velocity, L the characteristic length, t the experiment time and τ the mean residence time.

Bo and τ were automatically determined by fitting the derivative of the smoothed raw signals to Equation 1 (Bodenstein-Fit). The resulting functions were integrated and compared to the raw signal in order to check the quality of the fit.[26]

The experiments were carried out with two different pump types in order to obtain different flow characteristics. Flow properties derived by using a gear pump (SCHERZINGER 3000 3B) were assumed to be uniform. In contrast, the solenoid metering pump (PROMINENT GAMMA 5) delivered a pulsed flow. The stroke length of the PROMINENT pump was adjusted to enable a uniform pump frequency of 50-60 Hz at every mass flow rate.

For increasing mass flow rates a narrowed RTD was found for both flow characteristics. This trend was intensified by using the pulsed flow. In this case, from a mass flow rate of 1 g/s on, *Bo* is in typical regions in which near plug flow conditions can be assumed.

Heat Transfer Properties of the Tubular Reactor

For strongly exothermic reactions like a free radical polymerization, good heat

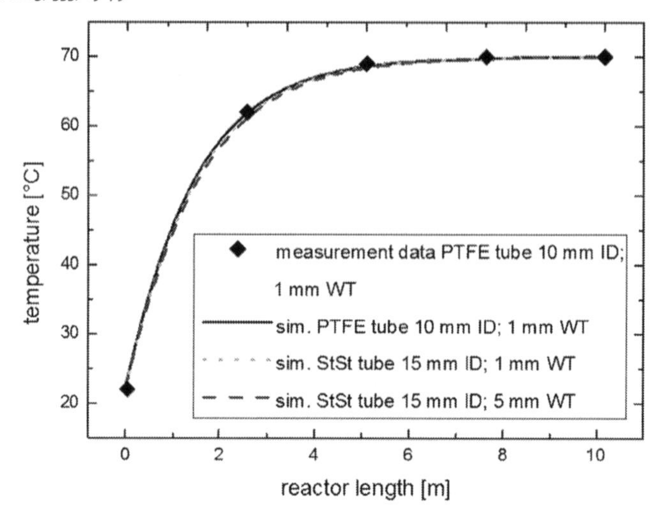

Figure 11.
Heat transfer properties of various tubular reactors. Measured temperatures for the 10 m helical coiled tube (◆) and simulated data for different PTFE and StSt tubes.[23]

transfer properties of the reactor are of prime importance to enable controlled and safe process conditions. The common construction material stainless steel (StSt) has a hundredfold higher thermal conductivity compared to PTFE. The heat transfer properties of the described helical coiled tubular reactor (*ID*: 10 mm; *WT*: 1 mm) with static mixers were determined by the measurement of the temperature of water running through the reactor which was installed in a thermostatic vessel at 70 °C. The theoretical temperature curves were determined through an iterative calculation of the Nusselt Number *Nu*.[27]

$$Nu = 15Re^{0.23}\Gamma^{-0.5} \tag{4}$$

$$Re = \frac{u \cdot d \cdot \rho}{\eta} \tag{5}$$

$$\mathrm{Pr} = \frac{\eta \cdot C_p}{\lambda} \tag{6}$$

$$\Gamma = \frac{L}{d} \tag{7}$$

Where u is the flow velocity, $d = ID$, ρ the density, η the dynamic viscosity, C_p the heat capacity, λ the thermal conductivity, and L the reactor length.

Figure 11 shows the comparison of the measured temperatures with the iteratively calculated values. Regarding tubes of identic *ID* and *WT*, a nearly perfect compliance of the measured and the calculated temperatures was found. In addition, linear StSt tubes with different *ID* and *WT* were simulated to find an appropriate scale-down approach.

Through this approach, it is possible to choose certain tube dimensions in order to keep the heat transfer properties identically while the reactor material is changed from StSt to PTFE in order to avoid fouling. It was found out that the used PTFE tube is a well-fitting scale-down approach for a linear StSt tube with an *ID* of 15 mm. Changing the *WT* of this StSt tube from 1 mm to 5 mm had no remarkable influence on the heat transfer properties (Figure 11). Further descriptions of such a scale-up/scale-down issue will be demonstrated in following publications.

Conclusion

A smart scale tubular reactor system for continuous emulsion copolymerization was constructed. The system can easily be operated in two different modes, isothermal and polytropic. The operation mode has a significant influence on the product properties, such as the MWD and the PDI.

Fouling on the reactor walls was prevented by using PTFE as reactor construction material. By using PTFE, a helical coiled structure of the tube and its partial equipment with Kenics static mixers was easily feasible. Through this combination of geometry and static mixing, it was possible to customize the flow pattern, that non-laminar flow behavior was obtained. This customized flow pattern led to near plug flow conditions. In addition, a positive influence of pulsing pumps on a narrowed RTD was found. Bodenstein Numbers up to 1355 for a production rate of 7.2 kg/h were obtained.

Another positive effect of the customized flow pattern is a reduced boundary layer thickness of the reaction mass on the reactor walls. Therefore, the heat transfer properties of the PTFE tube with an *ID* of 10 mm and a *WT* of 1 mm are comparable to linear StSt tubes with an *ID* of 15 mm and various values for *WT*. A replacement of linear StSt tubes by the described PTFE tubes, in order to prevent fouling, is possible by using the nearly perfect fitting scale-down approach.

Notations

Bo	Bodenstein Number
CFD	computational fluid dynamics
CSTR	continuous stirred tank reactor
GPC	gel permeation chromatography
ID	inner diameter
MWD	molecular weight distribution
Nu	Nusselt Number
PDI	polydispersity index
Pr	Prandtl Number
PTFE	polytetrafluoroethylene
Re	Reynolds Number
RTD	residence time distribution
SST	shear stress
StSt	stainless steel
WT	wall thickness

[1] A. Pashkova, L. Greiner, *Chemie Ingenieur Technik* **2011**, *83*, 1337.
[2] V. Hessel, I. V. Gürsel, Q. Wang, T. Noël, J. Lang, *Chemie Ingenieur Technik* **2012**, *84*, 660.
[3] EC FP7, *project COOPOL*, (NMP2-SL-2012-2808 27), **2012**.
[4] M. Escribà, *et al.*, *Green Chemistry* **2011**, *13*, 1799.
[5] A. K. Yadav, J. C. de la Cal, M. J. Barandiaran, *Macromolecular Reaction Engineering* **2011**, *5*, 69.
[6] D. A. Paquet, W. H. Ray, *AIChE Journal* **1994**, *40*, 73.
[7] K. Ouzineb, C. Graillat, T. McKenna, *Journal of Applied Polymer Science* **2004**, *91*, 2195.
[8] F. G. Lueth, *Diplomarbeit*, Universität Hamburg **2011**.
[9] EP0094443181 **1997**, BASF, invs.: H.-U. Moritz *et al.*
[10] S. Kim, S. J. Lee, *Experiments in Fluids* **2008**, *46*, 255.
[11] W. R. Dean, *Proceedings of the Royal Society A: Mathematical, Physical and Engineering Sciences* **1928**, *121*, 402.
[12] W. R. Dean, *Philosophical Magazine Series* **1927**, *7(7)*, 208.
[13] F. Jiang, K. S. Drese, S. Hardt, M. Küpper, F. Schönfeld, *AIChE Journal* **2004**, *50*, 2297.
[14] M. Apostel, W. Pauer, H.-U. Moritz, J. Kremeskötter, K.-D. Hungenberg, *Chimia Polymer Reaction Engineering* **2001**, *55*, 229.
[15] A. Georg, M. B. Däscher, *Chemie Ingenieur Technik* **2005**, *77*, 681.
[16] N. Elabbasi, X. Liu, S. Brown, *Comsol News* **2012**, 64.
[17] U. El-Jaby, G. Farzi, E. Bourgeat-Lami, M. Cunningham, T. F. L. McKenna, *Macromolecular Symposia* **2009**, *281*, 77.
[18] U. El-Jaby, T. F. L. McKenna, M. F. Cunningham, *Macromolecular Symposia* **2007**, *259*, 1.
[19] M. H. Pahl, E. Muschelknautz, *Chemie Ingenieur Technik* **1980**, *52*, 285.
[20] J. Schmidt, "*Fortschritte der Polymerisationstechnik XXXI*", H-. U. Moritz, Ed., Wissenschaft & Technik Verlag, Berlin **2010**.
[21] F. R. Menter, *AIAA Journal* **1994**, *32*, 1598.
[22] F. R. Menter, "*Zonal Two k-ω Turbulence Models for Aerodynamic Flows*", 24th Fluid Dynamics Conference **1993**, 1.
[23] F. Lueth, C. Retusch, W. Pauer, H.-U. Moritz, "Development of Milli-Reactor-Plants for Continuous Emulsion Polymerization", *IMRET 12*, **2012**.
[24] O. Levenspiel, *Chemical reaction engineering*, 3rd ed. John Wiley & Sons, Inc., Hoboken **1999**.
[25] U. Pallaske, *Chemie Ingenieur Technik* **1984**, *56*, 46.
[26] H. Schroeder, F. Lueth, *Forschungspraktikum*, Universität Hamburg, **2013**.
[27] W. Vauck, H. Mueller, "*Grundoperationen chemischer Verfahrenstechnik*";, 3rd ed. VCH Weinheim, Basel **1988**.

 DOI: 10.1002/masy.201300083

Polymerization of N-Vinyl Formamide in Homogeneous and Heterogeneous Media and Surfactant Free Emulsion Polymerization of MMA Using Polyvinylamine as Stabilizer

*J. Zataray, A. Aguirre, J.C. de la Cal, J.R. Leiza**

Summary: This work considers the homogeneous aqueous phase polymerization of n-vinyl formamide(NVF). Thus, the effect of temperature, initiator and monomer concentration in the kinetics and molar mass distribution (MMD) of the polyNVF produced was experimentally assessed. SEC-MALS analysis was misleading because anomalous elution was found due to interaction of the polyNVF chains with the column. This was solved by analyzing the polyNVF by asymmetric-flow field flow fractionation chromatography coupled with multi-angle light scattering and differential refractive index, AF4/MALS/RI. The second part of this work considered the synthesis of nanoparticles based on polyNVF. Two routes were explored. In the first one the inverse microemulsion photopolymerization of NVF was attempted and polyNVF dispersions in isopar M with solids content of 18 wt% and particle sizes in the range 50–70 nm with average molar masses of several millions were obtained. In the second route PolyNVF produced in homogeneous aqueous phase was hydrolyzed to yield polyvinyl amine, PVAm. The resulting water soluble polymers were used to produce polymethyl methacrylate, PMMA, nanoparticles by surfactant free emulsion polymerization initiated by tert-butyl hydroperoxide TBHP. Stable pH responsive PMMA cationic nanoparticles with amino functionalities in the surface were easily produced.

Keywords: cationic latexes; inverse microemulsion polymerization; poly(n-vinyl formamide); poly(vinyl amine); surfactant free emulsion polymerization

Introduction

Water-soluble polymers are widely used in a broad range of industrial products and processes including food, pharmaceuticals, cosmetics, personal care products, paints and other coatings, adhesives, paper making, biotechnology and water treatment.[1] Free radical polymerization (FRP) is often used to produce these polymers. The understanding of the mechanisms and kinetics involving the polymerization, is the key to the scientists and engineers to reduce the time required to develop new polymer products as well optimize existing processes. N-Vinylformamide (NVF) is a water-soluble monomer, which is a precursor to the amide and amine functional polymers and to other monomers, oligomers and functional polymers.[2] The best thing about NVF is its high reactivity: ready reaction under radical and base or acid conditions, and the wonderful mild hydrolysis under acid or base conditions even after polymerization.[3] PolyNVF has been proposed as a replacement for toxic acrylamide polymers for industrial use. The most appropriate method to produce

POLYMAT, Kimika Aplikatua Saila, Kimika Zientzien Fakultatea, University of the Basque Country UPV/EHU, Joxe Mari Korta Zentroa, Tolosa Hiribidea 72, 20018 Donostia-San Sebastián, Spain
E-mail: jrleiza@ehu.es

poly(vinylformamide) is FRP, although other methods like cationic[4] and anionic polymerization[5] exists. Even though the importance of the fundamental description of NVF polymerization there is only a few studies that focused in this subject. Schmidt et al. estimate the $kp/kt^{1/2}$ values for bulk FRP using differential scanning calorimetry.[6] A kinetic model based on free-volume theory was proposed by Gu et al.[7] to describe the kinetics of FRP in bulk and aqueous solution. The goal of the present work was twofold: First the kinetics and the mechanisms involved in the polymerization of NVF in aqueous media were analyzed. For this purpose, the effect of temperature, concentration of initiator and concentration of NVF in the kinetics and molar mass distribution (MMD) of the polyNVF produced was experimentally assessed. The second objective was to synthesize nanoparticle dispersions based on this polymer. Thus, inverse microemulsion polymeriation was used to produce very high molar mass polymers that can be used as flocculants. Furthermore, taking advantage of the mild conditions required to hydrolize PolyNVF to poly(vinyl amine), a surfactant free emulsion polymerization stabilized by the later was developed to produce pH responsive cationic poly-(methyl methacrylate) latexes.

Experimental Part

N-vinyl formamide, NVF, (Aldrich, Madrid, Spain, 98%, stabilized with 25 to 50 ppm Tempo/Tempol) was distilled under vacuum and stored at -10 °C. The free radical initiator 2,2'-azobis(2-methylpropionamide)dihydrochloride (AIBA) was used as received. Distilled water was used for the preparation of the polymerization solutions. Methyl methacrylate (MMA, Quimidrogra) and tert-butyl hydroperoxide (TBHP, Aldrich) were used as received.

Molar Mass Characterization

The polyNVF aqueous solution as withdrawn from the reactor, was added into methanol in order to precipitate the polymer. Afterwards, the polymer was dissolved in water and precipitated in methanol. This last step was repeated three times in order to ensure a polymer free of impurities and to remove the unreacted NVF (if any), which is liquid at room temperature, but with a boiling point of 210 °C. Then the polymer was left in the vacuum oven at 70 °C during 24 hours. Once the polymer was dried aqueous solutions at the required concentrations were prepared for injection in the AF4 equipment. The molar mass of the polymers was analyzed by AF4/MALS/RI. The equipment was composed by a LC20 pump (Shimadzu) coupled to a DAWN Heleos multiangle (18 angles) light scattering laser photometer equipped with an He-Ne laser ($\lambda = 658$ nm) and an Optilab Rex differential refractometer ($\lambda = 658$ nm) (all from Wyatt Technology Corp., USA).

Filtered toluene (HPLC-grade from Sigma-Aldrich, Madrid, Spain) was used for the calibration of the 90° angle scattering intensity. The detectors at angles other than 90° in the MALS instrument were normalized to the 90° detector using a poly (ethylene glycol) standard (PEG 21,030 g mol^{-1}, Đ = 1.07, Polymer Labs), which is small enough to produce isotropic scattering, at a flow rate of mixed eluent through the detectors of 1 mL min^{-1}. In addition, the same standard and conditions were used to perform the alignment (interdetector delay volume) between concentration and light scattering detectors and the band broadening correction for the sample dilution between detectors.

The analyses were performed at 35 °C and a mixed eluent water/acetonitrile (80:20 by weight) with 0.15 mol L^{-1} NaCl and 0.03 mol L^{-1} NaH_2PO_4 providing a pH 5.5 was used as mobile phase at a flow rate of 1 mL min^{-1}. The samples were injected without filtering. The recovery of the injected mass was calculated based on the dn dc^{-1} of poly NVF (0.159 mL g^{-1}) in DDI water.[8]

The separation in the AF4 equipment is achieved with no stationary phase, solely by a flow in an open channel where a perpendicular flow force is applied.[9] The channel consists of two plates joined together that are separated by a spacer. The bottom plate is permeable, made of a porous frit covered by a semi permeable membrane with a cutoff of 10 KDa. The membrane is permeable for the molecules of carrier, but impermeable for the polymer molecules (**Mw** > 10KDa) and therefore keeps the sample in the channel so that it is directed by a flow to the channel outlet. AF4 flow control was maintained with a Wyatt Eclipse 3 AF4 Separation System controller.

The AF4/MALS/RI data was analyzed by using the ASTRA software version 6.0.3. (Wyatt technology, USA). The absolute molar mass was calculated from the MALS/RI data using the Debye plot (with 1st order Zimm formalism for the polyNVF produced in solution polymerization and 2nd order Berry formalism for the polyNVF produced by inverse microemulsion polymerization).

Additional details of the AF4 analysis can be found elsewhere.[8]

TEM Analysis

The particle size and the morphology of the latexes were analyzed by means of transmission electron microscopy (TEM Tecnai™ G^2 20 Twin device at 200 kV (FEI Electron Microscopes)). The latexes were diluted to 0.5% of solids content and a drop was left to dry at 5 °C on the TEM grid. No staining was used in the analysis.

Flocculation Test

The efficiency of the polyNVF polymers as flocculants was assessed in a Turbiscan Labexpert apparatus (Formulaction). An aqueous solution of the polymer was added to a 0.38 wt% aqueous dispersion of colloidal silica (210 nm) placed in a cylindrical glass cell, which was scanned with a light source and the transmitted light measured. Upon addition of the flocculant, the dispersion became opaque, and then sedimentation of the flocs occurred leading to the formation of two separate regions: a clear phase at the upper part, which in most cases was basically devoid of particles, and a silica rich opaque phase at the bottom.

Aqueous Solution Polymerization of NVF

The homopolymerization reactions were carried out in a Reaction Calorimeter (RC1, Mettler Toledo, Barcelona, Spain) equipped with a stainless steel reactor with 1.5 L of capacity (HP60-Mettler Toledo, Barcelona, Spain). The reactor had a turbine type impeller with pitched blades, a Pt 100 sensor probe, an electrical resistance used to determine the heat transfer coefficient, and a pipe to withdrawn samples and for nitrogen purging.

In a typical experiment to synthesize polyNVF, an aqueous solution of the monomer was heated to the reaction temperature under constant stirring and nitrogen atmosphere. Once the reaction temperature reached the set-point value, polymerization was initiated by the addition of a known amount of a solution of AIBA initiator in distilled water.

The calorimetric conversion was determined by the integration of the heat generation rate, and the value of the enthalpy of polymerization of NVF ($\Delta H_{NVF} =$ -79.4 KJ/mol).

Inverse Microemulsion Polymerization of NVF

The challenge in this type of polymerization is to reduce the concentration of surfactant required to produce a thermodynamically stable microemulsion and at the same time to increase the content of the polymer in the aqueous nanodroplets. In this work it was assess the feasibility of producing high molar mass polyNVF dispersions by inverse microemulsion polymerizations.

In a typical inverse microemulsion polymerization an aqueous phase (36 wt%) containing the NVF monomer (50 wt% of the aqueous solution), and the organic phase (64 wt%) formed by Isopar M (46 wt%) and the mixture of surfactants composed by Arlacel 83 and Tween 80 (18 wt% in the whole formulation and HLB = 5.2) were mixed together to produce a thermodynamically stable microemulsion. An oil-soluble photoinitiator (1-hydroxycyclohexyl phenyl ketone, water solubility of 1.882 g/L) was used to initiate the polymerization in a UV chamber (model BS 03, Dr. Grobel UV-Elektronik GmbH) equipped with 20 UV lamps (wavelength range from 315 to 400 nm with a maximum intensity at 368 nm). The incident light irradiance (ILI) was measured using a radiometer UV sensor (which is part of the chamber) and two ILIs were used in the photopolymerization: 3.5 and 7 mW/cm^2. The photopolymerizations were carried out at room temperature in a jacketed quartz reactor after purging the microemulsion containing the initiator for 15 min with N_2.

Hydrolysis of PolyNVF to Produce Poly(vinyl Amine)

The poly(vinyl amine), PVAm, was synthesized by basic hydrolysis of polyNVF polymers[10] synthesized as described in the polymerization in aqueous phase section above. A solution of PolyNVF at 0.704 M was added to a jacketed reactor and purged for 30 minutes under mild N_2 flow. Reactor temperature was increased to 70 °C and a solution of sodium hydroxide (1.408 M) was injected to start the hydrolysis that was carried out for 6 hours. The kinetics of the hydrolysis reaction was followed by ^{1}H NMR and it was confirmed that the amido groups (at δ 7.8–8.1 ppm) fully disappeared during the 6 hour of reaction. The polyVAm solution was neutralized to pH = 6 using HCl (1N) and the solution was dialyzed to get rid of the sodium formiate formed during the hydrolysis reaction. The aqueous solution thus obtained were directly used in the surfactant free emulsion polymerizations of MMA.

Surfactant Free Emulsion Polymerization of MMA Stabilized by Poly(vinyl amine)

Polymerizations were carried out in a 125 ml jacketed reactor equipped with a turbine impeller rotating at 200 rpm and under a mild flow of N_2 at 80 °C. In a typical polymerization an aqueous solution of PVAm was introduced in the reactor and heated to 80 °C while purging with N_2. When the reaction temperature was reached the monomer (methyl methacrylate, MMA) was injected. The polymerization was started by adding an aqueous solution of terbutyl hydroperoxide, TBHP. Polymerizations were carried out using PVAm of different molar masses. The solids content, concentration of PVAm and TBHP concentration were varied to analyze the effect on the stability of the latexes and on the particle sizes achieved.

Results and Discussion

Aqueous Solution Polymerization of NVF

Table 1 presents the polymerization experiments carried out at various temperatures and concentrations of NVF and AIBA initiator.

Figures 1–3 show the evolution of the monomer conversion and the molar masses with respect to the three variables studied; namely, initiator concentration (Figure 1), monomer concentration (Figure 2) and temperature (Figure 3). Increasing AIBA concentration the polymerization rate increased (the slope of the conversion-time curve) and full conversion was achieved for all the initiator concentration employed. The molar mass decreased increasing initiator concentration as expected in solution polymerization.

Table 1.
Experimental conditions of the solution polymerization of NVF experiments carried out in the calorimetric reactor

Run	[NVF] (wt%)	[AIBA]x10^3 (M)	Temperature (°C)
1	9	1.47	70
2	9	1.84	70
3	9	2.21	70
4	9	2.54	70
5	9	2.94	70
6	6	1.47	70
7	12	1.47	70
8	15	1.47	70
9	9	1.47	60
10	9	1.47	80
11	9	1.47	90

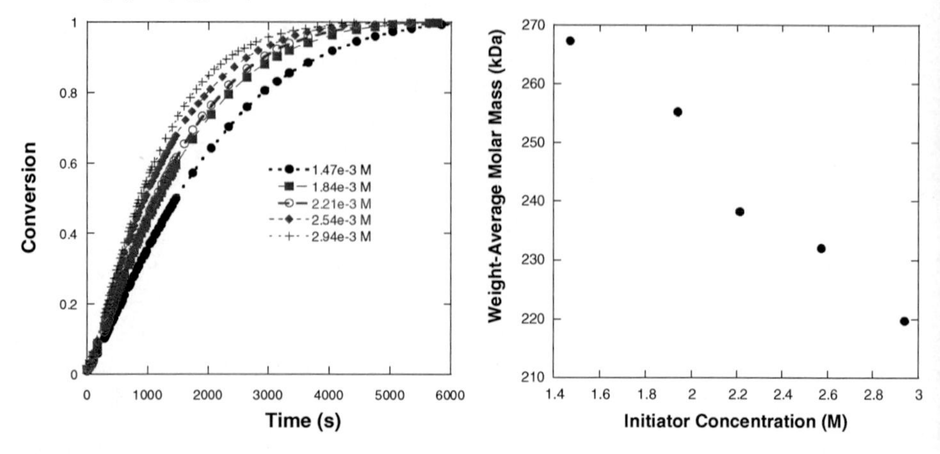

Figure 1.
Evolution of monomer conversion and molar mass for the solution polymerization experiments carried out at [NVF] = 9wt% and T = 70 °C varying AIBA initiator concentration.

For the experiments where the monomer concentration was varied the evolution of the conversion was not unique that is there was a dependence of the polymerization rate on the monomer concentration. This behavior was not expected for a first order reaction, in which for the same initiator concentration and temperature the conversion profiles should be the same. Also the polymerization rate increased decreasing the monomer concentration with little difference for the experiments at 12 and 15 wt% of NVF. This result can only be explained by assuming that the propagation rate coefficient depends on the monomer concentration. This was recently shown by Stach et al.[11] by PLP/SEC measurements that show that the propagation rate constant of NVF decreased by increasing the concentration of the monomer in aqueous solution. They proposed an Arrhenius equation with an additional monomer concentration dependence.

The molar masses increased with increasing the monomer concentration as it can be expected in solution polymerization.

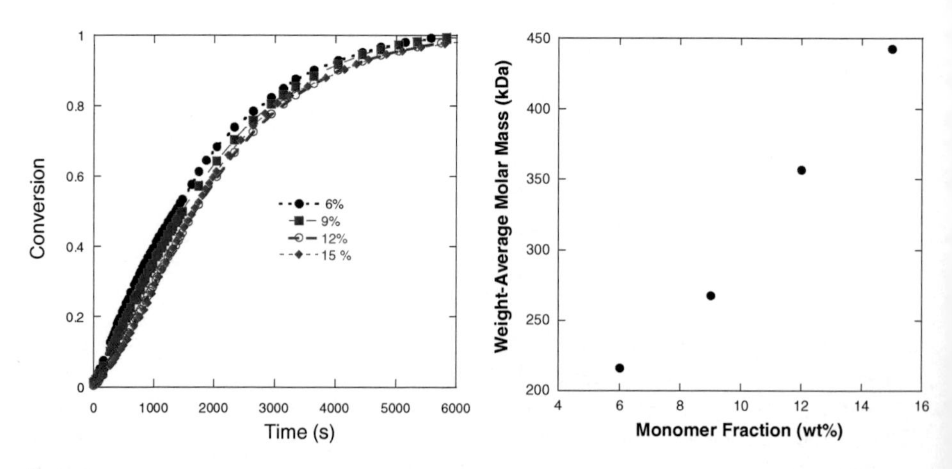

Figure 2.
Evolution of monomer conversion and molar mass for the solution polymerization experiments carried out at [AIBA] = 1.47×10^{-3} M and T = 70°C varying NVF concentration.

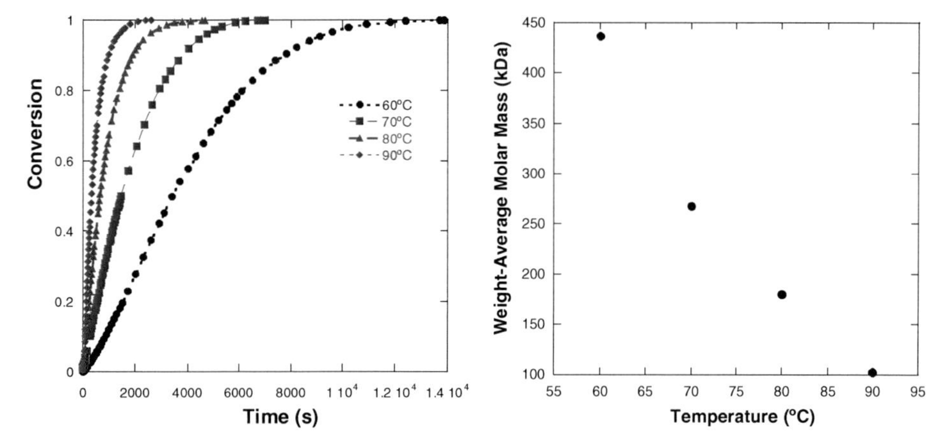

Figure 3.
Evolution of monomer conversion and molar mass for the solution polymerization experiments carried out at [NVF] = 9wt% and [AIBA] = 1.47×10^{-3} M varying reaction temperature.

At constant temperature, increasing concentration of monomer implies increasing the ratio ($k_p[M]/k_t[I]^{1/2}$) that controls the kinetic chain length. Therefore the effect of the monomer concentration on the molar masses fulfilled what it can be expected in solution polymerization.

Figure 3 presents the evolution for the effect of the temperature. Increasing the reaction temperature noticeably increased the polymerization rate and decreased the molar masses. Again this result is in good agreement with the classical theory of solution polymerization.

To sum up it was possible to obtain polyvinylformamide polymer solutions with molar masses in the range 100.000–450.000 Da by aqueous solution polymerization by conveniently varying temperature, initiator and monomer concentration. Some of these polymer solutions were used to produce the poly(vinylamine) used in the surfactant free emulsion polymerization experiments.

Inverse Microemulsion Polymerization of NVF

Monomers of the family of the acrylamide are widely used to produce flocculants that are extensively used in waste-water treatment applications.[12] The flocculants are produced by inverse microemulsion polymerization that allows producing ultra-high molar mass polymers due to the very small particles sizes achieved that favor an almost complete compartmentalization of the radicals. This feature minimizes bimolecular termination and hence leads to very large polymer chains.

Stable microemulsions at 36 wt% of dispersed aqueous phase (50 wt% being NVF monomer) were only achieved for surfactant concentrations between 15 and 20 wt% (based on the total formulation) and HLBs of the surfactant mixture between 4.5 and 6.0. The stable microemulsions were polymerized using different types of initiators. When redox initiators were used (TBHP/Ascorbic acid or TBHP/NaO-SO-CH-OH-COONa(Brugolite 7 or FF7)) the microemulsion did not remain stable and phase separation occurred. Stable polymerized microemulsion dispersions were only achieved using oil-soluble photoinitiators (PI) and initiating the polymerization with UV lamps. As it will be discuss below the faster the polymerization the easiest was to preserve the microemulsion during the polymerization.

Table 2 presents the formulations of the inverse microemulsion photopolymerizations carried out with different concentrations of initiator and irradiation power as well as the final particle size of the latexes produced.

Table 2.
Polymerization conditions and particle size obtained in the inverse microemulsion photopolymerizations carried out at room temperature with a concentration of the mixture of surfactants of 18wt% and a HLB of 5.25

Run	[PI]x10^3 (M)	Power (mW/cm^2)	Dp (nm)
IM1	1.47	3.5	65
IM2	1.47	7	80
IM3	2.21	3.5	52
IM4	2.21	7	54
IM5	2.94	3.5	51
IM6	2.94	7	51

Figure 4 presents the evolution of the conversion of the inverse microemulsion reactions for the three photoinitiator concentrations at the highest irradiation power (a similar trend was found for the lower irradiation power). In all cases full conversion was achieved and the polymerization rate increased with the concentration of photoinitiator and the irradiation power (not shown). For instance, full conversion was achieved in 15 minutes for [PI] = 2.94 M and an irradiation power of 7 mW/cm^2.

Figure 5 presents the pictures of the samples taken along the reaction for the experiments carried out at different photoinitiator concentrations for the highest power employed in the experiments (see

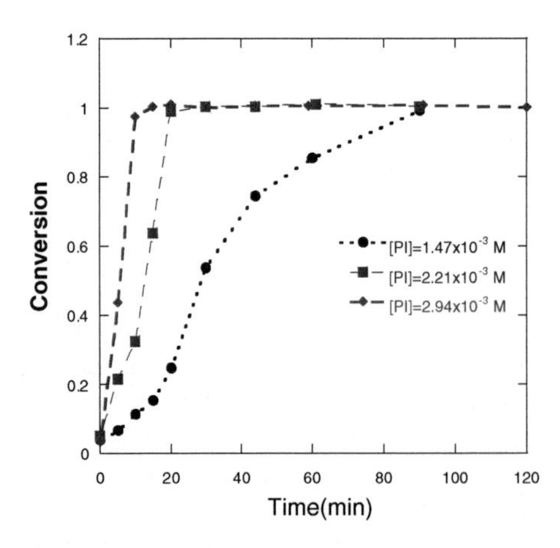

Figure 4.
Time evolution of the NVF conversion during the inverse microemulsion polymerizations carried out at room temperature and 7mW/cm^2.

Table 2). As it can be seen in the pictures the stability of the microemulsion was good (transparent dispersions) at the beginning of the polymerization in all the systems, but the slowest the polymerization rate the less stable was the microemulsion (it changed from transparent to turbid) during the polymerization. This is in agreement with the final particles sizes achieved as shown in Table 2. Relatively larger particle sizes were achieved for the lowest PI concentration. Indeed, the only polymerization in which unstable transitions were not observed (the samples were transparent during the whole polymerization) was the one carried out at the highest PI and irradiation power that is the fastest reaction as shown in Figure 4. This can be explained by the fact that the faster the polyNVF chains are formed in the nanodroplets, the better is the stability of the nanodroplet with respect to degradation and coalescence. Further work should be done to fully confirmed this hypothesis. According to these results true microemulsion polymerization conditions were only achieved in experiments IM4 and IM6.

Figure 6 shows the chromatogram of the AF4/MALS/RI analysis as well as the MMD in both the differential and cumulative forms for the experiment IM6. It can be seen that two clear populations of chains were produced. Note that elution in AF4 fractionation is opposite to that of SEC; namely, molar mass increases with elution time. Interestingly, the fraction of small molar masses is approximately 90% of the polymer and it is centered at 400,000 Da, whereas the high molar mass fraction (10% of the polymer) is centered at 80 million Da. The bimodal distribution is likely due to the extremely fast generation of radicals that causes two regimes of formation of chains: one at high concentration of PI that generates short molar mass chains because of the high radical flux to the particles favoring bimolecular termination and hence short chains to be formed. Note that termination by combination is the main termination mechanism of this monomer system as reported elsewhere[7,13] and also

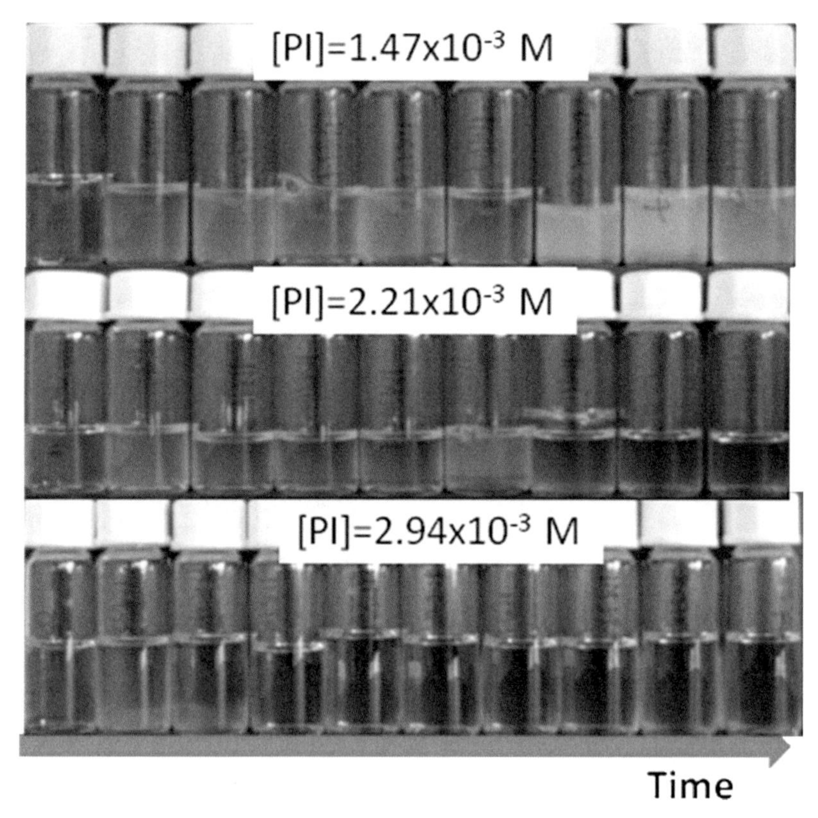

Figure 5.
Microemulsion samples taken from the reactor at different sampling times for the experiments carried out at 7mW/cm^2 with the photoinitiator concentrations indicated in the photographs.

found in this work because the dispersity index of the polyNVF polymers synthesized in solution polymerization were below 2. And the second regime at low concentration of PI, in which due to the small size of the particles and the substantially lower rate of generation of radicals produces extremely high molar mass chains although the percentage of those chains was modest (10 wt%).

The efficiency of the polyNVF polymer as flocculants was assessed by adding a solution of the polymer obtained in experiment IM6 to an aqueous dispersion of colloidal silica (210 nm and 0.38 wt%). It was found (not shown) that the solution of polyNVF by itself had very low flocculation efficiency likely because of the lack of charges in the structure of the polyNVF. However, when, in addition to the solution of polyNVF, a solution of the polyPVAm of low molar mass (obtained by the hydrolysis of a polyNVF synthesized in aqueous phase polymerization) was added to the silica dispersion the flocculation efficiency was substantially improved. Figure 6 shows the photograph of the silica dispersion before the addition of the polymers (polyNVF and cationic PVAm).

The combined effect of the cationic low molar mass PVAm and the high molar mass PolyNVF proved to be important to flocculate the dispersed silica particles.[14] The short cationic polymer reaches quickly the silica particles neutralizing the changes and aggregating them and the high molar mass polyNVF chains bridge-flocculates this aggregates to the bottom of the vial.

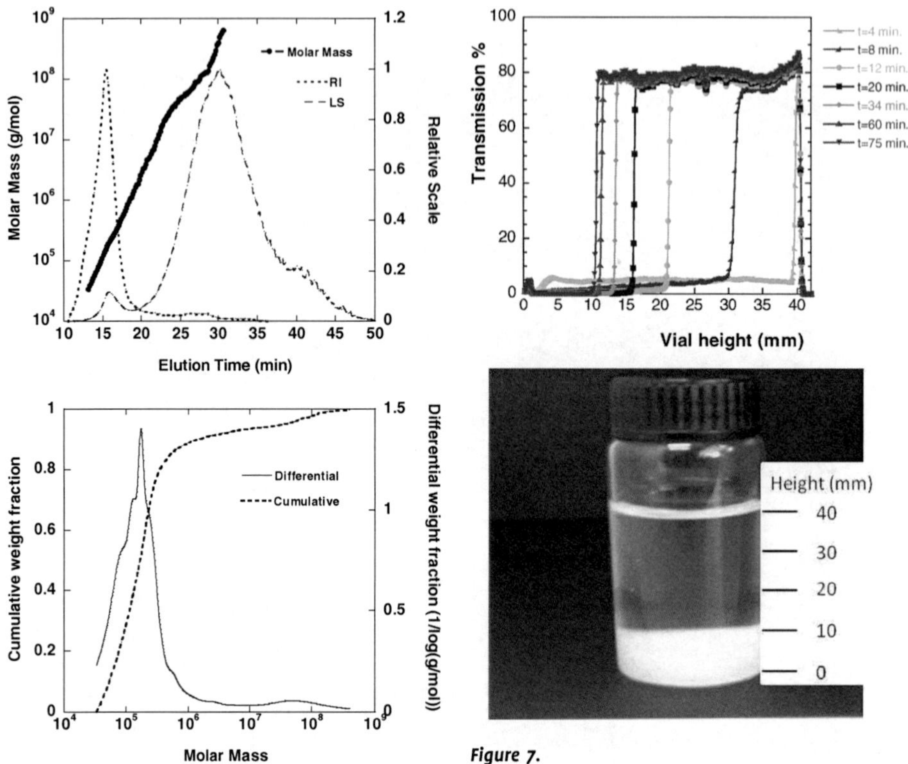

Figure 6.
(top)AF4/MAL/RI chromatogram and weight average molar mass vs elution time and (bottom) differential and cumulative molar mass distribution for sample IM6 on Table 2.

Figure 7.
(Top) Transmitted ligth vs time for a silica dispersion of 210 nm and 0.38wt%. (Bottom) Photograph of the vial after the addition of the polymer (PolyNVF plus PVAm) solution.

Surfactant Free Emulsion Polymerization of MMA Stabilized by PVAm

Pei Li et al. [15,16] pioneered the route to synthesize core-shell polymer nanospheres using water soluble polymers containing amino groups in absence of any surfactant. In this approach the initiator is a redox system formed by the oxidant, TBHP, and the reductant, the amine groups of the polyVAm. Thus, two radicals are formed; one in the amine groups of the polyVAm backbone, which has an slow motion because of the relatively high molar mass of the polymer chain, and a highly mobile and hydrophobic tertbutoxy one. The radicals in the PVAm backbone add MMA monomer dissolved in the aqueous phase and form graft PVAm-g-PMMA copolymer chains that are amphiphilic. These species aggregate and swell with monomer forming polymer particle precursors. On the other hand, the terbutoxy radicals can also add MMA from the aqueous phase forming oligoradicals of MMA or short chains of polyMMA that precipitate forming precursor polymer particles. The later radicals can also terminate in the aqueous phase, but likely they will be absorbed into the hydrophobic aggregates due to their hydrophobic character. The ratio PVAm/TBHP determines the concentration of radicals produced and the ratio PVAm/MMA affects the size of the particles achieved. Using this method Pei Li et al.[15,16] demonstrated that core-shell stable latexes could be obtained using a variety of amino containing natural and synthetic polymers (poletheleneimine, chitosan, casein, gelatin and others). A similar particle nucleation route was also proposed by the production of block copolymers of monohydroxy PEG (water soluble polymer) and NIPAM by radical polymerization using Cerium (IV) as redox initiation.[17,18] The Ce(IV) creates a redox pair with the CH_2-OH groups of the PEG that allows blocks of PNIPAM to be created in the end-groups. The block copolymers aggregated forming micelles to further polymerize the NIPAM monomer to yield stable particles.

Table 3 presents the experiments carried out in this work following this approach using PVAm polymers synthesized by the hydrolysis of polyNVF. PVAm's with three molar masses (217600, 91600 and 42800 Da) were used in the experiments. The ratios PVAm/THP and PVAm/MMA were varied in these experiments and the effect on the conversion and particle sizes produced was analyzed.

Figure 8 presents the evolution of the conversion for the experiments carried out with PVAm of decreasing molar mass (labeled R1, R2 and R3) and the same PVAm/TBHP and PVAm/MMA ratios (series 1, 2 and 3). For series 1 and 2 (Figures Figure 8a and 8b) the higher the molar mass of the PVAm the higher the conversions achieved and the higher the polymerization rate. This is likely related with the higher number of particles (smaller particle sizes) produced with the higher molar mass PVAm. There is no clear difference in particle size for the PVAm of 42800 and 91600 Da (approximately around 200 nm), but the particle size for the largest PVAm was in the range 110-125 and 125-130 nm for series 1 and 2, respectively (See Figure 9 for the TEM micrographs of the 1 series). The smaller particle sizes produced with the longer PVAm (at two different PVAm/TBHP ratios) should be related with the solubility of the grafted chains required to form aggregates. It seems that the longer the PVAm chain (above certain threshold value because there is no substantial difference for 42800 and 91600), the highest the number of aggregates and hence the higher the number of stable particles formed.

For the series 3, the ratio PVAm/MMA was doubled with respect to series 2 (with the same PVAm/TBHP) and also the total concentration of initiator was higher. In this case polymerization rates were substantially faster and particle sizes smaller than in series 2. This is because the amount of PVAm chains prone to graft polyMMA and hence form polymer particle precursors was higher and in addition the radical concentration was higher too. This led to very fast reactions where complete conversion was achieved in few minutes (see Figure 8c).

The stability of the latexes was very good even if surfactant was not used. Latexes up to solids content close to 20 wt% could be produced using slightly modified formulations with respect of those presented in Table 3. The stability of the dispersion comes from the cationic character of the amine groups of the polyvinyl amine that are protonated (and hence positively charged) at pH values

Table 3.
Polymerization conditions and final conversion and particles sizes of the latexes produced by surfactant free emulsion polymerization of MMA

Run	PVAm Molar Mass (Da)	PVAm/TBHP (wt/wt)	PVAm/MMA (wt/wt)	Final Conversion	Dp(nm) TEM*	SC(%) (PVAm+PMMA)
R1_1	42800	1387	1/4	0.85	180–200	5
R1_2	42800	222	1/4	0.92	200–220	5
R1_3	42800	222	1/2	1.00	185	6
R2_1	91600	1387	1/4	0.85	200	5
R2_2	91600	222	1/4	0.91	200	5
R2_3	91600	1387	1/2	1.00	120–130	6
R3_1	217600	1387	1/4	0.94	110–125	5
R3_2	217600	222	1/4	0.98	125–130	5

* The particle sizes provided are only indicative of the range.

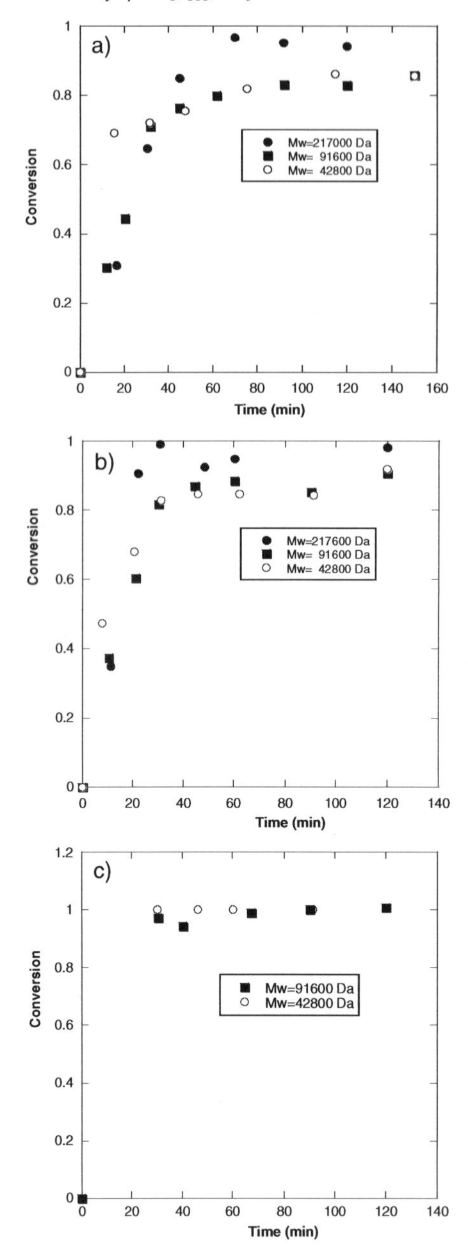

Figure 8.
Time evolution of conversion for the surfactant free emulsion polymerizations stabilized by PVAm. (a)PVAm/TBHP = 10 and PVAm/MMA = 1/4; (b) PVAm/TBHP = 2 and PVAm/MMA = 1/4;(c) PVAm/TBHP = 2 and PVAm/MMA = 1/2.

below 9.5. Figure 10a shows the ζ-potential of several latexes as a function of the pH. It can seen that the potential is in the positive range between 30-40 mV (which explains the good stability of these cationic latexes) and that when the pH of the continuous phase exceeded the value of 9–10, the ζ-potential abruptly decreased approaching to zero. This is very interesting because this imparts a pH responsive character to the latexes. The particle size shrinks at this pH as it can be seen in Figure 10b. In this figure the hydrodynamic particle size of the latexes was plotted as measured by dynamic light scattering(DLS). Note that the sizes measured by TEM were closer to the shrunk state of the particles, but still smaller because the hydrodynamic radius was measured by DLS. PVAm chains are positively charged at pH values below 9-10 and hence a polycation (positively charged polyelectrolite) is formed, which is fully soluble in the aqueous phase and hence the cationic PVAm arms are fully extended and swollen with water increasing the particle size from that measured in the dry state (TEM). When the critical pH is approached the amino groups deprotonated and the PVAm become hydrophobic and contracted over the surface of the particles expelling the water and notably reducing the particle size (between 2-3 fold depending on the reaction conditions).

Conclusion

N-vinyl formamide, a non-toxic isomer of acrylamide, was polymerized in homogeneous and heterogeneous conditions; namely solution and inverse microemulsion polymerization. The homogeneous polymerization behaved as expected from a classical theory of free radical polymerization with the exception that the monomer concentration affected the evolution of the conversion, which could be explained based on a recent report that demonstrated that the propagation rate coefficient of the NVF is inversely proportional to the concentration in the aqueous phase. For the polymerization in heterogeneous conditions, a formulation was found that provided stable microemulsions at 18 wt% solids. Polymerization of the microemulsions to produce

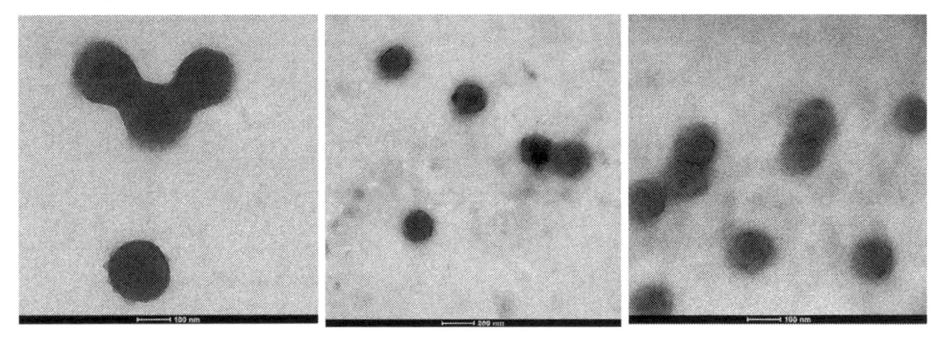

Figure 9.
TEM micrographs of experiments of series R1. (left) R1_1; (middle) R2_1 and (right) R3_1.

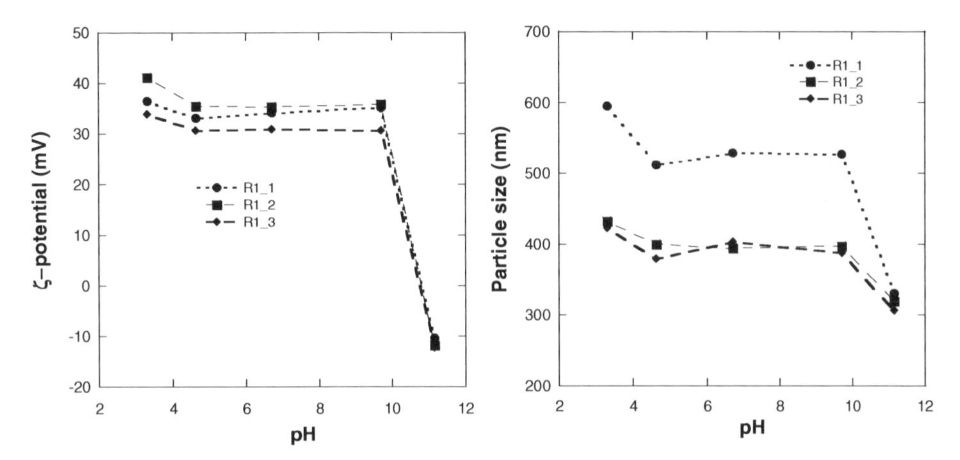

Figure 10.
ζ-potential (left) and particle size (right) for several latexes of the R1 series.

stable latexes was not straightforward because the microemulsion become unstable if the polymerization was not fast enough. Thus, it was found that photopolymerizations using oil-soluble photoinitiators and with a large irradiation power allowed to maintain true microemulsion polymerization conditions. Under these conditions stable latexes with particle sizes in the range of 50 nm and with a bimodal MMD were produced. The large molar mass mode of the distribution had average molar masses of 80×10^6 Da. Finally, cationic and pH responsive latexes of polyMMA containing amino functionalities at the surface were synthesized by surfactant free emulsion polymerization using poly(vinyl amine) as stabilizer. The PVAm was easily produced by hydrolysis of the polyNVF synthesized in the solution polymerization of NVF. These cationic nanoparticles are good candidates for biomedical applications.

Acknowledgements: Financial support from the Basque Government (GV-IT-303-10), Ministerio de Ciencia e Innovación (MICINN, Ref. CTQ2011-25572) and UPV/EHU (UFI 11/56) is gratefully acknowledged. The authors thank the sGIKer UPV/EHU for the electron microscope facilities of the Gipuzkoa unit.

[1] P. Williams, In *Handbook of industrial water soluble polymers*, Blackwell Publishing Ltd., **2007**.
[2] R. K. Pinschmidt, W. L. Renz, W. E. Carroll, K. Yacoub, J. Drescher, A. F. Nordquist, N. Chen,

Journal of Macromolecular Science-Pure and Applied Chemistry **1997**, *A34*(10), 1885–1905.
[3] R. K. Pinschmidt, *Journal of Polymer Science Part A-Polymer Chemistry* **2010**, *48*(11), 2257–2283.
[4] S. Spange, A. Madl, U. Eismann, J. Utecht, *Macromolecular Rapid Communications* **1997**, *18*(12), 1075–1083.
[5] R. Fikentscher, M. Kroener, *United States US* **1992**, *5155270*,
[6] C. U. Schmidt, K. D. Hungenberg, W. Hubinger, *Chemie Ingenieur Technik* **1996**, *68*(8), 953–958.
[7] L. Gu, S. Zhu, A. N. Hrymak, R. H. Pelton, *Polymer* **2001**, *42*(7), 3077–3086.
[8] J. Zataray, A. Aguirre, J. C. de la Cal, J. R. Leiza, *Journal of Chromatography A* **2013**, (submitted).
[9] S. Podzimeck, *Light Scattering, Size Exclusion Chromatography and Asymmetric Flow Field Fractionation: Powerful Tools for the Characterization of Polymers, Proteins and Nanoparticles*, Wiley ed., **2011**.
[10] L. Gu, S. Zhu, A. N. Hrymak, *Journal of Applied Polymer Science* **2002**, *86*(13), 3412–3419.
[11] M. Stach, I. Lacik, P. Kasak, D. Chorvat, A. J. Saunders, S. Santanakrishnan, R. A. Hutchinson, *Macromolecular Chemistry and Physics* **2010**, *211*(5), 580–593.
[12] G. Gonzalez, J. M. Ugalde, J. C. de la Cal, J. M. Asua, *Macromolecular Rapid Communications* **2009**, *30*(23), 2036–2041.
[13] S. Santanakrishnan, L. Tang, R. A. Hutchinson, M. Stach, I. Lacik, J. Schrooten, P. Hesse, M. Buback, *Macromolecular Reaction Engineering* **2010**, 4(8), 499–509.
[14] G. Gonzalez, J. C. de la Cal, J. M. Asua, *Colloids and Surfaces A-Physicochemical and Engineering Aspects* **2011**, *385*(1-3), 166–170.
[15] P. Li, J. M. Zhu, P. Sunintaboon, F. W. Harris, *Langmuir* **2002**, *18*(22), 8641–8646.
[16] P. Li, J. M. Zhu, P. Sunintaboon, F. W. Harris, *Journal of Dispersion Science and Technology* **2003**, *24*(3-4), 607–613.
[17] M. D. C. Topp, I. H. Leunen, P. J. Dijkstra, K. Tauer, C. Schellenberg, J. Feijen, *Macromolecules* **2000**, *33*(14), 4986–4988.
[18] K. Tauer, Polymer Nanoparticles with Surface Active Initiators and Polymeric Stabilizers. In: *Advanced Polymer Nanoparticles: Synthesis and Surface Modifications*, V. Mittal, (Ed., CRC Press, Taylor & Francis: Boca Ratón **2010**, pp 329–359.

 DOI: 10.1002/masy.201300090

Emulsion Polymerization Using Switchable Surfactants: A Route Towards Water Redispersable Latexes

Xin Su, Candace Fowler, Catherine O'Neill, Julien Pinaud, Erica Kowal, Philip Jessop, Michael Cunningham**

Summary: Colloidal latexes of polystyrene and poly(methyl methacrylate) have been prepared by emulsion polymerization using amidine-based switchable surfactants. Particles with sizes ranging from 50 nm to 350 nm were obtained. Destabilization of the latexes requires only air and heat which destabilize the latex by removing CO_2 from the system. The resulting micron sized particles can be easily filtered to yield a dry polymer powder and a clear aqueous phase. We have also developed a new benign means of reversibly coagulating anionic latexes by using "switchable water", an aqueous solution of switchable ionic strength. The addition of CO_2 and switchable water to an anionic latex can result in aggregation of the latex. Subsequent removal of CO_2 by sparging with air allows the aggregated latex to be redispersed and recovered in its original state.

Keywords: carbon dioxide; colloids; emulsion polymerization; radical polymerization; stimuli-sensitive polymers

Introduction

Latex destabilization is often carried out industrially using salts, acids or bases. A preferred greener approach would be to use a surfactant capable of being switched off when its stabilizing ability is no longer desired. From an environmental standpoint, the use of air as a coagulant is preferable to the currently used high concentrations of salts, acids or alkali. Carbon dioxide can be used as a trigger to switch a neutral amine or amidine into a charged form ("on" state).[1–16] Purging the system with inert gas or air will remove the CO_2 and switch the charged compound back into a neutral amine or amidine ("off" state) (Scheme 1). A few reversibly switchable surfactants have been developed but with economically or environmentally undesirable trigger agents. In no case has surfactant switching or recycling been proposed with a trigger as facile and reversible as simply exposing the solution to CO_2, the method that we have developed.[1]

We have prepared polystyrene and poly(methyl methacrylate) latexes by emulsion polymerization using cationic amidine-based switchable surfactants.[2–5] Particles with sizes ranging from 50 nm to 350 nm were obtained and the effect of factors such as initiator type, initiator amount, surfactant amount and solid content on the particle size and zeta potential of the resulting latexes have been examined. Destabilization of the latexes requires only air and heat which destabilize the latex by removing CO_2 from the system. The resulting micron sized particles can be easily filtered to yield a dry polymer powder and a clear aqueous phase. The switchability of aryl amidine and tertiary amine switchable surfactants has also been examined.[6] Despite the lower basicity of these compounds compared to alkylacetamidine switchable surfactants, it was found that amidinium and ammonium bicarbonates could be formed in sufficiently high enough concentrations to perform

Queen's University, Kingston, Canada
E-mail: michael.cunningham@chee.queensu.ca

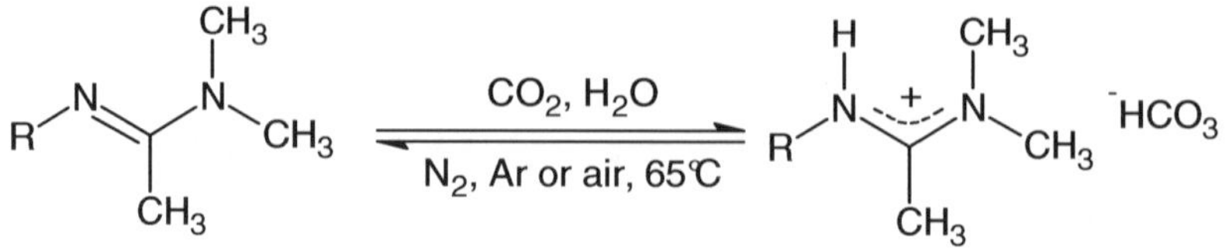

$$R-N=C(CH_3)-N(CH_3)_2 \underset{N_2,\ Ar\ or\ air,\ 65°C}{\overset{CO_2,\ H_2O}{\rightleftharpoons}} R-NH\overset{+}{=}C(CH_3)-N(CH_3)_2\ ^{-}HCO_3$$

Scheme 1.
Reaction of long chain alkyl amidines with CO2 to form amidinium bicarbonates.

emulsion polymerization of methyl methacrylate and stabilize the resulting colloidal latexes. However the coagulation rate is much higher in the case of the less basic aryl amidine and tertiary amine stabilized latexes.

We also report a new benign means of reversibly coagulating latexes by using "switchable water", an aqueous solution of switchable ionic strength.[8] The conventional anionic surfactant sodium dodecyl sulfate (SDS) is normally not stimuli-responsive when CO_2 is used as the stimulus but becomes CO_2-responsive or "switchable" in the presence of a switchable water additive. The addition of CO_2 and switchable water to a latex can result in aggregation of the latex. Subsequent removal of CO_2 by sparging with air neutralizes the amine DMEA and decreases the ionic strength allowing for the aggregated latex to be redispersed and recovered in its original state.

Results and Discussion

Alkyl Amidine Switchable Surfactants

Emulsion polymerizations of styrene and MMA were carried out using C12 amidine surfactants with a variety of azo-based free radical initiators to determine the stability and particle size of the resulting polymer latexes.[2–5] Table 1 shows the results obtained with the C12 amidine surfactant and each of the three initiators that were investigated. Latex generated using the cationic, water-soluble initiator VA-044 showed no visible signs of polymer settling after the removal of CO_2. Even after treatment of the latex with heat and air, followed by centrifugation for 10 min at 3500 rpm, the polymer particles remained suspended. Transfer of protons is expected to occur from the hydrochloride initiator to the surfactant resulting in sustained surface activity even in the absence of carbon dioxide (i.e. the HCl salts prevent the surfactant from switching off). Latexes formed using AIBN had large particles and very broad size distributions, and thus it was deemed unsuitable for further investigation. VA-061 is the neutral form of VA-044 and contains a cyclic amidine group that can react with CO_2 to yield a bicarbonate salt. The water soluble salt can then be used as the initiator for the polymerization reaction, and upon treatment of the resulting latex with air and heat, can be neutralized and the latex more easily aggregated.

Table 1.
Stability of polystyrene latexes formed using amidine switchable surfactants and different azo-based initiators.[a]

	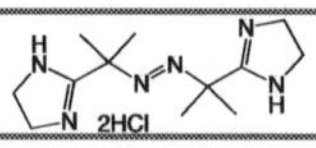 (structure, 2HCl)	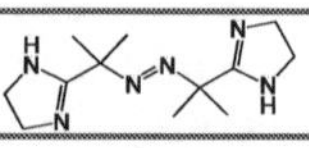 (structure)	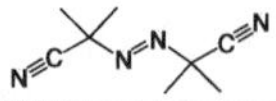 (structure)
Initiator Name	VA-044	VA-061	AIBN
Particle Size (nm)	82	65	>6000
Conversion (%)	93	81	–
Latex Stability[b]	Stable	Stable only in the presence of CO_2	Unstable

[a] Reactions were carried out for 5 h at 65 °C with N′-hexadecyl-N,N-dimethylacetamidinium bicarbonate.
[b] Stability was determined visually. Samples where no visible settling occurred for at least 2 weeks were deemed stable.

Table 2.
Variation in particle size and zeta potential of polystyrene latexes using different solid contents, surfactant concentrations and initiator concentrations.[a]

	Wt% Styrene	Mol% Surf[b]	Mol% Init[c]	Particle Size (nm) (PdI)[d]	Zeta Potential (mV)	Conversion (%)	Coagulum (%)
1	13.5	1.00	1.00	72 ± 0.1 (0.05)	51 ± 2	93	2
2	13.5	0.50	0.50	129 ± 1 (0.01)	45 ± 1	89	0
3	13.5	0.25	0.25	198 ± 2 (0.02)	49 ± 1	92	6
4	13.5	0.10	0.10	222 ± 2 (0.03)	44 ± 0.3	85	2
5	13.5	0.25	0.50	203 ± 3 (0.04)	59 ± 2	94	24
6	22.8	3.00	1.00	58 ± 1 (0.05)	31 ± 2[e]	–	–
7	23.5	1.00	1.00	61 ± 1 (1.2)	59 ± 6	100	2
8	23.5	0.50	0.50	85 ± 1 (0.06)	59 ± 1	95	3
9	23.5	0.25	0.25	203 ± 2 (0.03)	54 ± 1	98	5
10	23.5	0.10	0.10	178 ± 2 (0.02)	51 ± 1	94	7

[a] Reactions were carried out for 5 h at 65 °C under 1 atm of CO_2.
[b] Surfactant is *N*'-dodecyl-*N,N*-dimethylacetamidinium bicarbonate.
[c] With Initiator is the bicarbonate salt of 2,2'-azobis[2-(2-imidazolin-2-yl)propane].
[d] Values of PdI obtained from the Zetasizer ZS.

A series of PS latexes was prepared using the conditions shown in Table 2. The effects of solids content, initiator concentration, surfactant concentration on particle size and zeta potential were examined. Particle size was highly dependent on surfactant concentration; increases in particle size were observed when the surfactant concentration was decreased (entries 2 vs. 5 and 9 vs. 11). Simultaneously decreasing both the surfactant concentration and the initiator concentration also produced an increase in particle size (entries 1–4 and 7–10). This is expected behavior, as increasing the surfactant concentration increases the number of particles that can be stabilized, leading to smaller particle diameters. The particle size generally decreased with increasing wt% of styrene (at constant ratios of surfactant to monomer and initiator to monomer). Coagulum formation during polymerization was generally not a serious problem; most of the experiments produced <5% coagulum. Zeta potential measurements in all cases indicate that the latexes should be stable, as they have values greater than 25 mV.

Aryl Amidine and Tertiary Amine "Easy Off" Switchable Surfactants

While the latexes prepared using the C12 amidine surfactant yielded stable latexes that could be aggregated and redispersed using only air and gentle heat, long destabilization times (sometimes hours) were required. We then investigated the use of less basic aryl amidine and tertiary amine switchable surfactants with more facile switching (Scheme 2).[6] Less basic groups will switch off (neutralize) more

1a $C_{12}H_{25}$–N=C(CH$_3$)–N(CH$_3$)$_2$ ⇌ 1b [$C_{12}H_{25}$–NH=C(CH$_3$)–N(CH$_3$)$_2$]$^+$ HCO_3^-

2a $C_{10}H_{23}$–C$_6$H$_4$–N=C(CH$_3$)–N(CH$_3$)$_2$ ⇌ 2b [$C_{10}H_{23}$–C$_6$H$_4$–NH=C(CH$_3$)–N(CH$_3$)$_2$]$^+$ HCO_3^-

3a $C_{12}H_{25}$–N(CH$_3$)$_2$ ⇌ 3b $C_{12}H_{25}$–N$^+$H(CH$_3$)$_2$ HCO_3^-

Forward: CO_2, H_2O; reverse: Air, Heat

Scheme 2.
Aryl amidine and tertiary amine switchable surfactants.

easily, but may be more difficult to switch on (protonate). The reversibility of the conversion from uncharged base to cationic surfactant was demonstrated by bubbling CO_2 followed by argon through solutions of 2a and 3a in wet ethanol and measuring the change in conductivity of the solution. The conductivity increased almost immediately when CO_2 was bubbled through the solution and decreased again when sparged with Ar. The CO_2/Ar cycle was performed three times to demonstrate the repeatability of the switching process (Figure 1(a) and (b)). The application of Ar to 2b and 3b causes a rapid reduction in conductivity, and the original solution conductivity is restored after only 20 min, indicating that the surfactant is fully converted to the uncharged form.

These less basic switchable surfactants were successfully used in the emulsion polymerization of MMA. The resulting latexes were stable if kept under an atmosphere of CO_2. Upon CO_2 removal using a non-acidic gas, heat or a combination of both, the surfactant becomes uncharged and the latexes can be destabilized. These less basic surfactants offer a significant advantage over the previously developed switchable surfactants due to their ability to easily and rapidly revert to the uncharged forms, with a much lower zeta potential (approaching zero in some cases) than previously attainable. Both the aryl amidine and tertiary amine surfactants have similar basicities and yield similar results when used in emulsion polymerization. However, the long chain tertiary amine offers a clear advantage over the aryl amidine due to the lower cost and commercial availability of the amine. The greatest advantage of using 2b and 3b in emulsion polymerization is the ease of post-polymerization latex destabilization compared to the previously used C12 amidine 1b. Final zeta potential values were smaller in the case of these less basic surfactants due to the greater conversion of these surfactants to their neutral forms during the destabilization procedure.

Surfactant Free Switchable Latexes Made with DEAEMA

We investigated 2-(diethyl)aminoethyl methacrylate (DEAEMA) as a switchable co-monomer (Scheme 3), added at 0.54% with respect to monomer, for the surfactant-free emulsion polymerization of styrene initiated by the switchable azo-initiator VA-061, in order to prepare CO_2-responsive latexes (Scheme 4).[7]

When DEAEMA is fully converted to $DEAMA^{+}HCO_3^{-}$ before its addition to the reaction media by sparging the aqueous phase with CO_2, in-situ formation of surfactant is favored leading to more efficient nucleation, a higher number of particles and a fast rate of polymerization.

Monomer conversions of ~90% were achieved, with the formation of stable latexes with monodispersed particles having a Z-average diameter of ~300 nm. The latexes could be stored for weeks at room temperature while under a CO_2 atmosphere without showing any phase separation. Their characterization by DLS, TEM and SEM confirmed the monodisperse character of the particles which could also be noticed from the iridescence of the dried samples. The particle diameters were 230 to 300 nm depending on the monomer conversion. The narrow size distribution of particles can be explained by the formation of particles through a homogeneous nucleation process. To achieve coagulation, the latexes were first subjected to air bubbling at 40 °C and for 30 min. Once destabilized, the latexes could be easily filtered and dried under vacuum overnight to produce dried polymer powders. The dried powders were then redispersed in carbonated water to the same solid content (27 wt%) by CO_2 bubbling followed by ultrasound (sonication bath) for 10 min. All polymer particles were well dispersed and restabilized as attested by the recovery of the original zeta-potential. TEM and SEM (Figure 3) confirm the particles maintained their original size and shape. The ability of the polymer particles to be redispersed

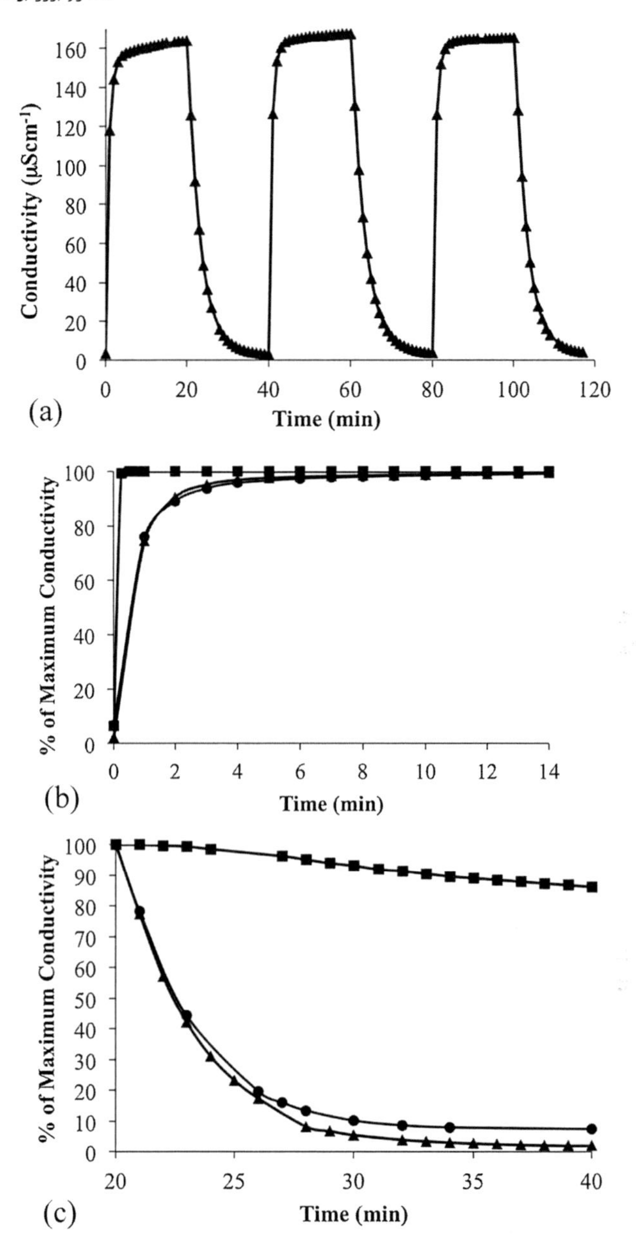

Figure 1.
The conductivity of solutions during CO2/Ar cycles in 20 mL ethanolic solutions containing 200 μL of water and 0.4 mmol of **2a** and the change in conductivity of wet ethanolic solutions of **1a** (■), **2a** (▲) and **3a** (●) at room temperature when (b) CO2 followed by (c) Ar are bubbled through the solutions.

from a dry-powder state to well-defined and stable latex is here a strong improvement as compared to our previous results where only destabilized latex solutions were able to be restabilized.

Reversible Aggregation of Anionic Latexes Using Switchable Water

We have developed a new method of reversibly breaking emulsions and latexes by using "switchable water", an aqueous

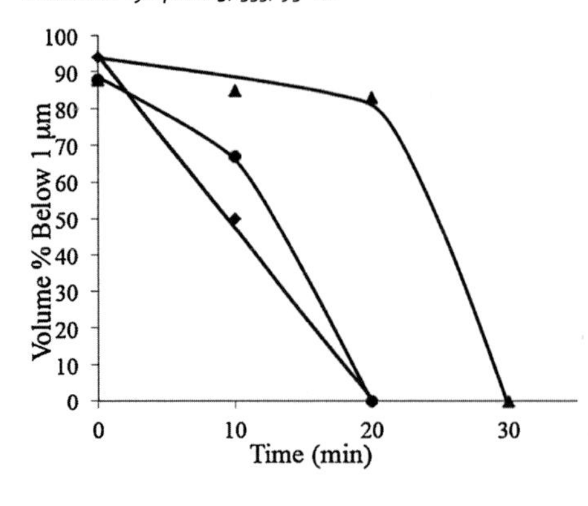

Figure 2.
Volume percent of PMMA particles below 1 μm as a function of time during destabilization using air at 65 °C (◆), 40 °C (●) and room temperature (▲) in a latex synthesized using surfactant **2b.**

solution of switchable ionic strength.[8] The conventional surfactant sodium dodecyl sulfate (SDS) is normally not stimuli-responsive when CO_2 is used as the stimulus but becomes CO_2-responsive or "switchable" in the presence of a switchable water additive. In particular, changes in the air/water surface tension and oil/water interfacial tension can be triggered by addition and removal of CO_2. It is found that a switchable water additive, N,N-dimethylethanolamine (DMEA), was an effective and efficient additive for the reversible reduction of interfacial tension and can lower the tension of the dodecane-water interface in the presence of SDS surfactant to ultra-

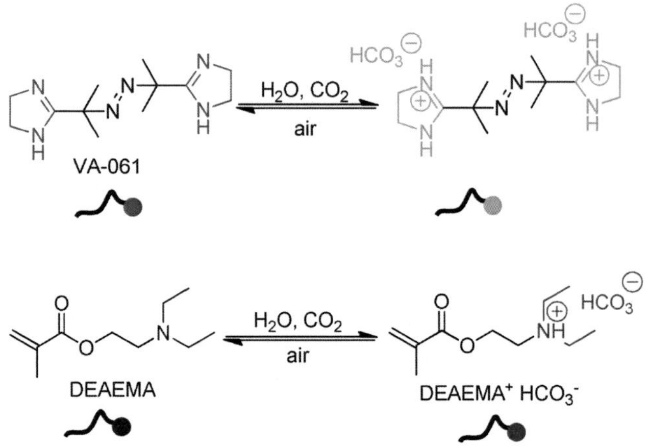

Scheme 3.
Switchable behavior of DEAEMA and VA-061 in the presence/absence of CO_2.

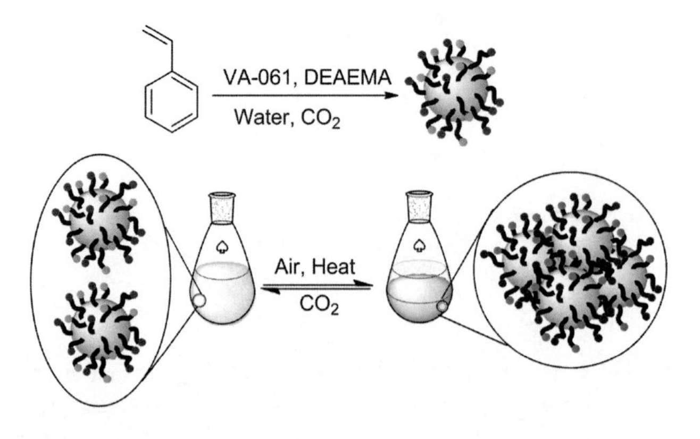

Scheme 4.
Preparation of a switchable polymer latex and reversible aggregation and redispersion of the latex triggered by removal and addition of CO_2.

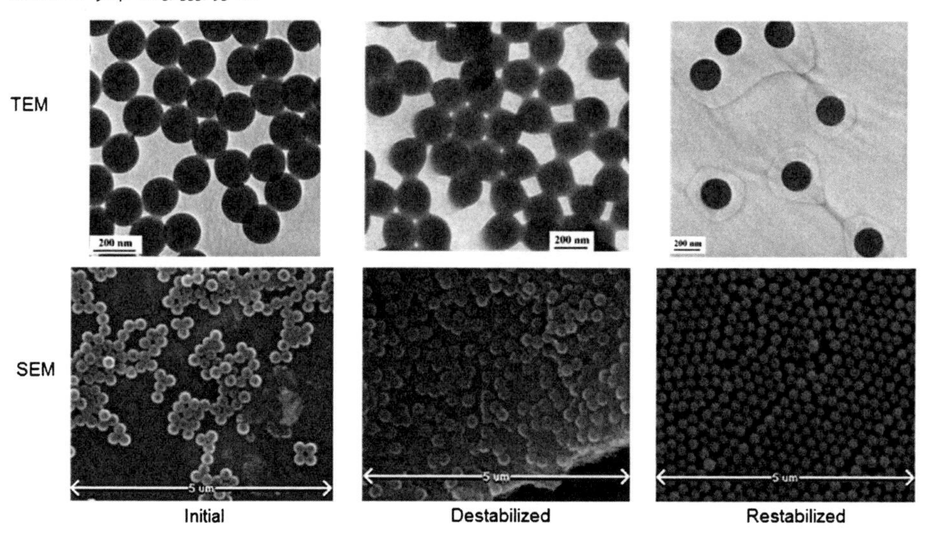

Figure 3.
TEM and SEM micrographs of latex after destabilization in the presence of Poly(DEAEMA) (middle) and redispersed after 10 min of sonication under CO_2 atmosphere (right).

low values at very low additive concentrations (Figure 4). Switchable water was can reversibly break a water-dodecane emulsion containing SDS as surfactant. Additionally, the addition of CO_2 and switchable water can result in aggregation of anionically stabilized latexes; the later removal of CO_2 neutralizes the DMEA and decreases the ionic strength allowing for the aggregated latex to be redispersed and recovered in its original state (Figure 5).

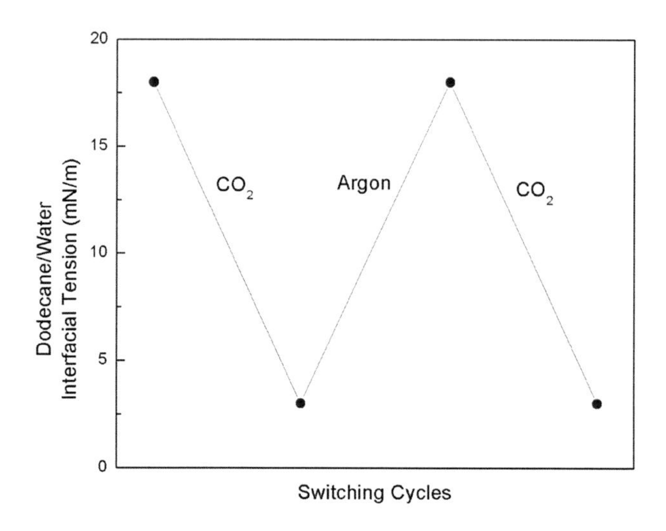

Figure 4.
The air/water surface tension (a) and the water/dodecane interfacial tension (b) for an aqueous solution of SDS (0.01 mol/L) with DMEA (20.0% v/v compared to water) at 25 ± 0.5 °C as a function of time during two cycles of sparging with CO_2 followed by argon.

Treatments	Original	Added 0.4 ml NaOH aq. (2.0 mol/L)	Added 5 mL DMEA	CO_2 10 min	Argon 10 min & Hand shaking
pH Value	2.3	9.7	11.0	6.5	10.8
Conductivity	2.0 Ms/cm	3.2 Ms/cm	3.5 Ms/cm	10.1 Ms/cm	3.6 Ms/cm
Particle size	101 nm	104 nm	103 nm	103,000 nm	102 nm
PDI	0.242	0.121	0.138	N/A	0.246
ζ -potential	-38.5 mV	-39.1 mV	-46.5 mV	-6.6 mV	-51.9 mV

Figure 5.
Reversible aggregation/redispersion behavior of a PS latex. To a PS latex (prepared using 15 mL of styrene, 0.50 g of SDS, 0.10 g of KPS and 50 mL of water) was added NaOH solution, DMEA (5 mL) and then CO_2 (flow rate of 90 mL/min at room temperature for 10 min); finally, the latex was treated with argon (flow rate of 90 mL/min at room temperature for 10 min).

Conclusion

Amidine or tertiary amine surfactants can be used to prepare stimuli-responsive "smart" polymer nanoparticles. Aqueous dispersions of these latexes can be aggregated by removal of CO_2 (using air, argon or any inert gas) and gentle heat. The aggregated latexes can be redispersed by introducing CO_2. The surfactant free polymerization of styrene in the presence of small amounts (0.54%) of DEAEMA and under CO_2 atmosphere allows the production of well-defined switchable latexes from commercially available compounds. The product latex is CO_2 stimuli-responsive and can be easily aggregated/redispersed using only air or CO_2. Even after being dried to a powder, the particles could be restabilized to restore the original latex. When the switchable water additive DMEA is present, the conventional surfactant SDS becomes stimuli-responsive, responding to CO_2 as a trigger in a manner that is reversible, repeatable, and controllable. Anionically stabilized PS latexes can be destabilized by CO_2 if DMEA is in the aqueous phase. The aggregated latexes are readily redispersed when CO_2 was removed. The use of the "switchable water" technique to make SDS respond to CO_2 as a trigger may be a general strategy for creating stimuli-responsive dispersions without the need to use switchable surfactants.

[1] Y. Liu, P. G. Jessop, M. Cunningham, C. A. Eckert, C. L. Liotta, *Science* **2006**, *313*, 958.
[2] C. I. Fowler, C. M. Muchemu, R. E. Miller, L. Phan, C. O'Neill, P. G. Jessop, M. F. Cunningham, *Macromolecules* **2011**, *44*, 2501.
[3] M. Mihara, P. Jessop, M. F. Cunningham, *Macromolecules* **2011**, *44*, 3688.
[4] X. Su, P. G. Jessop, M. F. Cunningham, *Macromolecules* **2012**, *45*, 666.
[5] C. O'Neill, C. Fowler, P. G. Jessop, M. F. Cunningham, *Green Materials* **2012**, *1*, 27.
[6] C. I. Fowler, P. G. Jessop, M. F. Cunningham, *Macromolecules* **2012**, *45*, 2955.
[7] J. Pinaud, E. Kowal, M. Cunningham, P. Jessop, *ACS Macro Lett.* **2012**, *1*, 1103.
[8] X. Su, T. Robert, S. M. Mercer, C. Humphries, M. F. Cunningham, P. G. Jessop, *Chem. –Eur. J.* (**2013**, in press).

[9] Q. Zhang, W.-J. Wang, Y. Lu, B.-G. Li, S. Zhu, *Macromolecules* **2011**, *44*, 6539.
[10] Q. Zhang, G. Yu, W.-J. Wang, H. Yuan, B.-G. Li, S. Zhu, *Langmuir* **2012**, *28*, 5940.
[11] Q. Zhang, G. Yu, W.-J. Wang, B.-G. Li, S. Zhu, *Macromol. Rapid Commun.* **2012**, *33*, 916.
[12] Y. Zhao, K. Landfester, D. Crespy, *Soft Matter* **2012**, *8*, 11687.
[13] V. Fischer, K. Landfester, R. Munoz-Espi, *ACS Macro Lett.* **2012**, *1*, 1371.
[14] D. H. Han, X. Tong, O. Boissiere, Y. Zhao, *ACS Macro Lett.* **2012**, *1*, 57.
[15] D. H. Han, O. Boissiere, S. Kumar, X. Tong, L. Tremblay, Y. Zhao, *Macromolecules* **2012**, *45*, 7440.
[16] Q. Yan, R. Zhou, C. Fu, H. Zhang, Y. Yin, J. Yuan, *Angew. Chem. Int. Ed.* **2011**, *50*, 4923.

DOI: 10.1002/masy.201300037

Micron-Sized Polymer Particles by Membrane Emulsification

*J.M.M. Simons, J.T.F. Keurentjes, J. Meuldijk**

Summary: Dispersions of micron-sized styrene droplets in water were produced by straight through microchannel arrays. The monomer droplets have been polymerized by suspension polymerization using an initiator system, which was only oil soluble. A significant amount of submicron particles was formed by secondary nucleation. The side reaction was emulsion polymerization, which was confirmed by the molecular weight distribution of the reaction product. Suppression of secondary nucleation was only partially possible by using the water soluble inhibitor $NaNO_2$. Increasing the monomer to water ratio before polymerization resulted in less emulsion polymerization. Molecular weight was found to be strongly dependent on the initiator concentration as well as on the inhibitor concentration. It was demonstrated that polymer particles with a diameter of about 10 μm can be produced using emulsification of the monomer by microchannel arrays prior to polymerization.

Keywords: emulsification; free radical polymerization; secondary nucleation; straight through microchannel arrays; suspension polymerization

Introduction

Monodispersed micron-sized polymer particles have various applications, *e.g.* support materials for enzymes,[1,2] and column packing for affinity chromatography.[3] The particle size distribution in suspension polymerization in stirred tanks is normally governed by droplet break-up and coalescence during the dispersion of the monomer(s) in water and during the first part of the polymerization. Particle size and particle size distribution strongly depend on impeller type and speed as well as on the stabilizer type and concentration, see *e.g.* Kalfas *et al.*[4] and Brooks.[5] Production of micron-sized polymer particles in the size range between 1 and 10 μm and a perfect control of the particle size distribution is difficult with conventional suspension polymerization in a stirred tank. Also emulsion polymerization combined with controlled coagulation is troublesome and time consuming. Rotor stator systems allow the preparation of monodispersed monomer in water dispersions with droplet sizes in the size range of 1–10 μm.[6] Note that Paine *et al.*[7] reported the production of micron-sized polystyrene particles in the desired size range with very smally size distributions using dispersion polymerization. Proper control of the particle size and particle size distribution can be expected when the monomer droplet formation (*i.e.* emulsification) and polymerization are separated in space.[8] Membranes[8,9] or straight through microchannel arrays,[10] see Figure 1, seem very suitable for the continuous and reproducible production of dispersions of micron-sized droplets of hydrophobic monomers in water (*i.e.* monomer in water emulsions) with a reproducible very narrow droplet size distribution. Membranes or straight through microchannel arrays allow a perfect control of the droplet size distribution

Department of Chemical Engineering and Chemistry, Laboratory of Chemical Reactor Engineering, Eindhoven University of Technology, P.O. Box 513, 5600 MB Eindhoven, The Netherlands
E-mail: j.meuldijk@tue.nl

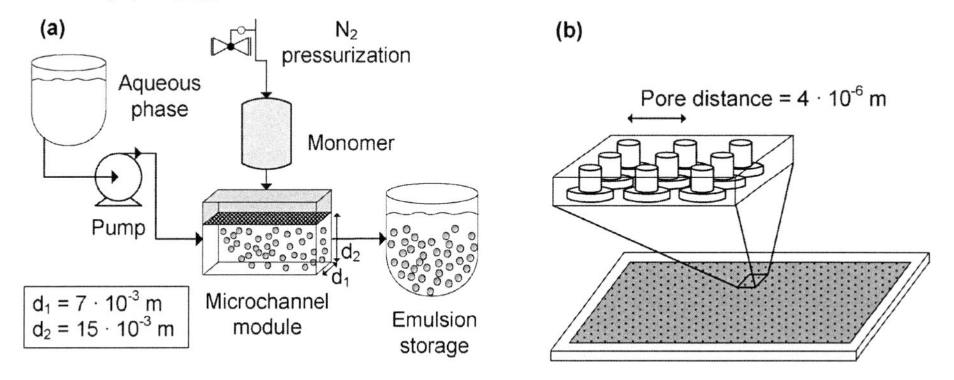

Figure 1.
Microchannel emulsification set-up (a) with microsieves, pore size ≥ 1.8 μm (b).[11]

under low shear conditions. Droplets of *very* sparingly water soluble monomers can then be transformed one-to-one into polymer particles by free radical polymerization in a separate reactor.

In the case where the monomer has a significant water solubility in the continuous water phase, *e.g.* styrene and methyl methacrylate, transfer of monomer (M) based radicals (IM_i^*) from the micron-sized droplet/particle phase to the continuous water phase may shift the locus of polymerization partially from inside the droplet/particle phase to the continuous water phase, see Figure 2.

These monomer based radicals may lead to secondary nucleated submicron particles. Also the presence of initiator based radicals I^* in the aqueous phase may lead to oligomer formation in the aqueous phase followed by nucleation. For the suspension polymerization of styrene, the secondary nucleated particles can dominate the polymerization process.[12]

In the work presented in this paper, asymmetric straight through microchannel arrays have been used for the production of a dispersion of monodispersed micron-sized styrene droplets in water (*i.e.* emulsification). The colloidal stability of the

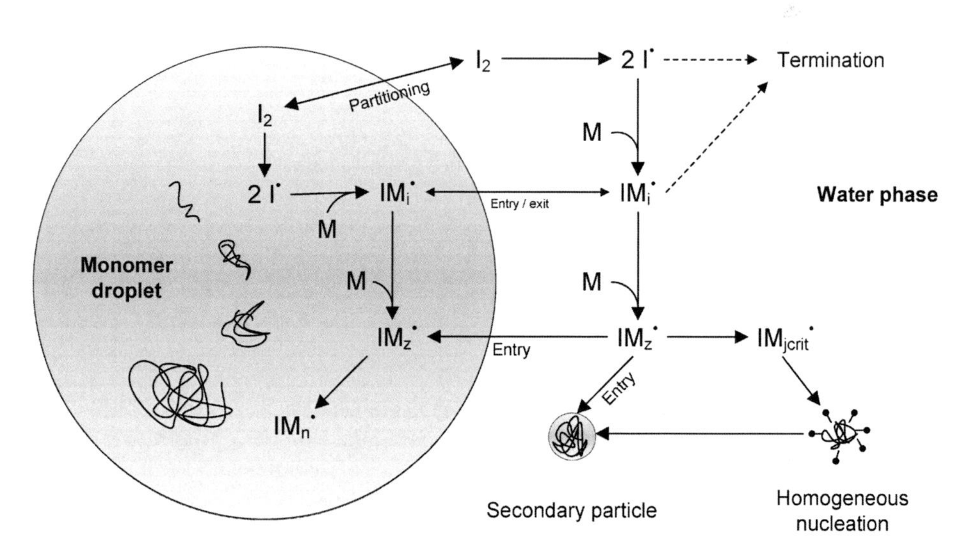

Figure 2.
Proposed mechanism leading to secondary particle formation. For styrene $i = 0 - 2$, $z = 3$, $n > 3$ and $j_{crit} = 4 - 5$.

resulting emulsions has been evaluated. Micron-sized polymer particles were produced by free radical polymerization in the monomer droplets. *i.e.* suspension polymerization in a separate reactor. The contribution of secondary nucleation has been evaluated by variation of the recipe (*i.e.* inhibitor concentration, initiator concentration and monomer to water ratio).

Experimental Part

Materials

Styrene (ReagentPlus, ≥ 99%), $NaNO_2$, polyvinylpyrrolidone (PVP) and lauroyl peroxide were obtained from Sigma-Aldrich (Steinheim, Germany). Styrene was distilled under reduced pressure prior to use. Sodium dodecyl benzene sulfonate (SDBS, ≥ 99%), was obtained from Merck (Hohenbrunn, Germany). Straight through microchannel arrays, type crocus and type tulip-4, respectively with pores of 1.8 and 4 μm, were obtained from Nanomi BV (Oldenzaal, The Netherlands).[13] Di(4-tert-butylcyclohexyl) peroxydicarbonate (Perkadox 16) was a kind gift from Akzo Nobel (Arnhem, The Netherlands).

Emulsification

Figure 1 shows the setup used for the emulsification with straight through microchannel arrays. Before emulsification, the straight through microchannel arrays were cleaned in a plasma asher (Emitech K1050X) using oxygen. Prior to use, all components were deoxygenated and stored under argon. The emulsification setup was treated with several vacuum – argon cycles and stored in an argon atmosphere. The water and monomer phase were prepared by recipes as listed in Table 1 and loaded into the aqueous and monomer phase vessel, respectively. The monomer/initiator mixture was pressed through the microchannel array using nitrogen. A gentle cross flow ($v_c = 0.1 \times 10^{-2}$ to 5×10^{-2} m/s) of the water/surfactant mixture, *i.e.* the continuous phase, carried the emulsion droplets from the microchannel module to the emulsion storage vessel. Proper emulsification with a narrow droplet size distribution was observed for applied pressures between 1 and 15 kPa.

Table 1.
Standard recipe for aqueous and to be dispersed phase prior to and after emulsification

To be dispersed phase		
Compound	Amount	
Styrene	5	[g]
Initiator	$C_{initiator}$	[mol/dm^3]
Continuous phase		
Compound	Amount	
Water	5×10^2	[g]
SDBS	0.174	[g]
Added after emulsification		
Compound	Amount	
$NaNO_2$	$C_{inhibitor}$	[mol/dm^3]
PVP	5	[g/dm^3]

Suspension Polymerization

$NaNO_2$ was used as an inhibitor in the water phase. PVP was used to agglomerate the excess of free surfactant molecules in the water phase.[14] Both $NaNO_2$ and PVP were added after emulsification to avoid influences on the emulsification process. Variations in pore activity of the microchannel array resulted in fluctuations of the throughput of the monomer phase. This caused variations in the monomer volume fraction Φ_M in the resulting emulsion, which is defined as:

$$\Phi_M = \frac{V_M}{V_{RM}} \text{ where } V_{RM} = V_W + V_M \quad (1)$$

V_{RM}, V_M and V_W are the volumes of the reaction mixture, the monomer phase and the water phase, respectively. For each emulsion, Φ_M was determined by measuring the styrene concentration in the sample by gas chromatography (Fision instruments, 8000/8160 series). To achieve a preset value for Φ_M, too concentrated samples were diluted with additional water phase. Samples that were too diluted were concentrated using a centrifuge (Heraeus Sepatech megafuge 1.0). Samples were centrifuged for 10 minutes at 4000 rpm in an argon atmosphere. After centrifugation, the calculated volume of supernatant was removed. The emulsion was transferred to a 2×10^{-3} dm^3 glass reactor. The polymerization was carried out with gentle stirring at 65 °C under argon for 24 hours. The conversion was calculated from the initial and final concentration of styrene, measured with gas chromatography.

Droplet and Particle Size Analysis

Monomer droplets and polymer particles were observed with an optical microscope (Axioplan 2 imaging, Carl Zeiss, Oberkochen, Germany), equipped with a digital camera (AxioCam Color, Type 412-312). Size distributions were obtained from images made with the digital camera by a custom-made routine written in Matlab R2006b (The Mathworks, Natick, MA, USA) and ImageJ.[15] The image analysis procedure consisted of some consecutive steps. First, the background was subtracted. The resulting image was binarized using a luminance threshold. The droplets and particles were separated from the noise by filtering for a luminance maximum. Further processing steps involved separation of connecting droplets and particles and filtering out objects with a circularity ratio below 0.8. From the resulting images, the droplet and particle areas were determined. Statistical analysis of the data provided the mean and standard deviation for the droplet and particle diameter distribution. Each distribution was based on the measurement of at least 9×10^3 droplets or particles.

Size Exclusion Chromatography

Molecular weight distributions of the polymer were measured with size exclusion chromatography (SEC). SEC was performed on a Waters GPC equipped with a Waters model 510 pump and a Waters model 410 differential refractometer. Two mixed bed columns (Mixed-C, Polymer Laboratories, 30 cm, 40 °C) were used. Tetrahydrofuran (Aldrich) stabilized with BHT was used as eluent. Narrow molecular weight polystyrene standards in the range of 600 to 7×10^6 g/mol were used to calibrate the system.

Results and Discussion

Emulsification

Membrane Performance

Figure 3a shows a typical picture of styrene droplets flowing from the microchannel

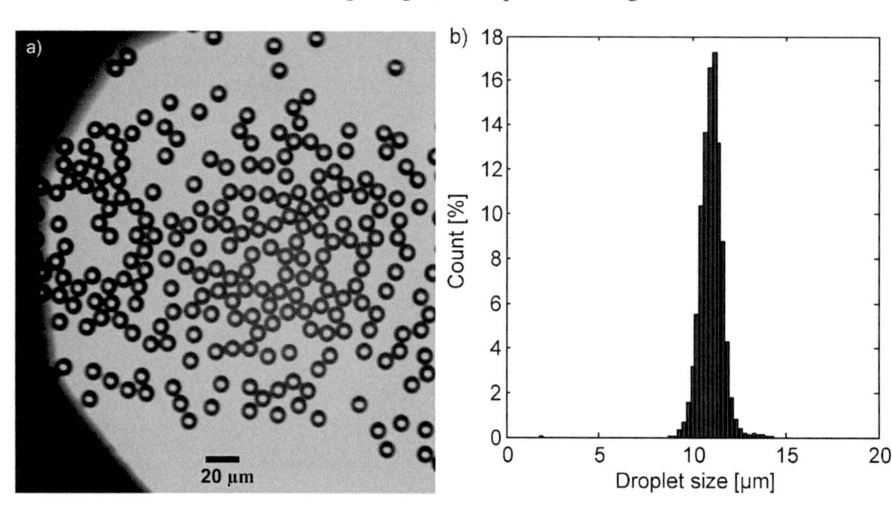

Figure 3.
a: Image of styrene droplets leaving the microchannel module with a membrane type tulip with a pore size of 4 µm. Velocity of the continuous phase, $v_c = 0.01$ m/s. b: Typical size distribution of styrene droplets produced with a straight through microchannel array.

module. The concentration of the surfactant (SDBS) in the fluid flowing parallel to the microchannel array was 10^{-3} mol/dm$^3_{water}$. Figure 3b shows the droplet size distribution, based on the measurement of over 11×10^3 droplets. The average droplet size as well as the droplets size distribution are not significantly dependent on the cross flow velocity v_c and the applied pressure (ΔP) in the cross flow velocity range $0.5 \times 10^{-2} < v_c < 5.0 \times 10^{-2}$ m/s and the pressure range $1 < \Delta P < 15$ kPa.

Figure 4 shows the relationship between the volume averaged droplet diameter ($<d_{droplet}>_V$) and the average pore diameter ($<d_{pore}>$) for the experiments with microchannel arrays, see also equation 2:

$$\langle d_{droplet} \rangle = c \times \langle d_{pore} \rangle \qquad (2)$$

In the range of the operating conditions used, c varied between 2.6 and 3.2, which is in line with the results of Charcosset *et al.*[16]

Required Surfactant Concentrations for Proper Droplet Formation

Maintaining a low surfactant concentration is a prerequisite to avoid formation of submicron particles, *i.e.* emulsion polymerization, during the suspension polymerization process. With increasing surfactant concentration, the probability for a free radical in the aqueous phase to participate in submicron particle formation increases. Therefore the lowest surfactant concentration required for proper droplet formation was determined. Figure 5 (left) shows the influence of the SDBS concentration in the continuous phase on the resulting droplet size distribution.

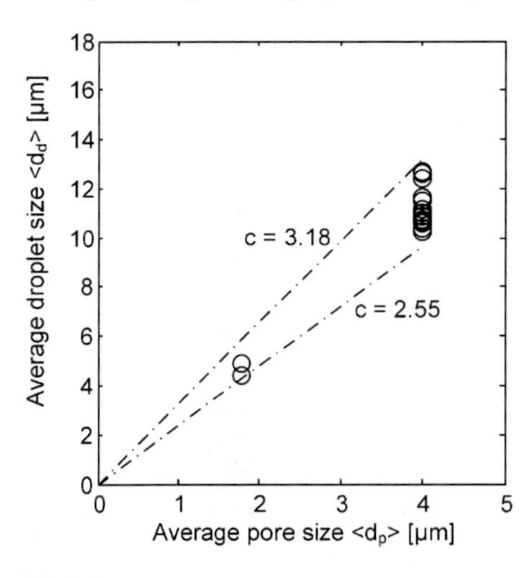

Figure 4.
Droplet sizes as a function of membrane pore sizes for straight through Nanomi microchannel arrays, type crocus, $<d_{pore}> = 1.8$ µm, and tulip, $<d_{pore}> = 4.0$ µm.

The concentration of SDBS does not seem to have a significant effect on $<d_{droplet}>_V$. The coefficient of variation (CV), however, decreases sharply with increasing SDBS concentration. For SDBS concentrations above 0.7×10^{-3} mol/dm^3, the CV does not change significantly with the SDBS concentration. Figure 5 clearly demonstrates that this SDBS concentration does not differ much from the critical micelle concentration (CMC) for SDBS in water, saturated with styrene, being 0.8×10^{-3} mol/dm^3 as derived from the styrene/water interfacial tension measurements. This value is slightly lower than the reported CMC values of SDBS in pure water, being 1.3×10^{-3} to 2.8×10^{-3} mol/dm^3.[17] This reduced CMC under the influence of monomer is referred to as the apparent CMC (CMC_{app}).[18] To obtain a direct relationship between the interfacial tension and the resulting droplet size distribution, evaluation of the dynamic interfacial tension is necessary. The dynamic interfacial tension is mainly influenced by surfactant migration rates and plays a dominant role at low surfactant concentrations.[19] The main conclusion that can be drawn from these results is that emulsification of styrene in water with SDBS using straight through microchannel arrays can only be done at surfactant concentrations higher than or around CMC_{app}. As a consequence submicron particle formation by micellar nucleation seems to be not completely avoidable during the polymerization of styrene droplets produced with microchannel arrays.

Stability of the Emulsion

The evolution of particle size distribution has been intensively studied for suspension

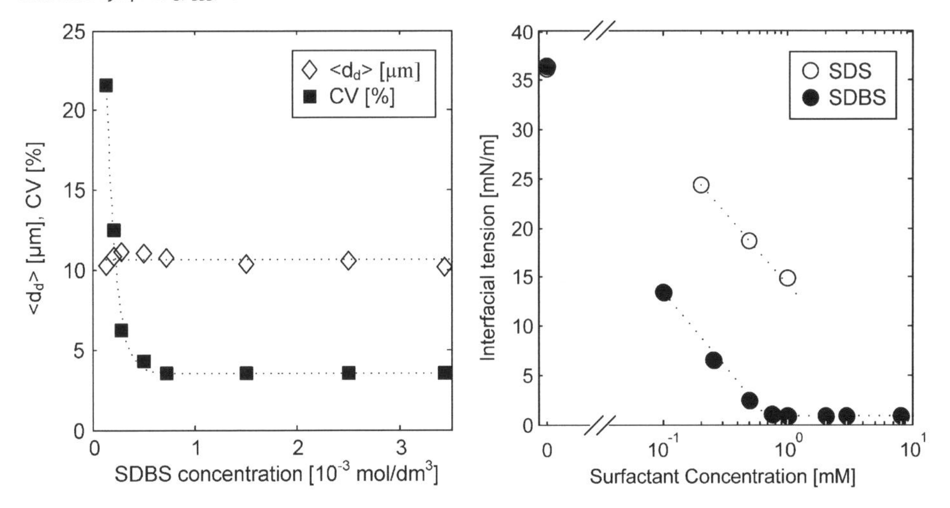

Figure 5.
Left: Mean droplet size and coefficient of variation (CV) for emulsification of styrene as a function of the aqueous concentration of SDBS after emulsification with. $CV = \frac{\sigma}{\langle d_{droplet} \rangle} \times 100\%$, where σ is the standard deviation. Straight through microchannel array: Nanomi, type tulip, $\langle d_{pore} \rangle = 4.0\ \mu m$. Right: Interfacial tension of water/styrene as a function of surfactant (SDBS) concentration.

polymerization processes.[20,21] The stability of the monomer emulsion is a key issue for the size distribution of the final polymer particles. The process in which large droplets grow at the expense of small ones and thereby reducing the total interfacial area is referred to as Ostwald ripening. The interfacial area is reduced via diffusional mass transfer from regions of high interfacial curvature to regions of smaller interfacial curvature. Figure 6 shows the broadening of droplet size distributions for a sample with $\Phi_M = 0.135$. The emulsion stored at 25 °C shows a significant increase of CV with time. This broadening of the droplet size distribution has no significant effect on the mean droplet diameter.

At a typical polymerization temperature, *i.e.* 65 °C, a similar trend was observed. The experiment was conducted for 4 hours and then stopped due to self-initiation of styrene.

Suspension Polymerization

Formation of Submicron Particles

The formation of submicron particles by secondary nucleation is to the best of the authors knowledge hardly reported in literature for suspension polymerization systems, although it is recognized that particle formation in such systems originates to a large extent from initiation in the water phase.[22] Focus on the formation of submicron particles became an important issue when emulsification and polymerization were conducted separately and a considerable part of the polymer was formed in these submicron particles. The secondary nucleation in such a system is expected to obey the mechanism illustrated in Figure 2.

The initiator I_2 as well as the the monomer based radical species (IM_j^*) partition between the monomer droplets and the water phase. Radical formation can therefore take place inside the monomer droplet as well as in the water phase. Inside the monomer droplet, reaction kinetics obey that for bulk polymerization, which is a common way to describe suspension polymerization. Monomer based radical species in the aqueous phase can among other things lead to submicron particles by homogeneous nucleation if the chain length i becomes equal to j_{crit}, the chain length at which precipitation occurs, see Figure 2. In the aqueous phase and in the submicron

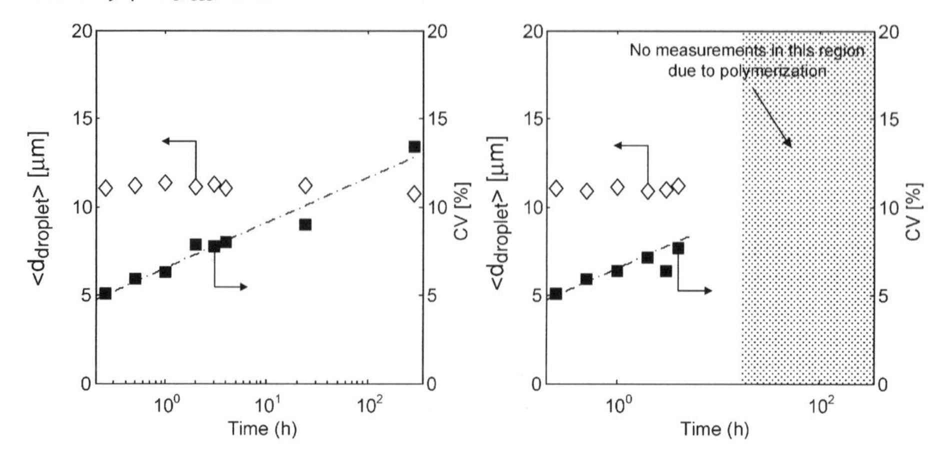

Figure 6.
Mean droplet diameter (<$d_{droplet}$>) and coefficient of variation (CV) as a function of time for styrene in water emulsions at 25 °C (left) and 65 °C (left). The broken lines are a guide to the eye.

polymer particles, the polymerization can be described by the kinetics of interval I and II of emulsion polymerization.[23–26] It is expected that Smith-Ewart case 1 kinetics is obeyed for the submicron polymer particles as the radical generation rate in the water phase is low due to the low water solubility of the initiator. Figure 7 shows a typical example of the molecular weight distribution of the reaction products in our work. The results in Figure 7 clearly demonstrate that suspension polymerization as well as emulsion polymerization occur simultaneously. The pseudo-bulk mechanism in the micron-sized particles leads to a weight averaged molecular weight, M_w, in the range of 10^4 to 10^5 g/mol. The polymerization in the submicron particles (*i.e.* emulsion polymerization) typically leads to a M_w of about 10^6 g/mol. Ma and Li reported that the water solubility of the initiator has a significant impact on the fraction of polymer present in the secondary nucleated particles.[12] These authors also reported that addition of a water soluble inhibitor such as $NaNO_2$ or diaminophenylene (DAP) suppressed the formation of submicron particles.

However, Ma and Li[12] reported that suppression of submicron particles formation was not clearly observed using hydroquinone. Under exclusion of oxygen hydroquinone only leads to retardation of submicron particles nucleation. In the final product the fraction of polymer in the submicron particles is not expected to differ significantly from that obtained with recipes without quinones.[27] The water soluble

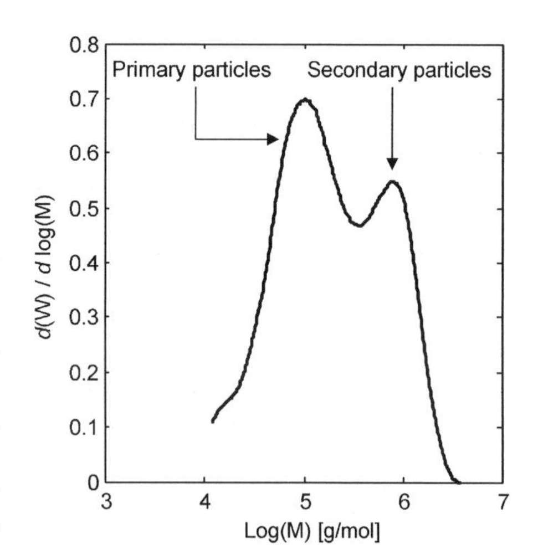

Figure 7.
Molecular weight distribution of polystyrene particles from a typical suspension polymerization experiment. Primary micron-sized particles and secondary nucleated submicron particles show distinct shoulders at a molecular weight (M) of 10^5 and 10^6 g/mol, respectively.

inhibitor increases the rate of chain stop for short chained radicals IM_i^* and I^* in the water phase, see the dashed lines in Figure 2. This prevents growing oligomers in the water phase from reaching chain lengths that allow them to become surface active. As a consequence, entry into submicron particles as well as homogeneous nucleation are suppressed. Ma and Li also observed that the addition of inhibitor also decreased the final monomer conversion from 100% to 76% in some cases.[12] So a recipe and operation strategy should be developed for which the inhibitor level is optimized so that secondary nucleation is suppressed as much as possible and simultaneously obtaining a high final conversion level.

Initiator Selection

The droplet size distribution of styrene in water emulsions broadens significantly over time, see Figure 6. The residence time of the droplets in the membrane module must therefore be kept as low as possible. Secondly, the time of polymerization has to be minimized to avoid broadening of the particle size distribution in the final product. A high rate of polymerization can be achieved by selecting an initiator with a high thermal decomposition rate coefficient, *i.e.* a low half life time, see equation 3.

$$R_p = k_p \times C_m \times \sqrt{\frac{f \times k_d \times C_{initiator}}{k_t}}, \qquad (3)$$

where k_p, C_m, f, k_d, $C_{initiator}$ and k_t respectively stand for the propagation rate coefficient, the monomer concentration at the locus of polymerization, the initiator efficiency, the rate coefficient for thermal initiator decomposition, the initiator concentration and the bimolecular termination coefficient.

The choice of initiator is therefore evaluated by the initiator half life time in addition to partitioning between the monomer and water phase. The initiator mixture used in this work consisted of 90 wt% Perkadox 16 and 10 wt% lauroyl peroxide. Perkadox 16 (half life time at 65 °C is about 1 hour) was chosen as a fast initiator to generate a high conversion increase during the first stage of the reaction. Lauroyl peroxide (half life time about one order of magnitude larger than that of Perkadox 16) was added to reach a high final conversion as Perkadox 16 is depleted in the final stage of the polymerization. Half life times of the initiators were borrowed from product information given by the suppliers.

Influence of Inhibitor

$NaNO_2$ suppresses polymerization reactions in the water phase. A side effect is the reduction of the rate of polymerization in the monomer droplets. Figure 8 (left) shows a decrease in molecular weight with increasing inhibitor content. This is due to a lower fraction of high molecular weight material in the sample formed by emulsion polymerization. Note that a lower radical concentration in the micron-sized particles leads to a higher molecular weight in these particles. Figure 8 (left) also shows that the overall conversion decreases with increasing inhibitor content, demonstrating that the suppression of polymerization by the $NaNO_2$ inhibitor is not restricted to the water phase.

The fraction of monomer that was not polymerized inside the droplets and participated in the formation of submicron particles was calculated from the initial droplet diameter and the diameter of the μm sized particles. The total mass of monomer, m_{mon}, before reaction can be expressed as:

$$m_{mon} = N_{droplet} \times \rho_{droplet} \times \langle V_{droplet} \rangle$$
$$\text{with } \langle V_{droplet} \rangle = \frac{\pi \times \left(\langle V_{droplet} \rangle_V \right)^3}{6} \qquad (4)$$

$N_{droplet}$, $\rho_{droplet}$ and $V_{droplet}$ are the total number of monomer droplets in the system, the density of the monomer phase and the volume of a monomer droplet, respectively. $<d_{droplet}>_V$ is the volume averaged droplet diameter. The total mass of the μm sized

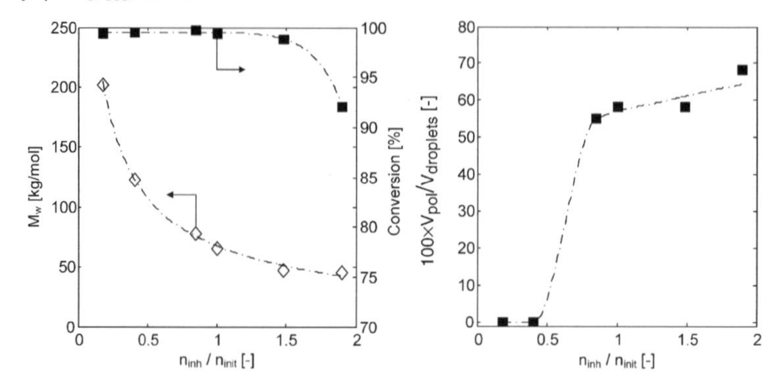

Figure 8.
Suspension polymerization of dispersions of micron-sized styrene droplets in water prepared by membrane emulsification: resulting properties as a function of the inhibitor to initiator molar ratio for: molecular weight (◇) and conversion (■) (left) as well as for the ratio of the total volume of particles originating from suspension polymerization and the initial droplet volume directly after emulsification (right). n_{inh} and n_{init} are the total number of moles of inhibitor and initiator in the system, respectively. For all samples $\Phi_M = 0.3$ and $C_{initiator} = 30 \times 10^{-3}$ mol/dm^3. Broken lines are a guide to the eye.

polymer particles, m_{pol}, follows from:

$$m_{pol} = N_{pol} \times \rho_{pol} \times \langle V_{pol} \rangle \quad \text{with } \langle V_{pol} \rangle = \frac{\pi \times \left(\langle d_{pol} \rangle_V \right)^3}{6} \tag{5}$$

N_{pol}, ρ_{pol} and V_{pol} are the total number of μm sized polymer particles, the density of these polymer particles and the volume of a polymer particle, respectively. $<d_{pol}>_V$ is the volume averaged particle diameter. Because the initiators have already been dissolved in the monomer before droplet formation, polymerization will take place in all droplets. During the first stage of the polymerization coalescence and redispersion of the micron-sized particles in the gently stirred reaction mixture is improbable. Together with the observation that the monomer droplets already show sufficient colloidal stability, see Figure 6, the assumption that $N_{droplet} = N_{pol}$ seems reasonable. The assumption $N_{droplet} = N_{pol}$ allows an estimation of the fraction of monomer that was polymerized in the μm sized droplets:

$$\frac{m_{pol}}{m_{mon}} = \frac{N_{pol} \times \rho_{pol} \times \langle V_{pol} \rangle}{N_{droplet} \times \rho_{droplet} \times \langle V_{droplet} \rangle} = \frac{\rho_{pol}}{\rho_{droplet}} \times \frac{\langle V_{pol} \rangle}{\langle V_{droplet} \rangle} = \frac{1}{0.85} \times \frac{\langle V_{pol} \rangle}{\langle V_{droplet} \rangle}, \tag{6}$$

with $\rho_{droplet} = 0{,}906$ kg/dm^3 and $\rho_{pol} \approx 1.07$ kg/dm^3.[28]

Figure 8 (right) shows the influence of $NaNO_2$ on the polymer particle to droplet volume ratio. At low inhibitor contents, the monomer is completely consumed by the formation of secondary particles. At higher inhibitor contents, the formation of submicron particles is suppressed. However, complete suppression can not be accomplished as the polymer particle to droplet volume ratio does not reach the value 0.85 even at high inhibitor contents.

The molecular weight of the polymer chains growing in the μm sized polymer particles can be influenced by changing the initiator concentration. Figure 9 shows the molecular weight as a function of the total initiator concentration. A lower initiator concentration allows polymer chains in the monomer droplets to grow to a larger degree of polymerization before termination. To maintain a constant molar ratio inhibitor/initiator, the total inhibitor concentration in the water phase was decreased for lower values of $C_{initiator}$. As a consequence of the lower absolute $NaNO_2$ concentration in the aqueous phase, the formation of submicron particles increases for decreasing values of $C_{initiator}$.

Finally, the influence of the monomer volume fraction in the emulsion on

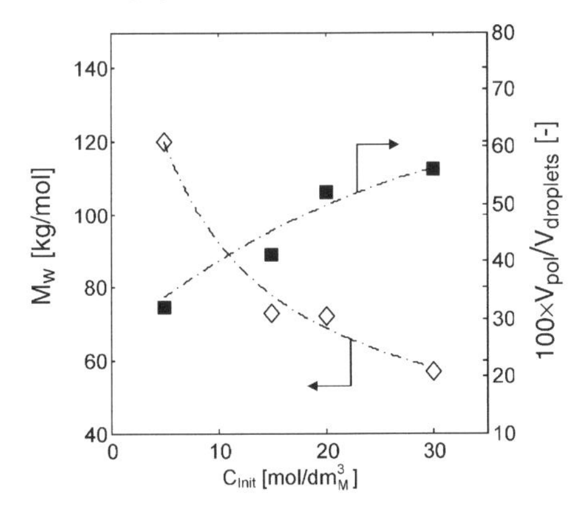

Figure 9.
Suspension polymerization of styrene in water emulsions prepared by membrane emulsification. Graphical relation between the initiator concentration and the molecular weight (◇). Graphical relation between the initiator concentration and the ratio of the total volume of the μm sized particles and the droplet volume direct after emulsification (■). All polymerizations were carried out with a molar ratio of the inhibitor and the initiator of one.

secondary nucleation was investigated. The results collected in Table 2 clearly demonstrate that secondary particle formation can be reduced by using polymerization recipes with high values for Φ_M. For a Φ_M value of 0.05, submicron particle formation was so dominant that no accurate value for $<d_{pol}>$ and the ratio of V_{pol} and $V_{emulsion}$ could be measured. The trend shows that for $\Phi_M = 0.3$ an optimum has not yet been reached, allowing further improvement for even higher values of Φ_M. Reaching these higher values of Φ_M was unfortunately not possible using our emulsification setup as very low cross flow rates of the water phase in the membrane module were required and monitoring of emulsification was not possible when using these very low cross flow rates.

Conclusion

Dispersions of micron-sized styrene droplets in water with a narrow droplet size distribution were prepared with straight through microchannel arrays. These dispersions were used as a starting point for suspension polymerization. It was shown that significant broadening of the droplet size distribution at room temperature occurs already on time scales of the emulsification process. Further broadening during the initial stage of polymerization is also significant. The formation of secondary nucleated submicron particles during suspension polymerization cannot be suppressed completely by using a water soluble inhibitor such as $NaNO_2$. The use of a water soluble inhibitor also has an effect on the polymerization in the μm sized particles, as it reduces overall conversion. It was found that Φ_M, *i.e* the ratio of the volume of the monomer phase (V_M) and the volume of the reaction mixture (V_{RM}), must be taken as high as possible. So a reduction of the total volume of the water phase in which secondary nucleation can occur will be accomplished. Further suppression of secondary nucleation accompanied by the production of high molecular weight material seems to be possible by using Φ_M values larger than 0.3. Note that only Φ_M values smaller than or equal to 0.3 were used in this work.

Table 2.
Influence of monomer to water phase ratio on resulting particle size. Average droplet diameter of the emulsion $<d_{emulsion}>$ was 10.4 ± 0.5 μm.

Φ_M [–]	$C_{inhibitor}$ [10^{-3} mol/dm^3_M]	n_{inh}/n_{init} [–]	$<d_{pol}>$ [μm]	$V_{pol}/V_{emulsion}$ [–]	Conversion [%]
0.05	2,1	1	–	–	36
0.05	12,9	6	–	–	23
0.1	4,3	1	6.8	28	75
0.1	12,9	3	7.4	36	46
0.25	10,7	1	8.5	54.0	99.5
0.30	12,9	1	8.7	58.4	99.7

Acknowledgement: The authors wish to thank Agentschap.nl for financial support in the framework of the MicroNed programme (project number BSIK033)
The authors owe many thanks to dr. G. Veldhuis (Nanomi BV, Oldenzaal, the Netherlands) for stimulating discussions and for the supply of straight through microchannel arrays.

[1] T. Hayashi, Y. Ikada, *Biotechnology and Bioengineering* **1990**, *35*, 518.
[2] S. Brahim, D. Narinesingh, A Guiseppi-Elie, *Journal of Molecular Catalysis B-Enzymatic* **2002**, *18*, 69.
[3] M. Tang, X. J. Cao, Z. Z. Liu, X. Y. Wu, D. Gance, *Process Biochemistry* **1999**, *34*, 857.
[4] G. Kalfas, H. Yuan, W. H. Ray, *Ind. Eng. Chem. Res.* **1993**, *32*, 1831.
[5] B. W. Brooks, "Free-radical Polymerization: Suspension" in: "*Handbook of Polymer Reaction Engineering*", Chapter 5 T., Meyer, and J. T. F. Keurentjes, Eds., Wiley - VCH Weinheim, **2005**.
[6] K. Urban, G. Wagner, D. Schaffner, D. Röglin, J. Ulrich, *Chem. Eng. Technol.* **2006**, *29*, 24.
[7] A. J. Paine, W. Luymes, J. McNulty, *Macromolecules* **1990**, *23*, 3104.
[8] S. Omi, K. Katami, A. Yamamoto, M. Iso, *J. Appl. Polym. Sci.* **1994**, *51*, 1.
[9] A. J. Abrahamse, R. van Lierop, R. G. M. van der Sman, A. van der Padt, R. M. Boom, *Journal of Membrane Science* **2002**, *204*, 125.
[10] J. M. M. Simons, L. M. Kornmann, K. D. Reesink, A. P. G. Hoeks, M. F. Kemmere, J. Meuldijk, J. T. F. Keurentjes, *Journal of Materials Chemistry* **2010**, *20*, 3918.
[11] J. M. M. Simons, "*Polymer Microparticles by Membrane Emulsification: Production and Applications*", PhD Thesis, **2011**, Eindhoven University of Technology, The Netherlands.
[12] G. Ma, J. Li, *Chemical Engineering Science* **2004**, *59*, 1711.
[13] G. Veldhuis, M. Girones, D. Bingham, *Drug Delivery Technol.* **2009**, *9*, 24.
[14] C. G. Bell, C. J. W. Breward, P. D. Howell, J. Penfold, R. K. Thomas, *Langmuir* **2007**, *23*, 6042.
[15] M. D. Abramoff, P. J. Magelheas, S. J. Ram, *Biophot. Int.* **2004**, *11*, 36.
[16] C. Charcosset, I. Limayem, H. Fessi, *Journal of Chemical Technology and Biotechnology* **2004**, *79*, 209.
[17] S. K. Hait, P. R. Majhi, A. Blume, S. P. Moulik, *Journal of Physical Chemistry B* **2003**, *107*, 3650.
[18] R. Q. F. Janssen, "*Polymer Encapsulation of Titanium Dioxide*", PhD Thesis, **1995**, Eindhoven University of Technology, The Netherlands.
[19] V. Schröder, H. Schubert, *Colloids and Surfaces, A: Physicochemical and Engineering Aspects* **1999**, *152*, 103.
[20] C. Kotoulas, C. Kiparissides, *Chemical Engineering Science* **2006**, *61*, 332.
[21] E. Vivaldo-Lima, P. E. Wood, A. E. Hamielec, A. Penlidis, *Ind. Eng. Chem. Res.* **1997**, *36*, 939.
[22] M. Nomura, K. Suzuki, *Ind. Eng. Chem. Res.* **2005**, *44*, 2561.
[23] W. D. Harkins, *Journal of the American Chemical Society* **1947**, *69*, 1428.
[24] W. V. Smith, R. H. Ewart, *Journal of Chemical Physics* **1948**, *16*, 592.
[25] I. A. Maxwell, B. R. Morrison, D. H. Napper, R. G. Gilbert, *Macromolecules* **1991**, *24*, 1629.
[26] R. G. Gilbert, "*Emulsion polymerization: A mechanistic approach*", Academic Press, London **1995**.
[27] M. F. Kemmere, M. J. J. Mayer, J. Meuldijk, A. A. H. Drinkenburg, *Journal of Applied Polymer Science* **1999**, *71*, 2419.
[28] J. Bandrup, E. H. Immergut, E. A. Grulke, "*Polymer Handbook*", Wiley Interscience, New York **1999**.

Monitoring Pyrrol Polymerization Using On-Line Conductivity Measurements and Neural Networks

Claiton Z. Brusamarello,[1] *Leila M. Santos,*[2] *Monique Amaral,*[2] *Guilherme M. O. Barra,*[3] *Montserrat Fortuny,*[2] *Alexandre F. Santos,*[2] *Pedro Henrique Hermes de Araújo,*[1] *Claudia Sayer**[1]

Summary: In this work, the chemical oxidative synthesis of polypyrrole was monitored through on-line conductivimetry. A group of reactions was carried out at two different temperatures (5 °C and 20 °C) with varying concentrations of pyrrole and oxidant agent ($FeCl_3$) to investigate the effect on conversion and conductivity. The relation of electrical conductivity with conversion was not straightforward. To overcome this, a neural network was proposed to predict conversion based on on-line conductivity measurements. In this way, the neural networks entry variables were: electrical conductivity, reaction temperature, initial oxidant and pyrrole concentrations. As optimization algorithms Levenberg-Marquardt and Descent Gradient with momentum term were evaluated. Results obtained by the neural networks based on on-line conductivity measurements showed a good agreement with off-line conversion data.

Keywords: artificial neural networks; conductivity; polypyrrole

Introduction

Conducting polymers have been attracting much attention due to their unique characteristics that combine the mechanical properties of conventional polymers with the electrical characteristics found in conducting materials as copper. In 1977 Hideky Shirakawa, in collaboration with professors Alan MacDiarmid and Alan Heeger, obtained the first conductivity polymer by doping polyacetilene by charge transfer reactions with an oxidizing or reducing agent. [1]

[1] Process Control Laboratory, Chemical and Food Engineering Department, Federal University of Santa Catarina - UFSC, Campus Universitário, 88040-900, Florianópolis, Santa Catarina, Brazil
E-mail: csayer@enq.ufsc.br

[2] Research and Technology Institute, Tiradentes University - UNIT, Av. Murilo Dantas, 300 49032-490, Aracaju, SE, Brazil

[3] Polymer and Composites Laboratory, Mechanical Engineering Department, Federal University of Santa Catarina – UFSC, Campus Universitário, 88040-900, Florianópolis, Santa Catarina, Brazil

Polypyrrole was first reported in 1979, becoming one of the most studied conducting polymers for diverse applications including batteries, electronic devices, electrochromic devices, optical switching devices, sensors and many other advanced technologies. [2,3] The skeletons of such polymers containing conjugated double bonds, exhibiting properties such as low ionization potential, high electron affinity and as a result, can be readily reduced or oxidized. [4,5]

Sensor technology for chemical and polymer industry is another field which has shown fast development in recent decades. [6,7] One of the driving forces for this growth was the need of sensors that are able to provide a safe and continuous process monitoring, providing on-line process information for optimization and control schemes to improve process safety, productivity, as well as, product quality. Polymerization processes are characterized by strong nonlinearities, long delays and great sensitivity towards impurities. In this way, many problems encountered in the control of polymerization reactors can be attributed to the lack of robust analytical instrumentation and sensors that make online measurements during the polymerization. [8] For example [9], the techniques available for monitoring the conversion are often complicated by the heterogeneity and the nature of the viscous polymerization systems and the non-linearity associated with the presence of more than one phase including polymer particles of different size. The development of sensors for on-line measurements during polymerization reactions requires a multidisciplinary effort: mathematical modeling and data processing, understanding and knowledge of the process and instrumentation.

Conductive polymers can be prepared with a wide range of properties. For example, chemical properties can be manipulated to produce materials capable of trapping simple anions, or to render them bioactive. Electrical properties can also be manipulated to produce materials with different conductivities, capacitance, or redox properties. [10] For these reasons it is extremely important to develop new tools to monitor these type of polymerizations.

Artificial neural networks (ANN) are inspired by the biological nervous system and learn from examples and are well known for being able of solving highly non-linear complex problems without prior knowledge on detailed mathematical model system by capturing relationships between input and output variables from a given pattern. [11] ANNs have been widely used to predict polymer properties and interpreting polymer characterization results, especially when the phenomenological models are extremely complex. [12,13] In this work an ANN was proposed to predict conversion based on on-line conductivity measurements during chemical oxidative pyrrole polymerizations. In this way, the neural networks entry variables were: electrical conductivity, reaction temperature, initial oxidant and pyrrole concentrations. As optimization algorithms Levenberg-Marquardt (LM) and Descent Gradient with momentum term (GDM) were evaluated. It was used 4 experimental sets to train and 1 experimental set to validate the data set.

Experimental Part

Reagents

Pyrrole (Py) (Sigma-Aldrich 98%) was vacuum distilled prior to use and stored in a refrigerator between 0 °C and 4 °C. The oxidant agent hexahidrated ferric chloride ($FeCl_3.6H_2O$; Sigma Aldrich) was used as received. Mili-Q water was used as reactional medium.

Pyrrole Polymerization

The experiments were carried out in a jacketed borosilicate glass reactor with total volume of 1000 ml and an internal diameter of 120 mm. A thermostatic bath provided heating/cooling water that is circulated in the jacket of the reactor, thus enabling temperature control of the reaction medium. A predetermined amount of $FeCl_3.6H_2O$ was dissolved in 50 mL of Milli-Q water. The solution was stirred for 30 min and then fed to the reactor. The Py/ $FeCl_3.6H_2O$ molar ratio was 1:2.3 (mol:mol). $FeCl_3.6H_2O$ molar ratio varied from 0.5M to 0.1M. A predetermined amount of pyrrole was added dropwise into the reactor already containing the $FeCl_3.6H_2O$ solution. The polymerization was carried out for 4h at two different temperatures (5 °C and 20 °C) at 300rpm. Table 1 summarizes the reactions carried out in this work.

At pre-established times (10, 20, 40, 60, 90, 120, 180, 240 minutes) aliquots were collected (±20 g) of the reaction medium with a syringe fitted with a pipe (±10cm long) attached to its tip, the sample was then diluted in a known mass of cold Milli-Q water (±3 °C) and stored in a 80 ml beaker. In sequence the withdrawn aliquots were weighed, vacuum filtered and washed with a copious amount of Milli-Q water until the effluent was colourless, as reported in literature [14,15,16]. The aim of this step was to remove the monomer and other

Table 1.
Recipes of pyrrole polymerizations.

	EXP 1	EXP 2	EXP 3	EXP 4	EXP 5
Water (g)	699.8	700.97	801.54	701.64	701.62
$FeCl_3.6H_2O$ (g)	94.94	94.87	54.31	47.9	19.12
Pyrrole (g)	10.22	10.81	5.95	5.31	2.13
$[FeCl_3.6H_2O]$ (M)	0.5	0.5	0.25	0.25	0.1
$FeCl_3.6H_2O$: Pyrrole (mol:mol)	2.3:1	2.3:1	2.3:1	2.3:1	2.3:1
Temperature (°C)	5	20	5	20	20

reagents present in the sample. The filtrate was dried in a forced convection oven at a temperature of 40 °C for a period of 24 hours. After this stage a fine powder is obtained, dark polypyrrole and gravimetric conversion (X) could be calculated by Equation 1.

$$X = \frac{\left(\frac{Ppy_{mass}}{RP_{mass} * Py_{add} * \left(1+0{,}33 * \left(\frac{MwCl}{MwPy}\right)\right)}\right)}{TM_{reac}} \quad (1)$$

where: Ppy_{mass} – Polypyrrole mass; RP_{mass} – Sample mass removed from reactor; Py_{add} – Pyrrole mass added to reactor; Mw_{cl} – Molecular weight of Chlorine; Mw_{Py} – Molecular weight of Pyrrole; TM_{reac} – Total mass in reactor; 0,33–a factor, because on each Cyanopropyne (C_4H_3N) there are 0.33 mol of Cl. [17]

The morphology of the polypyrrole particles was analyzed on a Jeol JSM-6390LV Scanning Electron Microscope (SEM)

On-Line Monitoring of Reaction Medium

The conductivity of the reaction medium was monitored on-line (every 10 seconds) using a S70 SevenMulti Mettler Toledo conductimeter, this equipment is able to monitoring the temperature, too. The equipment was coupled to a computer for data acquisition and processing through software LabX pH 2.1.

Artificial Neural Networks

MATLAB® 7.10 version was used as software to implement and simulate the neural networks models. Two different neural network models a feed-forward neural network and a cascade-forward with two different back propagation training algoritms (Levenberg-Marquardt (LM) and Gradient descent with momentum term (GDM)) were evaluated. The performance of feed-forward and cascade-forward were evaluated using Mean Square Error (MSE) and R^2.

Since one hidden layer with a sufficient number of neurons is enough to make any kind of mapping system [18] three-layer neural network models with only one hidden layer were used in this work. The training was carried out using 192 data sets for training and 48 data sets for testing. Each data set consists of 5 variables, four of which are considered input variables (electrical conductivity, reaction temperature, initial oxidant and pyrrole concentrations) and one output variable (pyrrole conversion).

Levenberg-Marquardt Algorithm

Like the quasi-Newton methods, the Levenberg-Marquardt algorithm was designed to approach second-order training speed without having to compute the Hessian matrix. [19]

The performance function for training feed-forward networks has the design of a sum squares and the Hessian matrix and gradient can be calculated by equation (2) and (3):

$$\mathrm{H} = \mathbf{J}^T\mathbf{J} \quad (2)$$

$$\mathbf{g} = \mathbf{J}^T\mathbf{e} \quad (3)$$

where **J** is the Jacobian matrix that contains first derivatives of the network errors with respect to the weights and biases, and **e** is a vector of network errors. The Jacobian matrix can be calculated using a back-propagation technique.

The Levenberg-Marquardt algorithm uses this approximation to the Hessian matrix in the following Newton-like update:

$$X_{k+1} = X_k - [\mathbf{J}^T\mathbf{J} + \mu\mathbf{I}]^{-1}\mathbf{J}^T\mathbf{e} \tag{4}$$

When the scalar μ is zero, this is Newton's method, using the approximate Hessian matrix. Scalar μ is decreased after each successful step (decrease of performance function) and only increased when a tentative step would increase the performance function, since Newton's method ($\mu = 0$) is faster and more accurate near an error minimum.

Gradient Descent with Momentum Term

The momentum rate is introduced in the gradient descent learning algorithm to attenuate oscillations in the iteration process when the minimum of the error function lies in a narrow valley.1[20]In this way the gradient of the error function is computed for each new combination of weights, but instead of just following the negative gradient direction a weighted average of the current gradient and the previous correction direction is computed at each step.

In standard back-propagation the input-output patterns are fed into the network and the error function E is determined at the output. When using back-propagation with momentum in a network with n different weights $w_1, w_2, \ldots, w_n$, the i-th correction for weight w_k is given by

$$\Delta w_k(i) = -\gamma \frac{\partial E}{\partial w_k} + \alpha \Delta w_k(i-1) \tag{5}$$

where γ and α are the learning and momentum rate respectively. In this work a momentum constant of 0.9 and a learning rate of 0.01 were used.

To minimize the effect of differences in parameter magnitude, the normalized variables were used for training and testing neural network model. In this way all variables were normalized to the dimensionless variables within a range of -0.9 to 0.9 using the MATLAB® function *mapminmax*. Another reason to make this transformation was the use of a sigmoid function ($f(x) = 1/(1–\exp(x))$) as activation function. The objective function is defined as the average error between the target output vector and the calculated model output vector of both training and testing data tests. In addition, the objective function at each training epoch is monitored to avoid overtraining and a method for improving generalization called early stopping was used in this work. In this technique data is divided into three subsets: 1) training set - used for computing the gradient and updating the network weights and biases; 2) validation set – used to determine the end of the training procedure, when the validation error increases for a specified number of iterations, the training is stopped and the weights and biases at the minimum of the validation error are returned and 3) test set - used to compare different models.[18] The dataset was randomly divided 60% for training, 20% for validation and 20% for testing. The effect of different numbers of neurons in the hidden layer (5, 10 and 15 neurons) was evaluated. Weights and biases were randomly initialized. The feed-forward (LM) and cascade-forward (CLM) networks with Levenberg-Marquard algorithm were trained with up to 100 epochs, for Gradient Descendent with momentum (GDM) the number of epochs was 2000.

Results and Discussions

According to Sulimenko [21] the change in electrical conductivity of the medium during aniline polymerization can have different causes: temperature, concentration and type of ionic species and polymer production. In this way, in order to verify if on-line conductivity measurements could be used to infer conversion during pyrrol polymerization the effect of different molar concentrations of $FeCl_3.6H_2O$ (0.5 M, 0.25 M and 0.1 M), keeping the oxidizing agent and pyrrole molar ratio constant, on the on-line conductivity of the medium and off-line polymerization conversion was evaluated for two different reaction temperatures 20 °C (Figure 1) and 5 °C (Figure 2).

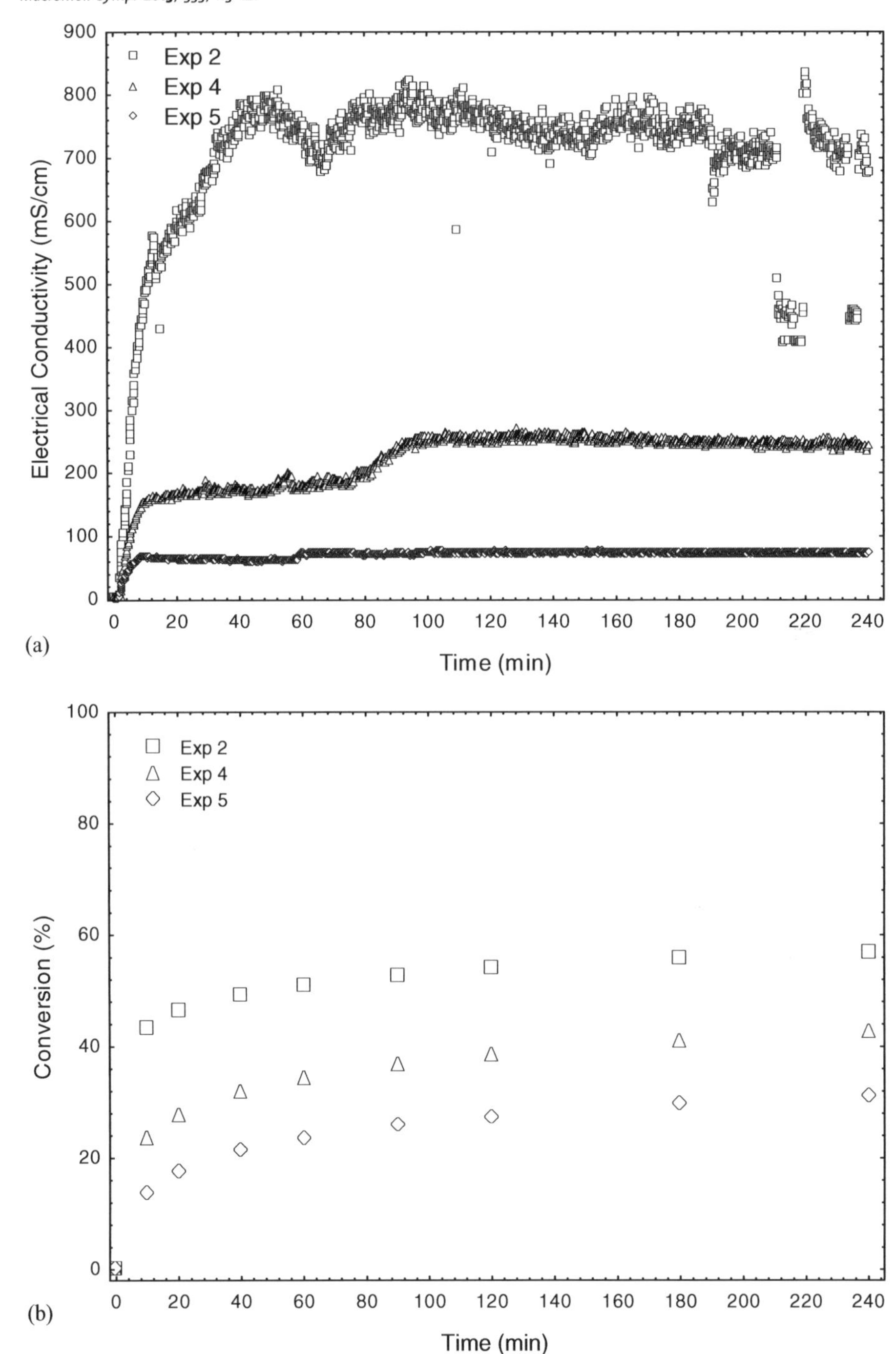

Figure 1.
Effect of different molar concentrations of Py and $FeCl_3.6H_2O$ on: (a) electrical conductivity of the reaction medium, (b) conversion of pyrrole, reaction temperature 20 °C, $FeCl_3.6H_2O$ molar concentrations: Exp 2 (0.5 M), Exp 4 (0.25 M), Exp 5 (0.1 M).

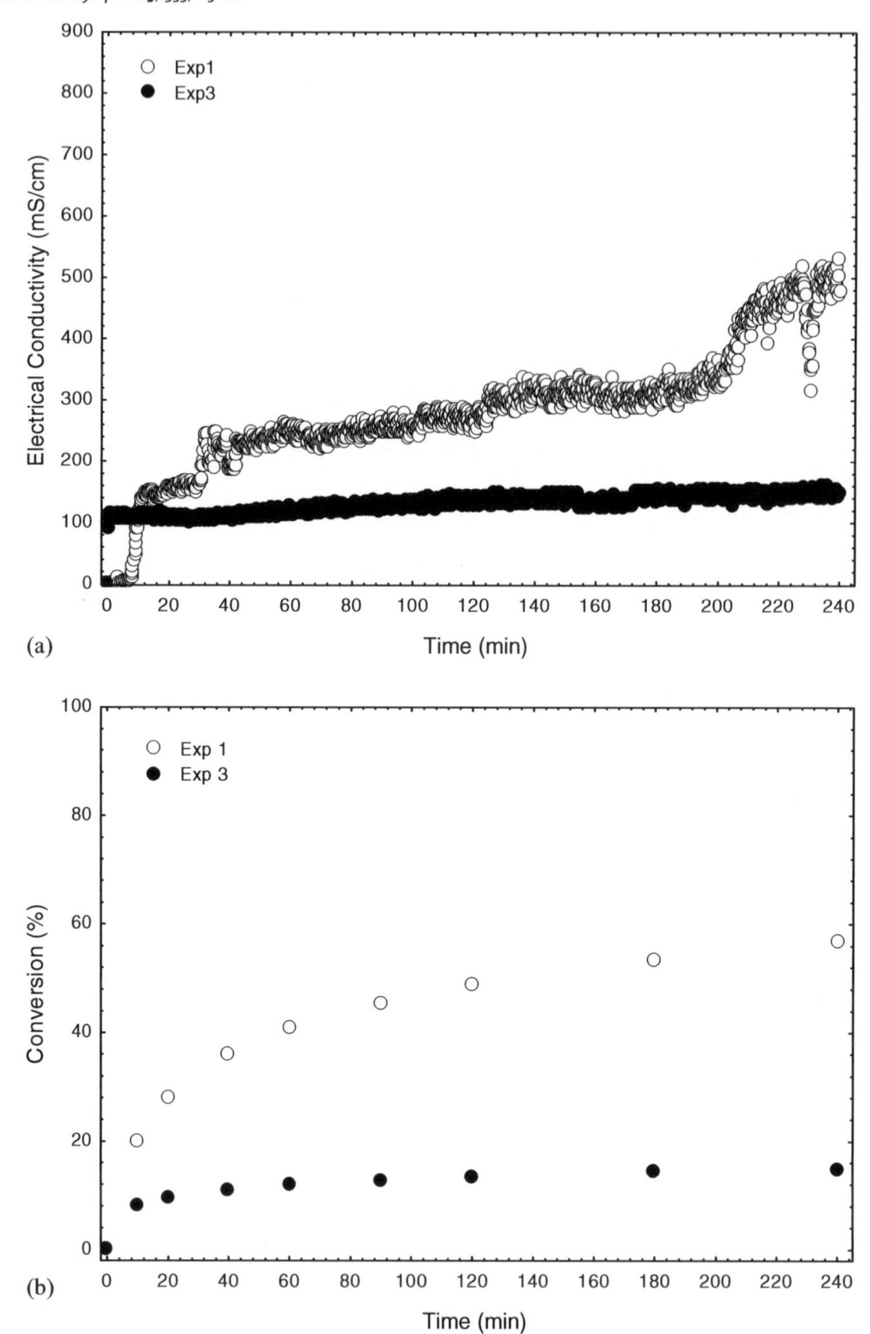

Figure 2.
Effect of different molar concentrations of Py and $FeCl_3.6H_2O$ on: (a) electrical conductivity of the reaction medium, (b) conversion of pyrrole, reaction temperature 5 °C, $FeCl_3.6H_2O$ molar concentrations: Exp 1 (0.5 M), Exp 3 (0.25 M).

As may be observed in results shown in Figure 1a, the electrical conductivity of the medium remains constant during the first reaction minutes indicating the presence of a short induction time, that has also been reported for the polymerization of aniline [21]. After this induction time, conductivity increases sharply due to the formation of polypyrrole until almost constant conductivity values are reached. Comparing off-line pyrrole conversion results (Figure 1b) with conductivity data (Figure 1a) the same major trends, steep increase followed by almost constant values may be observed. In addition, both conductivity of the reaction medium and pyrrole conversion increased with the concentration of the ionic species, but the effect on the former was more intense.

When reaction temperature was decreased from 20 °C (Figure 1) to 5 °C (Figure 2), the reaction rate became lower and final conversion and conductivity were also lower. Again the effect on conductivity was stronger since conductivity is affected by temperature and by the higher polymer concentration, as may be observed with the increase of conductivity with conversion.

Finally, in reactions Exp 2 (Figure 1a) and Exp 1 (Figure 2), that were carried out with a higher $FeCl_3.6H_2O$ molar concentration (0.5 M) and thus resulted in higher polymer contents, some oscillations in the conductivity measurements could be observed at the end of the experiments at around 220 min in Exp 2 and 230 min in Exp 1. These oscillations are attributed to the rearrangement and release o the polypyrrole film formed on the electrode during polymerization, thus affecting on line conductivity measurements.

During the polymerization polypyrrole precipitates forming aggregates with coral conformation of submicrometric spherical particles, as observed in the SEM image (Figure 3) of the polymer formed during Exp 3. According to literature several parameters can result in changes in polypyrrole conformation, as reaction temperature[22], solvent[23], co-dopant[24]. This precipitation during polymerization turns sampling and on-line monitoring more difficult.

Neural Network Performance

The performance of the feed-forward and of cascade forward neural networks using Levenberg-Marquard and Gradient Descendent with momentum term for predicting pyrrole conversion is presented in Table 2.

Figure 3.
SEM image of polypyrrole formed in polymerization Exp 3, Temperature 5 °C and 0.25M of $FeCl_3.6H_2O$.

Table 2.
Performance of feed-forward and cascade-forward neural networks for predicting pyrrole conversion.

Neural network	Neurons in hidden layer	Algorithm	MSE
Feed-forward	5	LM	0.00511
	10	LM	0.00485
	15	LM	0.0448
	5	GDM	0.0227
	10	GDM	0.00766
	15	GDM	0.00369
Cascade-forward	5	LM	0.00425
	10	LM	0.00928
	15	LM	0.00345
	5	GDM	0.01576
	10	GDM	0.00563
	15	GDM	0.00396

According to results shown in Table 2, when we compare results among feed-forward networks using both algorithms (LM and GDM), it is possible to see that the best result was achieved was with a feed-forward neural network with 15 neurons in the hidden layer with a GDM algorithm (MSE – 0.00369; R^2–0.0147).

Comparing the results obtained among the cascade-forward neural networks using networks with 15 neurons in the hidden layer also resulted in better results, with the cascade-forward with a LM algorithm leading to slightly better results (MSE – 0.00345; R^2–0.006).

Figure 4 shows a comparison between the predictions of the evolution of pyrrole conversion during polymerization Exp 1 using the developed neural network models and off-line gravimetric conversions. The cascade-forward neural network using a LM algorithm was the only one that was able to represent correctly the final almost constant conversion of reaction Exp 1. For all other models the increase of conductivity of Exp 1 around 200 min (Figure 2a) was reflected by an increase of the predicted conversion. This behavior is probably due to the relatively small data set used for training the neural networks.

Finally, results indicate that it is possible to combine on-line conductivity measurements of the reaction medium with neural network models to monitor on-line the conversion during pyrrole polymerizations. Nevertheless, since the conductivity is affected by several factors, robust

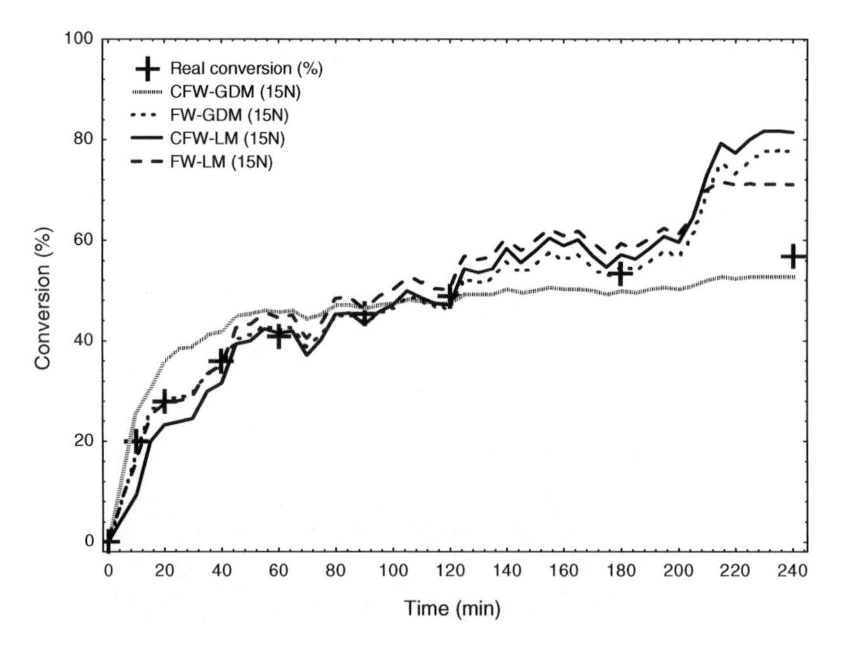

Figure 4.
Prediction of conversion of reaction Exp 1 using different neural network models.

estimation of conversion is not straightforward, especially based on a relatively small experimental data set.

Conclusion

The effects of reaction temperature (5 °C or 20 °C) and monomer and oxidizing agent concentrations on the evolution of conversion and conductivity of the reaction medium during pyrrole chemical oxidative polymerization were evaluated. Experimental data was used to verify if it is possible to train neural network models (feed-forward and cascade-forward) in order to predict pyrrole conversion based on on-line conductivity measurements. In this way, the neural networks entry variables were: electrical conductivity, reaction temperature, initial oxidant and pyrrole concentrations. As optimization algorithms Levenberg-Marquardt and Descent Gradient with momentum term were evaluated. The performances of the developed artificial neural networks were compared and all neural network models with 15 neurons in the hidden layer exhibited a good agreement with off-line pyrrole conversion data up 3 hours of reaction. Though the cascade-forward network using a Levenberg-Marquardt algorithm was the only one that was able to represent correctly the final almost constant conversion. These results indicate that it is possible to combine on-line conductivity measurements of the reaction medium with neural network models to monitor on-line the conversion during pyrrole polymerizations.

Acknowledgements: The authors thank CAPES - Coordenação de Aperfeiçoamento de Pessoal de Nível Superior and CNPq – Conselho Nacional de Desenvolvimento Científico e Tecnológico) for financial support and Laboratório Central de Microscopia Eletrônica (LCME) of Federal University of Santa Catarina for SEM analyses.

[1] A. Pron, P. Ronnou, *Prog. Polym. Sci.* **2002**, *27*, 135.
[2] S. Eofinger, W. J. Van Ooji, T. H. Ridgway, *J. Appl. Polym. Sci.* **1998**, *61*, 1503.
[3] T. A. Skotheim, R. L. Elsenbaumer, J. R. Reynolds, in: "*Handbook of conducting polymers*". (Ed., Marcel Dekker, New York **2007**, p. 1097.
[4] A. G. Macdiarmid, *Synth. Met.* **1997**, *84*, 27.
[5] D. Maia, M. De Paoli, O. L. Alves, A. J. G. Zarbin, S. Neves, S., *Quim. Nova* **2000**, *23*, 204.
[6] G. E. Fonseca, M. A. Dubé, A. Penlidis, *Macromol. React. Eng.* **2009**, *3*, 327.
[7] K. Petr, B. Gabrys, S. Strandt, *Comput. Chem. Eng.* **2009**, *33*, 795.
[8] G. Gattu, E. Zafiriou, *Chem. Eng. J.* **1999**, *75*, 21.
[9] F. Machado, E. L. Lima, J. C. Pinto, *Polímeros* **2007**, *17*, 166.
[10] G. G. Wallace, G. M. Spinks, L. A. P. Kane-Maguire, P. R. Teasdale, in: "*Conductive Electroactive Polymers Intelligent Polymer Systems*", 3 Ed., CRC Press, Boca Raton, Fl **2009**, p. 263.
[11] S. Anantawaraskul, M. Toungsetwut, R. Pinyapong, *Macromol. Symp.* **2008**, *264*, 157.
[12] Z. Zhang, K. Friedrich, *Compos. Sci. Technol.* **2003**, *63*, 2029.
[13] N. Sresungsuwan, N. Hansupalak, *J. Appl. Polym. Sci.* **2013**, *127*, 356.
[14] M. F. Planche, J. C. Thiéblemont, N. Mazars, G. Bidan, *J. Appl. Polym. Sci.* **1994**, *52*, 1887.
[15] H. V. R. Dias, M. Fianchini, R. M. G. Rajapakse, *Polymer.* **2006**, *47*, 7349.
[16] F. H. Hsu, T. M. Wu, *Synth. Met.* **2012**, *162*, 682.
[17] M. Omastová, M. Trchová, J. Kovářová, J. Stejskal, *Synth. Met.* **2003**, *138*, 447.
[18] G. Cybenko, *Math. Control Signals Syst.* **1989**, *2*, 303.
[19] H. Demuth, M. Beale, in: "*Neural Network Toolbox For Use With MATLAB*", Inc. Natick, MA **2002**.
[20] R. Rojas, in: "*Neural Networks – A Systematic Introduction*", Springer-Verlag, Berlin **1996**, p. 453.
[21] T. Sulimenko, J. Stejskal, I. Krivka, J. Prokes, *Eur. Polym. J.* **2001**, *37*, 219.
[22] J. Hong, H. Yoon, *J. Jang. Small.* **2010**, *6*, 679.
[23] J. Ouyang, Y. Li, *Polymer.* **1997**, *38*, 1971.
[24] M. Omastová, M. Trchová, J. Pionteck, J. Prokeš, J. Stejskal, *Synth. Met.* **2004**, *143*, 153.

DOI: 10.1002/masy.201300048

An In-Situ NMR Study of Radical Copolymerization Kinetics of Acrylamide and Non-Ionized Acrylic Acid in Aqueous Solution

*Calista Preusser, Robin A. Hutchinson**

Summary: An in-situ NMR technique has been developed to study the aqueous phase copolymerization of non-ionized acrylic acid (AA) and acrylamide (AM) under near-isothermal conditions at much higher monomer contents than previously reported in the literature. The composition data obtained over the entire conversion range provides a precise estimate of monomer reactivity ratios not available from low conversion data. The set of experiments, with initial monomer content in aqueous solution varied between 5 and 40%, were well-fit over the complete conversion range by $r_{AA} = 1.24 \pm 0.02$, and $r_{AM} = 0.55 \pm 0.01$. It was found that the rate of monomer conversion increases with increasing monomer concentration, a trend contrary to the known decrease in the AA and AM chain-end propagation rate coefficients.

Introduction

Water soluble polymers are an important family of products with diverse applications in personal-care (e.g., thickening agents for shampoos, hair gel, and other cosmetics, anti-flocculants in laundry detergents) and industrial (e.g., water treatment facilities, antiscalants in oil drilling) markets. Despite their commercial importance, the kinetics of these radical polymerizations are still not well understood. This paper will discuss the copolymerization of acrylic acid (AA) and acrylamide (AM) using an in-situ NMR technique at various monomer concentrations and monomer compositions, with the goal of improving the understanding of this complex system over a broader range of operating conditions.

Experimental pulsed-laser polymerization (PLP)[1,2] techniques have been applied to obtain reliable estimates of rate coefficients difficult to extract from conventional continuously-initiated reactions for radical polymerization in aqueous solution. Propagation rate coefficients, k^p, are estimated using PLP in combination with size exclusion chromatography (SEC) to measure chain growth that occurs during the dark periods between the laser pulses.[3] The PLP-SEC technique has been applied to the study of water soluble monomers including AA[4–6], AM[7–9], methacrylic acid (MAA)[10–12] and N-vinylpyrrolidone (NVP).[13] For all systems, the value of k^p increases as the monomer concentration is lowered, a result attributed to the reduced barrier to rotational motion in the transition state structure upon replacing monomer units with H^2O.[4,10,13,14] These theories have been confirmed by computational studies for AA[15] and AM.[16]

Similar to k_p, the termination rate coefficients, k_t, of AA,[17] NVP[18] and MAA[19,20] exhibit a dependency on the initial monomer concentration even at low conversion in aqueous solution, as measured using single pulse-PLP-electron paramagnetic resonance (SP-PLP-EPR)[17] and SP-PLP-near-infrared spectroscopy (NIR).[19] Studies with NVP[21] and MAA[22]

Department of Chemical Engineering, Dupuis Hall, Queen's University, Kingston, Ontario K7L 3N6, Canada
E-mail: robin.hutchinson@chee.queensu.ca

demonstrate that conversion profiles and molecular-weight data from continuously-initiated batch and semibatch polymerizations are well-represented using the concentration-dependent k_t and k_p values measured using PLP techniques. The modeling work for the MAA system has recently been extended to include the chain-length dependence of k_t measured using the SP-PLP-EPR technique,[20] which is especially important for systems with high levels of added chain-transfer agent.[23]

Intramolecular transfer to polymer, also referred to as backbiting, is a major side reaction that occurs during the polymerization of acrylates such as butyl acrylate (BA),[24] and also AA in aqueous solution.[17] Backbiting reactions occur when the polymerizing chain end curls back on itself to form a six-membered ring and abstracts a hydrogen atom from the polymer backbone with rate coefficient k_{bb}. A short-chain branch is formed when a monomer unit adds to the resulting midchain radical; as the rate coefficient for this monomer addition, $k_p{}^t$, is significantly smaller than k_p, the formation of midchain radicals leads to a reduction in the overall rate of conversion. The net effect of backbiting is an apparent reaction order of greater than unity when the variation of overall polymerization rate with respect to monomer concentration is examined. The rate coefficients for the backbiting reactions have been measured using SP-PLP-EPR[17,24,25] as well as PLP-SEC with varying frequency.[26] For AA k_{bb} and $k_p{}^t$ are also a function of monomer concentration, decreasing as monomer concentration in aqueous solution increases in a similar manner to k_p.[24]

The ultimate objective of this work is to develop an understanding of the copolymerization of AA and AM over a broad range of reaction conditions utilizing the rate coefficients measured using PLP techniques. In order to attain this goal, polymerization behavior at monomer concentrations higher than currently published in the literature are investigated. Conventional experimental methods using batch reactors that require mixing and sample collection are problematic, as the solution viscosity even at low monomer concentrations (~7 wt%) increases rapidly with monomer conversion[27] such that mixing and isothermal operation becomes difficult. Therefore the monomer concentrations in the literature are maintained below 5 wt% as a rule[28–30] to ensure good mixing of the reaction solution and isothermal operation. Limited data for AM homopolymerization is available to a maximum of 20 wt%,[27,31] and for AA a maximum of 30 wt% has been studied.[32] For AA/AM copolymerization, however, the monomer concentrations investigated are much lower, at less than 4 wt%.[33–35] Significantly higher monomer contents are utilized in industrial copolymerization to increase reactor productivity. At these higher concentrations, the variation of k_p with concentration, as well as the possible variation of diffusion-controlled k_t with conversion, may lead to rates of polymerization and comonomer incorporation that differ from behavior observed at low monomer concentration.

Several experimental methods of tracking monomer conversion on-line have been applied to water soluble polymerizations, including near-IR spectroscopy,[22,23] light scattering,[30,35] and NMR.[28,32,36] Due to the experimental constraints, namely the high monomer concentration, a method which requires a larger reaction volume and therefore mechanical mixing was not considered. The advantage of the in-situ NMR method is that monomer composition can be tracked in addition to the overall monomer conversion. In the literature in-situ NMR experiments for AA and AM homopolymerizations were successfully reported,[28,32,36] and these studies form the basis of the technique applied to AA/AM copolymerization. The group of Mahdavian has studied AM homopolymerization in aqueous solution[28] as well as various copolymerization in organic solution.[37–40] Three criteria for running in-situ NMR experiments were reported:[28] 1) a reaction rate that is less than the scanning rate, 2) solubility of the polymer in the solvent, and 3) distinguishable monomer/polymer peaks so that monomer composition and conversion can be determined. A fourth criterion, namely that isothermal conditions are maintained, should be added. As will be shown, these criteria are fulfilled for the copolymerization of AA and AM.

In this work, experimental procedures have been developed to study the isothermal batch copolymerization of non-ionized AA and AM. Monomer conversion profiles are presented from a set of experiments completed at 40 °C over a broad range of monomer concentrations and comonomer compositions. The data collected are used to estimate reactivity ratios and to develop an improved mechanistic understanding of the system.

Experimental Part

The monomers acrylic acid (99%, Sigma Aldrich) and electrophoresis grade acrylamide (99 + %, Sigma Aldrich) were used as received. Deuterated water (D_2O) (99.9%, Cambridge Isotope Laboratories Inc.) and 2,2-azobis(2-methylpropionamidine)dihydrochloride (V-50) (97%, Sigma Aldrich) were also used as is. Poly(acrylic acid) (pAA) (Sigma Aldrich) at an average molecular weight of 1800 g/mol was purchased and used without purification for development of the NMR procedures.

Stock solutions of monomer and initiator were prepared separately ahead of time, with the initiator stock solution of 4 wt% V-50 in D_2O made fresh weekly. The solutions were mixed at the appropriate ratios on the day of the experiment to the desired total monomer concentration and composition. The solutions were purged with nitrogen for 15 minutes to ensure complete mixing and solvation of the monomer in the D_2O, transferred to labeled NMR tubes and stored under refrigeration until polymerization. It was difficult to fully isolate the reaction mixture from air when transferring the solution to the 5 mm NMR tubes. However, a study by Cutié et al.[36] showed that while purging with nitrogen for the AA system reduces the inhibition time, it does not have an effect on the shape of the monomer conversion profile. Since the reaction mixture sits in the NMR machine for at least 4 minutes before the first scan is taken, an inhibition time is vital in order to ensure that a complete conversion profile is obtained. The inhibition times at 40 °C are between 15 minutes and 1 hour, depending on the AA content in the reaction mixture. The start time of the experiment is calculated by fitting the linear portion of the conversion profile and solving for the x-intercept.

The polymerizations are conducted using a Bruker 500 with TopSpin as the interface and a BVT3000 heating element. The NMR sample chamber is first increased to 40 °C using a flow rate of heated air of 535 L/h, and allowed to equilibrate for 20 minutes. The NMR tube containing the reaction mixture is then placed into the machine and allowed to heat for 4 minutes, at which time the instrument is shimmed and tuned to reduce the signal to noise ratio and obtain sharp peaks. Since peak positions and peak quality are temperature sensitive, it is necessary to perform the tuning and shimming when the reaction mixture has reached the reaction temperature. No changes are made to the standard ^1H NMR settings on the spectrometer (3.17 s acquisition time, 1 s relaxation delay, 6 μs dead time, 48.4 μs dwell time). Subsequent analysis showed good agreement between measured NMR monomer composition and conversion with values determined by alternate laboratory methods, indicating sufficient relaxation times. The first scan is taken approximately eight minutes after insertion of the NMR tube, and subsequent scans every two minutes afterwards until full conversion, with the total number of spectra collected ranging between 40 and 80.

Data Analysis

The NMR spectra were processed using the NMR software MestReNova 6.0. The spectra used for the following discussion were taken from a representative non-ionized experiment run at 40 °C with an initial AA monomer mole fraction (f_{AA0}) of 0.7, 20 wt% monomer and 0.217 wt% V-50 in D_2O. Figure 1 shows the ^1H NMR spectrum of the monomer solution and the peak assignment of the hydrogen atoms for AA and AM. Note that the exact positions of the peaks shift slightly relative to each other with monomer composition, overall monomer concentration and temperature; however, the overall shape and ordering of the peaks remains the same.

The monomer fractional molar composition can be calculated from the integrated area of the monomer peaks according to

$$f_{AM} = \frac{Area\,AM}{Area\,AM + Area\,AA} \tag{1}$$

Since one of the AA peaks overlap with an AM peak as can be seen in Figure 1, the peaks between 5.72 and 5.74 ppm for AM and between 5.90 and 5.92 ppm for AA were used. In order to test the validity of the measurement, the NMR values were compared to the "lab-measured" composition, as calculated by the measured masses of the reaction ingredients. As seen in Figure 2, the agreement is excellent. Based on 42 runs, the average relative error between the two values is 2%, with the absolute error ranging between 0 and 4%.

The AA and AM repeat units on the polymer backbones are similar and can therefore not be differentiated on the ^1H NMR spectra. The chemical shifts of the backbone hydrogen atoms are more shielded than those on the monomeric units; as can be observed in Figure 3, the CH peak is located between 2.59 and 2.09 ppm, while the CH_2 peak is between 2.07 and 1.16 ppm. The sharp peak at 1.46 ppm has been assigned to the methyl groups on the initiator, V-50. Although this peak is prominent on the spectra, the initiator levels cannot be directly measured due to the overlap with the CH_2 peak. However, by overlapping the spectra at different conversions in this region, a

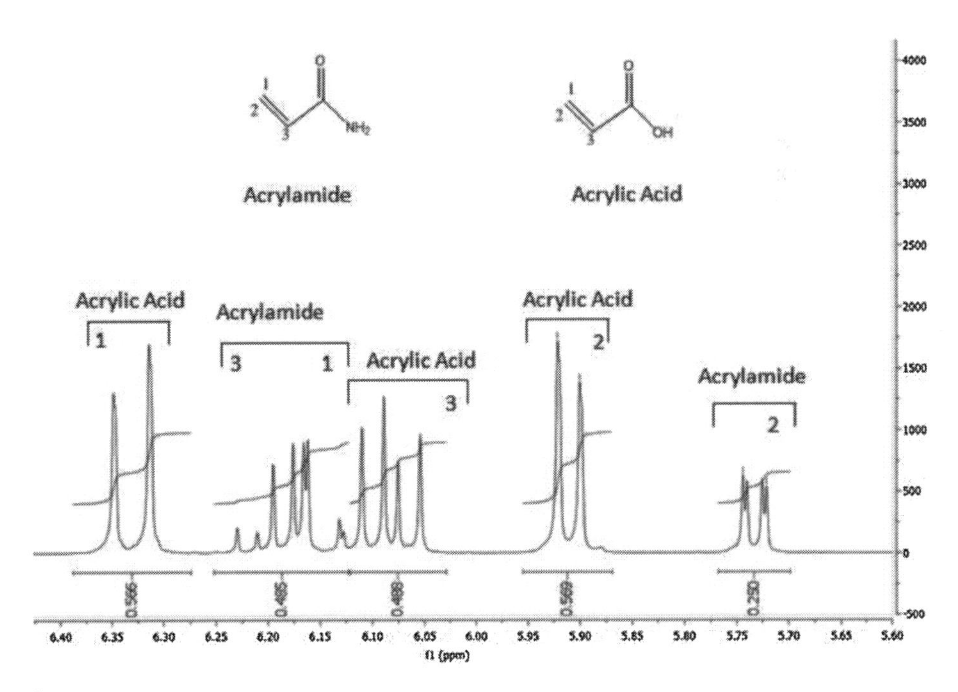

Figure 1.
^1H NMR peaks assignments for AA and AM monomer with $f_{AA0} = 0.7$ and 20 wt% monomer in D_2O with 0.217 wt % initiator at 40 °C.

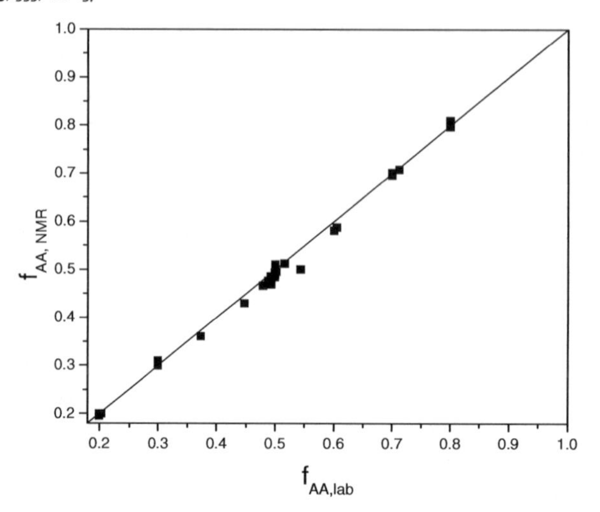

Figure 2.
Molar monomer composition of AA in the AA/AM mixture calculated in the lab compared to the monomer composition measured with NMR.

qualitative observation of the relative consumption of initiator can be made by comparing peak intensities. For this set of experiments, the intensity of the initiator peak is constant over the polymerization time of 1–2 h, as the half-life of V-50 at 40 °C is 100 h.

The fractional conversion of monomer is calculated using the following equation:

$$x = \frac{Area\,polymer}{Area\,polymer + Area\,monomer} \tag{2}$$

As with Equation 1, the areas of the polymer and monomer peaks are scaled to one hydrogen. The conversion measurements were verified by simulating specific conversions by adding pAA to monomer/D_2O mixtures. The agreement between the

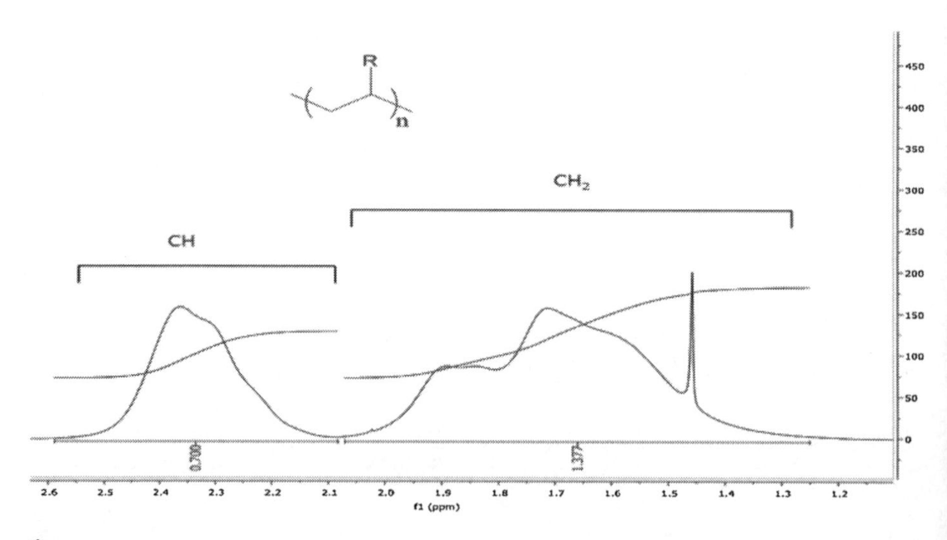

Figure 3.
^{1}H NMR peaks assignments for the polymer backbone with $f_{AAo} = 0.7$ and 20 wt% monomer in D_2O with 0.217 wt % initiator at 40 °C at 92% conversion.

'conversion' measured in the lab and that calculated from the NMR spectra is presented in Figure 4. The average relative error between the calculated monomer conversions is 6%, with the absolute error between 0 and 3% based on 13 data points.

Figure 5 shows the evolution of the NMR spectra collected over the course of an experiment. The solvent peak is locked at 4.7 ppm and taken as the reference peak for the integration in the analysis. One can clearly see with increasing time the disappearance of the monomer peaks between 5.5–6.5 ppm and the appearance and growth of the polymer peaks between 1.5–3.0 ppm.

Small peaks from diacrylic acid (DiAA) are observed between 4.37 to 4.31 ppm and 2.74 to 2.68 ppm, with each dimer peak representing two hydrogen atoms. These conclusions were confirmed with 2D NMR. DiAA is formed very slowly over time by a Michael addition reaction that is promoted with increasing temperature and humidity.[41] The fraction of DiAA in the acrylic acid can be calculated according to the following equation:

$$\% DiAA = \frac{Area\,DiAA}{\frac{Area\,DiAA}{2} + Area\,AA} \times 100 \quad (3)$$

Correcting for the consumption of two AA molecules to form a single DiAA molecule, the value represents the fraction of AA molecules in DiAA form. DiAA has a much lower reactivity than AA monomer and accumulates in the system with increasing conversion during polymerization. The DiAA level is monitored from reaction to reaction using the NMR spectra, and a fresh batch of monomer is purchased when it reaches 5%.

The experiments were run in randomized order and re-runs were periodically performed, with an example of the reproducibility shown in Figure 6. The monomer conversion profiles shown were from repeat experiments run during a period of over a month, with all of the reactions performed using different initiator stock solutions. The effect of varying initiator concentration on the monomer conversion rates was also examined and shown to give the expected behavior, in agreement with previous AA[32] and AM[27,28] homopolymerization studies.

Exotherm Experienced During the Polymerization

Due to the small reaction volume and the experimental NMR setup, it is impossible to directly measure the temperature of the reaction mixture. Instead, the measured temperature is of the air flowing through the cavity in which the NMR tube

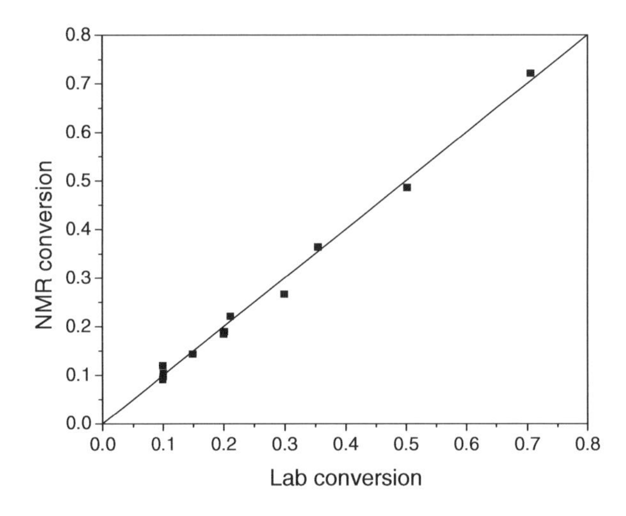

Figure 4.
Monomer conversion calculated in the lab compared to the monomer conversion measured with NMR.

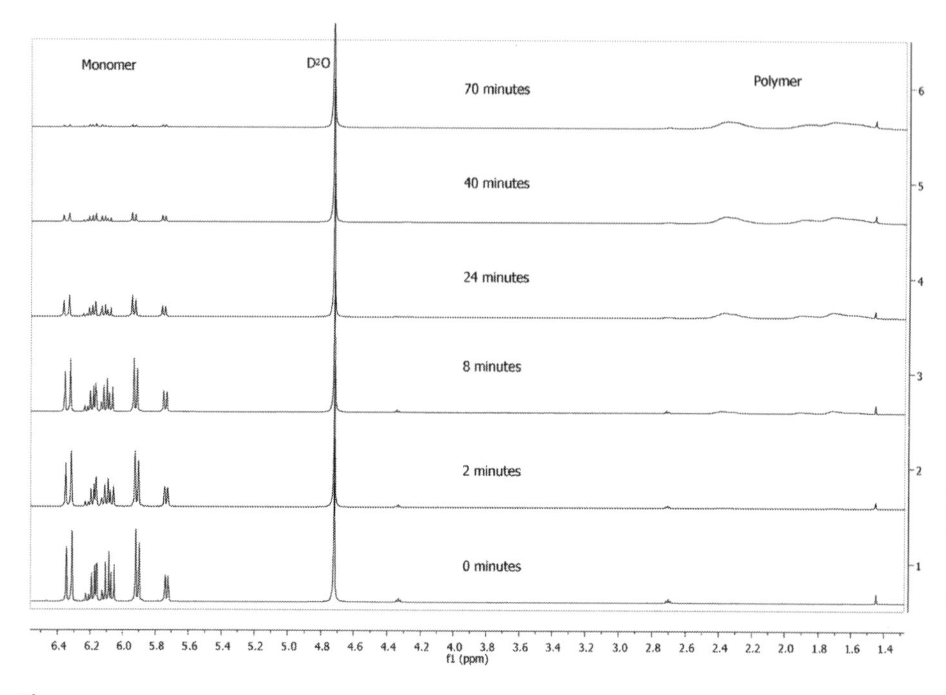

Figure 5.
Evolution of the ^{1}H NMR spectra from low to high conversion for an experiment run with $f_{AAo} = 0.7$ and 20 wt% monomer in D_2O with 0.217 wt% initiator at 40 °C.

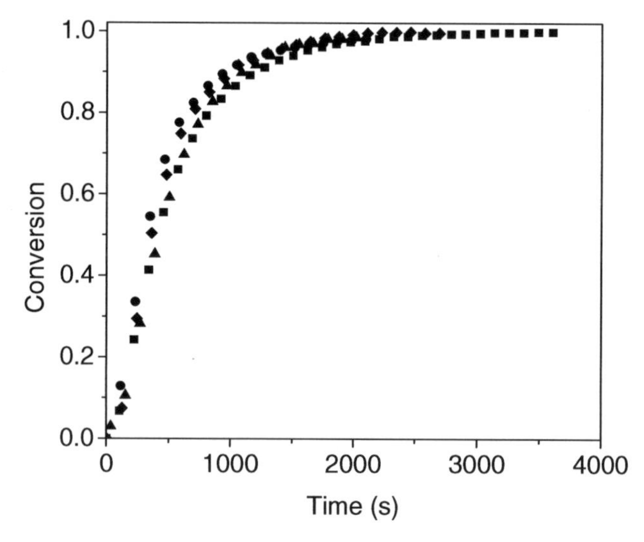

Figure 6.
Comparison of monomer conversion profiles for AM homopolymerization at 40 °C, 40 wt% initial monomer in D_2O, and 0.217 wt% V-50 for four experiments conducted over the period of one month with different initiator stock solutions.

resides. The reaction mixture in the 5 mm inner diameter NMR tube is generally filled to a height of about 3 cm, which corresponds to approximately 0.7 g of reaction mixture. The amount of heat released by the exothermic reaction increases with increased monomer concentration and thus good heat removal is essential. Therefore it is important that the air flow rate is at least 535 L/h at 40 °C; for higher temperatures a faster flow rate is advised. In order to monitor the extent of exotherm, the temperature of the reaction mixture can be inferred from the movement of the chemical shifts of the monomer peaks. In previous studies, the distance between the methylene and hydroxyl peaks in ethylene glycol has been used as an NMR thermometer, with the peak separation increasing by a chemical shift of 0.01 ppm/°C.[42] For systems with D_2O solvent, however, the OH groups are not visible due to the fast exchange of hydrogen with D_2O.[43] In the course of this investigation, it was determined that the position of monomer peaks were also sensitive to temperature and therefore systematic experiments were conducted by placing monomer/polymer mixtures of different monomer compositions and conversions without initiator into the NMR at 40 °C and then increasing the temperature to 50 °C in 1 °C interval steps, allowing the samples to equilibrate for 10 minutes after each step. As observed in Figure 7, the monomer peak position shifts by 0.01 ppm with every degree increase. Studies were conducted at different monomer compositions, conversions, and concentration, all showing the same trend.

To complicate the issue, however, a shift in monomer peaks is also observed with the presence of polymer, and therefore only a maximum temperature increase can be inferred. Figure 8 plots the shift in the peak position relative to the peak position at 0% conversion as a function of conversion for polymerizations run at 5, 20, and 40 wt% monomer and $f_{AA0} = 0.5$. Neglecting any peak shifts with the appearance of polymer and assuming a 0.01 ppm shift/°C, the maximum temperature increase varies between 1.5 and 3.5 °C. However, as the maximum rate of polymerization occurs at the start of the batch reaction, it can safely be assumed that the increase in polymer conversion contributes significantly to the movement of the peak position, and that the actual temperature increase is lower.

Results and Discussion

Overall Monomer Conversion

Experiments were conducted with non-ionized monomer and initial compositions of $f_{AA0} = 0.0$, 0.2, 0.5, 0.7, and 1.0 at initial

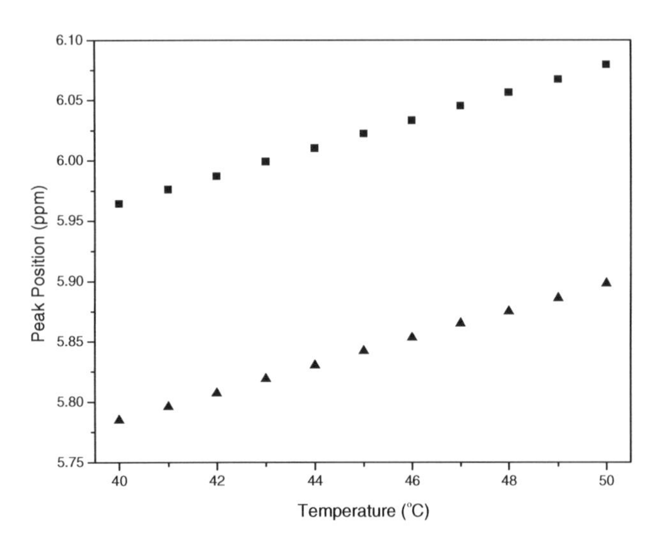

Figure 7.
AA (■) and AM (▲) monomer peak positions with 20 wt% monomer and $f_{AA} = 0.5$ as a function of temperature with no polymer or initiator present in the sample.

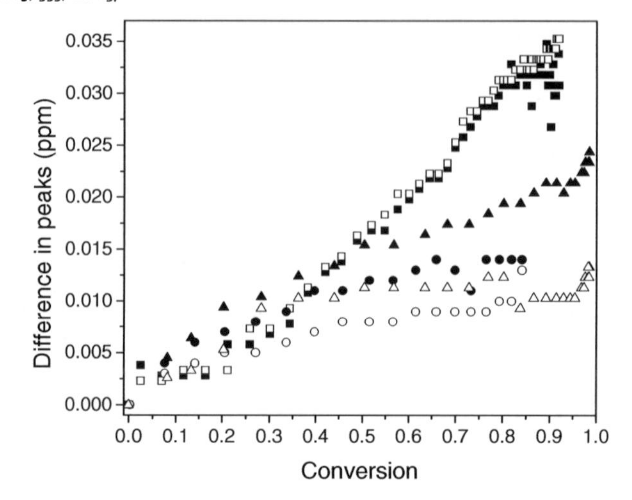

Figure 8.
Shift in the AA (filled symbols) and AM (empty symbols) monomer peak positions for experiments run at 5 (◆, ◇), 20 (■, □), and 40 (▲, △) wt% initial monomer concentration, $f_{AA0} = 0.5$, and 0.217 wt% V-50 at 40 °C. The difference in peak position is relative to the initial peak position at 0% conversion.

monomer concentrations of 5, 20, and 40 wt% in D_2O at 40 °C and 0.217 wt% V-50. Figure 9 demonstrates the influence of the initial monomer concentration on the measured monomer conversion profiles, while Figure 10 shows the effect of monomer composition. Figure 11 plots the initial rate of monomer conversion for all experiments, dx/dt, as estimated from the linear portion of the profiles between 0 and 40% conversion.

For all monomer compositions in Figure 9, including AA and AM homopolymerizations, the rate of monomer conversion increases with increasing monomer content. In most cases, the conversion profiles for 20 and 40 wt% monomer overlap, and are significantly faster than the profile measured with 5 wt% monomer. The faster rates for the 20 and 40 wt% experiments occur from the start of the polymerization, as seen by the initial rates of conversion in Figure 11. These results are in agreement with the previous AM[27,28,31] and AA[32,36] batch homopolymerization studies, and are indicative of an apparent reaction order with respect to monomer of greater than unity. The expected rate of monomer conversion (x) for FRP in a batch reactor is:

$$\frac{dx}{dt} = k_p(1-x)\sqrt{\frac{2fk_d[I]}{k_t}} \tag{4}$$

assuming the reaction is first order with respect to monomer concentration and also with respect to radical concentration. This relationship indicates that the absolute monomer concentration should have no effect on the initial rate of conversion, contrary to the observed trends in Figure 9 and 11. It is necessary to systematically examine possible explanations for the experimental deviations from expected behavior, starting with the propagation rate coefficient.

Published PLP-SEC data for AA and AM indicate a decrease in k_p with increasing monomer concentration,[4,7] as summarized in Figure 12. Assuming constant values for k_d (initiator decomposition), k_t (bimolecular radical termination), f (initiator efficiency) and $[I]$ (initiator concentration), one would therefore expect a decrease in the rate of conversion with increasing monomer concentration. This expected behavior has been observed in aqueous batch solution polymerizations of other water soluble monomers such as MAA[22] and NVP,[21] with the increased rate of conversion observed at lower

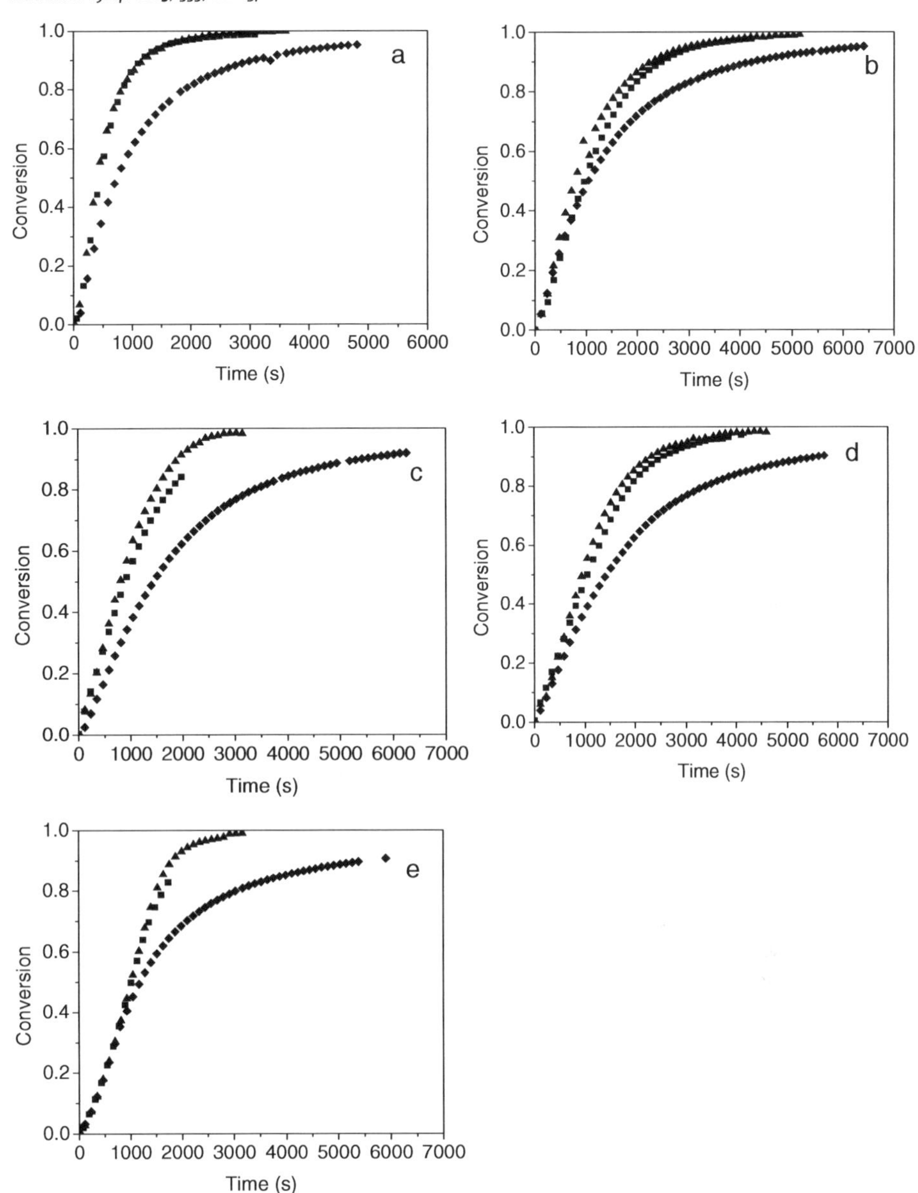

Figure 9.
Monomer conversion vs. time profiles comparing different initial monomer concentrations of 5 (◆), 20 (■), and 40 (▲) wt% for AA/AM batch copolymerization at 40 °C and 0.217 wt% V-50 for $f_{AAo} = 0.0$ (a), 0.2 (b), 0.5 (c), 0.7 (d), and 1.0 (e).

monomer concentrations well represented by models that account for the influence of concentration on k_p. However, for both AA and AM (as well as for their copolymerization), dx/dt increases with monomer concentration even though PLP-SEC studies indicate a decrease in k_p. Thus, the observed increase in the rate of conversion cannot be explained by k_p (indeed, is contrary to expected behavior), and must result either from changes in k_t or side reactions (backbiting) that

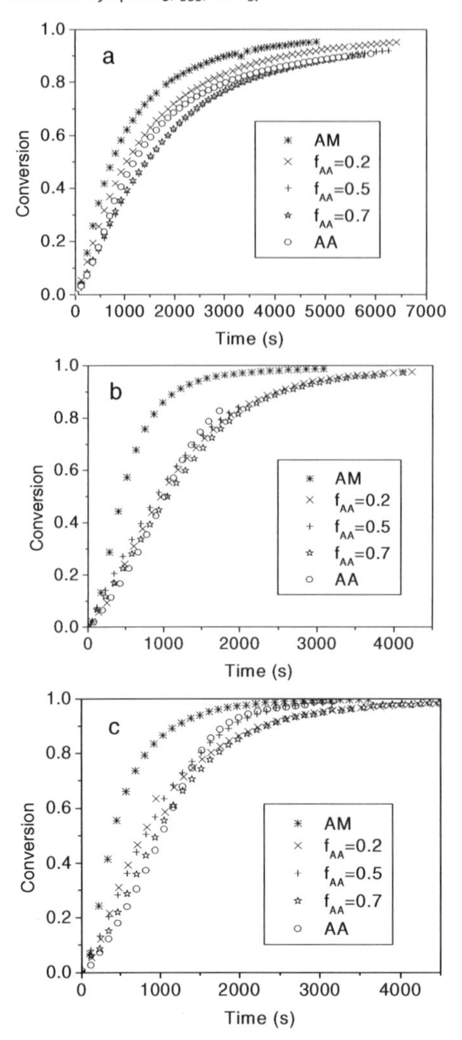

Figure 10.
Monomer conversion vs. time profiles for AA/AM batch copolymerization at 40°C and 0.217 wt% V-50 for 5 (a), 20 (b), and 40 (c) wt% monomer in D_2O with varying initial monomer compositions of $f_{AAo} = 0.0$, 0.2, 0.5, 0.7, and 1.0, as indicated in the figure legend.

influence the rate of monomer conversion significantly.

As seen in Equation 4, the rate of monomer conversion is also a function of k_t. Studies on monomers in aqueous solution such as AA,[17] MAA,[19] and NVP[18] have found that termination rate coefficients decrease with increasing monomer concentration. SP-PLP-EPR data on AA[17] show a decrease in the termination rate coefficient of two monomeric radicals at low conversion, $k_t^{1,1}$, by a factor of 6.5 as the monomer concentration increases from 10 to 50 wt%. (The termination rate for two polymer chains of length i is calculated by multiplying $k_t^{1,1}$ by a correction factor,[44] but the effect of monomer concentration on $k_t^{i,i}$ can be assumed to be the same.) By assuming that $(2fk_d[I])^{1/2}$ is constant, then the rate of conversion is proportional to $k_p/(k_t^{1,1})^{1/2}$. Using the available PLP-measured rate coefficients presented in Table 1, this ratio decreases from 2.6 to 2.4 as monomer

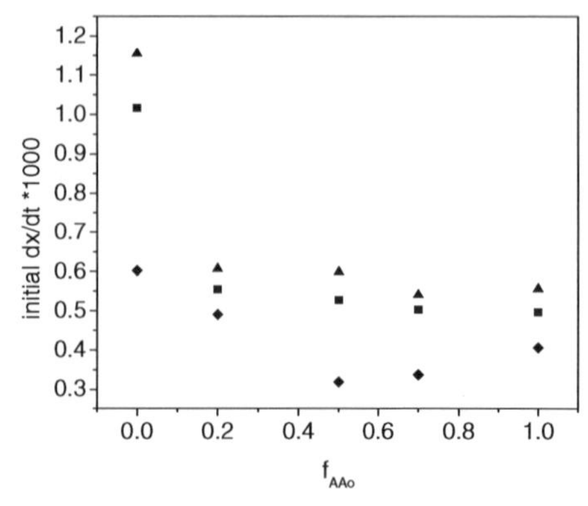

Figure 11.
Initial rate of monomer conversion, dx/dt, as a function of initial monomer composition for AA/AM batch copolymerization at 40 °C and 0.217 wt% V-50 with initial monomer concentrations of 5 (◆), 20 (■), and 40(▲) wt% in D_2O.

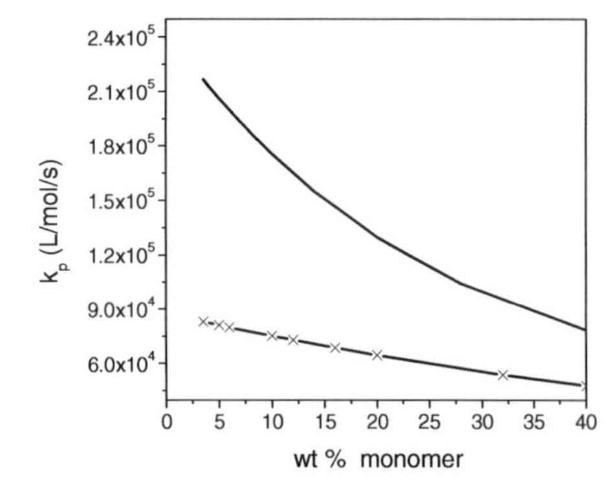

Figure 12.
Comparison of the chain-end propagation rate coefficient of AA (–)[4] and AM (-x-)[7] as a function of monomer concentration at 40 °C, as estimated by PLP-SEC.

Table 1.
Estimated individual and lumped rate coefficients for polymerization of non-ionized acrylic acid (AA) at 40 °C, with 10 and 50 wt% AA in aqueous solution.

Rate coefficient	10 wt% AA	50 wt% AA	Ref
k_p (L/mol/s)	1.76×10^5	6.46×10^4	[4]
$k_t^{1,1}$ (L/mol/s)	4.57×10^9	7.08×10^8	[17]
$k_p/(k_t^{1,1})$ (L/mol/s)$^{1/2}$	2.6	2.4	
k_p^t (L/mol/s)	87.5	24.5	[17]
k_{bb} (s^{-1})	423.6	199	[17]
k_p^{avg} (L/mol/s)	3.95×10^4	2.98×10^4	
$k_p^{avg}/(k_t^{1,1})$ (L/mol/s)$^{1/2}$	0.58	1.12	

concentration is increased from 10 to 50 wt % at 40 °C. Thus, even with a decrease in the termination rate coefficient, the rate of AA conversion is still expected to decrease slightly with increasing monomer concentration, contrary to the 35% increase observed experimentally between 5 and 40 wt% (see Figure 11).

Up to date data on the termination rate coefficient for AM are few. In a paper by Seabrook et al,[45] k_t is reported to be on the order of 10^7 L/mol/s for monomer concentrations of less than 1 wt% in aqueous solution at 50 °C. More recently, Schrooten has used the SP-PLP-NIR technique to measure a k_t on the order of 10^8 L/mol/s at 40 °C,[46] with the value decreasing as initial weight fraction of AM is increased from 20 to 50 wt%. This latter determination is more in line with the value required to fit the conversion rate profile for 5 wt% AM (Figure 9a); using the most recent PLP-SEC k_p data[7] a k_t value of 2.5×10^8 L/mol/s is required. However, in order to explain the increase in rate of conversion observed at the higher monomer levels of 20 and 40 wt %, a decrease in the k_t value by an order of magnitude would be required. This large shift may be plausible, as the k_t for NVP was found to decreases by over half an order of magnitude between 20 wt% and bulk.[18] More data and subsequent mechanistic modeling are required to develop a complete understanding of the termination rate coefficient of AM.

The above analysis indicates that it is difficult to explain the increasing rate of conversion observed with increasing monomer content observed experimentally by consideration of the recent PLP k_p and k_t data. Equation 4 is derived assuming that no side-reactions occur during polymerization. However, evidence has emerged that AA undergoes intramolecular chain transfer (backbiting), as summarized in the recent work of Barth et al.[17] As well documented for acrylates, these reactions decrease the overall observed propagation rate from the chain-end k_p value to yield an average propagation rate of:[26]

$$k_p^{avg} = k_p - \frac{k_p - k_p^t}{1 + \frac{k_p^t[M]}{k_{bb}}} \tag{5}$$

Using the rate coefficients from Barth et al.[17] summarized in Table 1, the values of $(k_p^{av}/k_t^{1,1})^{1/2}$ are 0.58 and 1.12 at 10 and 50 wt%, respectively. Thus, when the effect of backbiting on rate is accounted for, the predicted shifts in rate of conversion for AA are in agreement with the trends observed experimentally. Very recently, evidence of mid-chain radical formation, albeit at much lower rates, has been found for AM using the SP-PLP-EPR technique.[47] Backbiting in the AM homopolymerization system, along with a monomer concentration dependent termination rate coefficient, could combine to explain the increase in conversion rate with increasing monomer concentration in the AM homopolymerizations.

The analysis of the homopropagation rates of conversion for AA and AM can be extended to the copolymerization system. As shown in Figures 9 and 11, the initial rates of conversion for copolymerization are very similar to those for AA homopolymerization (but lower than for AM homopolymerization), independent of the monomer composition. It is likely that backbiting occurs with any amount of AA in the monomer mixture, leading to the same decrease in the initial rate of conversion at 5 wt% monomer relative to 20 and 40 wt% as observed in the homopolymerizations. Work is ongoing to measure the extent of short chain branch formation as a function of monomer composition and concentration using ^{13}C NMR. These data along with estimates of monomer reactivity ratios (see below) will be combined with k_p and k_t results from PLP to provide the kinetic coefficients necessary to model the set of conversion profiles obtained in this study.

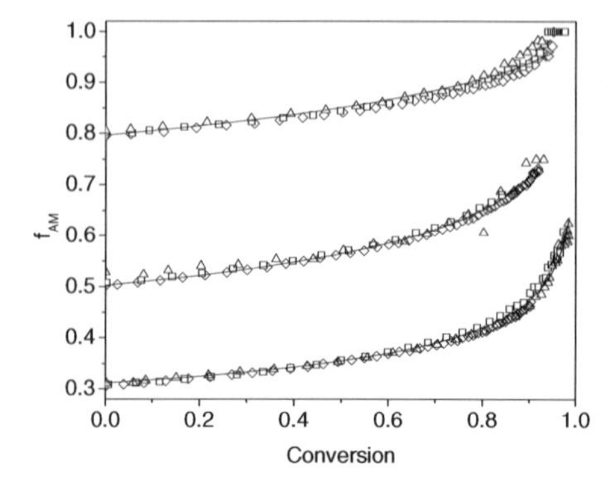

Figure 13.
Monomer composition as a function of conversion for all experiments at $f_{AA0} = 0.2$, 0.5, and 0.7 with initial monomer concentration of 5 (◇), 20 (□), and 40(△) wt% and 0.217 wt% V-50 at 40 °C, and the resulting fit (–) calculated with reactivity ratios $r_{AA} = 1.24 \pm 0.02$ and $r_{AM} = 0.55 \pm 0.01$, as estimated using the DNI method.

Copolymer Composition

In addition to the overall conversion profiles discussed above, the NMR experiments are used to track how monomer composition changes over the course of the batch experiment. Monomer composition is plotted as a function of conversion for the entire set of experiments in Figure 13. At all compositions examined ($f_{AM0} = 0.3$, 0.5 and 0.8), the monomer mixture becomes enriched in AM as the polymerization proceeds due to the preferential incorporation of AA into the copolymer. It is interesting to note that the composition drift is independent of the initial monomer concentration. The result suggests that the monomer reactivity ratios are independent of total monomer concentration. While commonly found for copolymerization in organic solution,[48] this result is not necessarily expected for AA/AM aqueous-phase copolymerization as the homopropagation rate coefficient of AA is a stronger function of monomer concentration than AM: as shown in Figure 12, the ratio of k_{pAA} to k_{pAM} increases from 1.6 at 40 wt% monomer to 2.6 at 5 wt%. With reactivity ratios defined according to the terminal model ($r_{AA} = k_{pAA}/k_{pAA\cdot AM}$ and $r_{AM} = k_{pAM}/k_{pAM\cdot AA}$), the data suggest that cross-propagation rate coefficients vary in the same fashion as the homopropagation values, such that r_{AA} and r_{AM} are invariant with monomer concentration.

The AA/AM reactivity ratios reported in the literature are determined using copolymer compositions measured in low conversion (<10%) experiments fit to the terminal model (TM),[33–35,49] where the instantaneous mole fraction of AM incorporated into the polymer (F_{AM}) is described by the Mayo-Lewis equation:[50]

$$F_{AM} = \frac{r_{AM} f_{AM}^2 + f_{AM} f_{AA}}{r_{AM} f_{AM}^2 + 2 f_{AM} f_{AA} + r_{AA} f_{AA}^2} \tag{6}$$

In order to fit the NMR data from this work using the same methodology, it is necessary to estimate F_{AM} from the consumption of monomer and the overall monomer conversion:

$$F_{AM}^{cum} = \frac{f_{AM0} - f_{AM}(1 - x)}{x} \tag{7}$$

Figure 14 plots the copolymer compositions estimated at conversions of 5 and 10%. The best-fit reactivity ratios, estimated using non-linear parameter estimation tools in Predici®, are $r_{AA} = 1.27 \pm 0.26$, and $r_{AM} = 0.54 \pm 0.21$. It is evident that significant uncertainty is introduced by the calculation of copolymer composition from

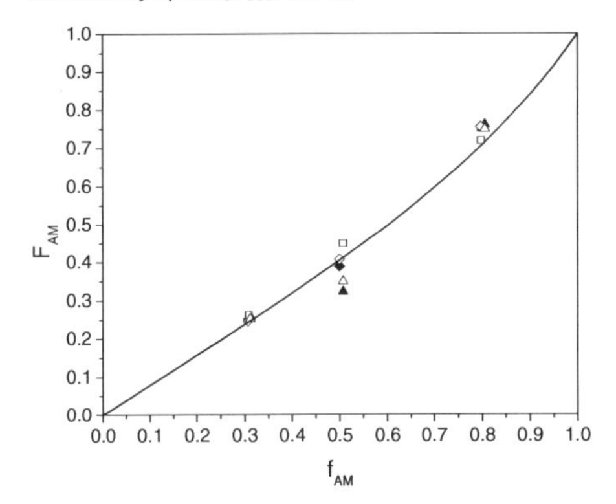

Figure 14.
Cumulative copolymer composition estimated at 5 (filled symbols) and 10 (empty symbols) % monomer conversion as a function of monomer composition with initial monomer concentration of 5 (◆,◇), 20 (■,□), and 40(▲,△) wt%, and the best fit (–) calculated from the combined dataset with $r_{AA} = 1.27 \pm 0.26$, and $r_{AM} = 0.54 \pm 0.21$.

the small change in monomer composition that occurs over the first 10% of monomer conversion.

To take advantage of the complete range of composition vs conversion data measured using the NMR technique (Figure 13), the reactivity ratios were also estimated using the integrated form of the Mayo-Lewis equation as derived by Meyer and Lowry (ML),[51] as well as the differential form of the balances coupled with direct numerical integration (DNI).[52] The DNI method relates overall monomer conversion, monomer composition, and the reactivity ratios according to:

$$\frac{df_{AM}}{dx} = \frac{f_{AM} - F_{AM}}{1 - x} \tag{8}$$

Here x is the total monomer conversion and F_{AM} is calculated using Equation 6. While slightly more computationally intensive (the differential equation is solved as part of the parameter estimation methodology), the DNI methodology is preferred as it avoids using any transformations and has a reduced error structure.[52] The reactivity ratios estimated using the DNI method are $r_{AA} = 1.24 \pm 0.02$ and $r_{AM} = 0.55 \pm 0.01$; values estimated using the ML integrated form of the equation are in excellent agreement, at 1.23 ± 0.02 and 0.58 ± 0.01 for r_{AA} and r_{AM}, respectively. Both estimates are in good agreement with those estimated using low conversion data ($r_{AA} = 1.27 \pm 0.26$, and $r_{AM} = 0.54 \pm 0.21$), but have a much lower associated error, as reflected by the reported 95% confidence intervals. The estimates provide a very good description of the drift in f_{AM} with overall conversion, as shown by the curves plotted in Figure 13. In addition, they are in reasonable agreement with the reactivity ratios of 1.48 and 0.54 for r_{AA} and r_{AM} reported by Rintoul and Wandrey[33] for experiments conducted with 4 wt% monomer. Work is ongoing to determine the reactivity ratios from polymerizations conducted at other experimental conditions.

Conclusion

The radical copolymerization of non-ionized AA with AM in aqueous solution was investigated at 40 °C using an NMR in-situ technique. The experimental methodology allows the study of monomer concentrations significantly higher than reported in the literature under near-isothermal conditions while simultaneously tracking changes in comonomer composition and overall monomer conversion. The composition data obtained over the entire conversion range provide a more precise estimate of monomer reactivity ratios than values estimated from low conversion data. The complete set of experiments, with initial monomer content in aqueous solution varied between 5 and 40%, were well-fit over the complete conversion range by $r_{AA} = 1.24 \pm 0.02$, and $r_{AM} = 0.55 \pm 0.01$, in good agreement with values reported in the literature for more dilute systems.

It was found that the rate of monomer conversion increases with increasing monomer concentration, a trend contrary to the known decrease in the AA and AM chain-end propagation rate coefficients with increasing monomer concentration. It is hypothesized that backbiting reactions, which exert a larger effect on polymerization rate as monomer concentration is lowered, significantly slow the conversion rate for both the AA and AA/AM polymerizations. Work is underway to measure the polymer molecular weights (using size exclusion chromatography) and branching levels (using ^{13}C NMR) of the samples obtained in the study, as well as applying the NMR technique to study copolymerization over a broader range of experimental conditions.

Acknowledgements: The authors thank Danielle Austin for experimental assistance, Prof. Michael Buback and Prof. Igor Lacík and their research groups for technical discussions and sharing of data throughout this ongoing collaboration, and the financial support from and technical interactions with researchers from BASF SE, Ludwigshafen.

[1] R. G. Gilbert, *Pure Appl. Chem.* **1992**, *64*, 1563.
[2] S. Beuermann, M. Buback, *Prog. Polym. Sci.* **2002**, *27*, 191.
[3] O. F. Olaj, I. Bitai, F. Hinkelmann, *Makromol. Chem.* **1987**, *188*, 1689.
[4] I. Lacík, S. Beuermann, M. Buback, *Macromolecules* **2003**, *36*, 9355.
[5] I. Lacík, S. Beuermann, M. Buback, *Macromol. Chem. Phys.* **2004**, *205*, 1080.
[6] I. Lacík, S. Beuermann, M. Buback, *Macromolecules* **2001**, *34*, 6224.
[7] L. Učňová, A. Chovancová, M. Stach, I. Lacík, manuscript in preparation.
[8] P. Pascal, M. A. Winnik, D. H. Napper, R. G. Gilbert, *Macromolecules* **1993**, *26*, 4572.
[9] S. A. Seabrook, M. P. Tonge, R. G. Gilbert, *J. Polym. Sci., Part A: Polym. Chem.* **2005**, *43*, 1357.
[10] S. Beuermann, M. Buback, P. Hesse, I. Lacík, *Macromolecules* **2006**, *39*, 184.
[11] S. Beuermann, M. Buback, P. Hesse, S. Kukučková, I. Lacík, *Macromol. Symp.* **2007**, *248*, 41.
[12] I. Lacík, L. Učňová, S. Kukučková, M. Buback, P. Hesse, S. Beuermann, *Macromolecules* **2009**, *42*, 7753.
[13] M. Stach, I. Lacík, D. Chorvát, M. Buback, P. Hesse, R. A. Hutchinson, L. Tang, *Macromolecules* **2008**, *41*, 5174.
[14] V. F. Gromov, N. I. Galperina, T. O. Osmanov, P. M. Khomikovskii, A. D. Abkin, *Eur. Polym. J.* **1980**, *16*, 529.
[15] S. C. Thickett, R. G. Gilbert, *Polymer* **2004**, *45*, 6993.
[16] B. De Sterck, R. Vaneerdeweg, F. Du Prez, M. Waroquier, V. Van Speybroeck, *Macromolecules* **2010**, *43*, 827.
[17] J. Barth, W. Meiser, M. Buback, *Macromolecules* **2012**, *45*, 1339.
[18] J. Schrooten, M. Buback, P. Hesse, R. A. Hutchinson, I. Lacík, *Macromol. Chem. Phys.* **2011**, *212*, 1400.
[19] S. Beuermann, M. Buback, P. Hesse, R. A. Hutchinson, S. Kukučková, I. Lacík, *Macromolecules* **2008**, *41*, 3513.
[20] J. Barth, M. Buback, *Macromolecules* **2011**, *44*, 1292.
[21] S. Santanakrishnan, L. Tang, R. A. Hutchinson, M. Stach, I. Lacík, J. Schrooten, P. Hesse, M. Buback, *Macromol. React. Eng.* **2010**, *4*, 499.
[22] M. Buback, P. Hesse, R. A. Hutchinson, P. Kasák, I. Lacík, M. Stach, I. Utz, *Ind. Eng. Chem. Res.* **2008**, *47*, 8197.
[23] N. F. G. Wittenberg, M. Buback, R. A. Hutchinson, *Macromol. React. Eng.* **2013**, DOI: 10.1002/mren.201200089
[24] J. Barth, M. Buback, P. Hesse, T. Sergeeva, *Macromol. Rapid Commun.* **2009**, *30*, 1969.
[25] J. Barth, M. Buback, *Macromolecules* **2012**, *45*, 4152.
[26] A. N. Nikitin, R. A. Hutchinson, M. Buback, P. Hesse, *Macromolecules* **2007**, *40*, 8631.
[27] T. Ishige, A. E. Hamielec, *J. Appl. Polym. Sci.* **1973**, *17*, 1479.
[28] A. R. Mahdavian, M. Abdollahi, H. R. Bijanzadeh, *J. Appl. Polym. Sci.* **2004**, *93*, 2007.
[29] G. T. Russell, *Macromol. Theory Simul.* **1995**, *4*, 549.
[30] A. Giz, H. Çatalgil-Giz, A. Alb, J.-L. Brousseau, W. F. Reed, *Macromolecules* **2001**, *34*, 1180.
[31] C. J. Kim, A. E. Hamielec, *Polymer* **1984**, *25*, 845.
[32] S. S. Cutié, P. B. Smith, D. E. Henton, T. L. Staples, C. Powell, *J. Polym. Sci., Part B: Polym. Phys.* **1997**, *35*, 2029.
[33] I. Rintoul, C. Wandrey, *Polymer* **2005**, *46*, 4525.
[34] W. R. Cabaness, T. Y. Lin, C. Párkányi, *J. Polym. Sci., Part A: Polym. Chem.* **1971**, *9*, 2155.
[35] A. Paril, A. M. Alb, A. T. Giz, H. Çatalgil-Giz, *J. Appl. Polym. Sci.* **2007**, *103*, 968.
[36] S. S. Cutié, D. E. Henton, C. Powell, R. E. Reim, P. B. Smith, T. L. Staples, *J. Appl. Polym. Sci.* **1997**, *64*, 577.
[37] A. R. Mahdavian, M. Abdollahi, *J. Appl. Polym. Sci.* **2007**, *103*, 3253.
[38] A. R. Mahdavian, M. Abdollahi, L. Mokhtabad, H. Reza Bijanzadeh, F. Ziaee, *J. Appl. Polym. Sci.* **2006**, *101*, 2062.
[39] A. R. Mahdavian, M. Abdollahi, L. Mokhtabad, F. Ziaee, *J. Macromol. Sci., Pure Appl. Chem.* **2006**, *43*, 1583.
[40] M. Abdollahi, B. Massoumi, M. R. Yousefi, F. Ziaee, *J. Appl. Polym. Sci.* **2012**, *123*, 543.

[41] A. Hartwig, R. H. Brand, C. Pfeifer, N. Dürr, A. Drochner, H. Vogel, *Macromol. Symp.* **2011**, *302*, 280.
[42] A. L. Van Geet, *Anal. Chem.* **1968**, *40*, 2227.
[43] D. L. Pavia, G. M. Lampman, G. S. Kriz, J. A. Vyvyan, *Introduction to Spectroscopy*, Brooks/Cole, Belmont CA (USA) **2008**, 656.
[44] J. Barth, M. Buback, G. T. Russell, S. Smolne, *Macromol. Chem. Phys.* **2011**, *212*, 1366.
[45] S. A. Seabrook, P. Pascal, M. P. Tonge, R. G. Gilbert, *Polymer* **2005**, *46*, 9562.
[46] J. Schrooten, PhD Thesis, U. of Göttingen, **2013**.
[47] H. Kattner, N. F. G. Wittenberg, M. Buback, personal communication, **2013**.
[48] K. Liang, R. A. Hutchinson, *Macromolecules* **2010**, *43*, 6311.
[49] A. Chapiro, *Eur. Polym. J.* **1973**, *9*, 417.
[50] F. R. Mayo, F. M. Lewis, *J. Am. Chem. Soc.* **1944**, *66*, 1594.
[51] V. E. Meyer, G. G. Lowry, *J. Polym. Sci. Part A* **1965**, *3*, 2843.
[52] N. Kazemi, T. A. Duever, A. Penlidis, *Macromol. React. Eng.* **2011**, *5*, 385.

 DOI: 10.1002/masy.201300043

On-Line Monitoring of Molecular Weight Using NIR Spectroscopy in Reactive Extrusion Process

Björn Bergmann, Wolfgang Becker, Jan Diemert, Peter Elsner*

Summary: Near infrared spectroscopy (NIR) is a versatile tool for process control, scientific research and quality management. Consequently it is our aim to utilize NIR technique for the understanding and following of reactive extrusion processes. As preliminary results it was shown the molecular weight of degraded, non-purified polyester can be measured and forecasted using purely NIR spectra analysis. In this paper we describe the close correlation of the received molecular weight values of partly glycolised Polyethylene terephthalate (PET) samples using reactive extrusion, measured using size exclusion chromatography (SEC) techniques and chemometric models basing on NIR measurements.

Keywords: molecular weight distribution; near-infrared spectroscopy (NIR); polyethylene terephthalate (PET); polyesters; reactive extrusion

Introduction

Controlling the molecular weight of a polymer is a crucial material property not only for the resulting material behaviour in use, but also to validate the stress which is brought on the material due to a non-optimized processing. [1,2]

Especially in reactive extrusion processes, a precise measurement of the resulting molecular weight reveals the effectiveness of the process. Polymerisations performed in an extruder or the precise reduction in molecular weight due to viscosity reasons in recycling processes are just two examples where the molecular weight of a polymer is modified in an extruder utilizing reactive extrusion. [3,4,5]

As state of art the molecular weight of a polymer is measured of utilizing different techniques, which all usually require some time for sample preparation, purification and measurement. Although first papers describing the correlation between molecular weight and NIR spectra have been published more than a decade ago,[6] this time saving technique has not been established in polymer engineering industry or reactive extrusion processes yet. Nevertheless the successful correlation between the molecular weight of several different, natural and artificial, polymeric systems and their molecular weight has been published. [6,7,8,9,10]

Today we present the measurement of the molecular weight of a non-purified polyester system using near infrared measurements. This leads way to the inline measurement of the molecular weight at an extruder dye, as it is shown near-infrared spectroscopy (NIR) applied here is a versatile tool for process control. [11]

Materials and Methods

Both, Polyethylene terephthalate (PET) grade T49H (received from INVISTA), and Ethylene glycol (received from Carl Roth) were used as received. The glycolysis was performed in a fully intermeshing

[1] Fraunhofer Institute for Chemical Technology - ICT, Joseph-von-Fraunhoferstraße 7, 76327 Pfinztal, Germany
E-mail: Bjoern.Bergmann@ict.fraunhofer.de

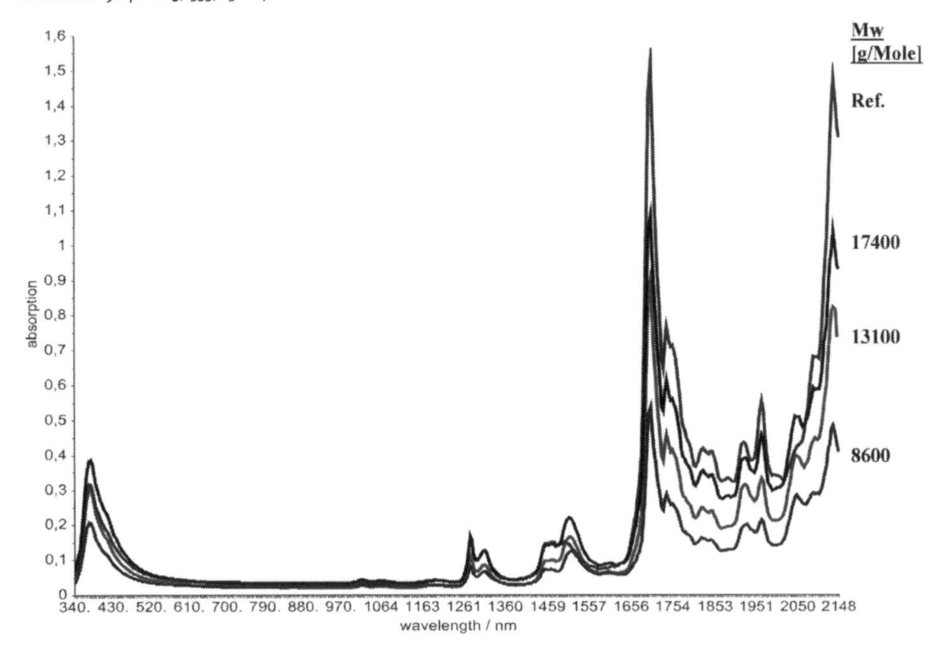

Figure 1.
UV/Vis and NIR spectra of three PET samples with varying molecular mass (indicated at the right). The reflection spectra were converted into absorption units via a Kubelka-Munk transformation.

"Coperion" twin screw compounder with a screw diameter of 32 mm and an L/D ratio of 52. The PET:Glycol ratio was varied between 1:0.7 and 1:0.1, the screw design provided several reaction zones at which the temperature was held at 320°C. By the means of throughput and rotating speed the residence time was varied in the range of 1–5 min.

Offline molecular mass measurements were performed in Hexafluoroisopropanol, using a commercial setup of "PSS" (LC 1100 Agilent), equipped with a refractive index detector. NIR spectra were recorded using a spectrometer system from Carl Zeiss MCS611 and MCS 612. The two spectrometers are interconnected and allow cascading measurements. The spectral measurements were performed in the ultraviolet / visible (UV/VIS) to the near infrared (NIR) spectral range (340 nm – 2100 nm wavelength). The spectral resolution $\Delta\lambda$ according to Raleigh criteria in the UV/VIS spectral range varies between 12 nm to 14 nm and in the NIR spectral range about 18 nm. Each sample was measured three times whereby each measurement consisted of 5 accumulations to reduce noise. The samples were measured in diffuse reflection with the reflection probe OMK 500 of Carl Zeiss. The irradiance source is a halogen lamp in the centre of the probe. The samples were placed on top of the probe in a petri dish in order to have reproducible measurement conditions. The spectra are shown in Figure 1.

Regression Model

A partial least regression model (PLS) was generated to correlate the UV/VIS and NIR spectra with the molecular weight of the samples. The reflectance spectra, which were converted into Kubelka-Munk units, were used for a one component model i.e. PLS1. There were no further data pretreatments of the spectra. The result of the PLS1 regression is shown in Figure 2.

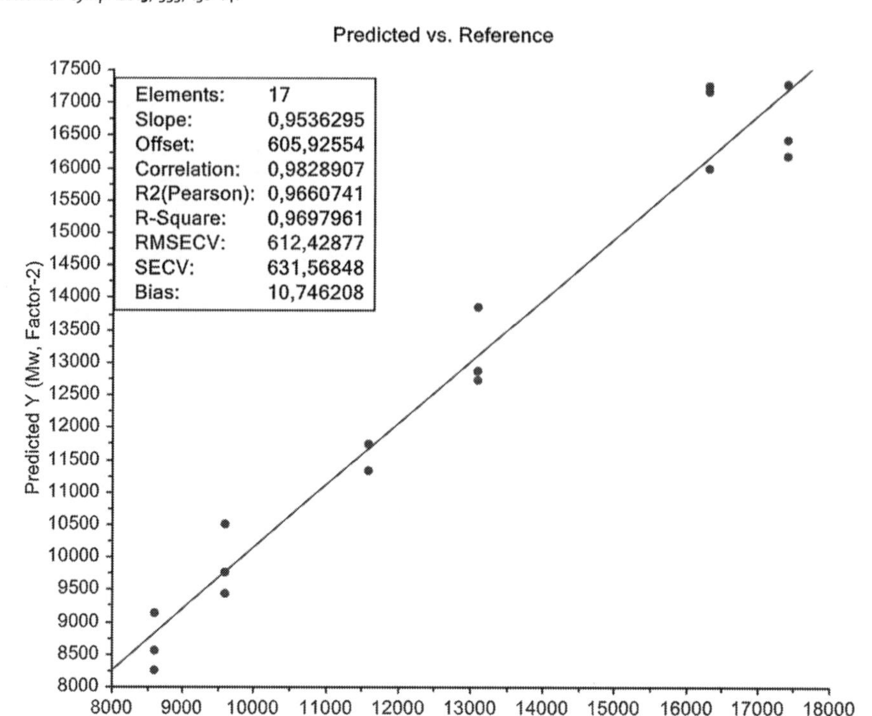

Figure 2.
Result of the PLS1 regression of the molecular weight with UV/VIS NIR spectra. The correlation coefficient is 0.983 and the RMSECV is 612.43 g/Mole.

The PLS1 model for the molecular weight prediction consists of two principal factors which indicate that, beside the varying molecular weight, another chemical or material factor causes spectral changes and hence is present in the polymer material. This factor originates from the varying production procedures and process conditions such as changed sheer energy induces using different screw configurations.

Table 1.
Molecular weight (Mw) of the PET samples together with some process parameters.

Sample Nr.	Glycol weight%	Residence time / min	Molecular weight g/Mole
1	10	2:40	8600
2	10	3:40	13100
3	20	1:50	17400

Conclusion

The correlation between spectral data and the molecular weight could be successfully shown. It could be shown a chemometric model can be applied to spectral data of Polyester, describing the molécular weight of the sample. The accuracy of the molecular weights as predicted using chemometric models is in good alliance with the molecular weights as measured using SEC techniques.

[1] T. Bartilla, D. Kirch, J. Nordmeier, E. Prömper, T. Strauch, *Adv. Polym. Technol.* **1986**, *6*, 339.
[2] L. J. Fetters, D. J. Lohse, D. Richter, T. A. Witten, A. Zirkel, *Macromolecules* **1994**, *27*, 4639.
[3] I. Alig, B. Steinhoff, D. Lellinger, *Measurement Science and Technology* **2010**, *21*, 062001.
[4] J. A. Biesenberger, C. G. Gogos, *Polym. Eng. Sci.* **1980**, *20*, 838.
[5] D. Carta, G. Cao, C. D'Angeli, *ESPR - Environ Sci. & Pollut. Res.* **2003**, *10*, 390.

[6] A. Cherfi, G. Févotte, *Macromol. Chem. Phys.* **2002**, *203*, 1188.
[7] A. Cherfi, G. Fevotte, C. Novat, *J. Appl. Polym. Sci.* **2002**, *85*, 2510.
[8] N. S. Othman, G. Févotte, D. Peycelon, J.-B. Egraz, J.-M. Suau, *AIChE J.* **2004**, *50*, 654.
[9] Q. Dong, H. Zang, A. Liu, G. Yang, C. Sun, L. Sui, P. Wang, L. Li, *J. Pharm. Biomed. Anal.* **2010**, *53*, 274.
[10] H. L. Pereira, F. Machado, E. L. Lima, J. C. Pinto, *Macromol. Symp.* **2011**, *299–300*, 1.
[11] T. Rohe, W. Becker, S. Kölle, N. Eisenreich, P. Eyerer, *Talanta* **1999**, *50*, 283.

 DOI: 10.1002/masy.201300088

Experimental Investigation of the Morphology Formation of Polymer Particles in an Acoustic Levitator

Robert Sedelmayer, Matthias Griesing, Annelie Heide Halfar, Werner Pauer, Hans-Ulrich Moritz*

Summary: Droplet polymerization of N-vinyl-2-pyrrolidone (NVP) and sodium acrylate (NaAA) was carried out in an acoustic levitator for different ambient temperatures and relative humidities. The resulting particle morphologies were compared to particles of two crystalline systems (mannitol and ammonium sulfate) and two disperse systems (silica and styrene - butyl acrylate dispersion), all obtained under similar drying conditions. NVP was found to form a higher amount of crystals with lower temperature and relative humidity, contrary to the behaviour observed for mannitol and ammonium sulfate. The processes of both, polymerization and drying of NaAA lead to similar morphologies at low temperatures and humidities, probably caused by the precipitation of NaAA during polymerization.

Keywords: acoustic levitation; drying; morphology; polymerization; spray

Introduction

Spray processes provide advantageous heat and mass transport properties considering the drying and polymerization of particles. Moreover, mild drying conditions are made possible, because the maximum temperature of a droplet is limited by the wet-bulb temperature as long as solvent is present. Spray processes like spray drying and spray polymerization are commonly used in the pharmaceutical, chemical or food industry.[1]

Today's research in spraying processes focuses largely on the development of powders consisting of particles with specifically designed properties. Powder characteristics like bulk density, redispersibility or flowability depend primarily on the size and morphology of particles and only to a minor extent on their chemical composition.[2,3]

An overview of possible particle morphologies is given by Charlesworth and Marshall.[4] This model describes the influence of the shell formation on the morphology of particles during the first period of the drying process. The morphology of highly porous particles does not depend on the drying temperature because the evaporation of liquid from the droplet is not constrained by the formation of the shell. Particles with a less porous shell develop cracks or "furry" outgrowths on the surface when the temperature during the evaporation process is below the boiling point. However, at temperatures higher than the boiling point they will break open. For shells that are impermeable for liquids, the resulting particle morphology is wrinkly and shrunken when temperatures during the evaporation process are below the boiling point or expanded and collapsed for conditions above the boiling temperature.

With regard to the eminent interest in morphology and morphology formation of particulate polymer products, this work presents a detailed analysis of two reactive systems. Polymerization inside a droplet was investigated for aqueous solutions of

Institute of Chemical Engineering and Macromolecular Chemistry, University of Hamburg, Bundesstr. 45, 21046 Hamburg, Germany
Fax: (+49) 40 42838 6008;
E-mail: robert.sedelmayer@chemie.uni-hamburg.de

N-vinyl-2-pyrrolidone (NVP) and sodium acrylate (NaAA) and compared to nonreactive crystalline and disperse systems. All of these systems are either actively produced by spray polymerization or spray drying or can serve as model systems for spray processes. Since processes inside of a spray tower are difficult to monitor, acoustic levitation is applied in this study as a method for observing single droplets for different temperatures and humidities with a focus on their morphology formation. Information obtained within this model system can be generalized to a realistic spray tower system.

Acoustic levitation allows the contactless positioning of droplets and particles in the nodes of a standing acoustic wave (see Figure 1). In order to generate such a standing acoustic wave, the distance between the sound source and the reflecting surface must equal a whole-numbered multiple of the half wavelength λ. In this standing acoustic wave axial forces compensate for the gravitational forces, while radial forces keep the droplet centered. This way, it is possible to observe a single droplet and all processes taking place in it.[5]

On the other hand it is known from literature that the acoustic field has non-negligible side-effects on the shape and the heat and mass-transfer processes of the droplet. The acoustic field leads to an inhomogeneous distribution of forces onto the surface of the droplet, causing its deformation to an oblate shape.[5] Furthermore the acoustic field leads to acoustic streaming around the droplet[6] which in turn has an impact on the effective rate of heat and mass transfer between the droplet and the gaseous phase. Additionally the acoustic streaming induces convective vortex flows inside of the droplet as reported by Yan *et al.*[7]. This leads to an increased mixing rate which should have an impact on the shell formation and its morphology and is typically not present in the droplets of a spray.

Being aware of these disadvantages, acoustic levitation is still the most feqasible metod for contactless positioning of small samples. Therefore it finds application in several research fields like analytical[8] and bioanalytical chemistry,[9–10] material sciences,[5,11–12] in pharmaceutical applications[13] or environmental sciences.[14–15] Priego-Capote gives an overview of recent research results in the field of acoustic levitation.[16]

Previous results address the formation of particle morphology of different substances under defined conditions in single droplets (*e.g.* aqueous solutions of salts,[2,17] silica suspensions,[2,18–19] aqueous sodium acrylate solutions[20] or *N*-vinyl-2-pyrrolidone.[21]

Experimental Part

The acoustic levitator used in this study consists of an ultrasonic sound source with a

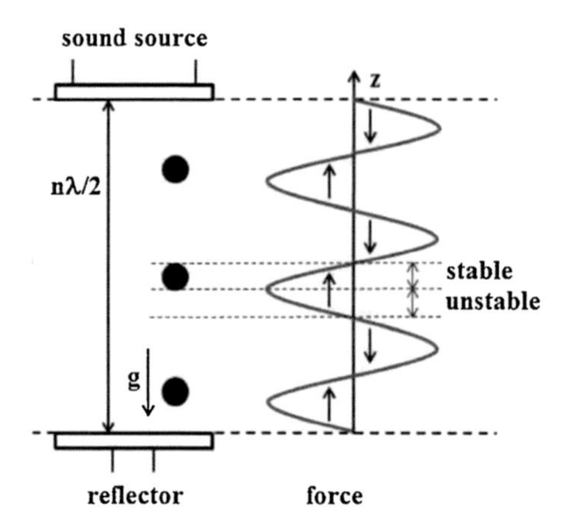

Figure 1.
Principle of acoustic levitation.

working frequency of 42 kHz placed opposite of a concave reflector. The ultrasonic source and the reflector are located inside of an insulated process chamber to control the desired reaction conditions. The temperature and the relative humidity can be adjusted by a continuous gas flow (here 0.02 m/s), which is run through a sinter plate at the bottom of the chamber. A mass flow controller (MFC 8711, Bürkert GmbH & Co. KG) regulates the continuous gas flow. The temperature inside of the chamber is continuously measured by thermo couple (type K, ES Electronic Sensors GmbH) and a Pt100 (ES Electronic Sensors GmbH). The relative humidity is monitored by using a humidity probe (HMT 337, Vaisala). A camera (PL-A741, PixeLINK) is used for the online monitoring of the droplet size. The particle surface is calculated online from the length of the major and minor axis. Scanning Electron Microscopy (SEM) (LEO Gemini 1525, Carl Zeiss AG) was used for analyzing the resulting particle morphology.

NVP was purchased from Acros Organics and VA-044 was purchased from Wako Pure Chemical Industries, Ltd. Mannitol PEARLITOL® 200 SD was purchased from Roquette Corporate and used as received. Ammonium sulfate was purchased from Merck KGaA and Evonik Industries AG donated the silica dispersion. The silica dispersion was diluted with 30 wt% distilled water. Styrene - butyl acrylate (SBA) dispersions (primary particle diameter 50 nm, 15 wt%) and sodium acrylate were synthetized in house and used with no further treatment.

Results and Discussion

Polymerization of N-Vinyl-2-pyrrolidone (NVP)

The polymerization of NVP was carried out in aqueous solution with an initial NVP content of 30 wt% and either 1 or 4 mol% of VA-044 as initiator. The ambient conditions were set to a temperature of either 70 or 95 °C and a relative humidity of 1 or 25%.

Due to the fact that solvent (H_2O) and monomer (NVP) evaporate during the process of polymerization, most of the drying curves exhibit three drying stages marked in Figure 2 by regions I-III. The first (I) drying stage is caused by the evaporation of only the solvent. First signs of shell formation are observed at the beginning of the second drying stage (II). At this point the droplet is roughly composed of equal amounts of water and NVP. The evaporation rate is reduced by 50%, thus indicating the formation of a porous or open shell. It is probably composed of large chunks of the polymer which are also visible in the SEM images (Figure 2b). In the third drying stage (III) no further reduction of the particle size can be observed because the shell formation is already completed. Therefore, the evaluation of the shadowgraphy images

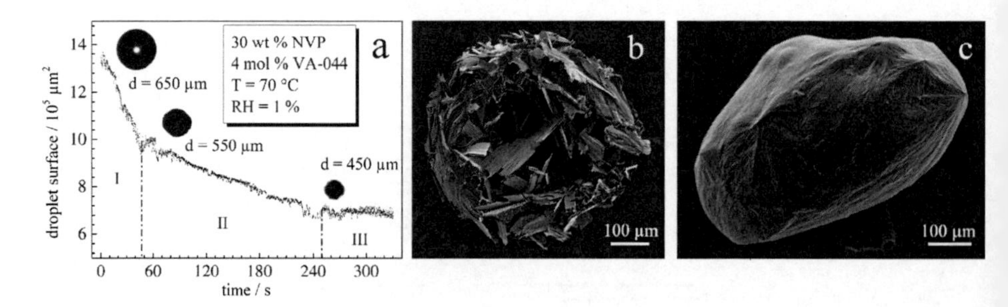

Figure 2.
Drying curve of an NVP/H_2O droplet presenting three drying stages during conversion into a PVP particle (a), SEM image of an PVP particle obtained at T = 70 °C and RH = 1% (b), SEM image of an PVP particle obtained at T = 95 °C and RH = 25% (c).

leads to a near-to-zero evaporation coefficient even though further drying of the particle is very likely. By contrast, the droplet polymerization at T = 95 °C and RH = 25% demonstrated no second drying stage. At these conditions the higher wet-bulb temperature ($T_{wb} \approx 64$ °C compared to $T_{wb} \approx 45$ °C for T = 70 °C and RH = 25%) leads to a higher initiator decomposition rate and thus to a faster polymerization process. Hence, the shell formation proceeds faster and is finished before all of the solvent has evaporated.

Figure 2 presents two SEM images of particles obtained at different temperatures and humidities. Particles polymerized at low temperatures and low humidities exhibit a structured morphology, consisting mainly of needle-shaped or flaky crystals (Figure 2 b). Contrary to that, high temperatures and high humidities led to smooth and amorphous particle surfaces (see Figure 2 c).

Figure 3 presents three contour graphs obtained from the evaluation of experimental results with respect to the *reduced evaporation coefficients* of the first (β_I) and second (β_2) drying stages and the *crystallinity index* of the particles. The *reduced evaporation coefficient* represents the evaporation coefficient of a drying stage (either β_I or β_{II}) divided by the evaporation coefficient of pure water at T = 95 °C and RH = 25%. The *crystallinity index* is based on the evaluation of SEM images and represents the fraction of the particle's surface covered by crystalline structures.

By using the design of experiments (DOE) approach for planning the experiments, it was possible to evaluate the results with statistical DOE analysis. This evaluation indicates that temperature and relative humidity have a significant influence on both drying coefficients and the morphology of such particles. As expected, higher temperatures and lower humidities lead to higher evaporation coefficients (Figure 3 a, b). Considering the influence of temperature and humidity on the morphology of the particles (compare Figure 3 c) an increase leads to an amorphous particle surface and lower grade of *crystallinity*. One explanation for this phenomenon could be a higher dissolution of primarily crystalized polymers on the surface due to the higher relative humidity. But as crystals can also be found at low temperature / high humidity conditions, this does not seem to be valid. It is more likely that the increase of the droplet temperature at high temperature / high humidity conditions leads to a higher polymerization rate, which in turn leads to an early shell formation and a suppression of crystal growth.

Polymerization of Sodiumacrylate (NaAA)

The polymerization of NaAA was carried out in aqueous solution with an initial NaAA content of either 16.25 or 8.75 wt% and either 0.85 or 0.55 mol% of VA-044 as

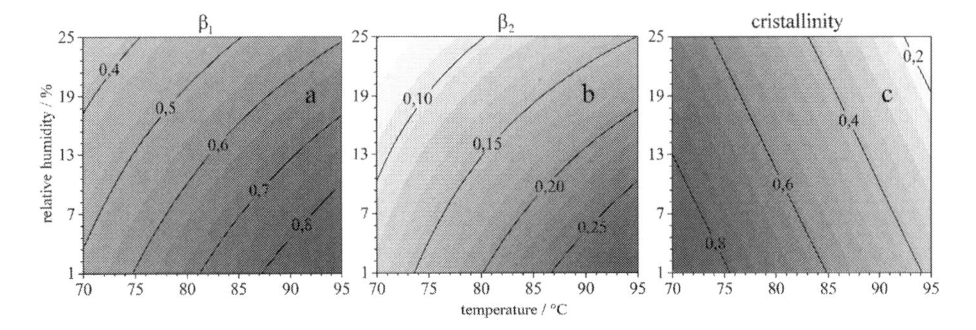

Figure 3.
Influence of relative humidity and temperature on the *reduced evaporation coefficient* of the first drying stage β_I (a), the *reduced evaporation coefficient* of the second drying stage β_2 (b) and the *crystallinity index* (c).

initiator. The ambient conditions were set to a temperature of either 69 or 86 °C and a relative humidity of 8 or 23%. In addition, aqueous solutions of NaAA were dried in the acoustic levitator.

Since NaAA is a nonvolatile component all drying curves exhibit only two drying stages (Figure 4 a, I and II) in comparison to the three drying stages of the drying curves of NVP (see Figure 2 a). The first drying stage (I) represents the direct evaporation of water from the solution in the droplet. The evaporation rate in the second drying stage (II) is near to zero because shell formation is completed. The particles obtained in this survey exhibited a great variety of structural elements on the surfaces. Among them were small and big creases, grooves, flakey structures, pores and many different types of indentations. Most of these structural elements can be observed without any correlation to a change of process parameters. However, particles polymerized at lower temperatures exhibit small creases and flakey structures while particles polymerized at higher temperatures present bigger grooves but smoother surfaces.

In Figure 5 SEM images of a dried NaAA (a) and two polymerized (b, c) PAA-Na particles are presented. The comparison of these pictures indicates that the surface structures of the dried NaAA particle are similar to the surface structures of the PAA-Na particle polymerized at lower temperatures. It seems reasonable that at mild drying respectively polymerization conditions precipitated NaAA governs the morphology of the particles due to its solubility limit of 35 wt% in aqueous

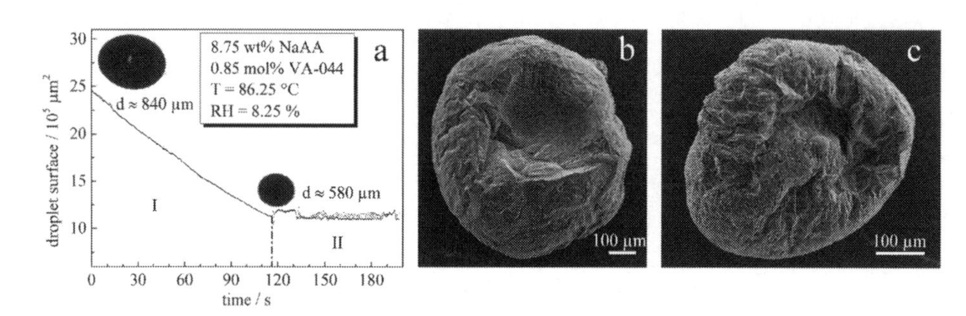

Figure 4.
Drying curve of an NaAA/H2O droplet presenting two drying stages during conversion into a PAA-Na particle (a), SEM image of an PAA-Na particle obtained at T = 69 °C and RH = 23% (b), SEM image of a PAA-Na particle obtained at T = 86 °C and RH = 23% (c).

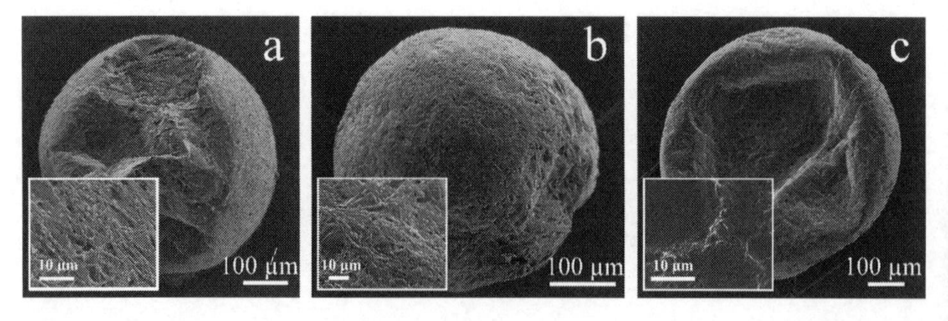

Figure 5.
SEM images of a NaAA particle dried at T = 77 °C and RH = 12% (a), a PAA-Na particle polymerized at T = 69 °C and RH = 8% (b) and a PAA-Na particle polymerized at T = 86 °C / RH = 8% (c).

solution. Precipitation should appear at first near to the droplet surface due to the evaporation of solvent. This leads to an increase of the local concentration of NaAA. Higher temperatures support higher conversion rates of NaAA to PAA-Na and therefore might reduce precipitation. This could be a reason for smoother surfaces of particles obtained under these conditions.

Morphology Formation of Aqueous Mannitol and Ammonium Sulfate Solutions

Aqueous solutions of mannitol with an initial mannitol content of either 5 or 15 wt% were dried at temperatures of 60, 90 or 120 °C and a relative humidity of either 1 or 30%. In addition, aqueous solutions of ammonium sulfate with an initial ammonium sulfate content of either 10 or 20 wt% were dried at temperatures of 60, 80 or 120 °C and a relative humidity of 1%.

SEM images in Figure 6 demonstrate the change of morphology of dried mannitol particles depending on the drying temperature. The amount of crystalline structures on the surface of the particle increases with higher drying temperatures. This behaviour is likely originating from more nucleation centers and a faster crystallization process at higher temperatures. All particles exhibit either hollow or collapsed surfaces. These results confirm those of Maas *et al.* [22] who investigated the surface structure of spray-dried mannitol particles. Supplemental experiments with RH = 30% did not reveal an influence of the relative humidity on the morphology formation.

Figure 7 a presents the influence of temperature, relative humidity and initial mannitol content on the evaporation coefficient of aqueous mannitol solutions. As expected, higher temperatures and relative humidities decrease the evaporation coefficient significantly. A higher initial mannitol

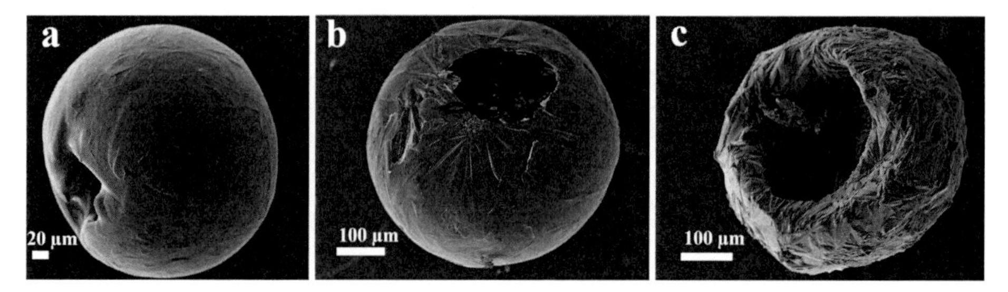

Figure 6.
SEM images of mannitol dried at RH = 1% and T = 60 °C (a), T = 100 °C (b) and T = 120 °C (c).

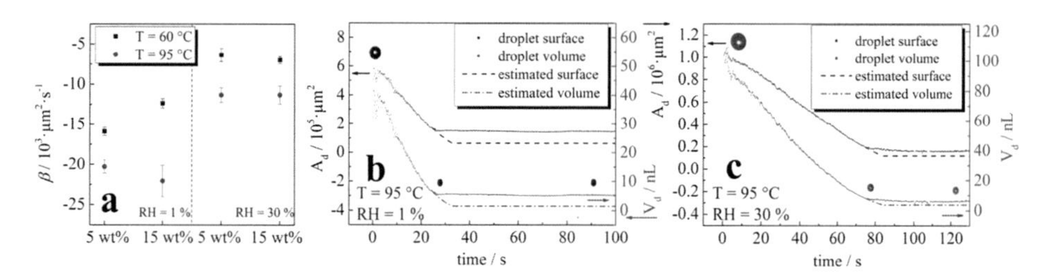

Figure 7.
Comparison of the mean average evaporation coefficient β for T = 60 °C and T = 95 °C at different RH and solid contents (a). Change of droplet surface A_d and volume V_d for 5 wt% mannitol during evaporation at T = 95 °C and RH = 1% (b) or RH = 30% (c). Corresponding shadowgraphy images are given for the starting point, for the change from first to the second drying stage and for the end of the drying process.

content also influences the evaporation coefficient, but only at mild drying conditions (T = 60 C, RH = 1%). This effect cannot be observed for higher temperatures or relative humidities.

Figures 7 b and c present drying curves for mannitol solutions with a content of 5 wt%, both measured and calculated. The drying was monitored at a temperature of 95 °C and relative humidities of 1 and 30%. A comparison of the measured and the calculated volumes in the second drying stage reveals another aspect of the drying behaviour. Lower humidities lead to bigger but hollow particles. With lower relative humidity, shell formation is completed earlier and excess solvent is then evaporated through holes or pores in the shell, leaving a hollow center in the particle. Particles dried at RH = 1% exhibited a four times larger volume than the calculated solid volume. On the contrary, particles dried at RH = 30% exhibited a nearly solid body.

As a comparison, single droplets of aqueous ammonium sulfate solutions were examined. Figure 8 presents SEM images of particles of ammonium sulfate dried at different ambient conditions. As observed for solutions of mannitol, the morphology changes with the increase of the operational temperature (compare Figure 6). At T = 60 °C a smooth outer surface with crystalline elements in the inside of the particle was observed (Figure 8 a). At T = 80 °C the SEM image (Figure 8 b) reveals a collapsed and rougher surface. This effect is even pronounced with a drying temperature of T = 120 °C (Figure 8 c). The corresponding SEM images present a non-uniform and collapsed particle structure (Figure 8, c1) with larger crystallites inside (Figure 8, c2). A detailed investigation of the effects of humidity and solid content on the morphology is ongoing.

Silica and Styrene - Butyl acrylate as Disperse Systems

As an example for the morphology formation of disperse systems a silica dispersion was dried at temperatures of either 23 or 100 °C and a relative humidity of 1%. Moreover, two different dispersions of styrene-butyl acrylate - copolymer were dried at temperatures of either 70 or 95 °C and a relative humidity of 1%.

SEM images of particles of silica dispersions dried at T = 23 °C as well as T = 100 °C reveal similarly smooth surfaces composed of nano-sized primary particles (Figure 9 a, b). Furthermore, for T = 100 °C crystalline structures were observed with shapes that vary from filigree needle to plump crystal (Figure 9 c). All these structures seem to grow from the inside of the particle through the shell but they also seem to grow through each other.

SBA dispersions examined in this work do not demonstrate a dependence of the evaporation coefficient on the monomer

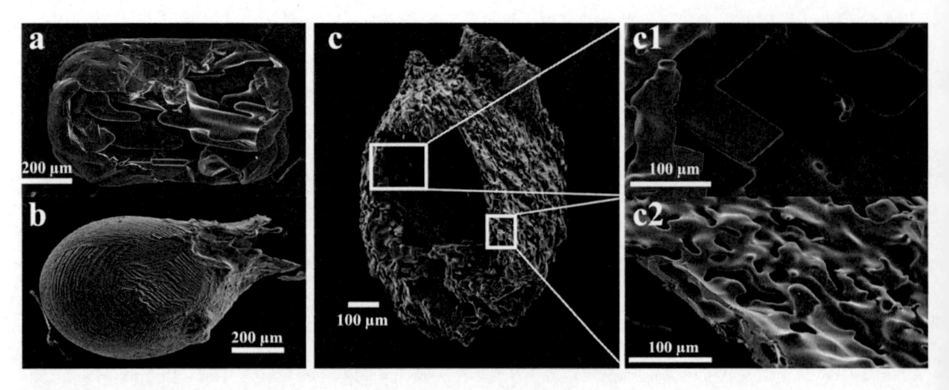

Figure 8.
SEM images of ammonium sulfate a) T = 60 °C, RH = 1%, 20 wt%, b). T = 80 °C, RH = 1%, 10 wt%, c) T = 120 °C, RH = 1%, 20 wt%.

Figure 9.
SEM image of silica particle dried at T = 23 °C and RH = 1% (a). Magnification presents particle shell with primary particles (b). Crystalline structures on particle dried at T = 100 °C and RH = 1% (c).

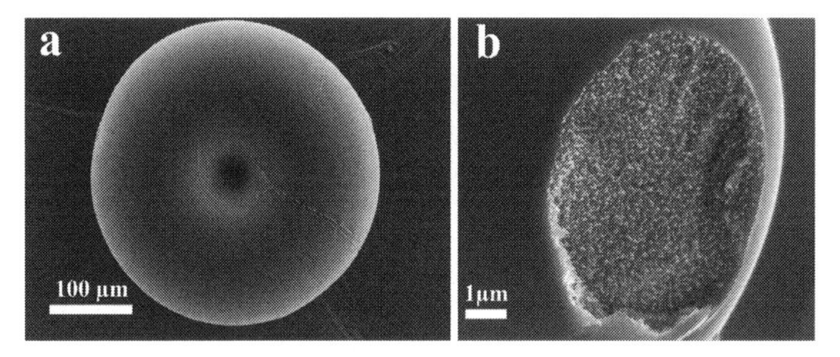

Figure 10.
SEM images of dried styrene – butyl acrylate copolymer particles (left: ratio S:BA = 20:80, T = 70°C, RH = 1%; right: ratio S:BA = 80:20, T = 95°C, RH = 1%).

ration of styrene - butyl acrylate (S:BA, either 80:20 or 20:80 (w,w)). With regard to the morphology, the polymer dispersions exhibited a smoother surface than the silica dispersion. Figure 10 reveals a view under the shell of a particle where primary particles (Figure 10 b) are visible. Further investigation to determine possible effects on the evaporation coefficient and morphology with lager primary particle size (300 nm vs. 50 nm) are ongoing.

Conclusion

In this study an acoustic levitator was used to examine the drying behaviour and morphology formation of two reactive aqueous monomer solutions, two aqueous solutions of salts and two disperse systems. It was demonstrated that acoustic levitation is an ideal tool to monitor the drying behaviour of single droplets. It gives access to information otherwise unobtainable in a spray tower system. Robust, yet non-complex measurements like shadowgraphy allow an easy and fast access to the droplet drying behaviour. If combined with morphology analysis by SEM morphological phase diagrams can be derived (see Figure 3 c).

The morphologies obtained in this study are in good agreement with the model of Charlesworth and Marshall.[3] Nearly all measurements were carried out below the boiling point of the solvent and all resulting morphologies fit into this class of the morphology model. However, the model is not detailed enough to classify all morphologies found in this survey, taking into account only a separation between temperatures above and below the boiling point. For more than one substance systematic differences of the morphologies were observed for different temperatures below the boiling point. Moreover, the model does not cover the influence of relative humidity on morphology formation.

The comparison between the two reactive monomer solutions reveals a very different behaviour. PVP forms defined crystalline structures and/or an amorphous shell, while NaAA exhibits a large variety of morphological forms. In this context the precipitation behaviour of NaAA has to be investigated more thoroughly.

The analysis of the PVP particle morphologies reveals an influence of the drying conditions on the crystallization, contrary to that of mannitol. While mannitol develops pronounced crystals with higher temperature and humidity, a lower temperature and humidity leads to an increase of crystalline structures for PVP. The drying behaviour of aqueous ammonium sulfate droplets confirm the behaviour found for mannitol, with respect to temperature changes. The deviating behaviour of PVP might be explained with the reactive nature of the system and the reaction conditions inside of the droplet, however, further investigations are necessary to fully understand the reason for this behaviour.

Acknowledgement: The authors are grateful to the German Research Foundation (DFG) for funding this research in the scope of SPP 1423 "Prozess-Spray" and Evonik Industries AG for donating silica dispersions. Furthermore we would like to thank R. Walter from the Biozentrum Grindel und Zoologisches Museum, Hamburg University for the SEM measurements.

[1] K. Masters, "Spray Drying Handbook". 5th ed. J. Wiley & Sons, New York **1991**, 309.
[2] P. Seydel, A. Sengespeick, J. Blömer, J. Bertling, *Chem. Ing. Techn.* **2003**, *75*, 714.
[3] M. Peglow, T. Metzger, G. Lee, H. Schiffter, R. Hampel, S. Heinrich, E. Tsotsas, in: "*Modern Drying Technology*", Vol. 2, E., Tsotsas, A. S., Mujumdar, Wiley - VCH, Weinheim **2009**, 187.
[4] D. Charlesworth, W. R. Marshall, *A. I. Ch. E. J.* **1960**, *6*, (1) 9.
[5] E. G. Lierke, *Acoustica.* **1996**, *82*, 220.
[6] E. H. Trinh, J. L. Robey, *Phys. Fluids* **1994**, *6*, 3567.
[7] Z. L. Yan, W. J. Xie, B. Wei, *Phys. Lett. A.* **2011**, *375*, 3306.
[8] E. Welter, B. Neidhart, *Fresenius J. Anal. Chem.* **1997**, *357*, 345.
[9] S. Santesson, M. Andersson, E. Degerman, T. Johansson, J. Nilsson, S. Nilsson, *Anal. Chem.* **2000**, *72*, 3412.
[10] D. D. Weis, J. D. Nardozzi, *Anal. Chem.* **2005**, *77*, 2558.
[11] E. H. Trinh, *Rev. Sci. Instrum.* **1985**, *56*, 2059.
[12] E. H. Trinh, H. Chaur-Jian, *J. Acoust. Soc. Am.* **1986**, *79*, 1335.
[13] S. Santesson, J. Johansson, L. S. Taylor, I. Levander, S. Fox, M. Sepaniak, S. Nilsson, *Anal. Chem.* **2003**, *75*, 2177.
[14] B. R. Wood, P. Heraud, S. Stojkovic, D. Morrison, J. Beardall, D. McNaughton, *Anal. Chem.* **2005**, *77*, 4955.
[15] P. Jacob, A. Stockhaus, R. Hergenröder, D. Klockow, *Fresenius J. Anal. Chem.* **2001**, *371*, 726.
[16] F. Priego-Capote, L. de Castro, *Trends in Anal. Chem.* **2006**, *25*, (9) 856.
[17] G. Brenn, T. Wiedemann, D. Rensink, O. Kastner, A. L. Yarin, *Chem. Eng. Techn.* **2001**, *11*, 1113.
[18] R. Mondragon, J. C. Jarque, J. E. Julia, L. Hernandez, A. Barba, *J Europ. Cer. Soc.* **2012**, *32*, 59.
[19] R. Kumar, E. Tijerino, A. Saha, S. Basu, *Appl. Phys. Lett.* **2010**, *97*, 123106.
[20] S. K. Biedasek, "Fortschritte der Polymerisationstechnik", Vol. XXVII, W&T-Verlag, Berlin **2009**.
[21] J. Laackmann, W. Säckel, L. Cepelyte, K. Walag, R. Sedelmayer, F. Keller, W. Pauer, H.-U. Moritz, U. Nieken, *Macromol. Sympos.* **2011**, *302*, 235.
[22] S. G. Maas, G. Schaldach, E. M. Littringer, A. Mescher, U. J. Griesser, D. E. Braun, P. E. Walzel, N. A. Urbanetz, *Powder Technology* **2011**, *213*, 27.

New Reactor for Polyester Polyols Continuous Synthetic Process

Like Chen, Zhenhao Xi, Zhen Qin, Ling Zhao, Weikang Yuan*

Summary: Based on the kinetic and thermodynamic equations, a comprehensive mathematical model for the continuous synthetic process of polyester polyols was developed, which was carried out in two innovational bubbling reactive distillation towers (BRDTs). In this new type of reactor, condensation reactions among oligomers of polyester polyols were accomplished efficiently and rapidly. Two bench BRDTs with the height of 2 m were applied for the continuous process of poly(ethylene adipate) (PEA). In the first continuous esterification operation at atmospheric pressure, linear oligomers were discharged from the bottom of the column, while water and ethylene glycol (EG) passed a few column trays and a packing section and reached the top of the column. Then water was discharged as byproducts and EG was sent back to the first tray for further reaction. In the second continuous polycondensation operated in the second BRDT under partial vacuum, PEA was discharged from the bottom of the column, while water and ethylene glycol passed a few column trays, they were discharged as byproducts. The influence of major operating conditions on reactor performance was also simulated. Simulation results provided a strategy for developing and optimizing this process.

Keywords: continuous operation; esterification; mathematical model; poly(ethyleneadipate); polycondensation

Introduction

Polyester polyols are macroglycols that are prepared by the polymerization of a glycol and a dicarboxylic acid or acid derivative via batch processes in large-scale operations. Polyester polyols of diverse specification and brands can produce polyurethane products by reaction with isocyanates. Global production capacity for polyurethane reaches 17 million metric tons in 2010.[1,2] Global demand for polyester polyols grows to 2.1 million metric tons in 2012, corresponding to an average growth of 8.4% per year.[3] In spite of the growing demand of poly(ethylene adipate) (PEA) and other polyester polyols, their continuous synthetic process has not been reported yet.[4]

Polyester polyols are produced commercially by batch processes in large-scale operations, which are generally configured with stirred tanks and have been disclosed in many patents.[5,6] The batch polyester polyols synthetic processes mainly consist of two stages: esterification and polycondensation. In the first stage, direct esterification of glycol and dicarboxylic acid in the absence of catalyst is carried out in a batch reactor under certain temperature and atmospheric pressure. As the distillation subsided, the pressure in the reactor will be reduced. If desired, catalyst can be used to accelerate the reaction rate, this is polycondensation stage. Typically, the components are reacted at temperature ranging from 130 °C to 230 °C, and are held at these temperatures long enough for the acid value drop to 0.04 $mol \cdot kg^{-1}$ or less. Batch processes are flexible for producing a variety of brands of polyester polyol products, but the product quality is unstable. Moreover, due to the limited heat and mass transfer, the reaction cycle is particularly long.

State Key Laboratory of Chemical Engineering, East China University of Science and Technology, Shanghai 200237, China
E-mail: zhaoling@ecust.edu.cn

Polymerization reaction for polyester polyols, from the reaction of two functional groups, is usually a reversible endothermic reaction with the formation of volatile byproducts.[7,8] The reaction and devolatilization are coupled in the synthetic process. In order to remove byproduct from the polymeric solution effectively, the synthetic apparatus should provide effective heat exchange and separation function. A heated counter-current sieve plate distillation column was used in continuous preparation of polyester prepolymer with the steps of esterifying unreacted acid and hydroxyl end groups.[9] In the heated column reactor, the reactants continuously flowed across multiple reactor plates, while inert gas flowed upward through the column reactor. The residence time for reaction was about 3 min. The reactor had no agitation. However, its operating flexibility was unsatisfactory, because of weeping or entrainment flooding.

In this study, an innovational BRDT reactor with or without the packing section[10] is used for PEA continuous esterification and continuous polycondensation processes, respectively. The BRDT contains several reaction trays with special construction. In these processes, molten linear oligomers flow through each tray from the top to the bottom by gravity, without agitation, and are discharged from the bottom of the column, while water passes the trays from the bottom to the top as a condensation byproduct. The flowing fluid on the trays has little dead zone due to the intensive bubbling effect of water, ethylene glycol (EG) vapor and a little inert gas for nitrogen protection, ensuring optimum intermixing.

A comprehensive mathematical model for the continuous synthetic process of PEA in the two series BRDTs is also developed. The influences of operating conditions on reactor performance are investigated.

Assumptions and Modeling

Polymer Segment Approach

In a polymerization process, polymers form in different chain lengths. A modeling approach that properly accounts for varying chain lengths is important to establish the framework for representing polymers in the reaction kinetics. There are two approaches: molecular species models and functional group models. Generally speaking, molecular species models are more comprehensive than functional group models in providing the information of product composition, but much more time is required for the calculation, because each unique polymer molecule is tracked as an independent component. In this study, the polymer segment approach,[11] which is a functional group approach, is used to establish the overall reaction network. Polymerization can be regarded as reactions between two reactive functional groups. The components in this reaction scheme are C_1 (-COOH), C_2 (EG), C_3 (-OH), C_4 (-COOC-), and C_5 (H_2O).

The chemical structure of PEA with P_n (number average degree of polymerization) can be expressed as follows:

$$\left.\begin{matrix}\text{H-}\\ \text{HOCH}_2\text{CH}_2\text{-}\end{matrix}\right] \left[\text{OOC(CH}_2)_4\text{COO}-\text{CH}_2\text{CH}_2 \right]_{P_n-1} \text{OOC(CH}_2)_4\text{COO} - \left[\begin{matrix}\text{-H}\\ \text{-CH}_2\text{CH}_2\text{OH}\end{matrix}\right. \tag{1}$$

The ratio of hydroxyl end groups to total end groups is:

$$\varphi = \frac{C_3}{C_1 + C_3} \tag{2}$$

where the atomic weights of H-, $HOCH_2CH_2$-, -CH_2CH_2-, -$OOC(CH_2)_4COO$- are 1.01, 45.06, 28.05,144.12, respectively. The relation between M_n and P_n can be given by

$$M_n = \frac{2000}{C_1 + C_3} \tag{3}$$

$$M_n = (P_n - 1)(144.12 + 28.05) + 144.12 + 2(1 - \varphi) \times 1.01 + 2\varphi \times 45.06 \tag{4}$$

$$P_n = \frac{M_n - 88.10\varphi + 26.03}{172.17} \tag{5}$$

Ester concentration C_4 can be expressed by

$$C_4 = \frac{C_1 + C_3}{2}(P_n - 1) \tag{6}$$

The carboxyl concentration C_1 represents the extent of esterification reaction, and the M_n represents the extent of condensation reaction in the continuous synthetic process.

Reaction Scheme

There are two primary reactions in PEA synthetic process: esterification (or water formation) and condensation (or trans-esterification).[12] Based on the hypothesis that the reactivity of functional group does not depend on the polymer chain length,[13,14] it can be assumed that the reactivities of acid end group on adipic acid (AA) and on polymer chain are the same while the reactivities of hydroxyl end group on EG and on half-esterified EG are different. The complete set of reaction is written as[15]

Esterification

Condensation

$$\underset{C_1}{-(CH_2)_4COOH} + \underset{C_2}{HO(CH_2)_2OH} \underset{k_2}{\overset{k_1}{\rightleftharpoons}} \underset{C_3}{-(CH_2)_4COO(CH_2)_2OH} + \underset{C_5}{H_2O} \qquad (7)$$

$$\underset{C_1}{-(CH_2)_4COOH} + \underset{C_3}{-(CH_2)_4COO(CH_2)_2OH} \underset{k_4}{\overset{k_3}{\rightleftharpoons}} \underset{C_4}{-(CH_2)_4COO(CH_2)_2OOC(CH_2)_4-} + \underset{C_5}{H_2O} \qquad (8)$$

$$\underset{C_3}{2-(CH_2)_4COO(CH_2)_2OH} \underset{k_6}{\overset{k_5}{\rightleftharpoons}} \underset{C_4}{-(CH_2)_4COO(CH_2)_2OOC(CH_2)_4-} + \underset{C_2}{HO(CH_2)_2OH} \qquad (9)$$

The associated reaction rates for these 3 reactions are [16]

$$R_1 = 2k_1C_1C_2 - k_2C_3C_5 \qquad (10)$$

$$R_2 = k_3C_1C_3 - 2k_4C_4C_5 \qquad (11)$$

$$R_3 = k_5C_3C_3 - 4k_6C_2C_4 \qquad (12)$$

For a simulation of synthetic process, the rate constant k_i ($i = 1\sim6$) and equilibrium constant K_i ($i = 1, 2, 3$) should be determined. The rate constants depend on temperature according to the Arrhenius equation

$$k_i = k_{0i}\exp\left(\frac{-Ea_i}{RT}\right) \quad (\text{for } i = 1,3,5) \qquad (13)$$

where k_{0i} is the preexpomential factor ($kg \cdot mol^{-1} \cdot h^{-1}$), Ea_i is the activation energy ($kJ \cdot mol^{-1}$), R is the gas constant ($J \cdot mol^{-1} \cdot K^{-1}$), and T is the temperature (K). The kinetic and thermodynamic parameters are given in Table 1.

Table 1.
Kinetic and thermodynamic parameters.[17,18]

No.	ΔS $J \cdot mol^{-1} \cdot K^{-1}$	ΔH $J \cdot mol^{-1}$	k_0 $kg \cdot mol^{-1} \cdot h^{-1}$	Ea $kJ \cdot mol^{-1}$
K_1	43.2	1.58×10^4		
K_2	50.4	2.01×10^4		
k_1			1.14×10^4	42.8
k_3	Esterification stage		2.46×10^5	54.6
k_5			2.24×10^5	60.1
k_1			2.00×10^3	30.5
k_3	Polycondensation stage		1.63×10^4	42.5
k_5			1.32×10^4	48.0

The complete set of reaction considered here is a reversible equilibrium reaction,[19] with $k_2 = k_1/K_1$, $k_4 = k_3/K_2$, and $k_6 = k_5/K_3$. For the temperature range considered, entropy and enthalpy change are constant in accordance with the thermodynamic equation [20,21]:

$$K_i = \exp\left(\frac{\Delta S_i}{R} - \frac{\Delta H_i}{RT}\right) \quad (\text{for } i = 1, 2) \tag{14}$$

$$K_3 = K_2/K_1 \tag{15}$$

where ΔS is the entropy change ($J \cdot mol^{-1} \cdot K^{-1}$), and ΔH is the enthalpy change ($J \cdot mol^{-1}$).

Phase Equilibrium

For the calculation of concentrations of volatile components (EG and water) in the liquid and vapor phases, a quasi-steady state assumption is used for the vapor-liquid equilibrium. The vapor-phase calculations for oligomeric components are not considered since the polymer is not volatile. The vapor phase is assumed to follow the ideal gas law:

$$p_i = P \cdot y_i \tag{16}$$

where P is the total pressure in the reactor, p_i and y_i are the partial pressure and mole fraction of volatile component i in the vapor phase, respectively. Using Raoult's law, the partial pressure of volatile components is

$$p_i = p_i^* \cdot x_i \tag{17}$$

where p_i^* is the saturated vapor pressure (Pa) of volatile component i, and x_i is the mole fraction of component i in the liquid phase. The vapor pressure data of water and EG are obtained from a pure component databank.[22,23]

for EG

$$\ln p_1^* = 79.3 - 10105/T - 7.5 \ln T + 7.3e^{-19} T^6 \tag{18}$$

for water

$$\ln p_7^* = 73.6 - 7259/T - 7.3 \ln T + 4.2e^{-6} T^2 \tag{19}$$

Mathematical Model

Figure 1 is the mathematical model diagram for PEA continuous synthetic process with two BRDTs, which are multistage bubble columns. Bubble columns are a class of multiphase reactor in which the discontinuous gas phase moves in the form of bubbles

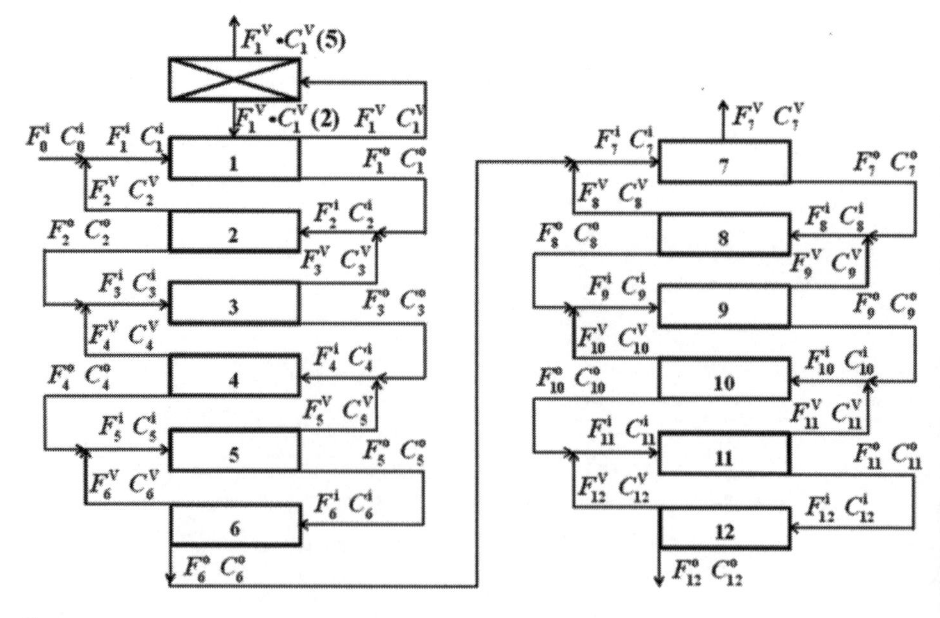

Figure 1.
Mathematical model diagram of PEA continuous synthetic process.

relative to the continuous liquid phase. They provide several advantages during operation and maintenance such as high heat and mass transfer rates, compactness and low operating and maintenance costs. Each of the 12 column trays can be considered as a perfect mixing reactor due to the intensive bubbling effect of water and EG and back mixing in both the phases.

The axial dispersion coefficient of the liquid phase D_{eL} can be calculated from Deckwer's correlations,[24] which show no influence of the liquid flow rate.

$$D_{eL} = 0.678 D_t^{1.4} u_{0G}^{0.3} \tag{20}$$

where D_t is the diameter of column tray (m), and u_{0G} is the superficial gas velocity ($m \cdot s^{-1}$).

Peclet number can be calculated from D_{eL}.

$$Pe_L = \frac{u_{0L} H}{(1-\varepsilon_G) D_{eL}} \tag{21}$$

where H is the height of column tray (m), and u_{0L} is the superficial liquid velocity ($m \cdot s^{-1}$), ε_G is the gas holdup of the column tray. ε_G can be estimated from u_{0G}.[25]

$$\varepsilon_G = \frac{u_{0G}}{0.3 + u_{0G}} \tag{22}$$

Figure 2 shows Pe_L of column tray under different superficial gas velocity. The Pe_L is small for the superficial liquid velocity is maintained to be lower than the superficial gas velocity by at least two orders of magnitude. The number of mixers of the cascade N can be calculated based on the relational expression of the series model and the axial dispersion model.[26]

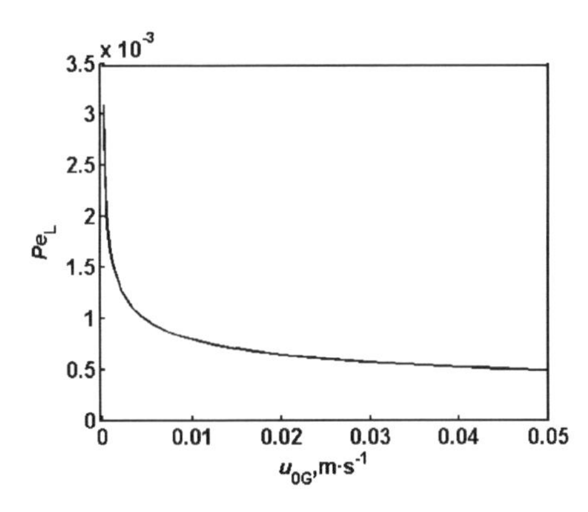

Figure 2.
Pe_L of column tray under different superficial gas velocity.

$$\frac{1}{N} = \frac{2}{Pe_L} - \frac{2}{Pe_L^2}(1-e^{-Pe_L}) \tag{23}$$

The number of mixers of the cascade N approximates 1, so each column tray can be considered as a continuous stirred tank reactor (CSTR).

The 12 column trays are assumed to be 12 CSTRs. The material balance equations of each component in a CSTR can be written as

$$m\mathrm{d}C_1^o/\mathrm{d}t = F^i C_1^i - F^o C_1^o + m(-R_1 - R_2) \tag{24}$$

$$m\mathrm{d}C_2^o/\mathrm{d}t = F^i C_2^i - F^o C_2^o + m(-R_1 + R_3) - F^V C_2^V + F^R C_2^R \tag{25}$$

$$m\mathrm{d}C_3^o/\mathrm{d}t = F^i C_3^i - F^o C_3^o + m(R_1 - R_2 - 2R_3) \tag{26}$$

$$m\mathrm{d}C_4^o/\mathrm{d}t = F^i C_4^i - F^o C_4^o + m(R_2 + R_3) \tag{27}$$

$$m\mathrm{d}C_5^o/\mathrm{d}t = F^i C_5^i - F^o C_5^o + m(R_1 + R_2) - F^V C_5^V + F^R C_5^R \tag{28}$$

where $C_1{}^i \sim C_5{}^i$ and $C_1{}^o \sim C_5{}^o$ represent the input and output concentrations of components in the reaction mixture, respectively, F^i and F^o represent the input and output flow rates, respectively, m is the mass of liquid phase, F^V and F^R are the vapor flow rate to the above tray and the reflux flow rate from the above tray, respectively. For the constant reactor volume, the following relation holds:

$$F^i + F^R = F^o + F^V \tag{29}$$

For the first CSTR, it is assumed that all vaporized EG is refluxed and all vaporized water is removed from the distillation column, *i.e.*, $F^V C_2{}^V = F^R C_2{}^R$ and $F^R C_5{}^R = 0$. This perfect separation simplifies the flash calculation with negligible errors. For the next five CSTRs, it is assumed that all vaporized EG and vaporized water are removed from the local distillation column to the above one, *i.e.*, $F^R C_1{}^R = 0$ and $F^R C_7{}^R = 0$.

For the seventh CSTR, it is assumed that all vaporized EG and water is removed from the top of BRDT because of vacuum effect, *i.e.*, $F^R C_2{}^R = 0$ and $F^R C_5{}^R = 0$. For the next five CSTRs, it is assumed that all vaporized EG and vaporized water are removed from the local distillation column to the above one, *i.e.*, $F^R C_2{}^R = 0$ and $F^R C_5{}^R = 0$.

For the steady-state simulation, the differential material balance equations are transformed into nonlinear algebraic equations.

Experimental Part

Materials

AA (>99.5%, AR grade), EG (>99.0%, AR grade), KOH (>85.0%, AR grade), Acetic anhydride (>98.5%, AR grade) and $Na_2S_2O_3 \cdot 5H_2O$ (>99.0%, AR grade) were supplied by Shanghai Lingfeng Chemical Reagent Co. Ltd. KI (>98.5%, AR grade), $KIO_4 \cdot 2H_2O$ (>99.0%, AR grade), Titanium Isopropoxide (TPT, >98%, CP grade) and Karl Fisher titration (5 mg water per milliliter) were supplied by Aladdin Industrial Corporation.

Apparatus and Process

We only have one BRDT, which has three main parts: tower shell, column trays, and packing section. Agitation and dynamic packing surface are avoided to prevent air leakage.

This bench BRDT with packing section was applied to PEA continuous esterification stage process, see Fig. 3(a). The mole feed ratio of EG to AA was 1.2, a paste of which was produced in the slurry kettle maintained at 95 °C and 1 atm. The material from the slurry kettle was supplied continuously to the first tray of BRDT by measuring pump, where the reaction temperature was controlled at 160 °C, which is above the melting point of adipic acid. The temperatures on next 5 trays were controlled at 180 °C, 200 °C, 220 °C, 230 °C, and

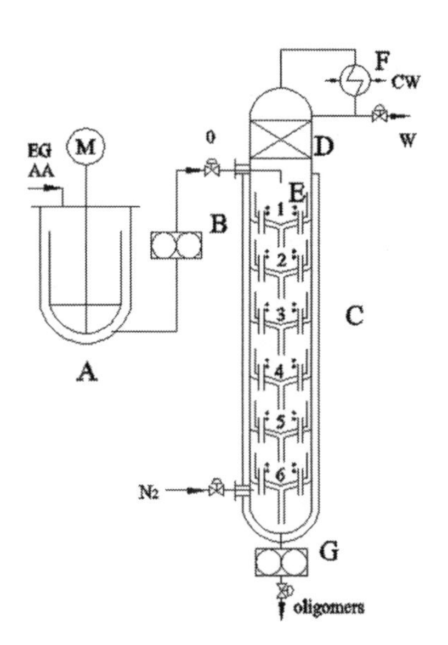

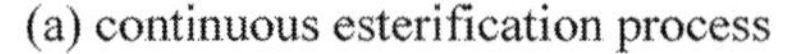
(a) continuous esterification process

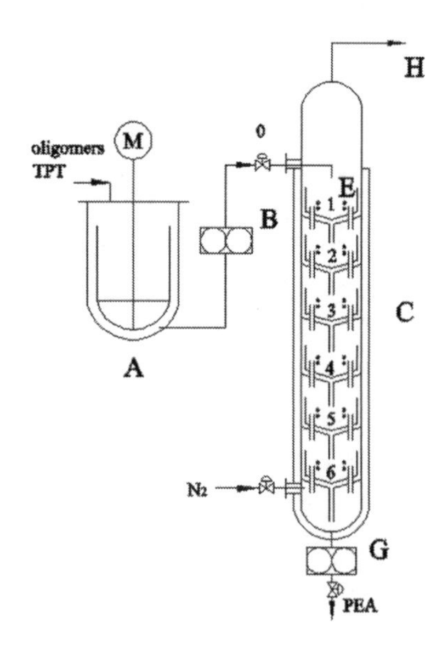

(b) continuous polycondensation process

Figure 3.
Flow diagram of bench BRDT with slurry kettle for PEA continuous synthetic process (A – slurry kettle; B, G – measuring pump; C – tower shell; D – packing tower; E – column trays; F – condenser; H - to vacuum system; M - motor; CW - condensed water; 0–initial input flow).

230 °C by temperature control devices. The homogeneous melt flowed down through 6 trays, and the product was discharged from the reactor by another measuring pump. The residence time on each tray was 30 min, and the total residence time in BRDT was about 3 h. Nitrogen was continuously bubbled through the reactor at 50 ml/min at the bottom of the reactor, and water and nitrogen passed the packing as a condensation byproduct.

This bench BRDT without packing section was applied to PEA continuous polycondensation process, see Fig. 3(b). The melt oligomers and 25 ppm TPT were mixed in the oligomers kettle maintained at 95 °C and 1 atm. The material from the oligomers kettle was supplied continuously to the first tray of BRDT by measuring pump, where the reaction temperature was controlled at 220°C (decrease the volatilization of EG with lower temperature). The temperatures on next 5 trays were controlled at 230 °C by temperature control devices. The homogeneous melt flowed down through 6 trays, and the product was discharged from the reactor by another measuring pump. The residence time on each tray was 30 min, and the total residence time in BRDT was about 3 h. Nitrogen was continuously bubbled through the reactor at 50 ml/min at the bottom of the reactor, and reaction gas and nitrogen passed the top of BRDT as a condensation byproduct.

Samples were taken from the bottom of the oligomers kettle and every tray when the acid value of product from the bottom of BRDT was constant.

Sample Analysis

The concentration of carboxyl (acid value, C_1) and that of hydroxyl (hydroxyl value, C_3) were determined from the samples by titration with KOH based on GB/T2708-1995 and GB/T2709-1995, respectively. The concentration of water (C_5) was determined by titration with METTLER V20 Karl Fisher Titrator based on GB/T12008.6-1989. The concentration of free EG (C_2) was analyzed according to literature.[27]

Results and Discussion

Verification of the Model

It is desirable to validate the simulation results using above model equations with experimental data. Figure 4 shows the comparison of simulated and experimental results. The calculating results fit the experimental data well. The average relative error of this model is anticipated to be smaller than 10%. As the reactions proceed, the concentrations of carboxyl C_1 decrease rapidly, while the concentrations of ester C_4 increase. Esterification and

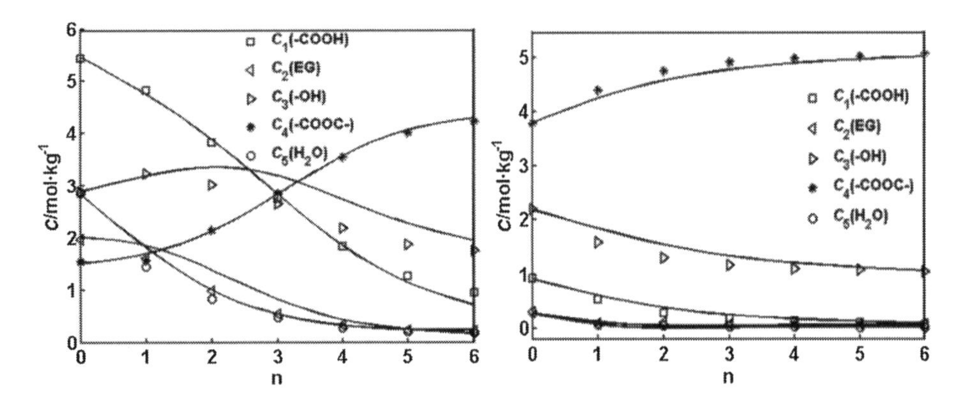

Figure 4.
Comparison of experimental and simulation results.

condensation are accomplished efficiently in the BRDTs under different stages.

Process Simulation

The model for PEA continuous synthetic process in BRDT reactor is very useful to predict the reaction results, although it is complicated because of the multi-reactor polymerization. Two of the main objectives in a PEA synthetic operation are to achieve high conversion of carboxyl (C_1<0.04 mol $\cdot$ kg^{-1}, even less than 0.02 mol $\cdot$ kg^{-1}) in short residence time and to obtain PEA products with certain molecular weight (about 2000). Thus we investigate the influence of residence time, monomer feed ratio, reaction temperature and pressure on the BRDT reactor performance using the model according to above experimental reaction conditions with the feedstock of AA and EG, see Fig. 1.

Effect of Residence Time

Figure 5 shows the effect of residence time of each column tray on the C_1 and M_n of the PEA products. As the residence time increases, the C_1 decreases. However, the effect is not significant for long residence time (more than 40 min) due to the chemical and phase equilibria. There is an extreme point for the M_n of PEA products. When the residence time is above about 35 min, the M_n will decrease a little as much more EG will join in the chain segments of PEA.

Effect of Monomer Feed Ratio

The monomer feed ratio is an important operating variable in the continuous synthetic process. Generally EG is in excess compared to AA so as to increase the conversion of AA, see Fig. 6. The decrease of C_1 is slower with the increase of feed ratio due to the chemical and phase equilibria. There is an extreme point for the M_n of PEA products. When the monomer feed ratio is below about 1.05, the M_n will decrease a little as the extent of reaction is lower with less EG. Moreover, the objective M_n of PEA products are about 2000, so the suitable monomer feed ratio is between 1.1~1.3.

Effect of Reaction Temperature

The reaction temperature of the 8~12th column tray for continuous polycondensation stage is an important operating variable in the continuous sythetic process. When the residence time and reaction

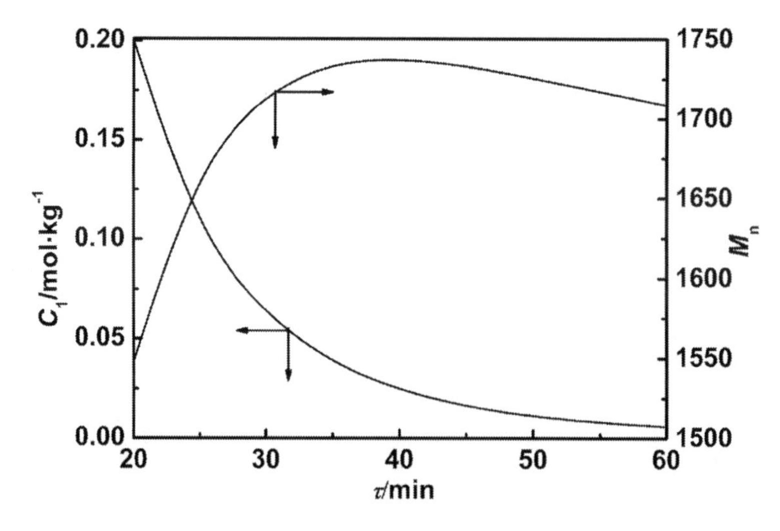

Figure 5.
Effect of residence time on C_1 and M_n of PEA products.

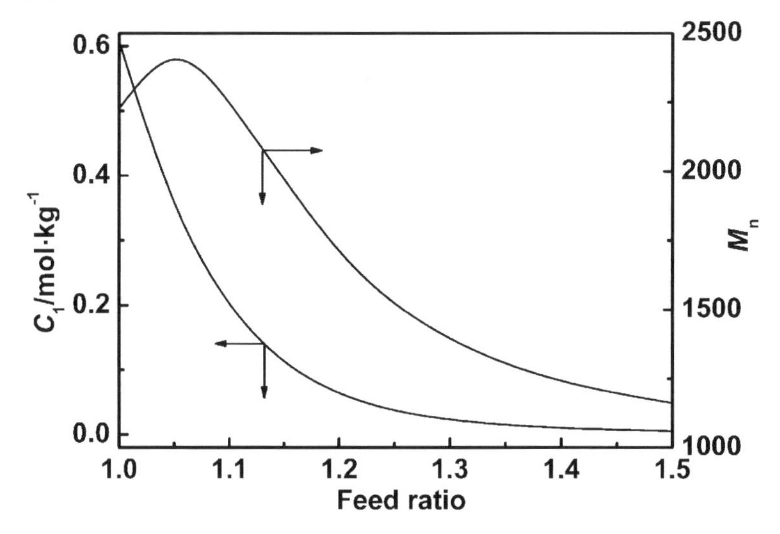

Figure 6.
Effect of monomer feed ratio on C_1 and M_n of PEA products.

pressure are kept constant, higher reaction temperature is favorable to increase the reaction rate and decrease the concentration of water in polymeric solution. From Fig. 7, it is shown that the conversion of acid end group and M_n of PEA increase as the reaction temperature increases. Although increasing the reaction temperature helps to decrease C_1, it will be more prone to thermal degradation for PEA when the reaction temperature is higher than 240 °C.[28] The ideal reaction temperature is between 230~240 °C.

Effect of Reaction Pressure

Esterification and condensation reactions for polyester polyols are reversible. The continuous polycondensation stage is generally operated under partial vacuum. Reduction of pressure in BRDT helps in increase the M_n of PEA as the

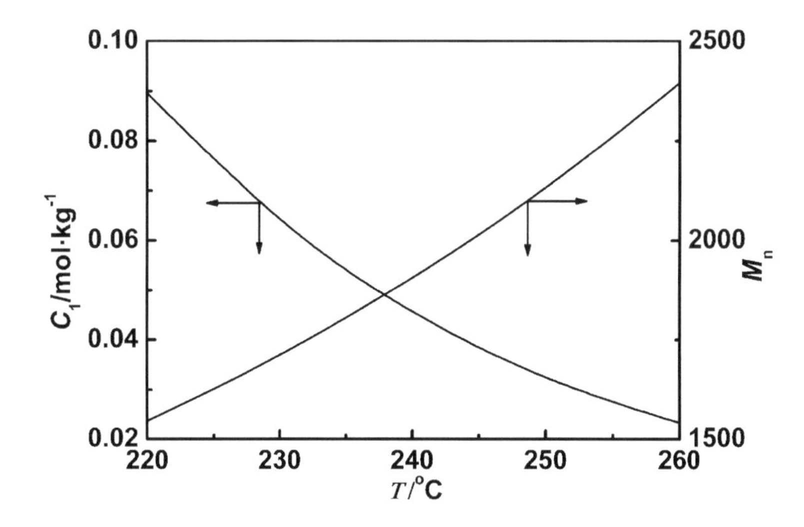

Figure 7.
Effect of reaction temperature on C_1 and M_n of PEA products.

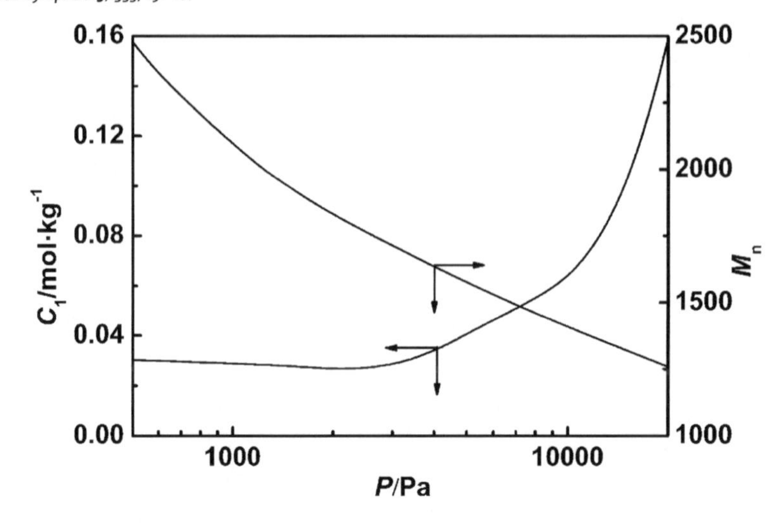

Figure 8.
Effect of pressure on C_1 and M_n of PEA products.

concentrations of EG and water in polymeric solution decreases rapidly (see Fig. 8 and 9). There is an extreme point for the C_1 of PEA, when the reaction pressure is below about 2000 Pa, the C_1 will increase a little as the reaction pressure reduces. The esterification reaction is the dominant reaction for low molecular weight synthesis. Because of the higher EG flow rate discharging from the top of BRDT ($F^V_7(2)/F^o_6(2)$) during the low pressure operation, the esterification reaction is reduced for the lower concentration of EG and hydroxyl in spite of the very little amount of water. The suitable reaction pressure is between 2000~10000 Pa.

Conclusion

Two bubbling reactive distillation towers were invented for continuous synthesis of polyester polyols. In these new types of reactors, Esterification and condensation

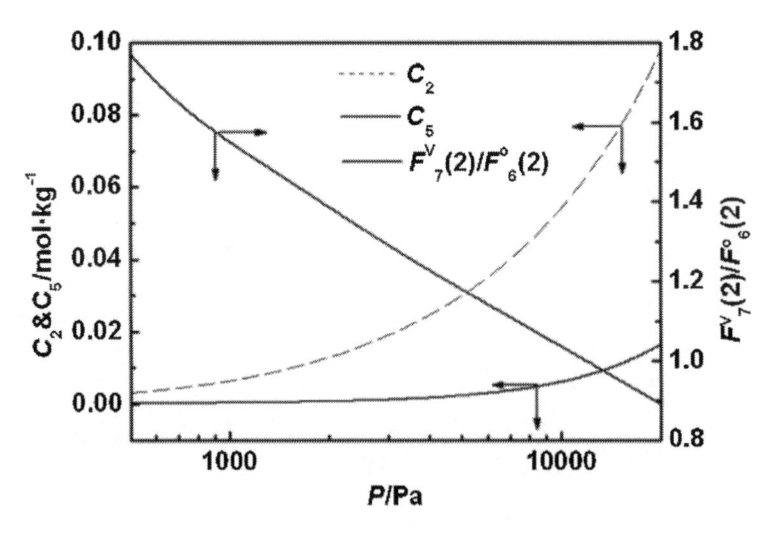

Figure 9.
Effect of pressure on C_2, C_5 of PEA products and $F^V_7(2)/F^o_6(2)$.

were accomplished efficiently and rapidly in these counter-current column reactors operated continuously under atmospheric or partial vacuum. Based on the kinetic and thermodynamic equations, a comprehensive mathematical model for the continuous synthetic process was developed. The influence of the major operating conditions, such as residence time, monomer feed ratio, reaction temperature and pressure, on reactor performance was investigated. Simulation results were in good agreement with experimental data, providing strategies for developing and optimizing this process. Furthermore, BRDT had wide operation flexibility as well as adjustable configuration parameters to meet different demands. This reactor was supposed to be a universal apparatus for many condensation processes.

[1] P. L. Xu, S. Q. Zhang, "*Handbook of Polyurethane Materials*", Chemical Industry Press, Beijing 2–6 **2011**.

[2] L. M. Zhu, Y. J. Liu, "*Polyurethane foam*", Chemical Industry Press, Beijing 219–232 **2005**.

[3] H. Chinn, U. Löchner, H. Mori, "Polyester Polyols", CEH Report **2012**.

[4] G. Vairo, M. D. Serio, "Process for the preparation of polyester polyols", WO Pat., 2008037400 **2008**.

[5] B. Thomas, M. Thomas, "Aromatic polyester polyols", US Pat., 2006069175 **2006**.

[6] W. Alexander, D. Matthias, T. Gerlinde, "Method for Producing Polyester Polyols of Polyvalent Alcohols", EP Pat., 1511787 **2005**.

[7] Z. H. Xi, L. Zhao, Z. Y. Liu, *Macromol. Symp.* **2007**, *259*(1), 10–16.

[8] W. Y. Tian, Z. X. Zeng, W. L. Xue, Y. B. Li, *Chinese J. Chem. Eng.* **2010**, *18*(3), 391–396.

[9] D. J. Lowe, W. Del, "Process of Making Polyester Prepolymer", US Pat., 5786443 **1998**.

[10] L. Zhao, Z. H. Xi, L. K. Chen, W. Z. Sun, "A method and device for continuous producing polyester polyols", CN Pat., 102432846 **2012**.

[11] T. Yamada, Y. Imamura, *Polym. Plast. Technol. Eng.* **1989**, *28*(7-8), 811–876.

[12] N. Luo, Z. C. Ye, W. M. Zhong, F. Qian, *CIESC Journal* **2010**, *61*(8), 1933–1941.

[13] P. Flory, *J. Am. Chem. Soc.* **1939**, *61*(12), 3334–3340.

[14] P. Flory, *J. Am. Chem. Soc.* **1937**, *59*(3), 466–470.

[15] K. Ravindranath, R. A. Mashelkar, *Polym. Eng. Sci.* **1982**, *22*(10), 610–618.

[16] K. Ravindranath, R. A. Mashelkar, *Polym. Eng. Sci.* **1982**, *22*(10), 619–627.

[17] W. J. Sun, L. K. Chen, Z. H. Xi, L. Zhao, "Reaction Equilibrium of Poly (ethylene adipate)". In: *6th Asia Pacific Chemical Reaction Engineering Symposium*, Beijing 58–59 **2011**.

[18] L. K. Chen, Z. H. Xi, L. Zhao, "Kinetic Model of Poly (ethylene adipate) Synthetic Process". In: *22th International Symposium Chemical Reaction Engineering*, Maastricht **2012**.

[19] T. Yamada, Y. Imamura, *Polym. Eng. Sci.* **1988**, *28*(6), 385–392.

[20] S. K. Gupta, A. Kumar, "*Reaction Engineering of Step Growth Polymerization*", Plenum Press, New York **1987**.

[21] H. K. Reimschuessel, B. T. Debona, A. K. S. Murthy, *J. Polym. Sci.* **1979**, *17*(10), 3217–3239.

[22] C. K. Kang, B. C. Lee, *J. Appl. Polym. Sci.* **1996**, *60*(11), 2007–2015.

[23] C. K. Kang, B. C. Lee, *J. Appl. Polym. Sci.* **1997**, *63*(2), 163–174.

[24] W. D. Deckwer, R. Burckhart, G. Zoll, *Chem. Eng. Sci.* **1974**, *29*(11), 2177–2188.

[25] J. B. Joshi, M. M. Sharma, *Trans. Inst. Chem. Eng.* **1979**, *57*, 244–251.

[26] K. R. Westerterp, W. P. M. Vanswaaij, A. A. C. M. Beenackers, "*Chemical reactor design and operation*", Netherlands University Press, Amsterdam 160–226 **1984**.

[27] Z. M. Liang, *China Food Additives* **2007**, (4), 160–162.

[28] Q. Z. Zhu, C. Zhang, S. Y. Feng, J. H. Chen, *J. Appl. Polym. Sci.* **2002**, *83*(7), 1617–1624.

DOI: 10.1002/masy.201300039

Quantification of Colour Formation in PET Depending on SSP Residence Time, Temperature, and Oxygen Concentration

Thomas Rieckmann,[1] Katharina Besse,[1] Fabian Frei,[2] Susanne Völker[3]*

Summary: A systematic approach was carried out to quantify the discolouration of different poly(ethylene terephthalate) (PET) samples such as virgin bottle-grade resins, bottle-to-bottle recycled chips, and bottle-to-bottle recycled flakes. The PET samples were treated in a fixed bed reactor in the solid-state. The possible influencing parameters such as residence time, temperature, and oxygen partial pressure were studied by analysing the CIELAB b*-coordinate (axis = yellow to blue) with a spectrophotometer at the end of the experiment. The PET samples were analysed once as chips respectively flakes retained from the experiments and again as plates obtained from a following standardised melting procedure. The combination of experiments and literature review results in the proposal of different discolouration mechanisms, which are influenced by temperature, residence time, and oxygen concentration. Using the design of experiments (DoE) approach, a mathematical model was compiled that can be used to predict the yellowing of the different PET grades.

Keywords: discolouration; PET; polyester; recycling; polymer degradation; solid-state polycondensation; SSP; yellowing

Introduction

PET has become one of the most important polymers during the last decades. It is used in a wide range of applications, especially in the food industry and the packaging sector. Due to this wide use of PET and growing public interest in ecological topics, a deposit system was introduced, e.g., in Germany. All of these factors entail an increasing amount of recycled PET. The recycled material should show the same material behaviour as the virgin one. Retaining most of the properties of PET can be easily achieved by various recycling systems. However, the property that is hard to control is the colour of the material. The virgin product is transparent and colourless or has a very bright white colour, but the recycling material becomes more and more yellow during the thermal treatment of reuse, which is an exclusion criterion for most consumers. Thermal degradation and discolouration are the main topics of research to achieve the aim of near virgin quality recycling materials.

There are a number of different mechanisms described in the literature concerning the discolouration of PET.[1–6] Most of them are based on molecules from degeneration reactions, which have a large impact on discolouration. The degradation of the PET chain, which means cleavage of the chain and reduction of molecular weight, can be triggered by two main effects. These are thermal degradation in an inert atmosphere and/or thermo-oxidative degradation with influence of air or oxygen. Side and follow-up reactions are much more

[1] Cologne University of Applied Sciences, Betzdorfer Strasse 2, 50679 Cologne, Germany
E-mail: thomas.rieckmann@fh-koeln.de
[2] Sika AG, Tüffenwies 16, 8048 Zurich, Switzerland
[3] 42 Engineering, von-Behring-Str. 9, 34260 Kaufungen, Germany

important for discoloration, because thermal degradation itself does not lead directly to discolouration.[1] Yet there are some side reactions that might have an influence, like vinyl end-group formation by chain scission of bound EG groups (bEG) or terminal EG-groups (tEG).[7]

It was determined that thermo-oxidative degradation, leading to enhanced discolouration, is faster than degradation in an oxygen free atmosphere and can be observed at lower temperatures.[6] However, there has been no systematic approach to this topic up to now.

Edge discovered hydroxylated therephthalic units as a precursor substance for colourant structures. Another possible pathway is oxygen triggered and followed by the formation of stilbenes and stilbene quinone (Edge [2,3]). A component that is generated in the first step of PET production is diethylene glycol (DEG) (1.0–2.5 wt % [7,8]), which functions as a co-monomer. It has been proven that the ether bond of DEG might be a starting point for thermal degradation.[4]

The examination of a slice of a degraded pellet shows that thermo-oxidative degradation is diffusion controlled.[4] Also, there is a difference in mass transfer and diffusion that is confirmed by a difference in degradation when the melting point of PET is reached.

There are some more research groups working on PET discolouration who have been publishing their results,[2,5] but up to now, an experimentally proven detailed understanding of how PET discolouration occurs in solid-state polycondensation (SSP) is not available.

Experimental Part

The temperature ranges (180°C to 220°C) of the experiments as well as residence times (1 to 4 h) and oxygen content (0 to 20 vol%) were chosen according to the process steps in PET production and recycling: drying, crystallization, solid-state polycondensation (SSP), and decontamination. Three different materials were chosen for the experiments: Virgin bottle-grade resin (Equipolymers, S93), bottle-to-bottle recycled chips (PET Vogtland GmbH), and bottle-to-bottle recycled flakes (Krones AG). This selection was made to determine the impact of the process conditions of different PET recycling processes.

The experimental setup (Figure 1) consists of a cylindrical glass tube as the SSP-reactor, which was placed in a former gas chromatography oven, because of its temperature uniformity and stability. The respective process gas mixture, composed of nitrogen 5.0 and synthetic air with 20 vol % O_2 and 80 vol% N_2 (Airliquid) is heated up in the oven before it enters the SSP reactor from the bottom through a mesh. Thermocouples are installed in the oven and in the reactor. With the flow control of both gases, the oxygen concentration in the process gas mixture is set to 0, 1, 10, and 20 vol % O_2, respectively. The oxygen concentration is checked in the outlet stream with an oxygen concentration analyser (MOCON portable O_2/CO_2 analyzer). By using a water column of 100 mm, where the outlet stream is led through, the overpressure in the system is set to 10 mbar. This is necessary to check the leak tightness of the system.

Sample Preparation

The sample is heated in the described experimental setup with a constant heating rate of approximately 3.0 K/min under a nitrogen (quality 5.0) atmosphere. When the set temperature is reached, the gas mixture can be varied, whereas the flow is kept steady at 0.15 m^3/h. At the end of the residence or treatment time, the sample is cooled rapidly by cold nitrogen gas and its colour is analysed.

From own prior experiments it is known that a non-coloured PET sample can show significant discolouration after remelting. Therefore, ca. 5.0 g of the treated sample were heated in a mould to 290°C under a thoroughly controlled nitrogen atmosphere.

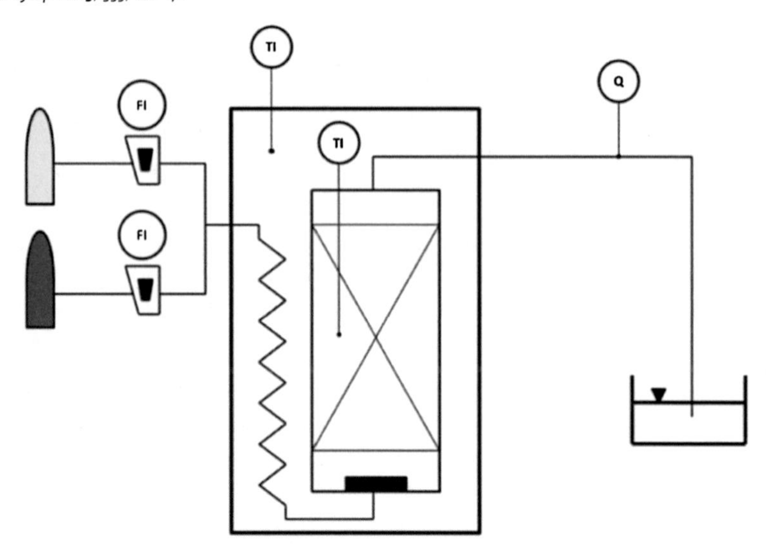

Figure 1.
Schematic of the experimental setup.

The samples were cooled down rapidly in demin. water, once the melting process was completed. This procedure resulted in a thin slice of opaque PET whose colour was also analysed.

To check the influence of dissolved oxygen in the polymer matrix, prior to the experiments some samples were treated under vacuum (7 days @ 65 °C) until the oxygen concentration was below the detection limit of the gas analyser.

Colour Analytics

The colour is measured with two different spectrophotometers (Konica Minolta CM-3600d and ColorLite sph900). All values of the CIELAB L^*-a^*-b^* space were reported. The b^* value is the most interesting parameter, as b^* describes the colour on the coordinate from blue (-100) to yellow (+100). A b^* value of approx. +2 can be identified as slightly yellow with the naked eye. A typical quality parameter of virgin PET bottle-grade chips is $b^* < 1$.

All 130 samples were measured five times with rearrangements between every measurement. In addition, a reference sample was measured regularly for calibration purposes ("white standard"). The overall treatment and analytical procedure are shown in Figure 2.

Results

To interpret the results, a look at the standard deviations of the colour measurements is essential. The deviations are highest for flake (0.5 in b^*), medium for recycled chips (0.2 for b^*) and lowest for virgin PET chips (0.1 in b^*). Knowledge of the standard deviations is necessary in interpreting the results of the measurements, so that their significance can be determined.

The virgin material was examined first. Here, the high thermal stability with and without oxygen atmosphere was considerable, as can be seen in Figure 3. Discolouration as a function of residence time was barely measureable at temperatures of 190 °C. A significant discolouration was measured at 200°C and continued to 210°C. Also, the oxygen concentration led to an increase in "yellowing" through the whole temperature range between 10 and 20% and became stronger at higher temperatures.

The behaviour of the bottle-to-bottle recycled chips was similar to the virgin

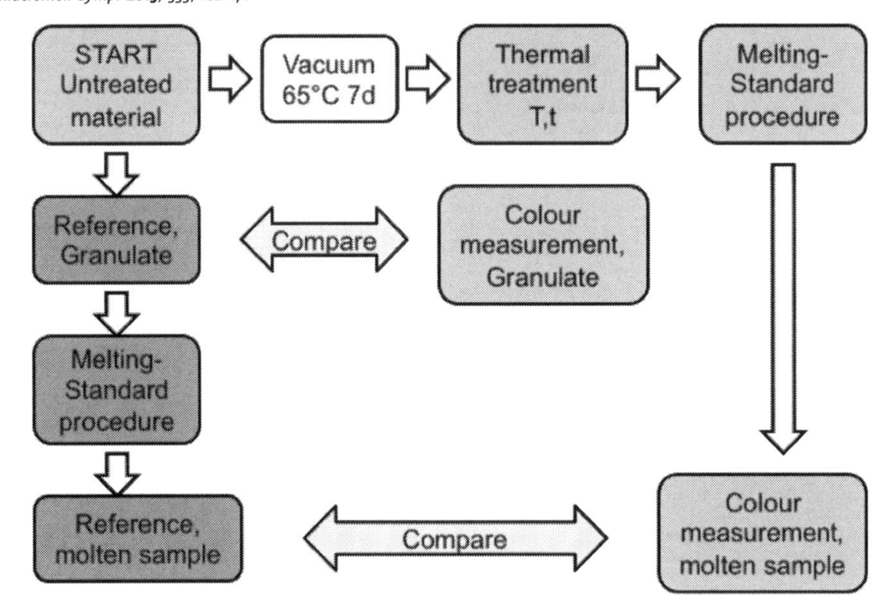

Figure 2.
Summary of the sample treatment for colour measurement. The vacuum treatment was only performed for selected samples.

material. However, the effects were much stronger, so the yellowing started to take place at more moderate conditions and reached values of $b^* > 5$ (Figure 4). The same effects of stronger yellowing by residence time or oxygen concentration can be observed. As the effects were much stronger, the material might be difficult to process and use, due to yellowing, for example, during drying.

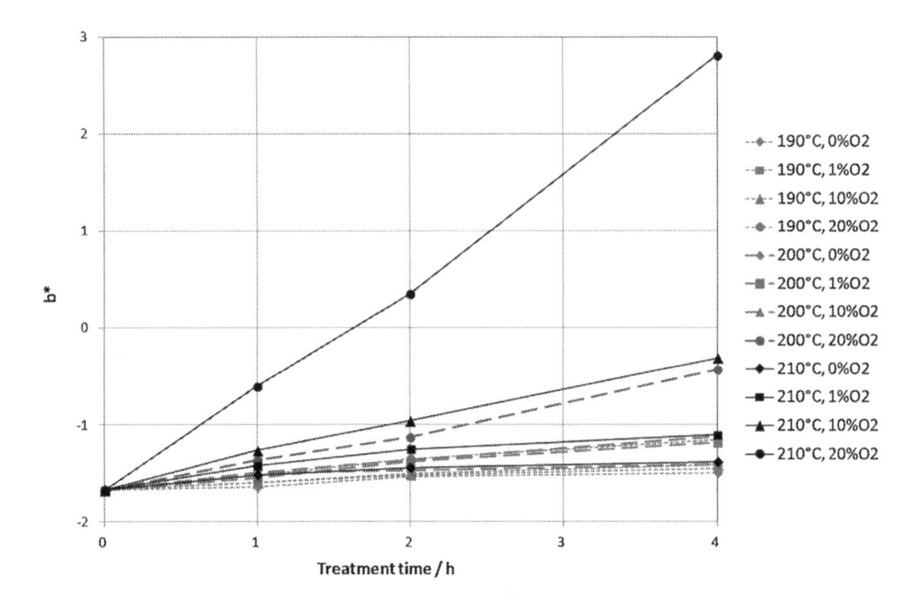

Figure 3.
Discolouration (b^*) of virgin PET chips vs. treatment time for different temperatures and oxygen concentrations.

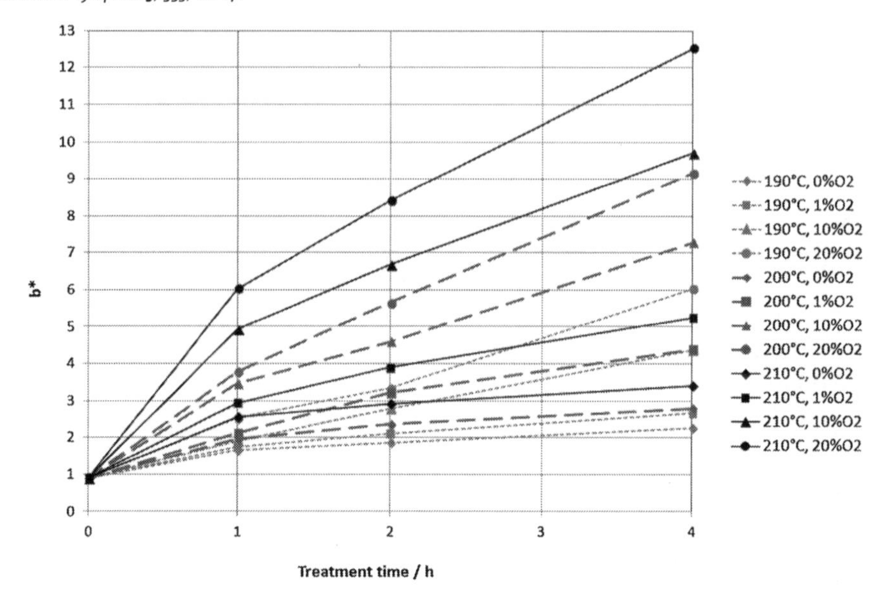

Figure 4.
Discolouration (b*) of recycled PET chips vs. treatment time for different temperatures and oxygen concentrations.

The PET-flakes showed the same impacts at higher values of influencing parameters, but a small statistic significance at lower ones, which made it difficult to interpret these measurements (see appendix, Fig. 11).

The idea of obtaining fewer outliers through melting the samples did not seem to be right. The measurements of the molten samples (e.g., recycled chips) showed the same trend as the original ones, but the standard deviation was larger for every sample. This effect was the largest for flakes. The virgin material improved its colour after melting (except under extreme conditions, see Figure 5). This colour improvement verifies the idea of an optimised temperature and oxygen concentration tolerance after melting.

The oxygen concentration had a strong influence on the yellowing mechanism during treatment time on every material: Virgin resin as well as recycled chips and recycled flakes.

The change of discolouration of recycled chips between 0 and 20% oxygen concentration was very steady. The highest gradient was observed between 0 and 1% oxygen concentrations (Figure 6). This supports the hypothesis that there is a mechanism change from thermal degradation to thermo-oxidative degradation, which supports the proposal by Jabarin et al.[6]

The same effects were observed in experiments with virgin material, where the discolouration was less than the discolouration of recycled chips. The gradient at the highest temperature up to the maximum oxygen concentration exceeded the start concentration gradient, but even under these experimental conditions, the effect of a mechanism change was detectable.

As proposed by MacDonald,[4] there might be an influence in discolouration, because of the presence of DEG. To examine this influence, two laboratory samples (courtesy by Dr. Otto from Lurgi Zimmer GmbH) with unusually high (3.7 weight %) and low DEG (0.36 weight %) contents were tested with our standardised method. However, in deviation from the other experiments, treatment temperatures of 200°C and 230°C were chosen as lower temperatures already have shown small impact on colour formation of virgin chips. The high DEG material responded to both

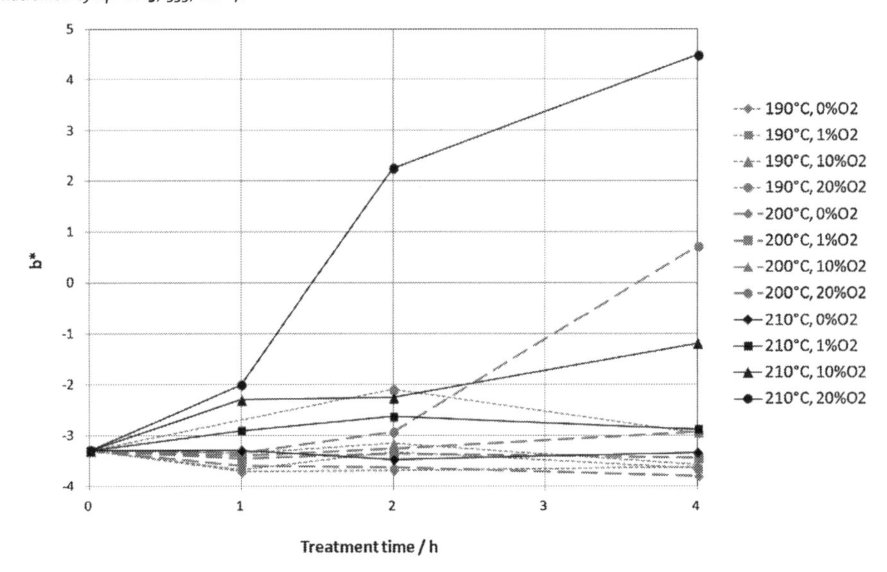

Figure 5.
Discolouration (b*) of virgin, molten PET chips vs. treatment time for different temperatures and oxygen concentrations.

thermal and thermal-oxidative treatment with significantly higher discolouration (Figure 8). The sample with low DEG content shows small colour changes even at 230°C whereas the sample with high DEG content is strongly affected by thermal and thermal-oxidative degradation.

Without oxygen at high temperatures and low DEG contents, there was no significant change in colour of the material as opposed to material with high DEG contents. This result confirms the proposals of MacDonald [4] and Ciolacu and Choudhury.[5]

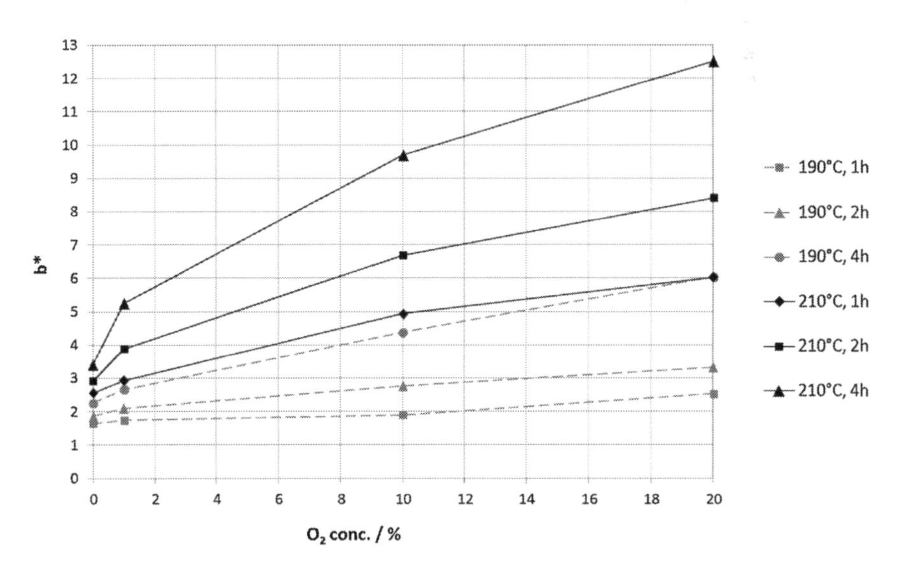

Figure 6.
Discolouration (b*) of recycled PET chips vs. oxygen concentration for different treatment times and temperatures.

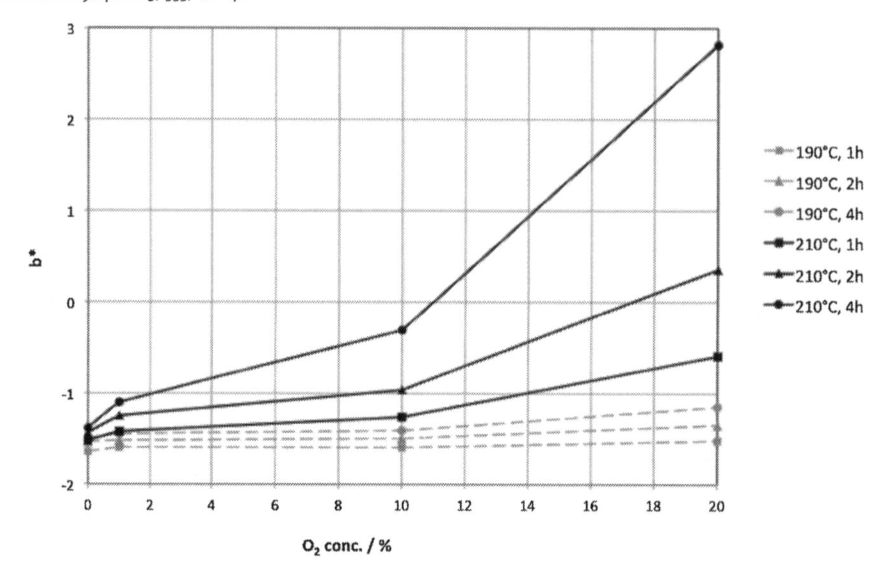

Figure 7.
Discolouration (b*) of virgin PET chips vs. oxygen concentration for different treatment times and temperatures.

As described above, there was a vacuum treatment of some samples before starting the experimental series in order to examine the influence of oxygen, which is dissolved in the solid PET matrix. The results of this experiments, as a comparison between non-treated and vacuum treated samples, revealed no significant reduction of yellowing by vacuum treatment. This led to the conclusion that the dissolved oxygen in the PET resin at saturated conditions of 0.2 bar does not contribute significantly to the discolouration of PET.

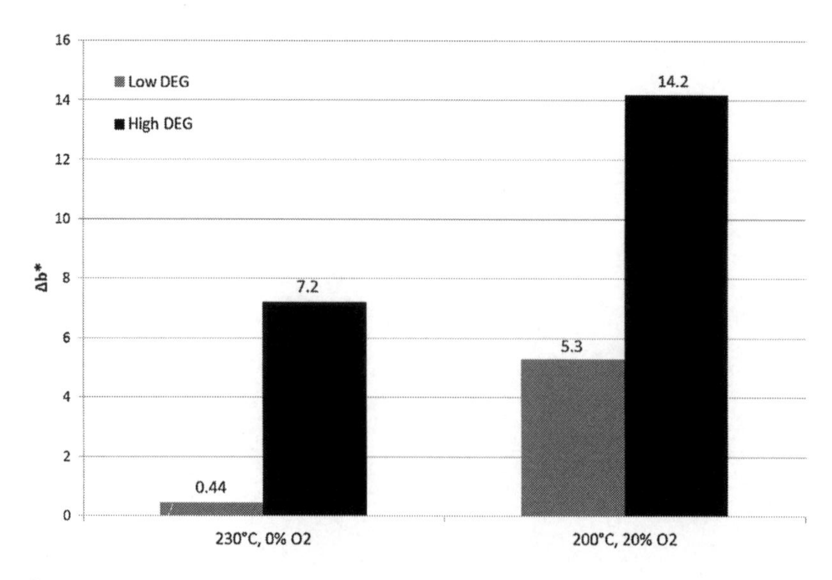

Figure 8.
Discolouration (b*) of PET samples with low and high DEG content without oxygen at 230 °C and with 20% oxygen at 200 °C.

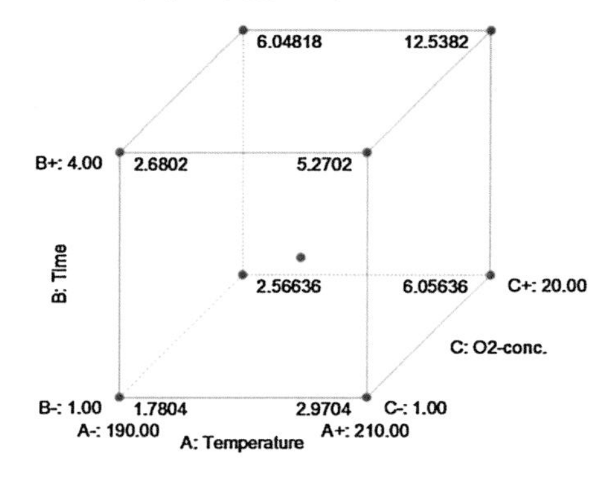

Figure 9.
Design of Experiments - two level factorial design for recycled PET chips.

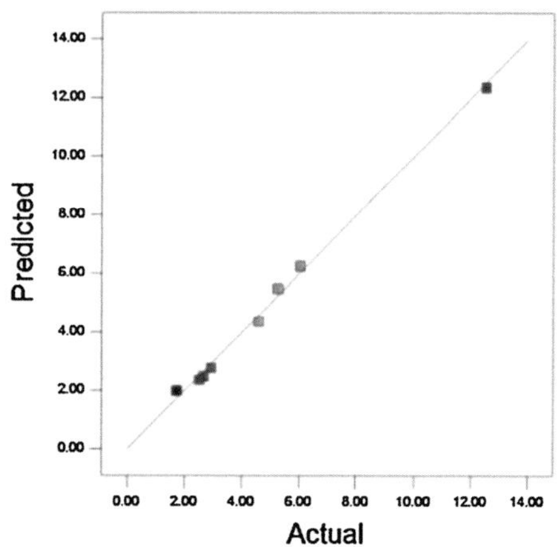

Figure 10.
Predicted b* vs. measured b*, 2^3 design for recycled PET chips.

Design of Experiments

The term "Design of Experiments" (DoE) is associated with a number of statistical methods, which have been developed to maximise the information that can be gathered with a minimum number of experiments. DoE is designed for multiple input factors and allows the identification of factor interactions, which would not be possible by varying only one factor at a time. The significant factors and interactions are summarised to an empirical equation, which allows the prediction of the response at a given set of parameters with adequate accuracy. The software "Design Expert" from StatEase Inc. was used to quantify the discolouration of PET.

The factorial design was limited to the different PET samples, but not linked among the materials, to enable a better description. A full factorial two level design was used for the experiments on virgin and recycled material. The upper and lower limits corresponded to the experimental set-up except for the lower range of oxygen concentration, which was 1%. This was due to the higher gradient at short residence times, which can only be described by a higher order model. There is one centre point used in the calculations that is the average of every range. The resulting design space is shown in Figure 9 as a 3-D – cube for the sample of recycled PET chips.

The result of the DoE calculations as a response of the model was checked by comparison between predicted and measured b* values, which show good agreement (Figure 10).

The response, i.e., the discolouration of the material, can be expressed by an equation comprising seven terms of different actual factors. The units for calculation are temperature T/°C, oxygen concentration O_2/vol% and time t/h:

$$b^* = 2.864 - 0.005 \cdot T - 1.589 \cdot O_2 - 6.859 \cdot t + 0.008 \cdot T \cdot O_2 + 0.037 \cdot T \cdot t + 0.059 \cdot O_2 \cdot t$$

The same method and design space was used for virgin material. Mathematically, there is a problem because of a prefix change in the response parameter b*. Therefore, a transformation had to be made to create a positive response throughout the design space. With this transformation, DoE yielded the following equation for the prediction of discolouration of virgin material:

$$\frac{1}{\sqrt{(b^* + 1.75)}} = 11.087 - 0.044 \cdot T + 0.132 \cdot O_2 - 1.053 \cdot t - 8.3E^{-4} \cdot T \cdot O_2 + 4.3E^{-3} \cdot T \cdot t$$

Conclusion

The undesired discolouration of PET was systematically investigated with the intention to specify parameters to allow a quantification of the overall discolouration in dependency of the most important parameters of influence. Up to now, a number of reaction paths and mechanisms towards different positively identified chromophores have been postulated in the scientific literature.

The results of the systematic experimental investigation of different materials (virgin chips, recycled chips, and recycled flakes) indicate a strong influence of temperature, residence time, and oxygen concentration of the process gas on discolouration. Dissolved oxygen is not affecting the colour at thermal treatment in solid PET. No significant difference in discolouring between molten samples and non-molten samples was found after thermal treatment in solid state. The experiments with different DEG content confirmed a strong impact of DEG on discolouration.

With Design of Experiments (DoE) it was possible to identify the parameter interactions and dependencies. It was determined that a linear two-level factorial design with nine experiments (eight plus one centre point) is capable of generating an adequate predicted response for the three main factors of influence (temperature, residence time, oxygen concentration), assuming that there is no mechanism change within the design space. The mathematical prediction of PET yellowing was possible for virgin and recycled chips. The standard deviations of the experimental data for recycled flakes were very high. Therefore it was not possible to calculate a model equation and only trends can be drawn from those experimental data.

The proposed model functions gained by the DoE approach do not base on physicochemical insights. It is not possible to extrapolate the response beyond the given design space and the validity of the model equations is restricted to the samples in connection with our experiments. If the virgin PET recipe, the recycling feedstock or the process conditions of the recycling processes change, the impact of temperature, residence time and oxygen concentration of the process gas might change also.

Appendix

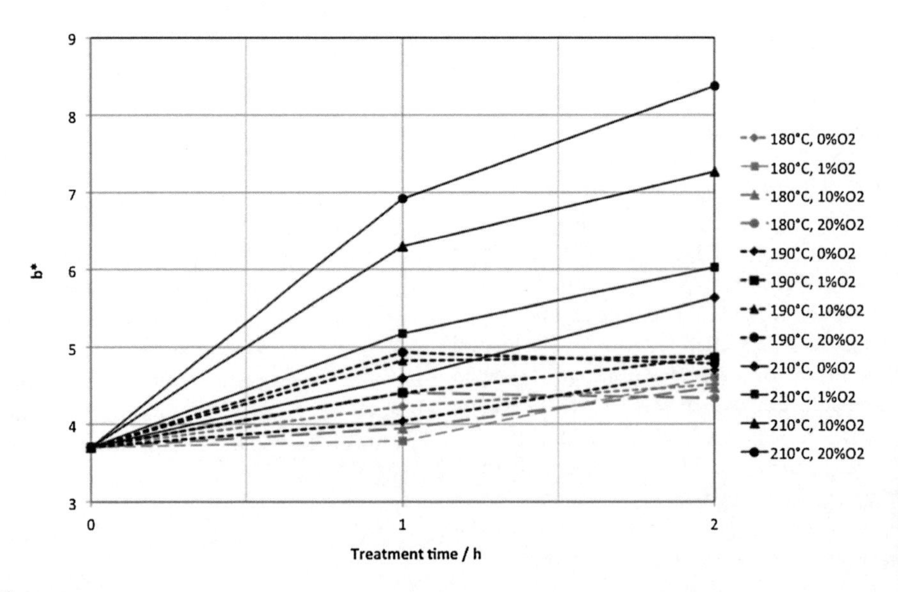

Figure 11.
Discolouration (b*) of recycled PET flake vs. treatment time for different temperatures and oxygen concentrations. Data with higher standard deviations at low changes of b*.

[1] H. Zimmerman, N. Kim, *Polymer Engineering and Science* **1980**, *20*, 680–683.
[2] M. Edge, R. Wiles, *Polym. Degrad. Stability*, **1996**, 53, 141–151.
[3] M. Edge, N. S. Allen, *Polymer*, **1995**, *36*, 227–234.
[4] W. A. MacDonald, *Polymer International*, **2002**, *51*, 923–930.
[5] L. Ciolacu, N. Choudhury, *Polym. Degrad. Stability*, **2006**, *91*, 875–885.
[6] S. A. Jabarin, E. A. Lofgren, *Polymer Engineering and Science*, **1984**, 24, 1056–1063.
[7] Th. Rieckmann, S. Völker, PET Polymerization - Catalysis, Reaction Mechanisms, Kinetics, Mass Transport and Reactor Design. in: *Modern Polyesters: Chemistry and Technology of Polyesters and Copolymers*, J., Scheirs, T. E. Long, Eds., John Wiley & Sons, **2003**.
[8] D. E. James, L. G. Packer, Effect of reaction time on poly(ethylene terephthalate) properties, *Ind. Eng. Chem. Res.*, **1995**, 34, 4049–4057.

 DOI: 10.1002/masy.201300052

Experimental and Modeling Study of Melt Polycondensation Process of PA-MXD6

*Zhenhao Xi, Like Chen, Yong Zhao, Ling Zhao**

Summary: The main polycondensation reaction of poly(m-xylene adipamide) (PA-MXD6) is a reversible reaction strongly coupled with mass transfer in melt polycondensation process. In this work, a realistic model for melt polycondensation of PA-MXD6 has been proposed taking in to account the kinetics data, equilibrium data and diffusion process for major by-product water in the melts. The characteristics of reaction and the effects of mass-transfer of polycondensation process have been studied by using stagnant film experiments. It is observed that the apparent rate of the polycondensation process increases with higher temperature, lower degree of vacuum and thinner film thickness with reduced specific interfacial area. Based on the experimental data, the model parameters including the kinetics data, equilibrium data and mass transfer coefficient of volatile have been estimated by the nonlinear least squares method. The model predictions are in a quite satisfactory agreement with the experimental data that all of the relative deviations are almost less than 2%.

Keywords: concentration of end groups; mass- transfer; melt polycondensation; model; poly(m-xylene adipamide)

Introduction

Poly(m-xylene adipamide) (PA-MXD6) is an aromatic polyamide prepared by the polymerization of m-xylylene diamine and adipic acid.[1,2] It is a semi-crystalline polymer which has the following molecular structure, as shown in Figure 1. Due to the slow crystallization and melting behaviour of PA-MXD6,[3] it shows good processability in the thermoforming and stretching operations. Due to its high mechanical strength, modulus, heat resistance, and thermal stability, PA-MXD6 is commonly blended with other polymers, such as PET, polyethylene, and polypropylene, and is suitable for engineering structural materials, especially when reinforced with glass fibre. Its good flow behavior, low moisture absorption, and chemical resistance enhance its usefulness. PA-MXD6 has excellent gas barrier properties, particularly low oxygen permeation which is much lower than that of other conventional polyamide resins, polyeaters, or poly(vinylidene chloride) (PVDC).[4,5] Due to its distinguished barrier properties, it is finding increasing use in multilayer food packaging applications such as flexible film and multilayer, stretch-blown bottles.[6] The availability of grades with different molecular masses permits use as components of polymer blends or multilayer films for retortable pouches, cup lids, flexible tubing, rigid bottles, etc.

The synthesis of PA-MXD6 is usually performed by direct polycondensation with two stages. In the early stage, low molecular weight oligomers are prepared by adding MXDA continuously to molten adipic acid under atmospheric pressure or a low steam pressure. Then the oligomers being exposed to a vacuum are polymerized in the second stage to a desired high molecular weight. As the main polycondensation

State Key Laboratory of Chemical Engineering, East China University of Science and Technology, 200237, Shanghai, China
E-mail: zhaoling@ecust.edu.cn

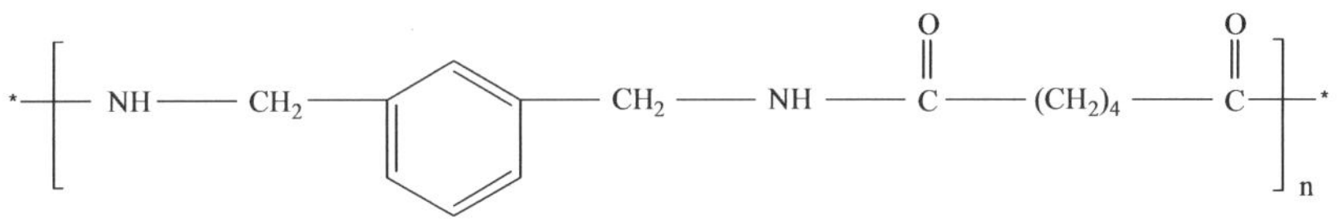

$$*\!-\!\left[\mathrm{NH}-\mathrm{CH_2}-\mathrm{C_6H_4}-\mathrm{CH_2}-\mathrm{NH}-\overset{\overset{\displaystyle O}{\|}}{C}-(\mathrm{CH_2})_4-\overset{\overset{\displaystyle O}{\|}}{C}\right]_n\!-\!*$$

Figure 1.
Repeating unit of MXD6.

reaction of PA-MXD6 is a reversible reaction, by-product (H_2O) has to be continuously removed from the highly viscous melt to promote chain growth reactions during the polycondensation.[7] Since the reaction and mass-transfer are coupled in polycondensation process, the process rate depends not only upon the chemical kinetics of polycondensation reaction but also upon the mass-transfer of volatile by-product through the melts.[8–10]

In spite of the commercial importance of PA-MXD6, there is little detailed knowledge about the polycondensation process especially the kinetics and mass-transfer data of polycondensation. The complex interconnection between the chemical reaction and mass-transfer problem in PA-MXD6 melt polycondensation process poses a serous challenge to make process more efficient and controllable. Generally the polycondensation rates are limited by the mobility of the reactive amino and carboxyl end groups of the polymer chains and by the diffusivity of water out of the melts; and hence, the molecular weight vary depending on the controllable operation conditions such as the temperature, pressure (vacuum degree), surface area of bulk melts and polycondensation time.[11,12]

In this article an overall apparent reaction kinetic model coupling reaction and mass transfer of the polycondensation process of PA-MXD6 was established. The characteristics of reaction and the effects of mass-transfer of the PA-MXD6 melt polycondensation process were studied by using stagnant-film experiments on bulk samples. Based on that, the parameters of the overall apparent reaction kinetic model were estimated, which could be used to simulate the polycondensation process of PA-MXD6.

Experimental Part

Materials

Adipic acid ($\geq$99.5%, analytical reagent) and sulphuric acid (98–100%, purity) were provided by Shanghai lingfeng chemical reagent Co., ltd. *m*-xylylene diamine (99.5%, polymer grade), o-cresol($\geq$99%, ReagentPlus®), chloroform ($\geq$99%, anhydrous), potassium hydroxide($\geq$85% KOH basis, pastilles) oxalic acid ($\geq$99%, anhydrous) and ethanol ($\geq$99.5%, anhydrous) were provided by Sigma-Aldrich.

The pre-polymer of PA-MXD6 was prepared with *m*-xylylene diamine and adipic acid in our laboratory for this study. The relative viscosity (η_r) of pre-polymer was 2.136, and the concentration of carboxyl end group was 0.0592 mol/kg.

Polycondensation

A typical example of stagnant-film experiments was executed as follows. The pre-polymer samples with different weight ($2.5 \sim 12$ g) were placed uniformly in the flat-bottomed thin- film reactors with same square size 10×10 cm. The paralleled reactors were connected to a public pipe that led to the nitrogen source with provision for applying vacuum. After the reactors were purged with nitrogen to remove all air, they were then heated in silicone oil bath to the desired temperature of 240–280 °C over the melting point of the pre-polymer. While holding at the temperature for 5 min, a vacuum from 50 Pa to 1500 Pa, adjusted by the thermocouple vacuum mometer, was applied with an oil rotary vacuum pump. After being exposed to the vacuum for a certain time, the samples were cooled down quickly together with the reactors. Figure 2 was the schematic of experimental apparatus.

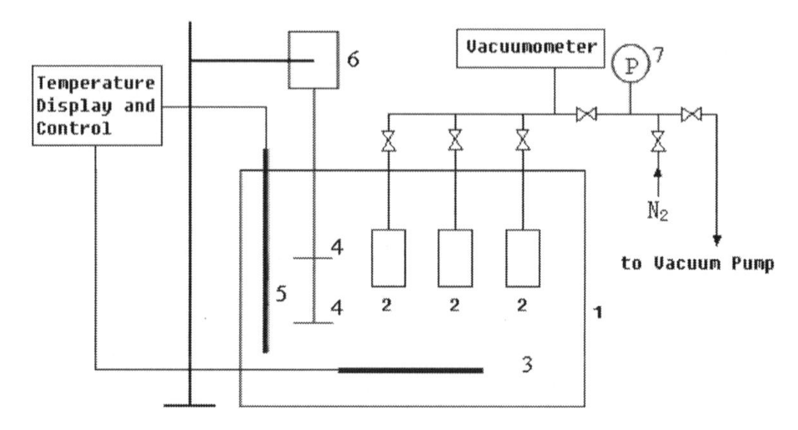

Figure 2.
Schematic of experimental apparatus: 1, silicone oil bath; 2, thin-film reactor; 3, electric heating elements; 4, agitator; 5, thermocouple; 6, motor; 7, pressure-and-vacuum gauge.

Before all subsequent processes, PA-MXD6 was shredded finely and then dried in a vacuum oven at 80 °C for 12h to minimize the possibility of hydrolytic degradation.

Concentration of Carboxyl End Groups

About 500 mg of PA-MXD6 were dissolved in 25 ml 70/30 (w/w) o-cresol/chloroform mixtures in the state of reflux. After dissolution, titration was performed through the use of a 0.05 mol/L potassium hydroxide ethanol solution. The titration agent was added at a rate of 0.20 ml/min that was controlled by a ZDJ-4B Automatic Potentiometric Titrator from INESA Scientific Instrument (Shanghai, China). Glass membrane electrode and standard calomel electrode also came from INESA Scientific Instrument. The pH curve was recorded on PC by specific Rex titration software. The following equation was used to calculate the concentration of carboxyl end groups.

$$[-COOH] = \frac{(V_1 - V_0) \times c}{m} \quad (1)$$

where $[-COOH]$ was the concentration of carboxyl end groups, mol/kg; V_1 was the titration volume for sample solution, ml; V_0 was the titration volume for blank test, ml; c was the contertation of the potassium hydroxide ethanol solution; and m was the weight of the PA-MXD6 sample, g.

Relative Viscosity

About 500 mg of PA-MXD6 were dissolved in 40 ml sulphuric acid (98–100%) and then the solution concentration was adjusted to about 10 g/L by 50 ml volumetric flask at 25 °C. The relative viscosity measurements of the solution were carried out on an Ubbelhode viscometer (SYP1003-IA), performed at 25 °C. An empirical linear correlation [eq. (2)] between average molecular weight M_n and relative viscosity η_r was used for further M_n calculation.[13]

$$M_n = K_1(\eta_r - K_2) \quad (2)$$

where $K_1 = 1.6 \times 10^4$ *g/mol*, $K_2 = 1.092$, and η_r is the relative viscosity.

Apparent Model of Polycondensation Reaction Rate

The development of a reaction scheme for polycondensation requies distinguishing funtional species from chemical ones. Funtional species are the entities directly involved in chemical rection like amino end-group, carboxyl end-group, or amide bond, in case of polyamides. However, only the reversible chain growth reaction between the functional end-groups, which is the main reaction during the polycondesation process, needs to be considered in our model to describe the change of the molecular weight of PA-MXD6. As the step-growth polycondensation of polyamides can be treated as a second-order reversibile reaction of carboxyl end-group (-COOH) and amino end-group (-N2H) to form amide bond (-CONH-), with the elimination of water (H2O), the thermodynamic rate and equilibrium constant are defined in terms of activities of the four components as follows:

~~~ PA-MXD6~~~ COOH + NH_2 ~~~ PA-MXD6~~~

~~~ PA-MXD6~~~CONH ~~~ PA-MXD6~~~ + H_2O

$$r = -\frac{d[-COOH]}{dt} = \frac{d[g]}{dt} = k \cdot \left([-COOH][-NH_2] - \frac{[A][g]}{K} \right) \tag{3}$$

where r is the intrinsic reaction rate, k is the rate constant of forward reaction, K is the reaction equilibrium constant, and $[-COOH]$, $[-NH_2]$, $[A]$ and $[g]$ represent the concentration of carboxyl end-group, amino end-group, amide bond and water respectively.

During the polycondensation process, the viscosity of the polymer melts typically increases by orders of magnitude with the reaction proceeding and the process becomes gradually limited by mass transfer. If the volatile by-product can be removed in time, the concentration of end-groups will decrease while the molecular weight of polymer will increase, and the chain growth reaction will go forward and reach the reaction equilibrium. In respect that the PA-MXD6 melts in thin-film reactor have huge specific interfacial area and are exposed to high vacuum, it can be assumed that the process can reach pseudo-steady state between the reaction and mass-transfer,[14,15]

$$N = k \cdot \left([-COOH][-NH_2] - \frac{[A][g]}{K} \right) \cdot W = K_L S \left([g] - [g]^* \right) \tag{4}$$

where $[g]^*$ is the interfacial concentration of volatile.

From above equation Eq. (4), $[g]$ can be converted to the expression:

$$[g] = \frac{k \cdot W \cdot [-COOH][-NH_2] + K_L \cdot S \cdot [g]^*}{K_L \cdot S + \frac{k \cdot W}{K} \cdot [A]} \tag{5}$$

If an overall apparent rate of PA-MXD6 melts polycondensation process is defined by the expression:

$$r_T = \frac{d[g]}{dt} = k_T \cdot \left([-COOH][-NH_2] - \frac{[A][g]^*}{K} \right) \tag{6}$$

it can be inferred that,

$$\frac{1}{k_T} = \frac{1}{k} + \frac{[A]/K}{K_L \cdot S_W} \tag{7}$$

where k is the temperature-dependent rate constant with an Arrhenius type $k = k_0 \cdot e^{-E/RT}$, S_W is interfacial area per unit weight of polymer, K_L is the overall mass transfer coefficient, and $[A]$ is equilibrium amido concentration which is nearly invariable after the Dp is over 30 ($[A] \approx 8.122 mol/kg$).

Assuming that the only end groups in PA-MXD6 are primary amine ($-NH_2$) and carboxylic acid (-COOH), the primary amine concentration $[-NH_2]$ can be calculated from the following equation [13]:

$$M_n = \frac{2 \times 10^3}{[-NH_2] + [-COOH]} \tag{8}$$

where M_n can be calculated through relative viscosity η_r using Eq. (2).

The interfacial concentration of volatile can be expressed as a thermodynamic functions related to the temperature and pressure as follows:

$$[g]^{*} = A\exp(-E_g/RT) \times P \quad (9)$$

where A is the pre-exponential factor, E_g is the activation energy, R is the universal gas constant, T is the melting temperature, and P is the pressure.

Results and Discussion

According to the overall apparent polycondensation rate model, the rate constant of forward reaction, the reaction equilibrium constant and the interfacial concentration of volatile are closely related to the processing conditions. Hence, in this experimental and modeling studies, the effects of polycondensation temperature, pressure, and film thickness on both the concentration of carboxyl end groups and the polymer molecular weight were investigated comprehensively.

Effect of Temperature

Figure 3 shows the effects of polycondensation temperature on the concentration of carboxyl end groups and the molecular weight of PA-MXD6 with 1.0 mm melts film thickness under 100 Pa. Notice that the concentration of carboxyl end groups decreases with the increasing polycondensation reaction temperature, while the molecular weight increases correspondingly. As the higher reaction temperature accelerates the diffusion rate of the chain end groups and increases the reaction rate of melt polycondensation, the polymer molecular weight increases more rapidly. It can also be seen that, over 75 min both of the carboxyl end groups and the molecular weight have little change with reaction time and even decrease at high temperature of 280 °C. That is due to the onset of thermal and oxygen-dependent degradation reactions especially for long time at high temperature.

Effect of Pressure

The effects of the pressure on the concentration of carboxyl end groups and the molecular weight of PA-MXD6 changed over time with 1.0 mm melts film thickness at 260 °C are shown in Figure 4. The pressure observably has an affect on the molecular weight and the melt polycondensation rate of PA-MXD6. As the lower pressure (higher vacuum degree) leads to a lower partial pressure of the volatile in the gas phase, the corresponding interfacial concentration of volatile in the melt is also smaller, which increases its concentration

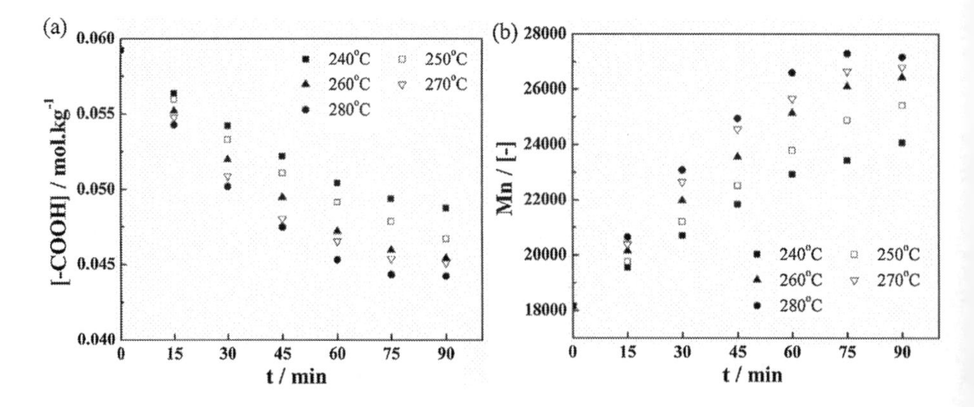

Figure 3.
Effect of temperature on concentration of carboxyl end groups and molecular weight, P = 100 Pa, d = 1 mm: (a) concentration of carboxyl end groups [-COOH]; (b) molecular weight M_n.

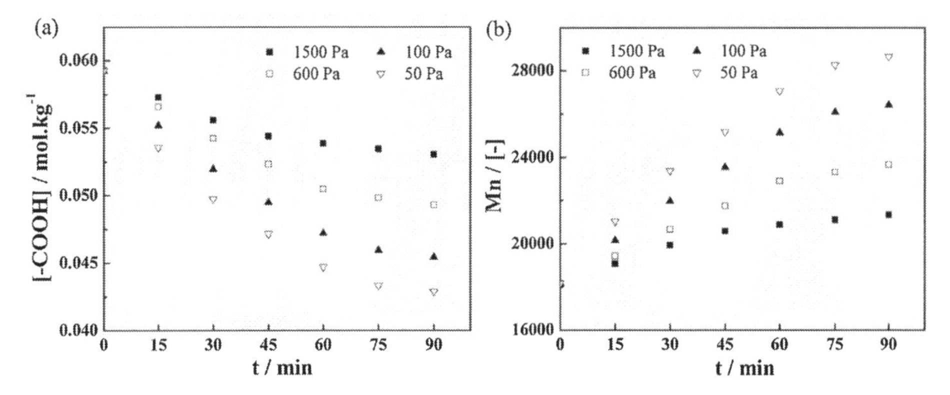

Figure 4.
Effect of pressure on concentration of carboxyl end groups and molecular weight, T = 260 °C, d = 1 mm: (a) concentration of carboxyl end groups [-COOH]; (b) molecular weight M_n.

gradient of the volatile from the bottom to surface in the film and accelerates the diffusion rate of the volatile. Thus, the overall reaction rate of melt polycondensation is improved and the polymer molecular weight increases more rapidly. It can also be observed that the overall reaction rate of melt polycondensation is seriously limited by the mass transfer when the pressure is over 100 Pa. Under this circumstances, it is difficult to get polymer with high molecular weight even with longer reaction time. Although low pressure is desirable from a mass-transfer point of view, industrial PA-MXD6 plants are limited to a mimimum pressure around 50 Pa due to the high equipment investment.

Effect of Melts Film Thickness

Figure 5 shows the effects of melts film thickness on the concentration of carboxyl end groups and the molecular weight of PA-MXD6 at 260 °C under 100 Pa. A decrease in the melts film thickness not only shortens the diffusion path of the volatile inside the melts but also increases the concentration gradient of the volatile from the bottom to surface in the film. The resistance to the diffusion of the by-product from the bulk melts to the film surface decreases pronouncedly with the film thickness decreasing. Meanwhile, the decrease in the film thickness increases the specific surface area for the diffusion into the gas phase. Therefore, it is expected that the melt

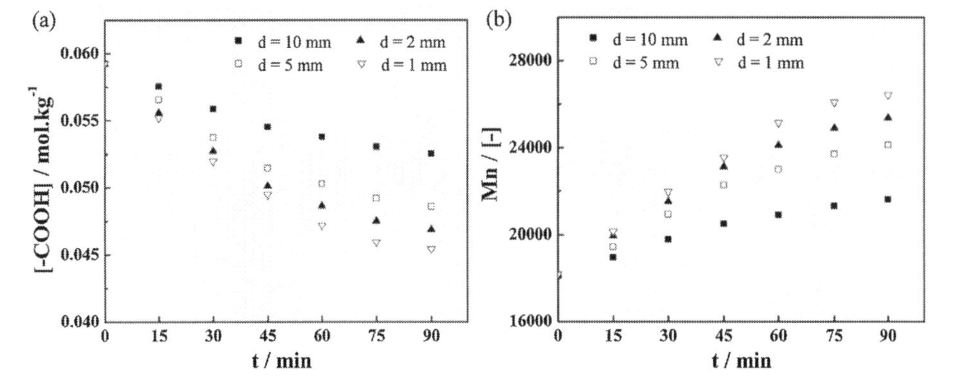

Figure 5.
Effect of film thickness on concentration of carboxyl end groups and molecular weight, T = 260 °C, P = 100 Pa: (a) concentration of carboxyl end groups [-COOH]; (b) molecular weight M_n.

polycondensation rate increases with the decreasing melt film thickness while the carboxyl end groups decreases correspondingly. This agrees with the results shown in Figure 5.

Parameter Estimation of Proposed Model

Using the experimental data reported above, the estimation of the kinetic data, equilibrium data and mass transfer coefficient of volatile via the nonlinear least squares (LSQ) method have been done by Matlab (The MathWorks, Inc.). With Matlab the integrator *ode45* and the optimisation routine fmins are implemented, and the objective function for the optimization is based on the concentration of carboxyl end-groups. The estimated results are summarized in Table 1.

$$f_{lsq} = \sum_{i}^{n} \left([-COOH]_i - [-COOH]_{c,i} \right)^2 < \varepsilon \tag{10}$$

where n is the number of experimental data, and ε is the fitness valves of objective functions.

The applicability of the apparent model of polycondensation reaction rate to the experimental values of [-COOH] in Figures 3–5 are shown in Figures 6–8, respectively. In these three figures each point represents an experimental value and the solid lines represent the apparent model predictions using the parameters above. From these

Table 1.
Parameters for the apparent model of polycondensation reaction rate

Parameter	Value	Units
A	8.92×10^{-3}	$mol.kg^{-1}.Pa^{-1}$
E_g	11.8	$kJ.mol^{-1}$
k_o	3.35×10^5	$kg.mol^{-1}.min^{-1}$
E	63.9	$kJ.mol^{-1}$
K	3.78×10^2	–
K_L	0.320	$m.min^{-1}$

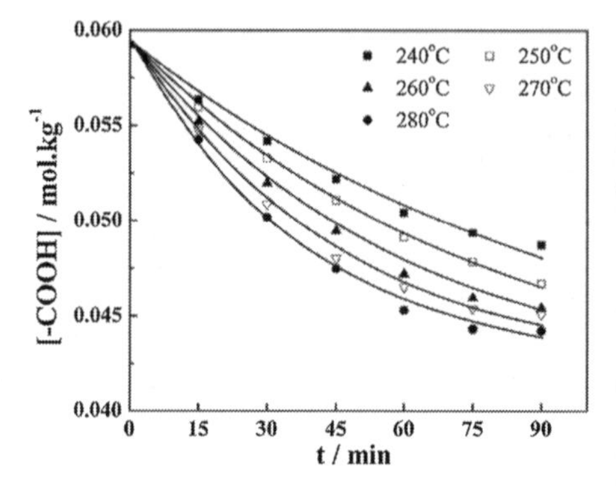

Figure 6.
Comparison between experimental data and model predictions for the concentration of carboxyl end groups, P = 100 Pa, d = 1 mm.

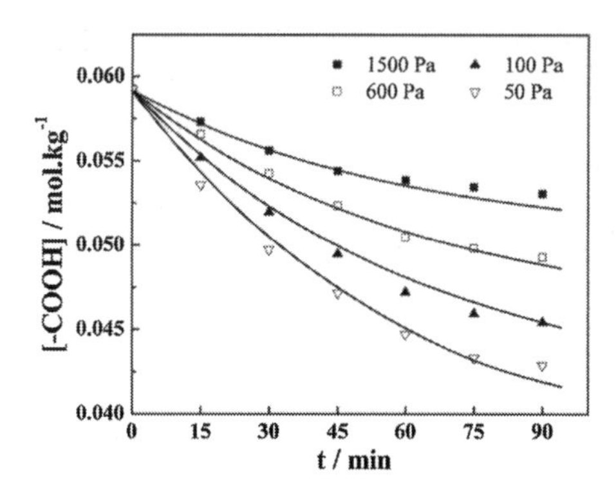

Figure 7.
Comparison between experimental data and model predictions for the concentration of carboxyl end groups, T = 260 °C, d = 1 mm.

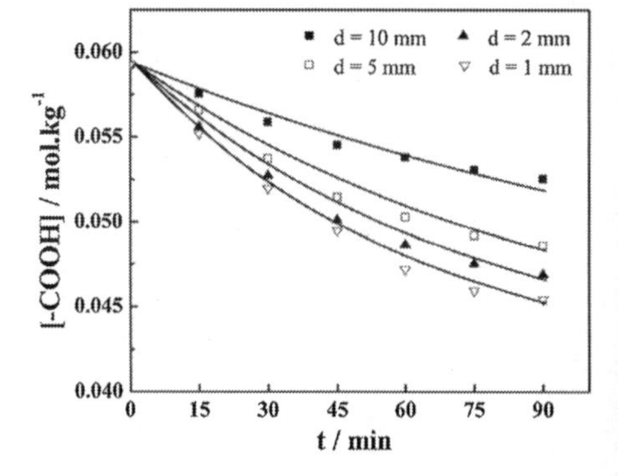

Figure 8.
Comparison between experimental data and model predictions for the concentration of carboxyl end groups, T = 260 °C, P = 100 Pa.

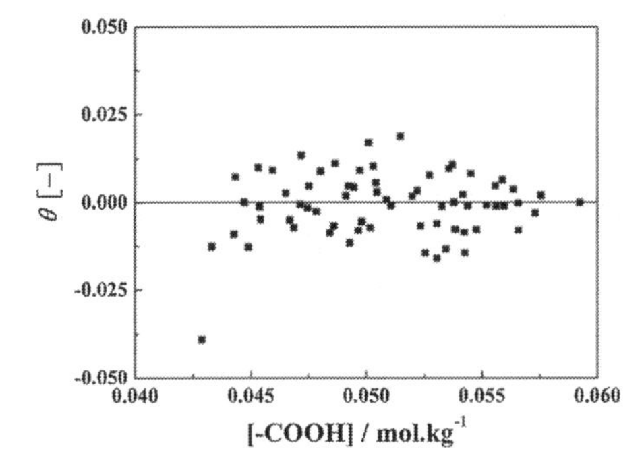

Figure 9.
Relative deviation of concentration of carboxyl end groups between experimental data and model predictions.

figures it is evident that the model predictions are in a quite satisfactory agreement with the experimental data. As is apparent, all of the relative deviations are almost less than 2%, as shown in Figure 9.

Conclusion

As the reversible polycondensation reaction and mass-transfer are strongly coupled in melt polycondensation process of PA-MXD6, the characteristics of reaction and the effects of mass-transfer in the process have been studied by using stagnant film experiments in this work. The apparent rate of the polycondensation process increases with higher temperature, lower degree of vacuum and thinner film thickness with reduced specific interfacial area.

A realistic model for melt polycondensation of PA-MXD6 has been proposed taking in to account the kinetics data, equilibrium data and diffusion process for major by-product water in the melts. The pseudo-steady state between the reaction and mass-transfer has been assumed in thin-film reactor. Based on the experimental data, the model parameters including the kinetics data, equilibrium data and mass transfer coefficient of volatile have been estimated by the nonlinear least squares method using Matlab. Then, the comparison between experimental data and model predictions have been done. The model predictions are in a quite satisfactory agreement with the experimental data that all of the relative deviations are almost less than 2%. The realistic model can be used to simulate and forecast the polycondensation process of PA-MXD6.

Acknowledgements: This research was supported by National Natural Science Fundation of China(General Program: 21176070) and the Fundamental Research Funds for the Central Universities.

[1] F. G. Lum, E. F. Carlston, *Ind. Eng. Chem. [J].* **1957**, (47), 1239.
[2] U.S. Pat. 2,997,463 (1961); 3,968,071 (1975); 4,433,136 (1984); 4,398,642 (1983); 6,596,803 (2003); Int. Pat. WO 98/58790 (1998).
[3] B. B. Doudou, E. Dargent, J. Grenet, *J. Therm. Anal. Calorim.*, **2006**, (85), 409.
[4] Mitsubishi Gas Chemical Catalog. "MX-Nylon", **2000**, 5.
[5] S. Seif, M. Cakmak, *Polymer*, **2010**, (51), 3762.
[6] Y. S. Hu, V. Prattipati, S. Mehta, et.al., *Polymer*, **2005**, (46), 2685.
[7] O. George, "*Principles of Polymerization*", New York John Wiley &Sons, **1981**.
[8] D. D. Steppan, M. F. Doherty, M. F. Malone, *J. Appl. Polym. Sci.*, **1987**, (33), 2333.
[9] Th. Rieckmann, S. Völker, *Chem. Eng. Sci.*, **2001**, (56), 945.
[10] J. Xie, *J. Appl. Polym. Sci.*, **2002**, (84), 616.
[11] S. K. Gupta, A. Kumar, "*Reaction Engineering of Step Growth Polymerization*", Plenum Press, New York **1987**.
[12] J. A. Biestenberger, D. H. Sebastian, "Principle of Polymerization Engineering". New York John Wiley & Sons, **1983**.
[13] B. Deroover, C. Coppens, J. Devaux, et. al., *J. Polym. Sci. Part A: Polym. Chem.*, **1996**, (34), 1039.
[14] L. Zhao, Z. Zhu, G. Dai, *Proc. 17th Int. Symp. Chem. React. Eng.*, Hong Kong, **2002**, MS# 0663.
[15] Z. Xi, L. Zhao, W. Sun, et. al., *J.Chem. Eng. Japan*, **2009**, (42), s96.

 DOI: 10.1002/masy.201300044

Energy and CO2 Savings: Systematic Approach and Examples in Polymer Production

Rolf Bachmann,[1] Christian Drumm,[1] Vijay Kumar Garg,[2] Jan Heijl,[2] Bert Ruytinx,[3] Johan Vanden Eynde,[2] Aurel Wolf*[1]

Summary: The chemical industry has set ambitious targets for the reduction of greenhouse gas emissions and energy consumption. In the present work, a systematic approach for a successful reduction of energy consumption and emissions is presented. A Structured Efficiency System for Energy (STRUCTese®) helps to identify, monitor and manage energy efficiency. The method fosters and accompanies energy saving measures as shown in the first two examples from polymer production: energy savings through operational efficiency and raw material savings with a new plant design. In addition, such a systematic approach pushes the development of innovative processes and provides a vision for an energy efficient future as shown in the third example: the use of CO_2 as a raw material in the copolymerization with epoxides.

Keywords: CO_2 reduction; energy efficiency; polymerization; reaction engineering; sustainability

Introduction

The chemical industry as an energy-intensive industry is motivated to reduce energy consumption and CO_2 emissions for two main reasons: rising concerns in companies, the public and scientific community about climate change or global warming, and the increasing fraction of energy in manufacturing costs.[1] Since energy represents a significant share of operating expenses at chemical plants, energy efficiency is presently the most effective and economic lever to sustainably lower energy consumption and CO_2 emissions.[2] On the other hand, energy efficiency is not a primary objective of chemical production – at least not prior to safety, throughput, yield, quality and overall equipment effectiveness (OEE). If a plant has already mastered these issues, then it is a real challenge to also address energy efficiency.[3] Therefore, successful energy and CO_2 reductions require a systematic approach, which ensures that energy efficiency is continuously and sustainably integrated at every level within the enterprise. Energy Management Systems (EnMS) are part of such an approach, establish the basis to increase energy efficiency in companies and organizations. Among other things, EnMS record the energy consumption, keep track of measures or investments in improving energy efficiency and support in target setting and decision making. The ISO 50001 standard defines standardized worldwide criteria for EnMS.[4] In order to benefit from tax reliefs in Germany it is necessary to have an EnMS according to ISO 50001 in place from 2013 on.[2] For the above reasons, Bayer started its climate program in 2007 and set targets for cutting greenhouse gas emissions in the period 2005 to 2020. In this connection, Bayer MaterialScience which has the highest energy consumption of the Bayer subgroups committed to reduce the specific greenhouse gas emissions by 40 percent. In

[1] Bayer Technology Services GmbH, D-51368, Leverkusen, Germany
E-mail: rolf.bachmann@bayertechnology.com
[2] Bayer Antwerpen NV, Antwerpen, Belgium
[3] Bayer Thai Company Limited, Rayong 21150, Thailand

this connection, the EnMS STRUCTese® (Structured Efficiency System for Energy), was developed as part of the sustainability strategy.

The benefits by using such a holistic energy efficiency approach are presented in this paper. The STRUCTese® methodology is briefly summarized and its basic ideas are introduced. In the following chapters several energy saving examples are given for polymer production processes, which also reflect different energy efficiency levels in the method, from operational improvements to energy efficiency investment projects to polymer research activities.

The first example describes continuous reduction of energy consumption over several years via improved plant operation and investments in energy efficiency.

The second example describes the optimization of the two-phase interfacial process for the production of polycarbonate. Despite the long history of optimizing this "old" process, several savings could be identified. One is a new reactor design for phosgene savings. Challenges are that changes in reactor design require an adjustment of process operations to meet the tight requirements on polymer quality (molecular weight distribution) and organic waste. Optimization has to take into account kinetics, distribution of reactants and solvents together with the built up and state of the dispersion.

The third example describes the development of a polymer with CO_2 as a copolymer together with propylene oxide and ethylene oxide which is used as a precursor for the production of polyurethanes. Apart from saving energy for the production of monomer, the CO_2 balance is improved by directly using CO_2 from power plants. A pilot plant was built and successfully produces polymer with good quality. The task now is to find the best production parameters.

Energy Efficiency Management System STRUCTese®

The EnMS STRUCTese® consists of several parts, which ensure idea generation, transparency and continuous focus on energy efficiency on all levels of an organization. An Energy Efficiency Check, which helps to identify all potentials for energy savings, is usually carried out as phase one of the STRUCTese implementation. It starts with a systematic analysis of overall energy consumption and energy distribution of a plant, followed by a collection and evaluation of improvement ideas with regard to technical feasibility, energy savings and profitability. The full range of optimization levels including energy & utility supply, raw materials, heat integration, equipment, operational improvements, process design improvements and buildings & facility (see Figure 1) is considered in the Energy Efficiency Check. A detailed description of the Energy Efficiency Check can be found elsewhere.[2]

In the second step, the Energy Loss Cascade, which is the key element of STRUCTese® is generated and implemented. It compares the current energy consumption of a production plant to several theoretical optimums under certain circumstances and breaks down and explains the differences between these energy levels in loss categories.[2] A simplified cascade is depicted in Figure 2. The Current Energy Consumption (CEC) on the right side of the cascade shows the measured actual energy consumption of the plant. The Operational Energy Optimum (OEO) represents the minimum energy consumption of the actual plant in case of optimum operation. The Plant Energy Optimum (PEO) is an important benchmark for the existing plant as it represents the minimum specific energy consumption to produce the desired product at the given site and available infrastructure. The way from the OEO to the PEO and therefore the loss code "suboptimal equipment" includes the investment measures from the Energy Efficiency Check. Finally, the Theoretical Energy Optimum (TEO) on the left side of the cascade reflects the best known process and the best infrastructure based on today's knowledge.

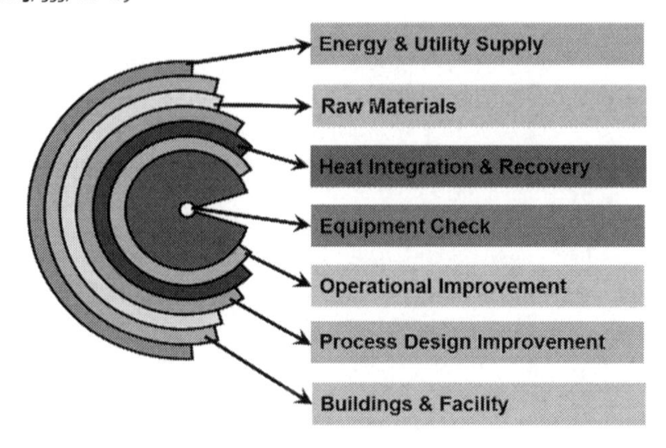

Figure 1.
Energy efficiency check.

The dynamic loss categories are usually identified by means of statistical data analysis of process data, while the energy levels and loss codes in the static part of the cascade (OEO, PEO, TEO) are simulated by means of a steady-state process model.

This approach achieves an unmatched level of transparency in the management of energy efficiency. The loss cascade is supplemented by further measures that create awareness for energy efficiency on all levels of the organization and integrate the topic into the daily workflow. A real-time energy efficiency monitor enables the plant operator to improve energy efficiency continuously through optimal process operation. Daily energy protocols summarize average energy consumptions and energy relevant process parameters of the last 24 hours. Key performance indicators that allow target setting and cross-plant comparison are derived from the energy levels and loss categories. With this integrated approach, the whole company is oriented towards energy efficiency from the upper management to the plant operator.[2]

To sum it up, the EnMS allows the detailed measurement and tracking of energy efficiency in contrast to measuring energy consumption alone. By visualizing different sources of energy efficiency losses, the key levers for energy efficiency, which

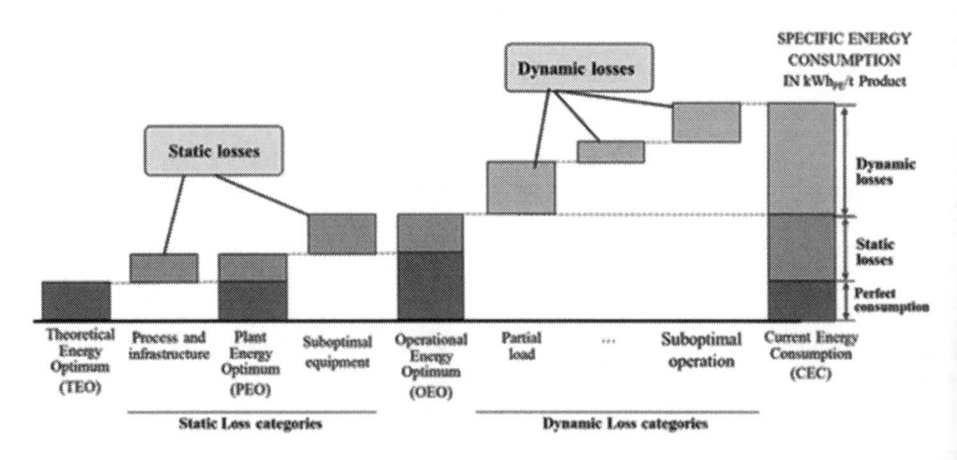

Figure 2.
Energy loss cascade.

can be investment projects or operational process improvements, are identified. A detailed description of STRUCTese® can be found in the original publication.[2]

In the following, three examples for improved energy efficiency in polymer plants are presented. One thing has to be made clear: these energy efficiency examples did not emerge from the EnMS and are not the sole merit of the STRUCTese® methodology. Just like the EnMS they are part of the systematic approach. However, the STRUCTese® methodology can accompany and foster energy efficiency measures as shown in the first two examples. In addition, the method pushes the development of new and innovative processes and provides a vision for an energy efficient future as shown in the third example.

First Example: – Operational Energy Efficiency

The first example shows how the continuous focus on energy efficiency can significantly reduce the specific energy consumption. Daily specific energy consumptions of a real chemical plant are depicted in Figure 3 for several years. The plant produces Bisphenol A (BPA) with an energy consumption of >200GWh per year. The specific energy consumption of the plant (in kWh primary energy per ton of product) is depicted over the load (t/h) of the plant. It becomes clear from the figure that the specific energy consumption was continuously decreasing from 2006 to 2011 down to around 150GWh per year. The energy consumption could be reduced more than 30% at full load over the years. Chemical plants usually have the best efficiency at maximum throughput, while the specific energy consumption is higher at partial load. This is also visible in the figure, where the specific energy consumption is twice at low load in the years 2006 and 2007. This increase due to partial load effect was also reduced considerably in the following years. The reduction of the energy consumption was achieved by means of a better process control, more efficient plant operation, better heat integration and investment projects in energy efficiency. For example process parameters were adjusted towards the energetically optimum, insulation was improved and vapor heat of condensation was used.

Such a success story was possible by a hard working plant team, which continuously focused on operational and design improvements. An Energy Efficiency Check was carried out in 2007 identifying

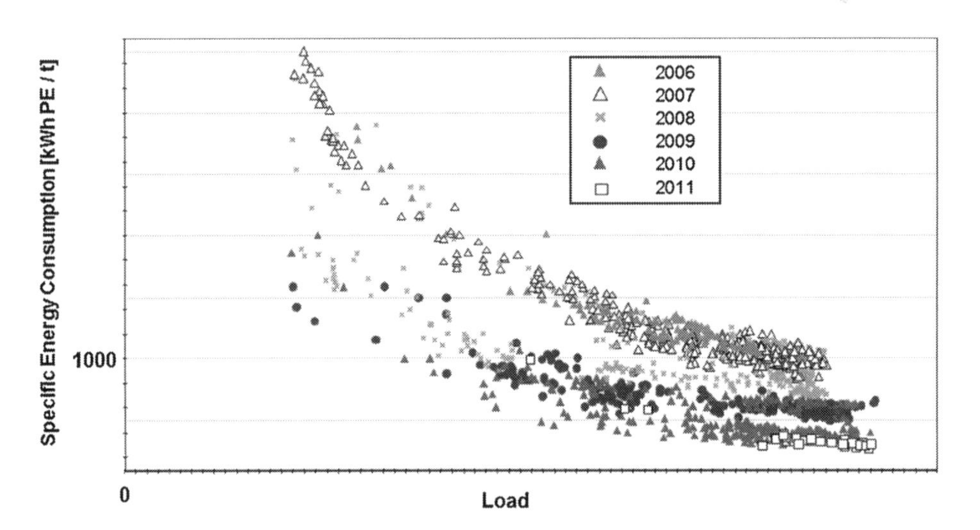

Figure 3.
Reduction of specific energy consumption for a polymer plant over several years.

35 energy projects with savings potentials of 41% in terms of energy costs. The STRUC-Tese® energy loss cascade and an online monitor were implemented in 2009. Since then, the plant personnel was supported by an online monitor, which visualizes the energy influencing parameters and gives real-time feedback on the energy efficiency of the plant to optimize the operation of the plant. An exemplary energy online monitor is shown in Figure 4. The plant manager and energy manager were supported by the energy loss cascade to focus on the most effective measures and to track the progress.

Example 2: Polycarbonate Production

The second example shows the production of polycarbonate via the interfacial condensation of phosgene and BPA. Polycarbonate is a polymer with superior mechanical properties over a wide temperature range together with high transparency and with many applications in the construction, automotive and electronic industries. Polycarbonates with the brand names Makrolon® and Apec® are two of Bayer MaterialScience's main products and the interfacial process accounts for roughly 1000 kt/a.

Production with this process started in the 1970ties and has since then undergone more than forty years of continuous optimization. This history of a mature process renders any substantial progress or optimization a special challenge. These are:

a) Limitations posed by existing plants. Apart from the requirement that changes to the current equipment should be minimal to keep costs low, there are legal limitations, production permits, specifications of waste water composition and quantity and the more.
b) Tight product specifications of well-known product. Any equipment changes are bound to affect the product and they have to be compensated by adjustments in the way the process is run. For polymers in general it is important to control all production stages, as the product once formed cannot be remedied.
c) The history of successful and unsuccessful optimization attempts. Many

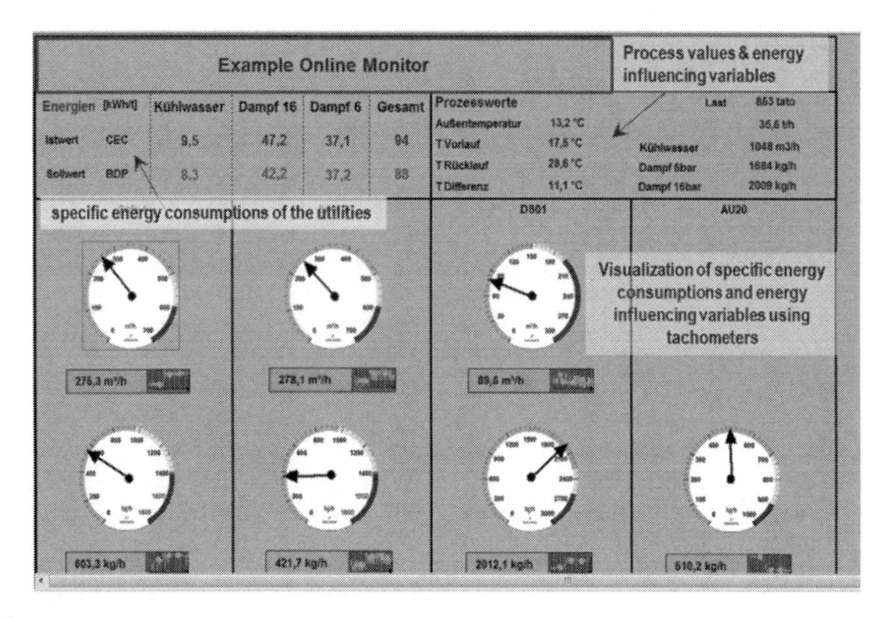

Figure 4.
Exemplary online monitor.

ideas are not completely new and arguments or even experience with a similar approach exist that a suggested improvement will fail. In this respect, the aforementioned systematic approach helps to identify the most promising items, quantify the potential, direct resources and get the commitment needed.

Here, the three stage interfacial process is described which is based on three main reactions:

Polycondensation of BPA and phosgene

$$nNa_2BPA + (n+1)COCl_2 \Rightarrow Cl - CO - [-BPA - CO]_n - Cl + (n+1)NaCl$$

Saponification of phosgene or the chloroformic acid end groups

$$COCl_2 + 4NaOH = Na_2CO_3 + 2NaCl$$

Molecular weight control via end capping with a monofunctional phenol

$$Cl - CO - [-BPA - CO]_n Cl + 2NaOPh \Rightarrow PhO - CO - [-BPA - CO]_n - OPh$$

The ideal product is completely end capped with no OH- and absolutely no Cl-end groups, with a tight molecular weight distribution (MWD) close to the ideal Schulz-Flory (SF) distribution.

In the first stage, an aqueous solution of BPA, an organic solution of phosgene, and caustic are mixed to produce oligomers. In the second stage chain terminator and the remaining caustic are added to ensure complete neutrality at the end of the reaction and in the last stage catalyst is added to drive the reaction to completion. Further production stages separate the aqueous and organic phase, remove the organic solvent and clean the aqueous phase. Alkaline conditions with a $pH > 10$ are required for the polycondensation reaction to proceed. Unfortunately, alkaline conditions also increase the saponification rates and hence increase the amount of phosgene surplus needed. Saponification, furthermore, consumes 4 NaOH molecules for every phosgene molecule. Phosgene and caustic consumption are two of the main energy factors in the production since they are produced from salt via electrolysis and any percent phosgene saved can be directly translated into energy savings.

The improvement measure to reduce the phosgene excess considers introducing a 4th reactor (prereactor) at the beginning of the reactor train where the organic and aqueous face are brought together through a nozzle to achieve a fine dispersion.[5] The nozzle ensures a high reaction rate and in the prereaactor the ratio of BPA / NaOH concentration is higher, favoring the main reaction. The general ideas were not new, but previous attempts had failed for a number of reasons. Prereactor and nozzle were designed using detailed measurements of the fast reaction of phosgene with BPA via stopped-flow experiments. The resulting reactor was small, could easily be fit into existing plants and investment costs were comparatively low. The new equipment, however, shifted the conversion and composition profile which affected product quality such as molecular weight distribution (MWD), end groups, as well as organic load in the waste water as well as process operability (separation of phases).

The origin for the shift in these parameters is found in the way, the reaction proceeds. As the name already implies, the reaction takes place at the interface between the organic phase and the aqueous phase. The effect of the interfacial polymerization on the molecular weight distribution is shown in

Figure 5 At intermediate stages the MWD is non-ideal and even differs for different end group combinations: OH-OH, OH-Cl or Cl-Cl, where Cl stands for a chloroformic acid and OH for a group of phenolic OH. In the final MWD some of these non-idealities such as excess oligomer content remain. For a good product, however, the MWD is expected to be close to ideal SF-distribution.

For the reaction to proceed, the two phases are brought together by shear forces applying mechanical energy to form a fine dispersion The dispersed phase can either be the organic or the aqueous phase. The

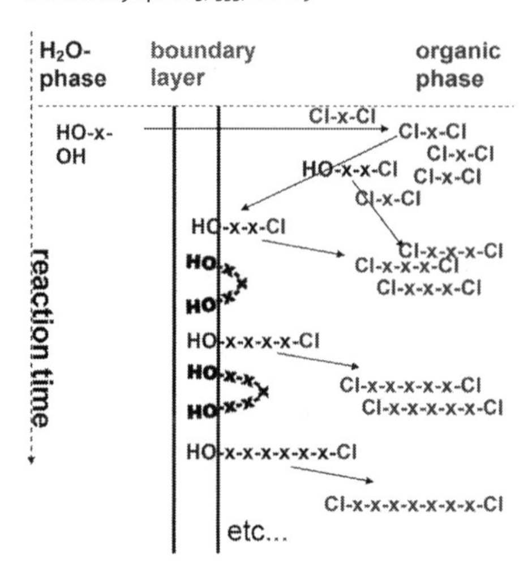

Figure 5.
Conception of the interfacial process.

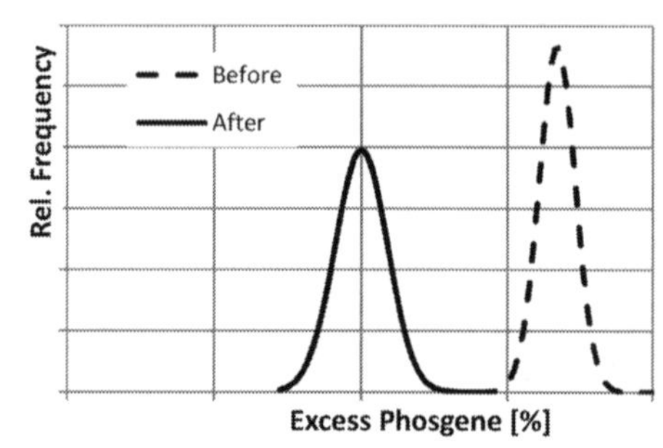

Figure 6.
Distribution of excess phosgene used, measured before and after the introduction of the prereactor.

phases tend to separate rapidly as the droplets coalesce. Therefore the dispersion energy has to be applied anew in regular intervals. During the reaction, phase inversion can occur and the dispersed phase may change up to two times depending on process parameters. As a result, apart from the reaction kinetics, transport to and from the interfacial surface, as well as type stability and dynamics of the dispersion play an important role. Here, it was found that surface creation was more important than actual surface area and diffusion is not the main way of transport to the surface.

These phenomena set out the framework for process optimization which requires a careful control of the dispersion along the reaction line, pH and chemical composition and hence point and form of addition of any feed stream. The equilibrium distribution of the different chemical species between the two phases as well as the dependence of the MWD on process parameters are well known. Kinetics, overall as well as detailed, have been measured several times. This information has been used in process models to predict the effect of process changes. However, the formation and dynamics of the dispersion and the transport processes connected with it elude full quantitative analysis and modeling.

The new reactor design was tested at a pilot plant and the optimal process parameters were determined. This leads to changes in the dosage profile of chain terminator, caustic, solvent and solvent concentration as well as adjustment of mixing / dispersion devices. The first lines are now successfully equipped with the new prereactor. The result on phosgene excess can be seen in Figure 6, which shows the distribution of phosgene excess obtained shortly after the implementation. The maximum has shifted to lower values as expected but also the width of the distribution has become broader. This last effect is due to the fact that experience with the new design is needed and becomes narrower with time.

Apart from phosgene savings, optimization leads to a reduction in organic load of waste water and a narrower molecular weight distribution. The new design will now be introduced at all sites. The expected overall energy savings corresponding to Cl2, CO and NaOH savings can be rougly estimated as 60 GWh/a after an implementation at all sites. Research is currently carried out on further phosgene reduction.

Example 3: CO_2 as Comonomer

The third example will focus on the leftmost part of the energy cascade: the design of a new process and / or introduction of a new product. A new chemical process promises the highest savings but also shows the highest demand on costs and investment. Challenges in the process design are different too. Neither product, chemical reaction nor process are (completely) fixed, the degree of freedom varying with each case. Product development has to ensure a suitable quality profile with a (potentially) high market demand. Reaction and catalysis need optimization and a new plant has to be designed.

The example chosen here is the copolymerization of epoxides with CO_2. The resulting polyethercarbonate polyols have a mix of properties comparable to polyether polyols and hence exhibit a vast potential for applications in the field of polymeric polyols, key building blocks for polyurethanes. Polyurethane production including foams, adhesives, coatings and fibres is a multi-million t/a industry.[6]

The most effective way to improve the CO_2 footprint and sustainability is probably the direct application of CO_2 as a substitute for fossil feedstock. The challenge using CO_2, however, is it's low reactivity and low energy content, which require substantial energy input and good catalysis to perform the reaction. In the process considered, the energy is provided by the epoxides and their heat of reaction in the polymerization process, which otherwise is usually wasted in cooling the process. The main overall copolymerization reaction is given by the following three steps:

Start with an alcohol, that may be multifunctional:

X-OH + (epoxide, R) ⟶ X–O–CH₂–CH(CH₃)–OH

Propagation

+ (epoxide, R) ⟶ X–O–CH(CH₃)–CH₂–OH

X-OH + CO2 ⟶ X–O–C(=O)–OH

There is a side reaction giving low molecular 5 ring carbonates.

(epoxide, R) + CO_2 ⟶ (cyclic carbonate, R)

Several catalysts can be applied for this reaction (see e.g.[7]&[8])). A good candidate was found to be the DMC catalyst (double metal cyanide: $Zn_3[Co(CN)_6]_2 \cdot a\, ZnCl_2 \cdot b\, L^1 \cdot c\, L^2 \cdot d\, H_2O$ with different Ligands L^1 and L^2, which is widely used for the polymerization of propylene oxide (PO) and ethylene oxide (EO).[9] CO_2 incorporation at the desired level greater 20% was achieved at pressures above 20 bar, however, reactivity decreased with increasing CO_2 pressure (Figure 7). Catalyst development was carried out to optimize reactivity, CO_2 incorporation, formation of side products, form of molecular weight distribution and sensitivity toward catalyst poisoning by reactants.[10]

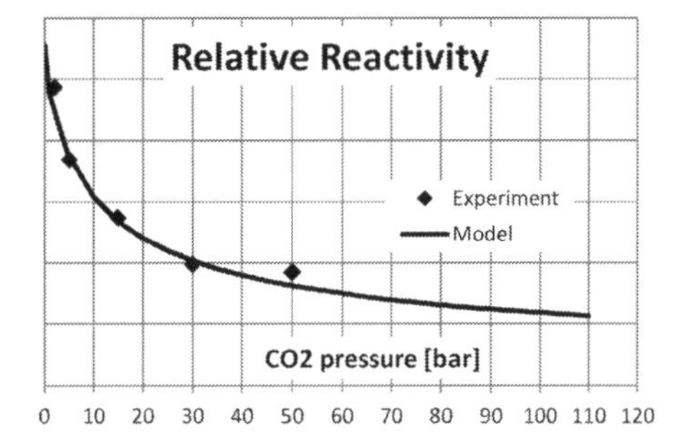

Figure 7.
Variation of relative pseudo first order reactivity of PO ($R(pCO_2)$ / $R(pCO_2 = 0)$) with CO_2 pressure.

The above mentioned quality parameters at the molecular level determine the suitability of a catalyst and at the same time form the criteria for the design and optimization of the process. There is a strong influence of process parameters such as catalyst level, dosage rate, residence time, temperature, temperature control and pressure on the quality parameters. The dependencies need to be quantified to design and optimize the process.

The homopolymerization of PO with DMC catalysis has been studied before and an unusual kinetic behavior of the polymerization has been found.[11] The reaction rate depends strongly on the molecular weight of the polymerizing species leading to the so called "catch-up" kinetics where the rate of polymerization is inversely proportional to molecular weight squared. The result is that small molecules grow much faster than larger ones resulting in a very narrow molecular weight distribution even in the case of a CSTR. This is employed in the so called CAOS process[12] where starter is continuously added during a semi-batch polymerization. A second feature is catalyst "poisoning" by low molecular weight starters, in particularly ones with a high functionality (for example 3 OH-groups). These starters reduce reactivity by forming a stable complex with the catalyst.

For the study of the copolymerization with CO_2 laboratory experiments were carried out in a continuous reactor as well as in semi-batch mode with wide variations of process parameters. Analysis of samples taken during or at the end of the reaction included ^{1}H-NMR spectroscopy and gel permeation chromatography (GPC) to obtain the concentration of free PO, cyclic carbonate and the chemical composition of the polymer and the MWD. PO concentration was monitored using in situ IR. In parallel, production runs were performed on a mini plant scale and analyzed accordingly. Apart from providing material for preparation and testing of polyurethane foams, they were a second source of valuable information on the kinetics and the overall behavior of the process. As in the case of homopolymerization, the results were unexpected compared to the usual copolymerization. CO_2 incorporation, selectivity, polydispersity and catalyst reactivity were found to be a function of molecular weight of the polymerizing species and catalyst concentration. In semi-batch runs, for example, CO_2 incorporatio varied with conversion.

To get a clear picture of the kinetics, two microscopic processes have to be considered separately: a) the transport of molecules to the catalytic particles and b) the chemical reactions at the catalytic sites. Heterogeneous catalyst particles are sparsely distributed in the melt or solution. For polymer molecules to grow they have to move to the catalyst sites and this diffusion process is responsible for the Mn^{-2} dependence of catch-up kinetics. At the catalyst site, the cycle of adsorption, reaction and desorption each time adds a new segment to the growing polymer molecule. Careful inspection of the polydispersity of the polymer allowed obtaining more information about these segments, and it was found that the formation of segments can be described in the usual context of polymerization kinetics with start, copolymerization and termination. It is the reaction at the catalyst site which determines the parameters of chemical composition (see Figure 8).

A further part of the kinetics is the formation and breakage of complexes of starter molecules with active catalyst sites. Last not least the thermodynamics of the system: the distribution of species between the gas and the liquid phase (solubility of CO_2 in the polymer solution, concentration of PO and cyclic carbonate in the gas phase[13]) as well as transport processes between the gas and the liquid phase are needed for a complete process description. A corresponding model was set up that describes copolymer formation and

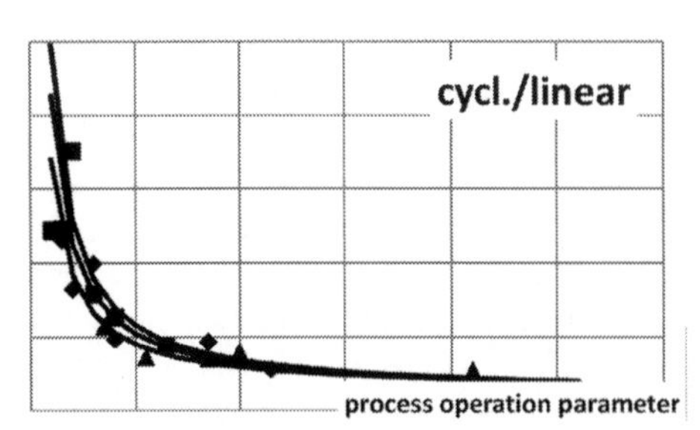

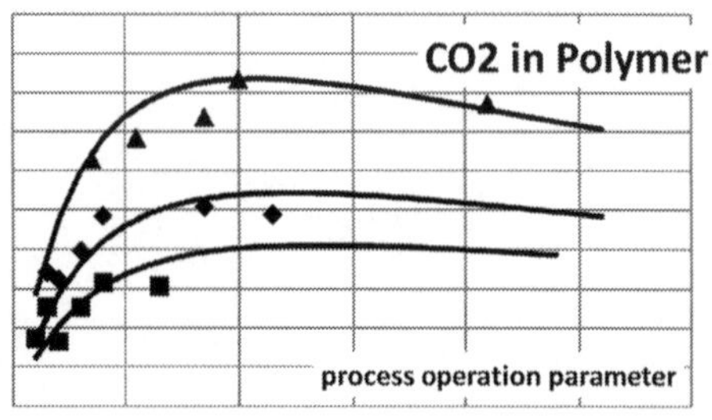

Figure 8.
Variation of the ratio cyclic / linear carbonates and CO_2 in polymer with temperature: Triangles 105°C, Diamonds 120°C, Squares: 130°C as a function of a process operation parameter.

chemical composition for different environment and equipment. It is used parallel to the experiments for the optimization of recipe, process parameters and equipment design.

Polyethercarbonate polyols thus produced were found to be easy to handle with standard equipment and fully suitable for further processing. The polyethercarbonate polyol based polyurethane foams exhibit similar basic physical properties[14] and thermal stability compared to conventional PU foams. These properties demonstrate that CO_2-based polyols are well suited for commercial polyurethane applications with a high potential for improving the carbon foot print.

Conclusion

The Energy Management System STRUCTese®, which allows the detailed measurement and tracking of energy efficiency in contrast to measuring mere energy consmpution was presented. Examples for three different ways of improving energy efficiency were shown: improvement by process operation, improvement by equipment and process changes and improvement by introduction of a new process. The shown examples are representative for countless energy efficiency measures, which form the systematic approach together with STRUCTese® to reach the ambitious targets for the reduction in green house gases and energy consumption. By the end of 2013, STRUCTese® will be implemented in 60 of the most energy-intensive plants. Bayer MaterialScience expects to realize an annual CO_2 emission reduction of 700,000 t by implementing the identified efficiency measures.

[1] H.-J. Leimkühler, in: "*Managing CO2 Emissions in the Chemical Industry*", H-. J. Leimkühler, (Ed., Wiley - VCH, Weinheim **2010**.

[2] C. Drumm, J. Busch, W. Dietrich, J. Eickmans, A. Jupke, Chem. Eng. Proc. **2013**, http://dx.doi.org/10.1016/j.cep.2012.09.009

[3] C.-H. Coulon, C. Drumm, in: "*Energieeffizienz – Standortkiller oder Innovationstreiber*", M. Siwek,, N. Wolfs, (Eds., T. A. Cook Research, Berlin **2011**.

[4] International Organization for Standardization. Win the energy challenge with ISO 50001, International Organization for Standardization, Geneva, **2011**.

[5] EP 2098553. **2009**, *Bayer* AG, invs.: W. Ebert, Chr. Kords, R. Bachmann, L. Obendorf, J. Eynde vanden B. Ruytinx, J. Seeba.

[6] K. Uhlig, *Polyurethan-Taschenbuch, 3. Auflage*, **2006**, Carl Hanser Verlag, München, Wien

[7] J. Donald. Darensbourgh, Making Plaxtics from Carbon Dioxide: Salen Metal complexes as Catalysts for the Production of Polyacarbonates from Epoxides and CO2. *Chem. Rev.* **2007**, *107*, pp 2388–2410.

[8] A. Gerrit, Luinstra: "Poly(Propylene Carbonate), Old Copolymers of Propylene Oxide and Carbon Dioxide with New Interests: Catalysis and Material Properties." *Polymer Reviews* **2008**, *48*, pp 192–219.

[9] U.S. Patent 5,470,813. (1993) inv. : B. Le-Khac.

[10] J. Langanke, A. Wolf, J. Hofmann, K. Müller, M. A. Subhani, T. E. Müller, W. Leitner, and C. Gürtler, (2013) CO_2 as sustainable feedstock for polyurethane production, Angew. Chemie 2012, in preparation.

[11] J. F. Pazos, K. McDaniel, E. Browne, K. Haider, IMPACT™ - The Next Generation of Polyol Technology: Unique Kinetics API Polyurethanes Technical Conferrence **2006**, Salt Lake City September 25-27, Conference Proceedings, 8–15.

[12] U.S.Patent 5,689,012. (1997) invs. : J. F. Pazos and T. Shih.

[13] J. M. S. Fonseca, R. Dohrn, A. Wolf, R. Bachmann, *Fluid Phase Equilib.* **2012**, *318*, 83–88.

[14] H. G. Pirkl, B. G. G. Kleczewski, CO2-based Polyols: A Future New Product Class for the Flexible Foam Market? *UTECH Europe exhibition program* 17. April **2012**.

DOI: 10.1002/masy.201300089

Semi-Batch Copolymerization of Propylene Oxide and Carbon Dioxide

*Benjamin Nörnberg, Carmen Spottog, Anne Rahlf, Revaz Korashvili, Claas Berlin, Gerrit A. Luinstra**

Summary: A nanoscopic heterogeneous catalyst system based on zinc glutarate is used to prepare poly(propylene carbonate) (PPC) from CO_2 and propylene oxide (PO). The catalyst was exposed to a defined humid atmosphere. The water uptake resulted in a deactivation to a minimum of 25% of the original activity, depending on time of exposure and water concentration. In addition, the progress of the copolymerization at constant pressure was monitored by measuring the CO_2 uptake. It is shown that the rate is linear with time after an initial phase and that this rate is in linear correlation with the amount of catalyst. The copolymerization was also performed in several modes of discontinuous addition of PO. The productivity of the catalyst was in the range of 1.6–2.0 kg PPC/g Zn, the highest productivity reported for the zinc carboxylate catalyst system so far. High molecular weights of around 220 kg/mol (M_w) and low polydispersities of 2.2–2.5 were achieved in all experiments.

Keywords: carbon dioxide; catalysis; kinetics; poly(propylene carbonate)

Introduction

Thermodynamically stable carbon dioxide can be copolymerized with alkylene oxides in a spontaneous process. It is one of the few exergonic reactions of CO_2 yielding products of commercial interest. The alternating copolymerization of propylene oxide (PO) and CO_2 gives access to the (substantial) alternating copolymer poly(propylene carbonate) (PPC), a compostable and biodegradable compound with a CO_2 content of over 40% by weight. Packaging, toys, agricultural and medical applications are only a few potentially interesting commercial applications of the versatile polymer.[1] The lack of a catalyst system that is productive enough to yield a stable polymer with a low ash content which does not require further processing *and* which can be prepared at a reasonable effort has so far impeded the large-scale use of PPC. At present, PPC has found limited application, e.g., as sacrificial binder in ceramic processing, which is, among other factors, due to the relative high production costs compared to commoditiy plastics. These high production costs can also be attributed to the separation process of polymer and catalyst residues.[2]

In this paper, we present studies on a nano-scaled zinc glutarate catalyst for the PO/CO_2 copolymerization with a productivity in the range of 2–3 kg PPC/g Zn.[3] This range is approaching the level necessary for simplifying the preparation of PPC ($\sim$ 10 kg PPC/g Zn). The catalyst can be prepared from cheap starting materials - glutaric acid, zinc nitrate hexahydrate and hexadecylamine. The latter can easily be recycled in the catalyst synthesis.

Institut für Technische und Makromolekulare Chemie, Universität Hamburg, Bundesstr. 45, 20146 Hamburg, Germany
Fax: +49 40 42838 6008
E-mail: luinstra@chemie.uni-hamburg.de

Experimental Part

Materials

Zinc nitrate hexahydrate (98% purity, *Sigma-Aldrich*), glutaric acid (99.8% purity,

Molekula), hexadecylamine (95% purity, *Merck KGaA*), carbon dioxide (99.995% purity, *Linde-Gas*) and propylene oxide (99.9% purity, *GHC Gerling, Holz & Co.*) were all used as received.

Synthesis of Zinc Glutarate

3.00 g (10 mmol) zinc nitrate hexahydrate and 1.26 g (9.5 mmol) glutaric acid were dissolved in 150 mL ethanol. After addition of 10.0 g (4.14 mmol) hexadecylamine, the mixture was stirred for 15 hours. Subsequently, the viscous mass was filtered off, washed three times with 50 mL ethanol and dried at 60 °C.

Copolymerization of CO_2 and PO

Copolymerization experiments were conducted in 300 mL and 2000 mL stainless steel autoclaves (*Parr Instruments Company*, model *4560* and *4520*) equipped with propeller stirrers. The reactor vessel was loaded with zinc glutarate catalyst, sealed, evacuated and pressurized with 1 MPa of CO_2. PO was subsequently added with a HPLC pump (*HPD Multitherm 200, BISCHOFF Analysentechnik und –geräte*). At the reaction temperature of 60 °C, the pressure was increased to 3 MPa. The pressure was kept at that level, CO_2 uptake was monitored by a mass flow controller. Copolymerization was terminated by cooling to room temperature, releasing the pressure and degassing to remove surplus propylene oxide. The resulting product was dissolved in acetone. The viscous solution was cast into an evaporating dish and dried in a dynamic vacuum to constant weight.

Water Saturation of Zinc Glutarate

Atmospheres with different levels of relative humidity were created inside of desiccators over saturated salt solutions of potassium carbonate (43%), sodium chloride (75%) and potassium sulfate (97%). Samples of 200 mg zinc glutarate were stored in the desiccator for various periods of time. After 1, 2, 5 and 9 days, the copolymerization activity of the catalyst was tested in a four-hour polyreaction under CO_2 pressure of 3 MPa (T = 60 °C, 50 mL PO). Each experiment was conducted twice, average values of the activity are reported.

Characterization

Product distribution and polymer composition were determined by ^{1}H-NMR spectroscopy on an *Avance Ultrashield-400* spectrometer from *Bruker* at room temperature. Spectroscopic data of PPC and common byproducts are given in various reports.[4–6] For sample preparation, 25 mg of the crude polymer were dissolved in 0.7 mL $CDCl_3$ containing 0.03 vol% TMS. Weight average molecular weights (M_w) and dispersity (M_n/M_w) were determined against PS standards by size exclusion chromatography. The measurements were carried out at room temperature with THF as eluent over a set of two columns (2×5 μm *Polypore, Varian*) on a chromatograph equipped with a *RI 101* refractive index detector from *Shodex*. For sample preparation, 3 mg of the polymer were dissolved in 1 mL THF and filtrated over a 2.4 μm syringe filter.

Results and Discussion

Catalyst Synthesis and Activation

DFT calculations on the mechanism of the zinc glutarate catalyzed copolymerization of PO and CO_2 suggest that polymer growth is solely taking place on the crystal surface.[7] A high catalyst activity can, thus, be achieved with a large surface area and a high number of active sites. Accordingly, efforts were directed at synthesizing zinc glutarate particles with small dimensions and a high surface-to-volume ratio.[8] A nano-scaled zinc glutarate catalyst can be prepared in three steps as depicted in Scheme 2.[9] The X-ray diffraction pattern of nanoscopic zinc glutarate is the same as for zinc glutarate prepared from $Zn(OH)_2$/glutaric acid or from ZnO/glutaric acid.[10] The identical crystal structure is expected to be present as published on single crystals[11,12] or a powder sample of zinc glutarate.[13] A pronounced line broadening of the reflexes is observed that originate from both crystal defects and small crystallite dimensions. Further research about the synthesis and morphological properties of nanoscopic zinc glutarate is in progress and will be reported upon.[14]

The Effect of Humidity on the Catalytic Activity

The importance of water coordination at the catalyst surface in the PO/CO_2 copolymerization was addressed by exposing the dry catalyst after synthesis to atmospheres with a defined humidity of 43, 75 and 97% saturation. The activity of the catalyst was subsequently measured under standardized copolymerization conditions (T = 60 °C, p = 3 MPa, t = 4 h) after catalyst exposure times of 1, 2, 5 and 9 days. Figure 1 depicts the resulting activity as a function of exposure time. One day of storage decreases the initial activity, depending on the humidity, to about half of the dry catalyst. The same observation are made after two days of catalyst exposure to humidity with a small further decrease in activity. The activity of the catalyst is reduced to 25% after 9 days of storage in a 97% water-saturated atmosphere. From these results, it can be anticipated that the catalyst is little sensitive to water and can be handled without inert atmosphere for several hours if the ambient humidity is 43% or lower, i.e. without a substantial loss of activity, e.g., while loading a reactor, and, that dry conditions are necessary for a high activity.

Kinetics of the Copolymerization

The kinetics were investigated by performing bulk polymerizations with various amounts of catalyst in PO under a CO_2 pressure of 3 MPa. The PO conversion was calculated from the yield of crude product, taking the product distribution (PPC and cyclic propylene carbonate cPC) into account as determined from an ^{1}H NMR spectrum. Figure 2a shows this conversion and the corresponding CO_2 uptake as measured by a mass flow controller as a function of the reaction time. The reaction has several phases of rates. The fast uptake of 800–1000 mL in the first 30 minutes is due to both the reaction and the saturation of PO with CO_2. Subsequently, a characteristic phase of linear CO_2 uptake is observed. It was found, that the conversion of PO has the same linear rate as the CO_2

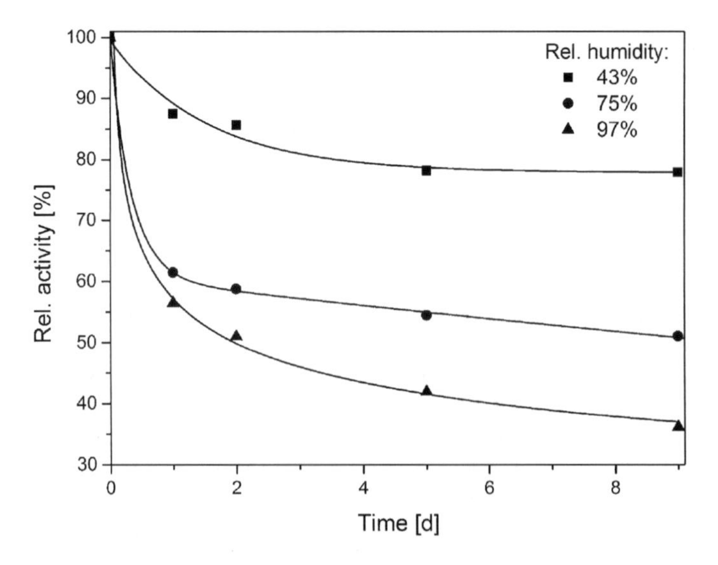

Figure 1.
Relative catalyst activity as a function of time during storage at different levels of humidity. (Reaction conditions for copolymerization: p = 3 MPa, t = 4 h, T = 60 °C, 200 mg zinc glutarate, 50 mL PO).

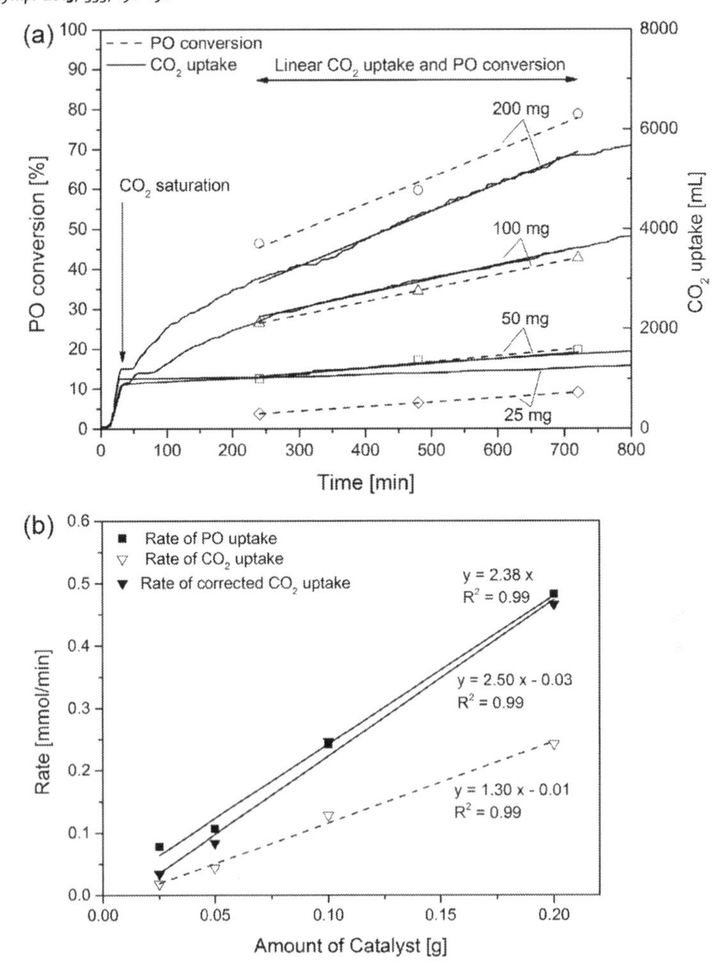

Figure 2.
a) PO conversion as a function of time for catalyst amounts of 25, 50, 100 and 200 mg. b) Rate as a function of the amount of catalyst. Reaction conditions: p = 3 MPa, T = 60 °C, 50 mL of PO.

uptake by determining the product yield over the relevant interval. The rate constants were determined and showed also a linear dependence on the amount of catalyst (Figure 2b). As expected, the observed CO_2 uptake is lower than the factual CO_2 consumption. The amount of CO_2 dissolved in PO decreases with the decrease of the PO quantity. The concomitant CO_2 release was calculated from the solubility in the medium. Figure 3 depicts the mole fraction of CO_2 as a function of pressure and temperature.[14] At the relevant conditions (p = 3 MPa, T = 60 °C), the mole fraction is about 0.33. Since each mole of CO_2 is fixing one mole of PO into PPC (and cPC), 0.48 mole of CO_2 are released from a saturated solution. In consequence, for each mole conversion of PO, 0.52 mole of CO_2 are entering the reactor via the mass flow controller. The corrected rate of CO_2 uptake (2.50 mmol/g catalyst · min) is thus in good agreement with the rate of PO conversion (2.38 mmol/g catalyst · min). A turnover frequency of 28.6 h^{-1} was calculated from the mean value of both slopes.

The rate law for the copolymerization is, thus, as follows (equation 1):

$$-\frac{d[PO]}{dt} \approx \frac{d[PPC]}{dt} = k[\text{catalyst}] \qquad (1)$$

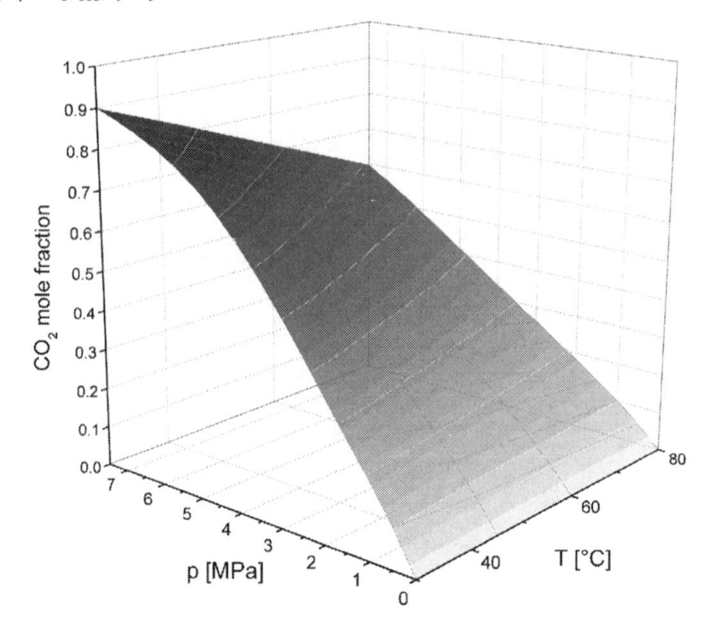

Figure 3.
Mole Fraction of CO_2 in pure PO as a function of pressure and temperature. The mole fraction at employed copolymerization conditions (T = 60 °C, p = 3 MPa) is 0.33.[14]

Integration yields equation 2:

$$n_{PPC} = n_{PPC,0} + k \cdot n_{cat} \cdot t \tag{2}$$

where $n_{PPC,0}$ is the amount of PPC formed prior to the linear phase of reaction.

Byproduct Formation

Cyclic propylene carbonate (cPC) is a common byproduct of the copolymerization of CO_2 and propylene oxide under the action of many transition metal catalysts (Scheme 1), which is difficult to remove. The byproduct can be either formed directly on the catalyst surface or by thermal degradation of PPC. The cPC content of the reaction product appears to depend on the amount of catalyst and on the reaction time. The concentration of cPC increases from 2.9% after 4 hours to 6.5% after 24 hours in case of 25 mg of catalyst. Copolymerizations with a higher amount of 200 mg of catalyst give cPC between 3.8% and 6.9% in the corresponding time (Figure 4). The time dependence of the byproduct formation leads to the hypothesis that in the early stage of the copolymerization, one type of catalytic center is present that catalyzes the formation of PPC, which then decomposes to cPC with a more or less fixed rate constant. The rate of decomposition increases relative to the rate of copolymerization when the concentration of PO decreases. This is faster at higher concentration of catalyst and is attributed to the decomposition of the PPC.

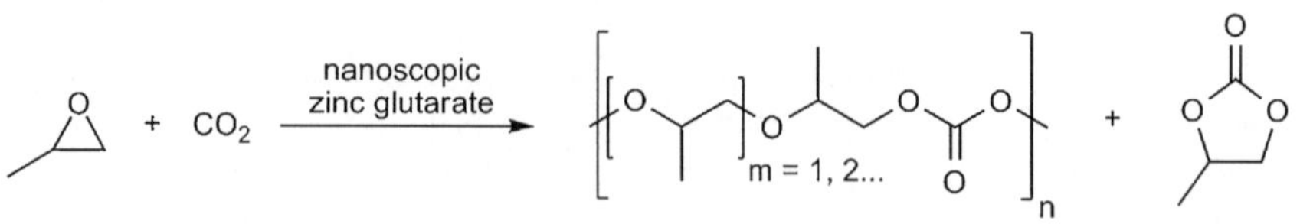

nanoscopic zinc glutarate
+ CO_2
m = 1, 2...
n

Scheme 1.
Alternating copolymerization of PO and CO_2 catalyzed by nanoscopic zinc glutarate.

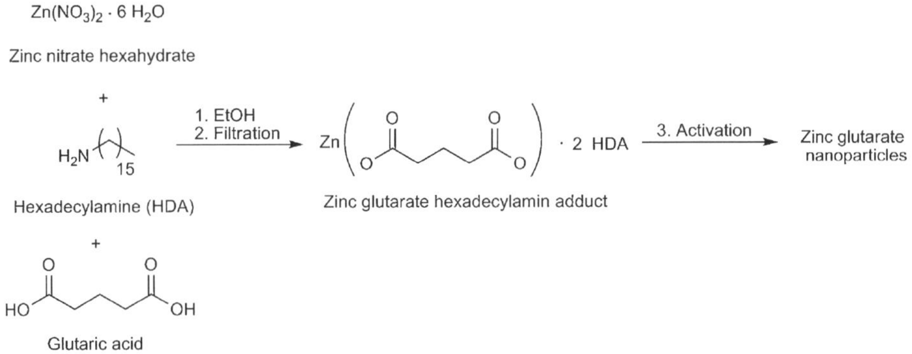

$Zn(NO_3)_2 \cdot 6\,H_2O$

Zinc nitrate hexahydrate

+

Hexadecylamine (HDA)

+

Glutaric acid

1. EtOH
2. Filtration

Zinc glutarate hexadecylamin adduct $\cdot$ 2 HDA

3. Activation

Zinc glutarate nanoparticles

Scheme 2.
Synthesis of zinc glutarate nanoparticles in a three-step method.[8,9]

Semi-Batch Polymerization with Respect to PO

The homopolymerization of PO to PPO by base or double metal cyanide catalysis is frequently realized in a semi-batch operated reactor. The concentration of PO is held at a low level to keep the heat content of the chemicals in the reactor at a low level. It also ensures that a homogenous product quality can be reached.[15] This mode of operation would be attractive for the preparation of PPC, too. The discontinuous batch PO/CO_2 copolymerization mediated by nanoscopic zinc glutarate was, therefore, evaluated as a first step. Reactions were carried out in a 2000 mL stainless steel reactor ($p_{max} = 350$ bar). The copolymerization was started ($p = 3$ MPa, $T = 60\,°C$) from 200 mL of PO and 800 mg of zinc glutarate. A total of 400 mL of PO was added along various protocols, the total time of reaction was 72 h. Table 1 summarizes the results.

A carbonate selectivity in the polymer backbone of around 92% is achieved in all experiments. The PPC has a high molecular weight of around 220 kg/mol and a low polydispersity. The productivities of 1.6–2.0 kg PPC/g Zn are easily reached.[9] The highest PO conversion of 70% is achieved in the batch reaction of 600 mL (Exp. No. 1). Addition of PO in 8 portions

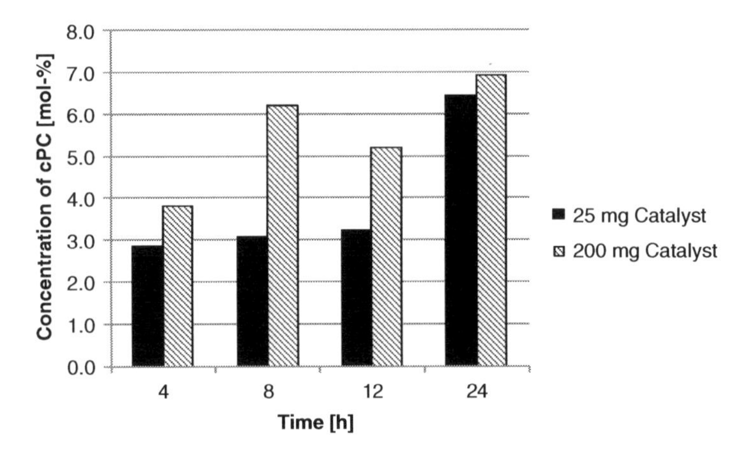

Figure 4.
Concentration of propylene carbonate (cPC) in mol% in the product mixture as a function of time for catalyst amounts of 25 and 200 mg.

Table 1.
Copolymerization of PO and CO_2 in discontinuous semibatch.

Exp. No.	Dosage points h	PO conv. %	Productivity kg_{PPC}/g_{zinc}	cPC mol%	Carbonate Selectivity mol%	M_w kg/mol	*D*
1	–	69.9	2.0	10.4	91.5	208	2.1
2	1/2/3/4/5/6/7/8 (50 mL)	69.7	2.0	12.0	91.8	234	2.5
3	0-8 (400 mL)	65.8	1.8	13.7	91.1	226	2.2
4	8/16 (200 mL)	64.2	1.8	10.8	92.3	224	2.3
5	24/48 (200 mL)	58.2	1.6	13.4	92.5	210	2.2

(Reaction conditions: CO_2 pressure = 3 MPa, T = 60 °C t = 72 h, 600 mL PO, 800 mg of zinc glutarate).

of 50 mL during the first eight hours of reaction (Exp. No. 2) basically gives the same results. PO addition of 200 mL after 8 and 16 h resp. leads to a total conversion of 64%, and after 24 and 48 h resp. to a lower conversion of 58%. All these results are consistent with the batch behavior with a catalyst - depicted in Figure 2a -, under the assumption that the catalyst does not deactivate.

Conclusion

A new nano-scaled zinc glutarate was evaluated as catalyst in the copolymerization of CO_2 and PO. Water deactivates the catalytic surface. The deactivation is slow and incomplete after 9 days in a 97% water-saturated atmosphere. The kinetics of the bulk polymerization of PO and CO_2 show a first order dependence on the catalyst amount with a rate constant of 28.6 h^{-1} in the early phase of the copolymerization. Discontinuous addition of PO is possible and has no major impact on the process.

Abbreviations

cPCcyclic propylene carbonate
HDA hexadecylamine
PO propylene oxide
PPC poly(propylene carbonate)
PPO poly(propylene oxide)

[1] G. A. Luinstra, E. Borchardt, *Adv. Polym. Sci.* **2012**, *245*, 29.
[2] www.empowermaterials.com.
[3] R. Korashvili, B. Nörnberg, N. Bornholdt, E. Borchardt, G. A. Luinstra, *Chem. Ing. Tech.* **2013**, *85*, 1.
[4] G. A. Luinstra, *Polym. Rev.* **2008**, *48*, 192.
[5] M. H. Chisholm, D. Navarro-Llobet, Z. Zhou, *Macromolecules* **2002**, *35*, 6494.
[6] M. J. Byrnes, M. H. Chisholm, C. M. Hadad, Z. Zhou, *Macromolecules* **2004**, *37*, 4139.
[7] S. Klaus, M. W. Lehenmeier, E. Herdtweck, P. Deglmann, A. K. Ott, B. Rieger, *J. Am. Chem. Soc.* **2011**, *133*, 13151.
[8] R. Korashvili, *Dissertation*, Universität Hamburg, **2012**.
[9] R. Korashvili, A. K. Brym, J. Zubiller, G. A. Luinstra, WO Patent 2013034489 (A1), **2013**.
[10] M. Ree, J. Y. Bae, J. H. Jung, T. J. Shin, *J. Polym. Sci., Part A: Polym. Chem.* **1999**, *37*, 1863.
[11] M. Ree, Y. Hwang, J. S. Kim, H. Kim, G. Kim, H. Kim, *Catal. Today* **2006**, *115*, 134.
[12] Y. Q. Zheng, J.-L. Lin, H.-L. Zhang, *Z. Kristallogr. NCS* **2000**, *215*, 535.
[13] D. J. Darensbourg, *Chem. Rev.* **2007**, *107*, 2388.
[14] R. Korashvili, B. Nörnberg, G. A. Luinstra, in preparation.
[15] C. Berlin, *Dissertation*, Universität Hamburg, **2013**.
[16] S. D. Gagnon, *Polyethers, Propylene Oxide Polymers*, Kirk-Othmer Encyclopedia of Chemical Technology, **2000**.

Effect of Six Technical Lignins on Thermo-Mechanical Properties of Novolac Type Phenolic Resins

Juan D. Martínez, Jorge A. Velásquez*

Summary: Lignin has been studied as an alternative to phenol in phenolic resins, due to economic and environmental reasons. The possibility of partially replacing phenol in phenol–formaldehyde (PF) resins by different types of lignins and analyze the lignin functional groups influence on the thermo-mechanical properties of the resins, has been explored. Novolac phenolic resins were prepared with six lignins: five lignins from black liquor of soda pulping bagasse and a commercial ammonium lignosulphonate from *pinus taeda* (LSA). These lignins have different percentages of functional groups (OH aromatic carbonyls and methoxyl) and polysaccharides. The kinetic study found that modified resins elaborated with lignin LSA exhibit the highest values of activation energy (148.66 and 148.92 kJ mol^{-1} in models of Flynn-Wall-Osawa and Friedman respectively), resulting to be the most stable of modified resins. Finally, this same resin has the highest value of the modulus of elasticity (MOE) of 8.02 GPa, even higher than unmodified phenolic resins. Therefore, LSA lignin, showed the best results in friction phenolic resins, this would imply the stability is within the bio-polymer and attributable to the linkage between PF and lignin.

Keywords: composites; kinetics (polym.); lignin, novolacl; stiffness

Introduction

Studies have been conducted to find modifiers to partially replace phenol in novolac resins, which is an expensive material which comes from a fossil resource and therefore non-renewable.[1–7] Some authors show that PF resins have been replaced by up to 40%, but it is known that the greater the degree of substitution, the greater the properties of the final product deteriorate.[8,9] To improve the modification of resins, lignins has been modified by demethylation, hydroxymethylation, fractionation and phenolation, the latter being one of the best to get a good performance in novolac type phenolic resins.[2,7,10–12] Lignin is presented as an attractive alternative because it generally comes from a residue.[13] Although it is a complex material, its structure has phenolic hydroxyl groups that may react in similar way as does the phenol with formaldehyde in phenolic resins.[14–16]

After cellulose, lignin is the most abundant organic substance in the plant world. It is a three-dimensional natural macromolecule, randomly arranged, composed of phenylpropane units with hydroxyl and methoxyl groups, molecules grouped hydroxyphenyl (H), Guaiacyl (G), and Syringyl (S).[17,18] Industrially, it is obtained as a byproduct pulping processes and represents an economical and abundant raw material. Depending on the source and extraction methods, the physical - chemical characteristics of the lignin and its use in polymer formulations can vary widely.[19]

Numerous investigations have been carried out in the formulation of lignophenolic resins.[14,16] However, due to the structural complexity of the lignin and their polymerization processes is desirable to

Facultad de Ingeniería Química, Universidad Pontificia Bolivariana, Cq. 1 No 70-01 B 11-250, Medellín, Colombia
E-mail: juandavid.martinez@upb.edu.co

explain the lignins effect on thermo-mechanical properties of the novolac resin.

Thermogravimetry (TGA) is a useful tool for studies on thermosetting resins pyrolysis.[20] Kinetic parameters may be obtained from the mass loss data supplied by this test. The most accepted velocity expression for the study of the degradation of polymers is given by the equation (1).[20]

$$\frac{d\alpha}{(1-\alpha)} = \frac{A}{\beta} \cdot e^{\left(\frac{-E}{RT}\right)} \cdot dT \qquad (1)$$

Where α is the fractional conversion at temperature T, R is the universal gas constant, and β is the reaction rate. A and E are the pre-exponential factor and the activation energy respectively. For a particular degradation step, α is calculated using the equation (2)

$$\alpha = \frac{W_0 - W_t}{W_0 - W_f} \qquad (2)$$

Where W_o, W_t and W_f corresponding to the initial mass of the degradation step, the mass at any time and the residual mass respectively.

Some of the methods proposed for phenolic resins are those of iso-conversion, which use multiple heating rates. They can be differential or integral when using the general equation of velocity or an integrated equation respectively.[6] Some integrals models are given by Flynn - Wall – Ozawa [21,22] and Friedman.[23]

Flynn, Wall and Ozawa developed a method which uses the approximation of Doyle (1965),[6] without knowing the reaction order and thus can determine the activation energy. Thus, the equation (3) is obtained.

$$\log\beta = \log\frac{AE}{R \cdot g(\alpha)} - 2.315 - \frac{0.457 \cdot E}{R \cdot T_\alpha} \qquad (3)$$

Where T_α is the temperature at a given conversion. The activation energy is obtained from the slope of the graph of $\log \beta$ vs. $1/T_\alpha$.

Friedman's method is based on the degree of conversion link with the inverse of temperature. It requires knowledge reaction rates depending on the temperatures and heating rates for the same degree of conversion, as shown in equation (4).

$$\ln\frac{d\alpha}{dt} = \ln f(\alpha) + \ln A - \frac{E}{R \cdot T_\alpha} \qquad (4)$$

By plotting ln $(d\alpha / dt)$ vs. $1/T_\alpha$ for different degrees of conversion, it is possible to determine the value of the slope and then obtain the activation energy.

Lignin incorporation in phenolic resins allows to obtain polymers with properties very similar to the no lignin polymers. The most affected property is the flow because this property is directly dependent on the polymer mesh forming during the reaction. Despite this, either the amount or the type of lignin does not affect flow, if the polymer has lignin there is always low flow. In industrial applications could be necessary to alter the pressing conditions in the composites building.[6,24]

The major components needed to form friction materials are resins, fillers and reinforcing material. In this case, the matrix are the resins of study, barium sulfate used as filler and for environmental reasons instead of asbestos, glass fiber is used as reinforcing material.[25–27] Resins are one of the most important ingredients in friction materials because they bind the ingredients firmly and allow them to contribute effectively to the desired performance. Ideally, there should be no significant deterioration in the function of the binder when the brake is operated under diverse conditions. Consequently, a friction material's thermal stability, its capacity to retain mechanical properties, and its ability to hold its ingredients together under adverse conditions all depend on the resin.[28]

Tests have been conducted in three point bending, which refers to the ability of a material to withstand forces applied to its longitudinal axis between its points of support, which is a combination of tensile, compressive and shear. It is performed to measure the behavior basically stress - strain.[29] From the bending test it is possible to obtain the modulus of elasticity (MOE), which refers to the rigidity or resistance of a material to the elastic deformation.[29] It was also possible to obtain the modulus of flexural strength (MOR) which is defined as the ability of a material to resist deformation under load. Furthermore represents the greater strain undergone by the material at rupture.

This paper explores the possibility of partially replacing phenol in PF resins by different types of lignins and analyzes the lignins functional groups influence in the thermo-mechanical properties of the resins. Here, the thermal properties were studied through the degradation temperature and the activation energy obtained from thermogravimetric tests. The mechanical properties were studied from the bending test at three points calculating the modulus of elasticity (MOE) and rupture (MOR). Finally, in order to analyze statistically the differences provided by the functional groups of lignin, an analysis of variance (ANOVA) was performed using STATGRAPHICS®. Thus, comparing the samples with statistically significant differences and conclude whether or not there is a real effect of such variations. This study uses the method of Fisher's least significant difference (LSD).[30]

Material and Methods

Materials

The six technical lignins are a lignin from Colombian sugar cane bagasse (LBC), ammonium lignosulphonate from pine (*pinus taeda*) (LSA) supplied by Borregaard LignoTech, lignin from Brazilian sugar cane bagasse pretreated with sulfuric acid (LBB), mixture LBC and LSA (50% by weight) (LBCLSA), and two lignins from chemical treatment and purification of LBC, LBCF and LBCP respectively. Phenol, curing agent hexamethylenetetramine (HMTA) and formaldehyde (50% aqueous solution) were supplied by Interquim S.A. The glass fiber and barium sulfate to make composites were supplied by Mafriccion and Reco respectively.

Physico Chemical Characterization of Lignins

Acid Hydrolysis and Klason Lignin: This is based on Klason's method.[31] The acid soluble fraction, is determined by ultraviolet spectroscopy.[32] Hemicellulose and cellulose were determined from a mass balance of hydrolyzed monomers by high performance liquid chromatography (HPLC).[33]

Phenolic Hydroxyls (Spectroscopic Method): This method is described by Wexler (1964).[34] A standard solution is prepared by weighing approximately 5 mg lignin of known moisture content and was dissolved in a 10 ml dioxane (96%). An aliquot of this stock solution, was diluted in 1:10 proportion of dioxane and water 1:1 (v/v) and adjusted to pH 13 by addition of 1 N NaOH. A control was prepared with the same dilution but pH adjusted to 1 by the addition of 1 N HCl. Then, a UV scan, using a PERKIN ELMER *Precisely*. Lambda 25 UV/VIS Spectrometer.

Total Hydroxyls: It is performed through a modification of the method described by Barnett (1982).[35] In a stoppered test tube 0.03 g dry lignin was mixed with 0.24 ml reagent (pyridine/acetic anhydride 10:3). The mixture was placed in a heater overnight at 65 °C, after which the mixture was transferred to an Erlenmeyer flask after adding 15 ml acetone and 15 ml distilled water. The mixture was then left for 1 hour to assure the destruction of any residual acetic anhydride. Acetic acid formed in the reaction was titrated against a standard solution of 0.1 N NaOH with phenolphthalein indicator.

Determination of Carbonyl Groups: This method was developed by Alder and Marton in 1959 and reported by Lin (1992).[36] It involves differential absorption measurements that take place when carbonyl groups are reduced at the benzylic alcohol corresponding with sodium borohydride. 2 ml of 0.2 mg/ml lignin solution in dioxane 96% was mixed with 1 ml 0.05 M sodium borohydride solution. The mixture was left in darkness for 45 hours at ambient temperature and after, its absorbancy in the

200–400 nm band determined against an unreduced lignin solution at the same concentration. The purpose was to distinguish carbonyl groups in two lignin models structures: coniferyl aldehyde and acetoguaiacon.

Instrumental Analyses

FT-IR Spectroscopy: Samples were previously dried overnight in an oven at 60 °C and then for 2 hours at 105 °C. KBr pellets were prepared and compacted at a pressure between 10 and 12 Tnf.cm^{-2} in vacuum. Each spectrum was recorded over 32 scans, in the range from 4000 to 400 cm^{-1} with a resolution of 8 cm^{-1}. FT-IR measurements were performed in a Perkin Elmer FT-IR System Spectrum GX.

Thermogravimetric Analysis (TGA): TGA measurements were performed in a *Mettler Toledo* TGA/SDTA 851e in nitrogen atmosphere (50 ml.min^{-1}) at a heating rate of 10 °C/min over the temperature range from 30 °C to 800 °C. Runs at different heating rates (10, 15, 20, 25 and 30 °C/min) were performed to degradation kinetics study in all resins.

Resin Production

Modified phenolic novolac resins were produced with each lignin (PFLBC, PFLSA, PFLBB, PFLBCLSA, PFLBCF and PFLBCP), where phenol in resins has been partially substituted by lignin (10% w/w). To compare two resins without lignin were produced. A reference novolac resin (PF) and PFB which is a resin with a default of phenol 10% (wt). Finally, a resin (PF) was produced with the addiction of 10% LBC to see the effect as a filler (PF + LBC). All resins have been synthesized according to the typical procedure used for commercial novolac in the industry, with the only difference of a previous lignin mixing step in addition for modified systems. Those prepolymers have been cured with 10 wt % of HMTA.

Composites

The resin (50 wt %), glass fiber (40 wt %), and barium sulfate (10 wt %) were mixed, then stacked between two polished steel plates to form plaques of 150 mm diameter and 3.2 mm thickness. Technique used was hot forming in hydraulic press for 15 minutes at 180 °C and 12 MPa. After, 5 specimens were made of each composite of 12.6 mm wide and 94 mm long.

Flexural Strength: It was performed according to standard ASTM D 790-07.[37] The tests were performed in a universal machine Instron 5582 with a ratio of specimen thickness (d) and distance between supports (L) of 16 to 1 and a load cell of 1 kN. In order to statistically analyze the differences presented by the functional groups of lignin on the flexural tests, was realized an analysis of variance (ANOVA) using STATGRAPHICS®.

Results and Discussion

Table 1 shows the content of cellulose, hemicellulose, lignin and ash for samples. LBC has a 25.47% total carbohydrate and compared with other samples, it has the highest value. This may be due to the formation of lignin-carbohydrate complex during precipitation of the lignin.[38] LBB has the lowest carbohydrate (0.59%), because this lignin has been pre-treated with sulfuric acid which hydrolyses the existing polysaccharides. LSA has 17.47% of carbohydrate and since it is an ammonium lignosulfonate, the amount of lignin is 100% soluble in acid, whose properties make it ideal for emulsions of asphalt and phenolic resins.[39]

The FT-IR spectrum of all lignins are shown in Figure 1. Samples show bands at

Table 1.
Mass balances

Sample	Cellulose (%)	Hemicellulose (%)	Lignin (%)	Ash (%)
LBC	1.35	24.12	66.44	2.88
LBB	0.00	0.59	90.78	1.03
LSA	3.40	14.07	79.46	0.00
LBCLSA	2.36	19.62	75.18	1.20
LBCF	1.67	21.32	80.37	0.69
LBCP	1.76	22.35	82.89	0.79

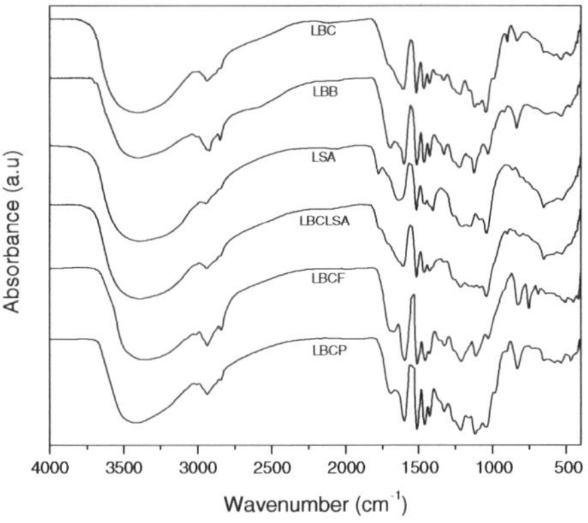

Figure 1.
FT-IR spectra of lignins.

1600, 1510 and 1424 cm^{-1} corresponding to aromatic ring vibrations of the phenylpropane structures. Absorption in the region of 3400 cm^{-1} is assigned to the aliphatic and aromatic OH groups. Bands between 2942 and 2849 cm^{-1} and 1460 cm^{-1}, are related to the vibration of C-H bonds of CH_2 and CH_3 groups of the side chain.[40] The low intensity of the bands at 1424, 1330 and 1126 cm^{-1} shown by LSA, can be attributed to process of obtaining lignosulfonates, which is the result of chemically degraded materials. All lignins have S typical bands (1330, 1260 and 833 cm^{-1}) and G typical bands (1260, 1151 cm^{-1}), such as expected.[17,40,41] LSA does not present vibration at the band 1330 cm^{-1}, and it was found that this lignin is the guaiacyl type, which was anticipated, given that the feedstock source was *pinus taeda* and after lignosulfonate pulping, the guaiacyl estructures of the native pine were retained, which may be important during resin synthesis.

Lignins composed of G units have free C5 position (*ortho* to the phenolic hydroxyl) and thus the ring is susceptible to reaction with formaldehyde.[42,43] However, in lignins composed of S units, C3 and C5 positions are linked with methoxyl groups, resulting in low reactivity with formaldehyde. This can promote high noncovalent interactions between the lignin making it behave like a rigid macromolecule and with high steric hindrance.[42] This may result in formulations of lignin modified phenolic resin, a significant decrease of the final properties. From this point of view, type G lignins are more suitable for resin formulations.

Table 2 shows the percentages of some functional groups present in the lignins. LBB has the highest content of phenolic hydroxyl followed by LBCF and LBCP. This is consistent with what has been analyzed from FT-IR for these three lignins. Results for total hydroxyl and carbonyl have very low rates. Carbonyl are found in extremely small concentrations in native lignins and significant concentrations in lignins that have suffered degradation or chemical processes.[17] However, it is unreasonable to draw conclusions with these low amounts.

Table 3 presents the percentage of loss in thermal degradation of the resins.[6,20,44,45] At study the thermal stability of resins in friction materials application requires that the degradation is low at temperatures below 300 °C.[9] By analyzing the mass loss until the material reaches 300 °C,

Table 2.
Functional groups of the lignins.

Lignin	Phenolic Hidroxyls (%)	Total Hydroxyls (%)	Carbonyl Groups (%)	
			X*	Y**
LBC	5.36 ± 0.002	3.74 ± 0.20	0.036 ± 0.00	0.144 ± 0.002
LBB	6.49 ± 0.00	2.59 ± 0.23	–	0.025 ± 0.006
LSA	3.31 ± 0.003	2.30 ± 0.17	0.013 ± 0.001	0.090 ± 0.003
LBCLSA	4.26 ± 0.00	3.35 ± 0.34	0.003 ± 0.002	–
LBCF	6.79 ± 0.02	4.54 ± 0.08	0.028 ± 0.00	0.094 ± 0.00
LBCP	5.79 ± 0.002	3.24 ± 0.17	0.046 ± 0.00	0.141 ± 0.00

* Coniferyl aldehyde structures
** Acetoguaiacon structures

Table 3.
Thermal degradation of the resins.

Resin	Mass-loss (%)			Residue at 800 °C (%)
	300 °C	450 °C	650 °C	
PF	7.31	28.23	45.81	50.36
PFB	7.50	29.66	48.56	47.38
PF + LBC	9.54	22.74	43.77	51.35
PFLBC	7.85	27.88	47.45	48.39
PFLBB	7.89	29.32	47.98	47.74
PFLSA	7.25	25.77	44.90	51.02
PFLBCLSA	10.27	30.62	49.04	46.73
PFLBCF	7.36	27.17	46.80	49.09
PFLBCP	8.68	28.19	47.24	47.85

differences arise due to moisture, degradation of cellulose, hemicellulose or lignin that are not incorporated in the polymer mesh, can lead to decreased thermal stability.[46] Therefore, the most stable is PFLSA given the mass loss even lower than PF. This would imply the stability is within the bio-polymer and attributable to the linkage between PF and lignin.[47,48]

The maximum temperature degradation rate is about 500 °C.[46,49] There is a remarkable difference with PF + LBC that exhibits a lower loss compared to the other. This resin has a 10% lignin as a filler and hence the loss is mainly due to this associated lignin fraction. Therefore, even if lignin is incorporated at the process initial, it does not replace the phenol stoichiometrically, nor does it act as an inactive load.

In the study of the kinetics of thermal degradation, Table 4 shows the activation energy. PF presents higher values (177.27 and 157.97 kJ.mol^{-1}), so is the one with the best thermal performance. PFLSA is the modified resin that exhibits higher values of the activation energy (148.66 and 148.92 kJ mol^{-1}). So that loses less mass by the effect of temperature and is the modified resin greater thermal stability. This result is consistent with other similar studies.[6]

Table 4.
Activation energy for resins

Resin	Activation energy (kJ mol-1)	
	Flynn-Wall-Osawa	Friedman
PF	177.27	157.97
PFB	148.12	127.42
PF + LBC	130.15	139.71
PFLBC	147.85	140.38
PFLBB	129.26	108.14
PFLSA	148.66	148.92
PFLBCLSA	112.80	67.59
PFLBCF	139.41	117.67
PFLBCP	147.90	132.72

Figure 2 presents the average values of the modulus of elasticity (MOE) associated with least significant difference intervals. These results are similar to those reported for similar systems.[20,25,50,51] There was a P value less than 0.05, which proves that there is statistically significant difference between the means of the MOE for each type of resin with a 95% confidence. It was observed that PFLSA has the greatest value of MOE (8.02 GPa), even greater than the unmodified resin (PF and PFB). This fact can be explained considering that lignin can serve as a compatibilizer between the matrix and the reinforcing, improving interface properties.

Figure 3 shows the average values of the module of resistance (MOR). The P value was less than 0.05, so there are significant differences between the results. This MOR values are quite low compared with those obtained with other composite materials.[25,50] This is due to the high percentage of binder 50% (values ypically used below 30%) and a low impregnation of the fibers. The PF resin has the best performance for this property, while the molar ratio change (in the case of PFB) and lignin modification decreases.

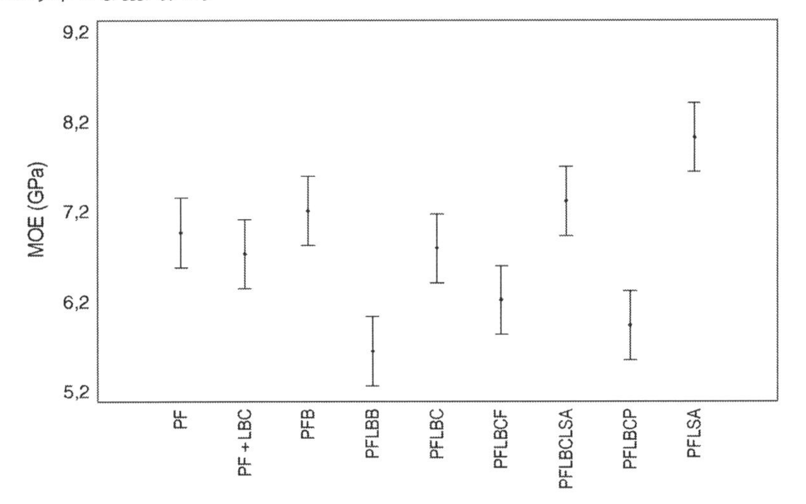

Figure 2.
Median value for MOE associated with 95.0 percent least significant difference intervals.

Conclusion

Bagasse lignins are GS type, which is consistent because they come from monocotyledonous. They present greater steric hindrance, chemical complexity, and less available OH. These facts prevent the formation of linear chains and retard the formation of methylene bridges, affecting the reactions with formaldehyde in the condensation process of the resins. LSA has little syringil moieties, which may explain the good performance of resins made from this lignin.

Thermogravimetric study found that PF + LBC has greater mass loss than PFLBC. Lignin LBC added at the beginning of the polymerization reaction is incorporated in some form into the phenolic mesh and functional complexity can affect the curing reactions.

Kinetic study found that PFLSA is the resins modified which exhibits higher values of the activation energy (148.66 and

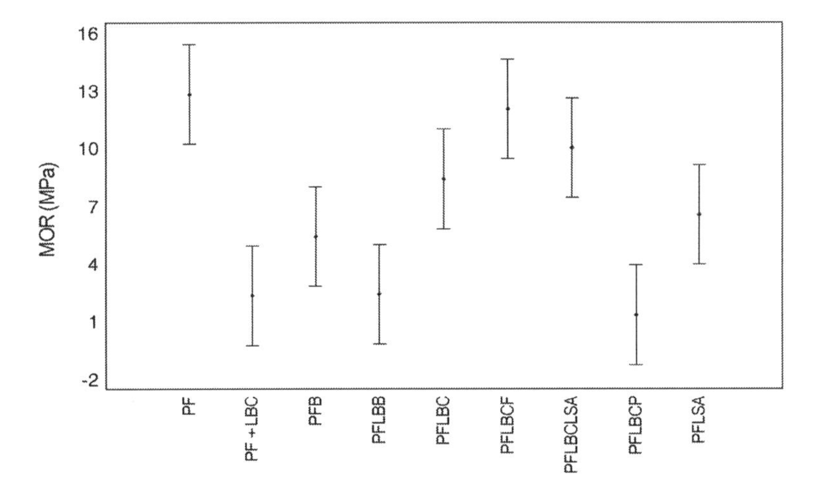

Figure 3.
Median value for MOR associated with 95.0 percent least significant difference intervals.

148.92 kJ mol-1 depending on the model of Flynn-Wall-Osawa and Friedman respectively). This shows that it is the one with the best thermal performance.

From the study of the mechanical properties, PFLSA has the highest value of the MOE (8.02 GPa), even higher than unmodified resins (PF and PFB). This fact shows that lignin can serve as a compatibilizer between the matrix and the reinforcing, improving interface properties. Furthermore, the chemical structure of LSA, allows a better arrangement of the polymer, in turn increased the stiffness. Moreover, the resin PFLBB, has the lowest value of the MOE (5.64 GPa) and considering also its low thermal performance, it is found that the low amount of carbohydrates, negatively influences the resins.

A low carbohydrate content of lignin, has a negative effect on the development of composite materials. Finally, it was determined that a degraded lignin, mainly composed by guaiacyl units such as lignin LSA, exhibits better behavior of phenolic resins as a modifier of friction.

[1] J. D. Martinez, J. Velásquez, W. Ramírez, P. Gañán, *Scientia et Technica*, **2007**. *XIII*, 683–688.
[2] M. V. Alonso, M. Oliet, F. Rodríguez, J. García, M. Gilarranz, J. Rodriguez, *Bioresource Technology*, **2005**. *96*, 1013–1018
[3] F. Dos Santos, A. Curvelo, *Polímeros: Ciencia e Tecnología*, **1999**. vol. Ene/Mar, pp. 49–58 Enero/Marzo
[4] J. D. Martinez, S. Pavone, J. Velásquez, P. Gañan, J. Cruz, in: *Congreso Iberoamericano de Investigación, CIADICYP*, Santiago y Valdivia - Chile, **2006**, p. 33.
[5] J. D. Martinez, C. Gómez, D. Restrepo, P. Gañan, *Suplemento de la Revista Latinoamericana de Metalurgia y Materiales*, **2009**. 3, S1, pp. 1173–1179 .
[6] J. M. Perez, *Universidad Complutense de Madrid, Madrid*, **2008**.
[7] A. Tejado, Ciencias Químicas, Escuela Universitaria Politécnica de San Sebastián, San Sebastián, **2007**.
[8] N. AMA, M. A. Yousef, K. A. Shaffei, A. M. Salah, *Pigment & Resin Technology* **1999**. *28*, 143–148.
[9] J. D. Martínez, J. Velásquez, G. Quintana, J. Cruz, P. Gañan, in *Proyecto Colciencias. Cod. 12100814227. Contrato 441-2003*, ed, **2006**.
[10] M. Olivares, J. A. Guzmán, B. Amadel, A. Natho, I. Ibañez, *Apuntes de Ingeniería*, **1987**. 30, 21–42.
[11] A. Toledano, A. García, I. Mondragon, J. Labidi, *Separation and Purification Technology*, **2010**. *71*, 38–43.
[12] P. Benar, A. R. Gonçalves, D. Mandelli, U. Schuchardt, *Bioresource Technology*, **1999**. *68*, 11–16 4//
[13] J. Velasquez, C. Barrera, B. Zapata, *Ingeniería Química*, **1999**. *356*, 182–188.
[14] J. M. PÉREZ, F. Rodriguez, M. V. Alonso, M. Oliet, J. M. Echavarría, *BioResources*, **2008**. 2, 270–283.
[15] A. Tejado, C. Peña, J. Labidi, J. M. Echeverria, I. Mondragon, *Bioresource Technology*, **2007**. *98*, 1655–1663 .
[16] M. V. Alonso, M. Oliet, J. M. Pérez, F. Rodríguez, in: *Congreso Iberoamericano de Investigación en Celulosa y Papel - Ciadicyp- Córdoba*, España, **2004**, 1–8.
[17] D. Fengel, G. Wegener, *Wood: Chemistry, ultrastructure and reactions* (Ed., Berlín: Walter D, Gruyer **1984**.
[18] K. V. Sarkanen, C. H. Ludwing, *Lignins Ocurrence, formation, estructure and reactions*, Washington Wiley - Interscience, **1971**.
[19] N.-E. El Mansouri, *Universitat Rovira i Virgili, Tarragona*, **2006**.
[20] S. Betancourt, *Universidad Pontificia Bolivariana, Medellín*, **2010**.
[21] J. H. Flynn, L. A. Wall, *Journal Research Natural Bureau Standard Section A - Physicochemical A*, **1966**. *70A*, 487–523.
[22] T. Osawa, *Bulletin of Chemical Society of Japan*, **1965**. *38*, 1881–1886 .
[23] H. L. Friedman, *Journal of Polymer Science Part C - Polymer Symposium*, **1964**. *6*, 183–195.
[24] U. S. Hong, S. L. Jung, K. H. Cho, M. H. Cho, S. J. Kim, H. Jang, *Wear*, **2009**. *266*, 739–744.
[25] Z. Da-Peng, *Journal of reinforced plastics and composites*, **2008**. *0*, 1–12.
[26] H. Jang, J. S. Lee, J. W. Fash, *Wear*, **2001**. *251*, 1477–1483 .
[27] D. Hull, *Materiales compuestos* (Ed., Barcelona: Reverté S.A, **1987**.
[28] J. Bijwe, N. Nidhi, B. K. Majumdar, *Wear*, **2005**. *259*, 1068–1078 .
[29] W. Callister, *Fundamentals of Materials Science and Engineering* (Ed., New York John Wiley & Sons, Inc, **2001**.
[30] D. Montgomery, G. Runger, *Probabilidad y estadísticas aplicadas a la ingeniería* (Ed., México Limusa-Wiley, **2010**.
[31] "TAPPI standard 175 T-13," ed.
[32] E. Maekawa, T. Ichizawa, T. Koshijima, *Journal of Wood Chemistry and Technology*, **1989**. *9*, 549–567.
[33] W. E. Kaar, L. G. Cool, M. M. Merriman, D. L. Brink, *Journal of Wood Chemistry and Technology*, **1991**. *11*, 447–463.
[34] A. S. Wexler, *Analytical Chemistry*, **1964**. *36*, 213–221.
[35] C. Barnett, K. Loferski, C. Wartman, in: *Department of Forest Products* (Ed., Madison, **1982**.
[36] S. Lin, C. Dence, *Methods in Lignin Chemistry*, New York Springer - Verlag, **1992**.

[37] "ASTM Standard D 790, 2007. Standard test Methods for Flexural Properties of Unriforced,Reinforced Plastics,Electrical Insulating Materials," ed: ASTM International,West Conshohocken,PA, **2007**.
[38] G. Quintana, *Universidad Pontificia Bolivaraina, Medellín*, **2009**.
[39] Melbar Productos de lignina LTDA.
[40] C. BOERIU, D. Bravo, R. J. A. Gosselink, J. E. G. Van Dam, *Industrial Crops and Products*, **2004**. *20*, 205–218.
[41] Ayo. Tejado, *Bioresource Technology*, **2007**. *98*, 1655–1663 .
[42] L. G. Wade, *Química Orgánica* (Ed., Madrid: Pearson Education S.A., **2004**.
[43] A. Gardziella, L. A. Pilato, A. Knop, *Phenolic Resins* (Ed., Berlín: Springer, **2000**.
[44] D. A. Kourtides, W. J. Gilwee, J. A. Parker, *Polymer Engineering and Science*, **1979**. *19*, 24–29.
[45] Y. Chen, Z. Chen, S. Xiao, H. Liu, *Thermochimica Acta*, **2008**. *476*, 39–43 9/30/
[46] M. Křístková, P. Filip, Z. Weiss, R. Peter, *Polymer Degradation and Stability*, **2004**. *84*, 49–60.
[47] S. Aradoaei, R. Darie, G. Constantinescu, M. Olariu, R. Ciobanu, *Journal of Non-Crystalline Solids*, **2010**. *356*, 768–771 4/1/
[48] H. D. Rozman, K. W. Tan, R. N. Kumar, A. Abubakar, Z. A. Mohd. Ishak, H. Ismail, *European Polymer Journal*, **2000**. *36*, 1483–1494 .
[49] L. Costa, L. R. di Montelera, G. Camino, E. D. Weil, E. M. Pearce, *Polymer Degradation and Stability*, **1997**. *56*, 23–35 4//
[50] C. P. Reghunadhan, *Progress in Polymer Science*, **2004**. *29*, 401–498.
[51] M. R. Orpin, *Proc Instn Mech Engrs*, **1995**. *209*, 19–23.

 DOI: 10.1002/masy.201300040

A Theoretical and Experimental Kinetic Investigation of the ROP of L,L-Lactide in the Presence of Polyalcohols

Konstantina Karidi,[1] *Prokopios Pladis,*[1] *Costas Kiparissides**[1,2,3]

Summary: The present work deals with the experimental and theoretical investigation of the ring-opening polymerization of L,L-lactide with stannous octoate as a catalyst for the production of high molecular weight polymers with improved physical and mechanical properties. The polymerization of L,L-lactide was experimentally studied at various temperatures and different monomer to initiator concentration ratio values. Subsequently, co-initiators with different number of hydroxyl groups (i.e., 1,4-butanediol, glycerol, di-trimethylolpropane and poly-glycidol) were utilized to obtain branched or star shaped polymers. A comprehensive mathematical model was developed based on a detailed kinetic mechanism to predict the dynamic evolution of monomer conversion and molecular weight averages (i.e., Mn, Mw of PLLA). The predictive capabilities of the mathematical model are demonstrated by a direct comparison of model predictions with literature and new experimental data on monomer conversion as well as number and weight average molecular-weight of PLLA.

Keywords: kinetics; L,L-Lactide; polyalcohol; ring-opening polymerization

Introduction

Polylactide (PLA) or polylactic acid is a recyclable and compostable polymer. It is considered to be the best representative of biodegradable polymers since its constituent monomer (lactic acid) can be produced from renewable sources (i.e., derived from corn, sugar beets, etc.). Its physical and mechanical properties can be altered via the copolymerization of lactic acid with suitable co-monomer (e.g., ε-caprolactone, glycolic acid, etc.). PLA is a thermoplastic polymer with excellent mechanical properties and non- toxic degradation by-products. PLA has found a great number of end-use applications, including packaging, resorbable medical implants, and microspheres for drug delivery systems, etc.[1–5] The common requirement for all these applications, with the exception of drug delivery carriers, regards its high molecular weight.

Currently, there are two main manufacturing routes employed for the production of PLA. The first one involves the direct polycondensation of lactic acid. The main drawback of this technology is the requirement for the continuous removal of the condensation water with the aid of a suitable solvent, under high vacuum and high temperature conditions. Moreover, the molecular weight of PLA produced by this process is in general low. The second production route employed by the industry is the ring-opening polymerization (ROP) of L,L-lactide that leads to the synthesis of high molecular weight polymers. The main advantage of this method is ta wide range of linear and branched polymers (i.e., of

[1] Chemical Process and Energy Resources Institute/ Centre for Research & Technology Hellas, P.O. Box 60361, Thessaloniki 57001, Greece
[2] Department of Chemical Engineering, Aristotle University of Thessaloniki. P.O Box 472, Thessaloniki 54124, Greece
[3] Department of Chemical Engineering, The Petroleum Institute, Abu Dhabi, UAE
E-mail: cypress@cperi.certh.gr

different molecular weights and polymer chain microstructures) can be obtained in the presence of a suitable co-initiator (i.e., mono-, bi- and multi-functional alcohols). Moreover, the fact that PLLA is produced in melt does not require the use of any environmentally hazardous solvents.

The present study deals with the experimental and theoretical investigation of the ROP of L,L-lactide in the presence of $Sn(Oct)_2$ for the production of high molecular weight PLLA. Stannous octoate is a very efficient initiator that results in PLAs with a low degree of racemisation (even at high temperatures) and has low toxicity. Thus, it is accepted by the FDA (Food and Drug Administration) [6] for biomedical applications of PLA.

In a number of experimental studies,[7–11] it was found that by varying the co-initiator type and concentration (i.e. the number of hydroxyls in the alcohol) polylactides with different chain microstructure characteristics and molecular weights could be synthesized. In particular, in the presence of mono- and bi-functional alcohols, linear polymer chains were produced while in the presence of polyalcohols with more than two hydroxyl groups, star-shaped and branched polymers were synthesized. Thus, by selecting a suitable polyalcohol as co-initiator, the molecular structure (i.e., linear branched, comb-like and/or star-shaped) of polylactides can be affected.[7–14]

In the present study, the ROP of of L,L-lactide was experimentally investigated at high temperatures to obtain linear and branched PLLAs of high molecular weight. Co-initiators with different number of hydroxyl groups (i.e., 1,4-butanediol, glycerol, di-trimethylolpropane (DTMP) and polyglycidol) were utilized to obtain branched or star shaped polymers. A number of kinetic experiments were carried out, at two different polymerization temperatures (i.e., 160 °C and 180 °C), by varying the value of the monomer to initiator molar ratio (i.e., [M]/[I] equal to 5000, 10000 and 20000) when no co-initiator was used, and the value of the monomer to co-initiator molar ratio (i.e., [M]/[CI] equal to 2000, 5000 and 10000 in the case of 1,4-butanediol, glycerol, DTMP and 12500 to 200000 in the case of polyglycidol) when the molar ratio of monomer to initiator was kept constant (i.e., [M]/[I] = 10000 in the case of 1,4-butanediol, glycerol, DTMP and 5000 in the case of polyglycidol). In addition, a comprehensive mathematical model is developed based on a detailed kinetic mechanism of the ROP of L, L-lactide to predict the dynamic evolution of the monomer conversion and molecular weight developments (i.e., M_n and M_w of PLLA). The proposed kinetic mechanism for the ROP of L,L-lactide comprises a series of elementary reaction steps, including the activation of initiator molecules, chain initiation, propagation, chain transfer to water and to octanoic acid, transesterification, esterification and chain scission reactions.

Experimental Part

It is well-known that the ROP of L,L-lactide is extremely sensitive to the presence of reactive impurities and, thus, it is difficult to control its polymerization rate and molecular weight developments. As a result, careful removal of all the potential impurities from the various reagents employed in the recipe is a prerequisite for the effective control of the polymerization rate and molecular weight distribution of polylactide.

Commercial $Sn(Oct)_2$ is known to be contaminated by impurities, including water and octanoic acid, which are difficult to completely remove, even on repeated distillations. However, shortly after distillation, newly formed acid is being formed. This indicates that even when no alcohol or acid is added to the system, there exists some amount of impurities that can not be removed from the catalyst. Thus, the water traces in the $Sn(Oct)_2$ are responsible for the formation of the active initiating species. In the present study, all other reagents (i.e., L,L-lactide, solvents, e.t.c.) were systematically dried and purified leaving the stannous octoate impurities as

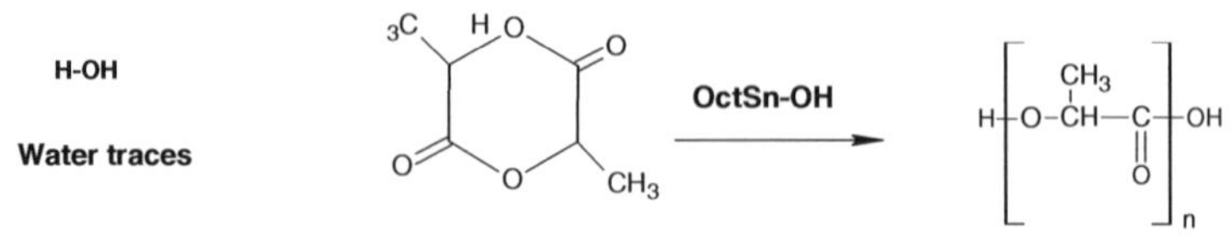

H-OH

Water traces

OctSn-OH

Scheme 1.
The synthetic route of PLLA in the absence of co-initiators.

the only source of co-initiating water species.

Linear and star-shaped high molecular weight polylactides were produced in the absence or presence of co-initiators, respectively, according to the reaction Schemes 1 and 2.

Synthesis of Polylactides

The freshly recrystallized and sublimated L, L-lactide was weighted in 5 ml Schlenk flasks under a continuous flow of nitrogen, followed by the addition of the specified amounts of initiator (i.e., $Sn(Oct)_2$ in toluene) and co-initiators (i.e., 1,4-butanediol, di-trimethylolopropane in acetone and glycerol, polyglycidol in DMF) solutions. The addition of the initiator and co-initiator solutions to the Schlenk flasks was carried out with the aid of flame dried glass syringes, under continuously flushing nitrogen conditions, to ensure a strictly anhydrous environment. In the experimental studies without the use of a co-initiator, the molar ratio of monomer to initiator was varied in the range 5000–20000 while it was kept constant (i.e., [M]/[I] = 10000) in all experiments using co-initiators whereas the

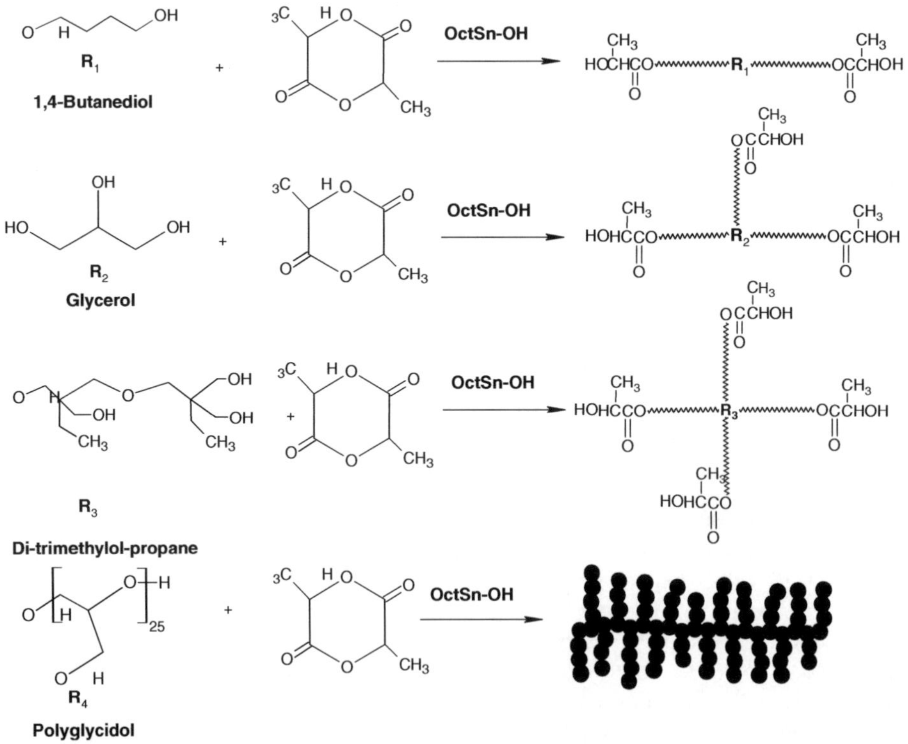

R1
1,4-Butanediol

R2
Glycerol

R3
Di-trimethylol-propane

R4
Polyglycidol

OctSn-OH

Scheme 2.
The synthetic route of PLLA in the presence of co-initiators (i.e., 1,4-butanediol, glycerol, di-trimethylolopropane and polyglycidol).

molar ratio of monomer to co-initiator was varied in the range of 2000–10000. In the case of polyglycidol (linear polymer with 25 hydroxyl groups) the monomer to co-initiator ratio was equal to 5000. The flasks were then sealed under nitrogen and immersed into a thermostated oil bath. The polymerization was carried out at two different temperatures (i.e., T = 160 and 180 °C). At predefined times, the polymerization was stopped by quenching (i.e., by placing the flask into an ice bath). The polymer produced, at a specified time, was recovered by dissolving the total flask content (i.e., polymer, unreacted monomer, etc.) in chloroform, followed by polymer precipitation in excess of cold methanol. The precipitated polymer was then dried under vacuum before its analysis by GPC, DSC and TGA.

Mathematical Modeling of the ROP of the L,L-lactide

Based on a series of previous investigations,[15–17] the ring opening polymerization of L,L-lactide in the presence of $Sn(Oct)_2$ was assumed to follow a coordination insertion mechanism, comprising the set of elementary reactions that is presented in Table 1:

The above kinetic mechanism is sufficiently general and includes a number of reaction steps, namely, initiator activation, chain initiation, propagation, chain transfer reactions (i.e., chain transfer to water, to octanoic acid and to monomer), cyclization, esterification, transesterification and chain scission reactions as well as reactions leading to the formation of specific end-groups.[16,18,19] In the present study, the method of moments was employed to recast the infinite system of dynamic molar species balance equations into a low-order system of differential equations for the leading moments of the "live" and "dead" NCLDs that can be easily solved. The method of moments is based on the statistical representation of the average molecular properties of the polymer chains in terms of the leading moments of the number-chain-length distributions (NCLDs) of the "live" and "dead" polymer chains.

Results and Discussion

It is commonly accepted that the ROP of L,L-lactide in the presence of $Sn(Oct)_2$ proceeds via a coordination insertion mechanism. Accordingly, the initiator reacts with the hydroxyl bearing species to form an alkoxide, which is considered to be the actual initiation species. Apart from their chain initiating role, the hydroxyl groups can also act as chain transfer agents. As a result, the ROP of L,L-lactide is very sensitive to the hydroxyl groups concentration. Thus, it is of paramount importance to closely control their concentration since they largely affect the polymerization rate and molecular weight of PLLA.

To elucidate the effect of monomer to initiator molar ratio on the polymerization rate and the MWD of PLLA without the use of a co-initiator, a series of experiments were carried out for different values of the [M]/[I] ratio at two temperatures (i.e., 160 and 180 °C). It should be noted that the polymerization reactions were conducted at conditions (e.g., temperature and molar ratios) similar to industrial ones. Figure 1 shows the effect of the monomer to initiator molar ratio (i.e., [M]/[I] = 5000, 10000 and 20000) on the dynamic evolution of the L,L-lactide conversion at 180°C. The discrete points represent the experimental measurements obtained from NMR analysis while the continuous lines denote the respective model simulations. As can be seen, as the amount of initiator concentration increases (i.e., the value of the molar ratio [M]/[I] decreases) the polymerization rate increases. Note that, at longer polymerization times, the monomer conversion reaches a limiting conversion that corresponds to the equilibrium monomer concentration determined by the reversible propagation reaction at the specified polymerization temperature. In Figures 2, model predictions (continuous and dashed lines)

Table 1.
Mechanism reactions of the ROP of L,L-lactide in the presence of water traces.

Initiator Activation	$I + H_2O \underset{k_{a'}}{\overset{k_a}{\rightleftarrows}} I^* + OA$
Chain Initiation	$I^* + M \overset{k_{pi}}{\longrightarrow} R_1$
Propagation	$R_{n-1} + M \underset{k_d}{\overset{k_p}{\rightleftarrows}} R_n$
Chain Transfer to Water	$R_n + H_2O \underset{k_{tw'}}{\overset{k_{tw}}{\rightleftarrows}} I^* + D_n$
Chain Transfer to Octanoic Acid	$R_n + OA \underset{k_{ta'}}{\overset{k_{ta}}{\rightleftarrows}} I + D_n$
Chain Transfer to Monomer	$R_n + M \overset{k_{tm}}{\longrightarrow} R_1 + D_n$
Esterification Reactions	$R_n + D_x \underset{k_{est'}}{\overset{k_{est}}{\rightleftarrows}} R_{n+x} + H_2O$
	$D_n + D_x \underset{k_{c'}}{\overset{k_c}{\rightleftarrows}} D_{n+x} + H_2O$
Intermolecular Transterification Reactions	$R_n + R_x \overset{k_{se}}{\longrightarrow} R_{n+p} + R_{x-p}$
	$R_n + D_x \overset{k_{se}}{\longrightarrow} R_{n+p} + D_{x-p}$
Intramolecular Transterification (Formation of Cyclic Chains with Stannous)	$R_n \overset{k_{ts}}{\longrightarrow} OA + CID_n$
Formation of Cyclic Chains (Scission)	$D_n \overset{k_{ssd}}{\longrightarrow} D_{n-s} + CD_s$
Ester End-group Formation	$D_n + OA \overset{k_e}{\longrightarrow} B_n + H_2O$

The symbols I, I^*, M, OA, R_n, D_n, CID_n, CD_n, B_n denote the initiator (stannous octoate), the stannous alkoxide, the monomer (L,L-lactide), the octanoic acid, the "live" and "dead" polymer chains, the cyclic chains (with and without stannous octoate), the "dead" polymer chains with an ester end-group molecular species, respectively. The subscript "n" denotes the degree of polymerization.

are compared with experimental measurements (discrete points) on number and weight average molecular weights for three values of the molar ratio [M]/[I] (i.e., 5000, 10000 and 20000, respectively). Note that the present GPC values for M_w are similar to those reported in the literature.[8,20,21] As can be seen, the molecular weight of PLLA

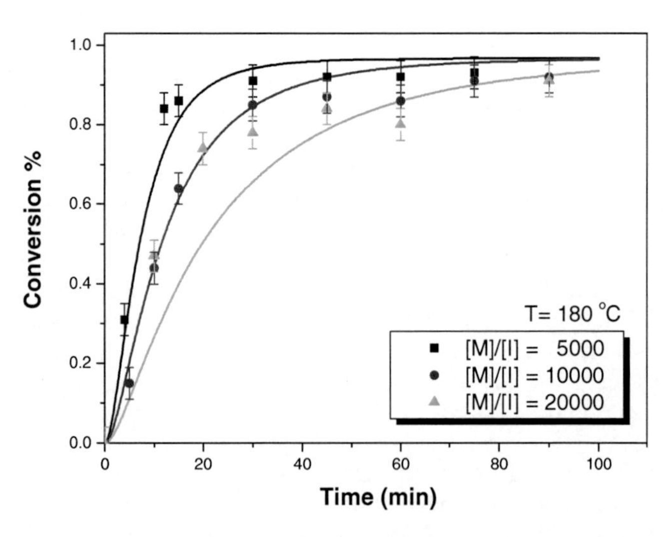

Figure 1.
Model predictions and experimental of L,L-lactide conversion measurements (Temperature: 180°C; [M]/[I] = 5000, 10000 and 20000).

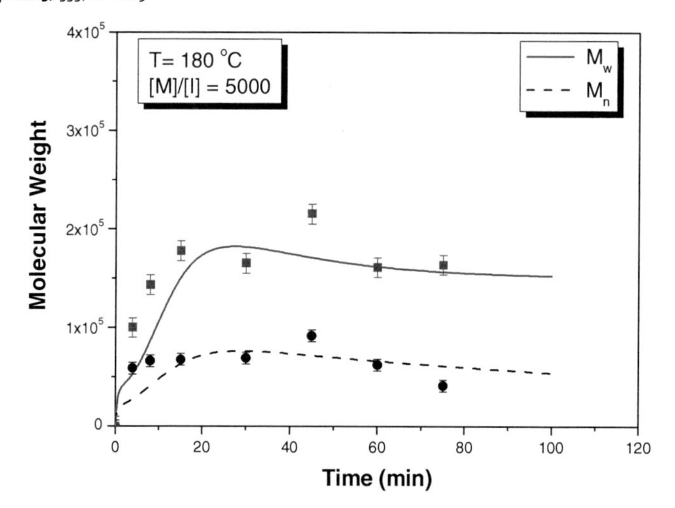

Figure 2.
Comparison of model predictions with experimental measurements on number and weight average molecular weights (Temperature: 180°C; [M]/[I] = 5000).

initially increases with polymerization time up to a maximum value. The early formation of carboxyl end groups results in an increase of the esterification reaction rates (see Table 1). As a result, from the early stages of polymerization, the ring-opening polymerization of L,L-lactide without the use of a co-initiator deviates from the "ideal living" polymerization behaviour (i.e., the number average molecular weight does not exhibit a linear dependence with monomer conversion).

Figure 3 shows the effect of the co-initiator type on the dynamic evolution of monomer conversion at two different temperatures (i.e., 160°C and 180°C in the insert Figure). It is apparent that at high polymerization temperatures (i.e., 180 °C), the co-initiator type does not significantly affect the polymerization rate since all potential hydroxyl groups exhibit similar reactivities due to lower melt viscosities. In Figure 4, the effect of 1,4-butanediol concentration at 180°C on the time evolution of the weight average molecular weight (M_w) of PLLA is depicted. It can be seen that the weight average molecular weight decreases as the co-initiator concentration increases. Similarly to the results of Figure 2, the M_w initially increases with time up to a maximum value. Subsequently, the M_w decreases due to the dominant role of intramolecular transesterification reactions resulting in a decrease of chain length (i.e., due to chain scission).

It is well known that the glass transition temperature (T_g: 53°–64°C) and melting point (T_m: 145–186°C) of semi-crystalline PLAs are important for determining the processing behaviour for various applications. Figure 5 shows the effect of the DTMP concentration on the Differential scanning calorimetric measurements. As it can be seen the increase of DTMP concentration leads to reduction of the glass transition temperature (from 61.8 to 42.9 °C) and to a small reduction of the melting temperature (from 176.7 to 168 °C). Similar behaviour has been observed for all the branched polylactides synthesized in this study indicating the significant effect of the polymer structure on the PLLA thermal properties.

In Figure 6, the TGA diagram of PLLA (i.e., synthesized at 160 °C for a monomer to initiator molar ratio equal to 10000 and a monomer to DTMP molar ratio equal to 2000) is shown. It can be seen that for the linear PLLA (i.e., DTMP = 0) the polymer decomposition temperature is around 350 °C. When DTMP was used as co-initiator the star-shaped PLLA exhibited a different

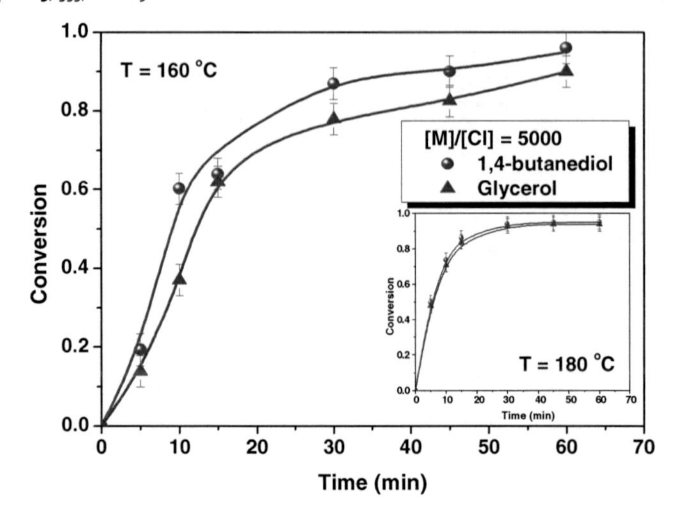

Figure 3.
Evolution of the monomer conversion with respect to time at the two different temperatures in the presence of 1,4-butanediol and glycerol. Polymerization conditions ([M]/[I] = 10000, [M]/[CI] = 5000).

degradation profile due to the branched polymer chain architecture. As can be seen, the polymer chain degradation started at lower temperatures than in the case of the linear polymers. In particular, the polymer degradation profile consisted of two parts. This 'two-part' decomposition profile is due to the initial breakage of the labile bonds between the L,L-lactide and DTMP units (first part), followed by the breakage of the lower molecular weight polylactide chains. From the shape of the thermograms it can also be concluded that the synthesized polymer is of high purity while the residual mass, at the end of the temperature scan, is almost zero.

Figure 7 shows the effect of the polyglycidol (PGL) on the dynamic evolution of monomer conversion at 160°C. In all the experiments the monomer to initiator ratio

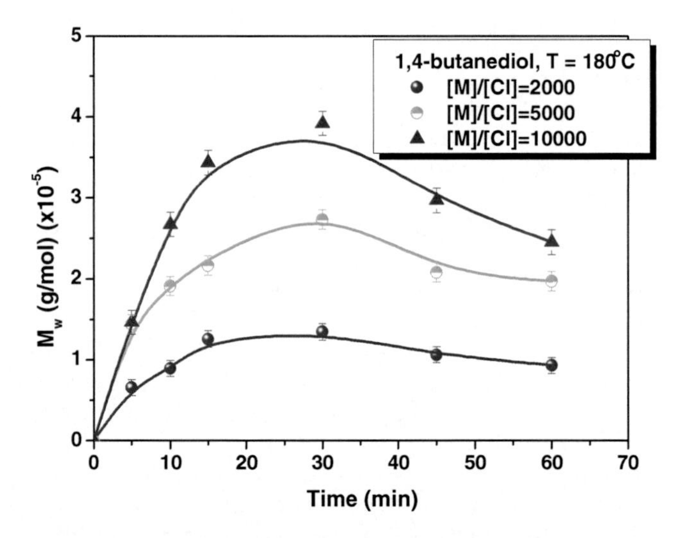

Figure 4.
Variation of the weight-average molecular weight evolution with time at 180°C for different values of the monomer to 1,4-butanediol molar ratio:Polymerization conditions ([M]/[I] = 10000).

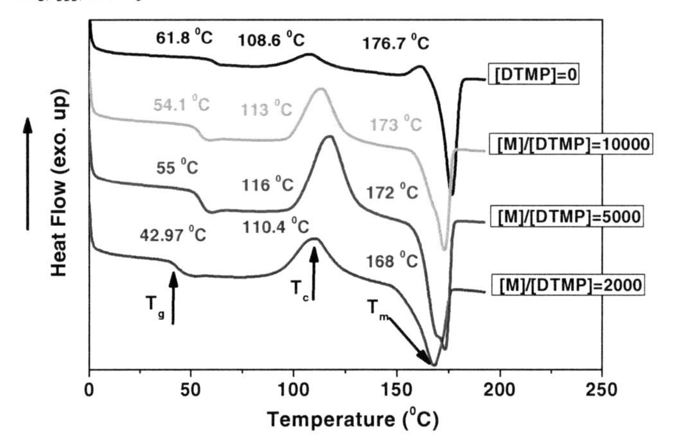

Figure 5.
DSC measurements of PLLA samples synthesized at 160°C for different values of the monomer to DTMP molar ratio.

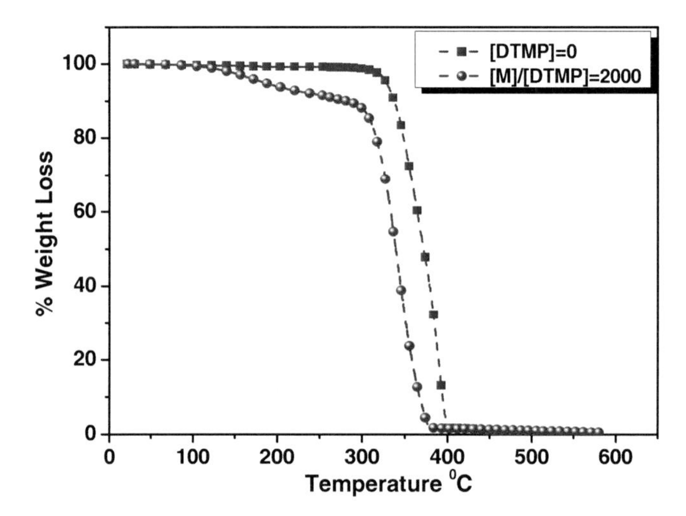

Figure 6.
Comparison of TGA measurements of PLLA samples synthesized at 160°C with and without DTMP co-initiator: Polymerization conditions ([M]/[I] = 10000, [M]/[CI] = 2000).

was equal to 5000. It is apparent that the increase of the polyglycidol concentration (i.e., low [M]/[CI] ratio decrease the L,L-lactide monomer conversion probably due to steric effects). This also results in lower polymerization rates. Finally in Figure 8, the effect of the polyglycidol co-initiator on the MWD distribution of the synthesized PLLAs is shown. As can be seen, in the presence of polyglycidol as co-initiator, the MWD of PLLA is clearly bimodal. The low molecular weight peak can be attributed to the linear polymer chains initiated by the water impurities and the high molecular weight peak is a multi-branched polymer initiated by the hydroxyl groups of the co-initiator.

Conclusion

In the present study, a comprehensive experimental and theoretical kinetic investigation of the polymerization of L,L-lactide

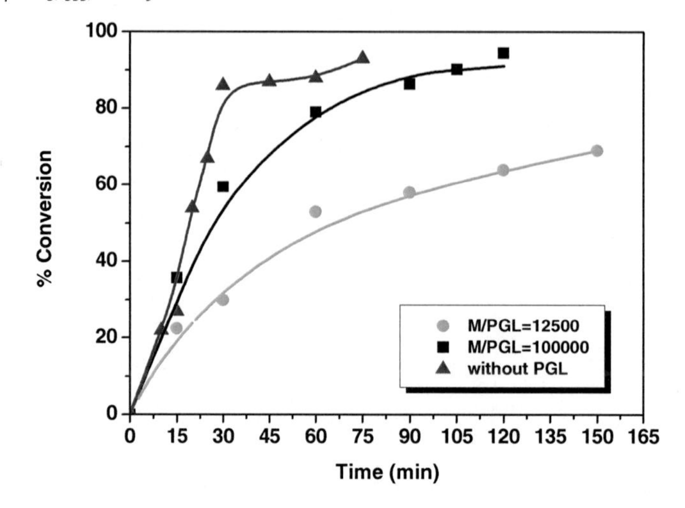

Figure 7.
Evolution of the monomer conversion with respect to time at the two different temperatures in the presence of polyglycidol. Polymerization conditions ([M]/[I] = 5000, T = 160 °C).

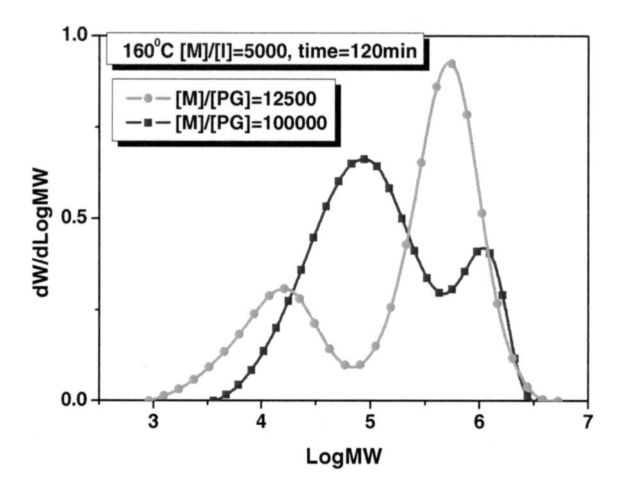

Figure 8.
Molecular weight distribution of the PLLA synthesized with polyglycidol as co-initiator for different values of the monomer to co-initiator molar ratios: Polymerization conditions (T = 160°C, [M]/[I] = 5000).

in the presence of stannous octoate as initiator was carried out at two different temperatures (i.e., 160 and 180°C) and three different values of the monomer to initiator molar ratio ([M]/[I] = 5000, 10000 and 20000). Moreover, the kinetic investigation of the polymerization of L,L-lactide in the presence of stannous octoate as initiator and 1,4-butanediol, glycerol, di-trimethylolpropane and polyglycidol as co-initiators was also carried out at two different temperatures (i.e., 160 and 180°C) and different values of monomer to co-initiator molar ratios. The produced polymers were characterized with regard to their molecular and thermal properties. It was found that the polymerization temperature, the monomer to initiator molar ratio and the monomer to co-initiator molar ratio have a significant effect on the polymerization rate and the molecular properties (i.e., M_n, M_w and MWD) of the synthesized PLLAs.

[1] W. Hoogsteen, A. R. Postema, A. J. Pennings, *Macromolecules* **1990**, *23*, 634.
[2] M. Vert, G. Schwarch, J. Coudane, *J. Macromol Sci Pure Appl Chem* **1995**, *A32*, 787.
[3] P. Mainil-Varlet, R. Rahm, S. Gogolewski, *Biomaterials* **1997**, *18*, 257.
[4] M. Vert, S. Li, G. Spenlehauer, P. Guerin, *J. Mater Sci Mater Med* **1992**, *3*, 432.
[5] R. Duncan, J. Kopecek, *Adv. Polym. Sci.* **1984**, *57*, 51.
[6] H. R. Kricheldorf, A. Serra, *Polym. Bull.* **1985**, *14*, 497.
[7] E. S. Kim, B. C. Kim, S. H. Kim, *J. Polym. Sci. B* **2004**, *42*, 939.
[8] H. Korhonen, A. Helminen, J. V. Seppala, *Polymer* **2001**, *42*, 7541.
[9] T. Biela, A. Duda, S. Penczek, K. Rode, H. Pasch, *J. Polym. Sci A* **2002**, *40*, 2884.
[10] Y. L. Zhao, Q. Cai, J. Jiang, X. T. Shuai, J. Z. Bei, C. F. Chen, F. Xi, *Polymer* **2002**, *43*, 5819.
[11] S. H. Kim, Y. K. Han, Y. H. Kim, S. I. Hong, *Makromol Chem* **1992**, *193*, 1623.
[12] I. Barakat, P. Dubois, R. Jerome, P. Teyssie, E. J. Goethals, *Polym Sci A* **1994**, *32*, 2099.
[13] A. Breitenbach, T. Kissel, *Polymer* **1998**, *39*, 3261.
[14] J. L. Eguiburu, M. J. Fernandez-Berridi, J. San Roman, *Polymer* **1996**, *37*, 3.
[15] R. Mehta, V. Kumar, H. Bhunia, S. N. Upadhyay, *J. Macrom. Sci. Part C Polym. Rev.* **2005**, *45*, 325.
[16] A. Kowalski, A. Duda, S. Penczek, *Macromolecules* **2000**, *33*, 7359.
[17] Y. Yu, G. Storti, M. Morbidelli, *Macromolecules* **2009**, *42*, 8187.
[18] K. Majerska, A. Duda, S. Penczek, *Macromol. Rapid Commun.* **2000**, *21*, 1327.
[19] S. Penczek, A. Duda, A. Kowalski, J. Libiszowski, K. Majerska, T. Biela, *Macromol. Symp.* **2000**, *157*, 61.
[20] S. H. Hyon, K. Jamshidi, Y. Ikada, *Biomaterials* **1997**, *18*, 1503.
[21] D. R. Witzke, R. Narayan, *Macromolecules* **1997**, *30*, 7075.

 DOI: 10.1002/masy.201300041

Stereoselective Condensation of L-Lactic Acid in Presence of Heterogeneous Catalysts

Jennifer Marina Raase, Karl-Heinz Reichert, Reinhard Schomäcker*

Summary: Melt phase condensation of L-lactic acid with heterogeneous catalysts was performed at 0.5 mbar and 180 °C for 1 to 5 h, aiming at enantiopure products. The products were characterized by DSC, HPLC, ^{1}H-NMR spectroscopy and GPC. Zirconium sulfate tetrahydrate was found to be a suitable catalyst with respect to its selectivity and activity. The activity of this catalyst depends on the calcination conditions. By supporting the catalyst on an acidic material with a very high specific surface area, the catalytic activity can be strongly increased without losing selectivity. Products with nearly 100 mol% L-lactic acid units were obtained even at high conversion and catalyst concentration.

Keywords: activity; condensation of L-lactic acid; heterogeneous catalyst; homogeneous catalysis; stereoselectivity

Introduction

Biodegradable polymers such as polylactic acid (PLA) are recognized as important materials for packaging and medical applications.[1,2] Production of PLA by condensation reaction is attracting more and more attention as an alternative industrial route, due to the fact that this process is simpler and cheaper than the ring-opening polymerization of dilactide which is carried out today. Acidic catalysts are able to accelerate the condensation of lactic acid in solution and melt phase. In literature, the use of strong and weak acids such as sulfuric and phosphoric acid as well as numerous organometallic compounds based on germanium, tin, iron, aluminum, titanium or zinc was reported. Especially tin(II) octoate and titanium(IV) butoxide are proven to be suitable catalysts for the condensation of lactic acid, but these Lewis acids show a high degree of racemization of the products.[3] PLA with more than 15 mol% of D-lactic acid units is not crystalline.[3] Products with lower melting points and low mechanical strengths result.[4] With the objective of getting enantiopure products, suitable heterogeneous acid catalysts, e.g. heteropoly acids and sulfated zirconia, were studied. With regard to the condensation of lactic acid, zirconium sulfate tetrahydrate turned out to be an active and selective catalyst.

Institut für Chemie, Technische Universität Berlin, Straße des 17. Juni 124, 10623 Berlin, Germany
E-mail: j.m.raase@gmail.com

Experimental Part

Materials

L-lactic acid (LA) was purchased as 90 wt% aqueous solution from Purac with an optical purity of at least 99% according to the manufacturer. Zirconium sulfate tetrahydrate (ZST, 99.99%) was supplied by ABCR. The silica SIPERNAT® 310, 350 and 2200 were received from Evonik (SIPERNAT® 310: median pore diameter $d_P = 2$ nm, particle size $d = 8.5$ µm, surface area $A = 700$ m^2/g, absorbs 2.65 mL/g; SIPERNAT® 350: $d_P = 100$ nm, $d = 4.5$ µm, $A = 55$ m^2/g; SIPERNAT® 2200: $d_{P,max} = 20$–30 nm, $d = 320$ µm, $A = 190$ m^2/g). Triethylamine (TEA, 99%) from Fluka was distilled before use. The following materials were used without treatment: chloroform-d1 (99.8%) from

Deutero, hydrochloric acid (37%) and acetic acid (HOAc, 99%) from ROTH, sodium hydroxide and anhydrous magnesium sulfate from Fluka. ROTISOLV® HPLC-grade solvents water, ethanol (EtOH) and chloroform with 1 vol% EtOH were purchased from ROTH. D-lactic acid reference solution (90 wt%) was supplied by Uhde Inventa-Fischer.

Characterization Techniques

Number average molecular weights were determined by ^{1}H-NMR (nuclear magnetic resonance) spectroscopy and with respect to polystyrene standards by gel permeation chromatography (GPC) with two PLgel 5 μm Mixed D columns (300 × 7.5 mm) and a refractive index detector. 200 μL of a 3 mg sample/mL chloroform solution were injected. 1 mL chloroform/min was used as eluent flow rate. The analyses were done at room temperature for two samples of each product.

For NMR spectroscopy, 90 mg of each sample were dissolved in 1 mL chloroform-d1. The NMR spectroscopic measurements were done on a Bruker Aspect 2000 (200 MHz) spectrometer with 128 scans. The peak integration was performed with the program MestReC. Baselines were corrected manually.

Scanning electron microscopy (SEM) was carried out on a JEOL 7401F instrument equipped with a Bruker-AXS quantax 4010 detector for element analysis by energy dispersive X-ray (EDX) spectroscopy. Images were recorded using 10 kV as electron accelerating voltage in SE mode.

For transmission electron microscopic (TEM) measurements, the FEI Tecnai G^{2}20 S-TWIN microscope with an electron accelerating voltage of 200 kV was applied. An EDAX r-TEM SUTW detector was used in the case of EDX analysis.

Chemical analysis of zirconium and sulfur content was performed using an ICP-OE (inductively coupled plasma optical emission) spectrometer Varian 715-ES (wavelength $\lambda_{Zr} = 344$ nm, $\lambda_S = 182$ nm).

X-ray powder diffraction (XRD) was done on an X' Pert PRO MPD diffractometer (PANalytical) in reflection mode using CuK_α radiation (wavelength $\lambda = 0.154$ nm). The diffractograms were recorded from 5 to 80° (2θ) with step scan of 0.013°.

Differential scanning calorimetric (DSC) measurements were carried out on a Perkin-Elmer DSC-7 between 25 and 200 °C with a heating rate of 10 °C/min. The degree of crystallization was determined from the area difference of the melting and cold crystallization peak from the first scan, given by the program Pyris Series, in relation to the reference value 93.6 J/g of poly-L-lactic acid.[5]

High performance liquid chromatographic (HPLC) analyses were made on a HPLC system with an isocratic pump, a guard column (30 × 4.6 mm) and a chiral Astec CHIROBIOTIC® TAG column (250 × 4.6 mm) from Sigma-Aldrich at 25 °C. The mobile phase A was a mixture of EtOH/0.1 vol% TEA/0.2 vol% HOAc (pH 6) obtaining a higher efficiency of the column.[6] Water was used as mobile phase B. 10 μL of a solution consisted of 2 mg sample/mL EtOH was injected into the HPLC system. The flow rate of the mobile phase (A/B 83/17) was 0.2 mL/min. UV-Vis detector Dionex UVD170S with 206 nm was used.

Catalyst Preparation

ZST was calcined at different temperatures between 25 and 750 °C for 1 to 3 h. Various silica supports (SIPERNAT® 310, 350 and 2200) were pretreated at 500 °C for 1 h. Silica with different loadings between 7 and 70 wt% were prepared as described below. Loading is defined as mass ratio (1)[7] and was determined by ICP-OES detecting the flushed mass of ZST during filtration.

$$\text{Loading} = \frac{m_{ZS}}{m_{ZS} + m_{Sio_2}} \qquad (1)$$

The desired amount of ZST in aqueous solution corresponding to the pore volume of the support SIPERNAT® 310 (2.65 mL/g) was impregnated on silica using the incipient wetness method.[8] The precursor solution of ZST was added dropwise to a thin layer of support material. Obtaining a loading of 30 wt%, diffusion-controlled impregnation is described: 2.23 g of support material emulsified in 11.2 mL water were impregnated with an aqueous precursor solution of 1.19 mg ZST. The suspension was stirred for 24 h at room temperature. After impregnation, the water was removed with a rotary evaporator. Performing another variant of diffusion-controlled impregnation, catalysts were prepared in the same way as before, but 20% less water volume and more ZST were used in each case. Instead of eliminating water by using a rotary evaporator, the water and surplus ZST were removed by vacuum filtration followed by washing with 10 mL distilled water and drying in air at room temperature.

Condensation of L-Lactic Acid

In a 250 mL round-bottom flask with two indentations on the wall, 45 g of a 90 wt% L-lactic acid aqueous solution were evaporated using a rotary evaporator with a rotation speed of 100 rpm. The water was removed under reduced pressure (50 mbar) at 80 °C. A uniform suspension of the calcined catalyst in the dehydrated lactic acid was prepared by an ultrasonic treatment before starting the reaction. The condensation of L-lactic acid was performed at 160 to 200 °C under reduced pressure (0.5 mbar) and 100 rpm for 1 to 5 h. Selecting 200 °C for condensation, a large amount of dilactide was formed. Low molecular weight product was synthesized at 160 °C condensation temperature. Running the condensation at 180 °C, 0.5 mbar and 100 rpm with 0.3 wt% catalyst for 5 h turned out to be a convenient procedure for condensation of lactic acid getting enantiopure oligomers with a relatively high molecular weight. This is defined as standard procedure in the following section. The pressure and temperature profiles of the condensation reaction are given in Figure 1. After reaction, the reaction mixture was cooled down in air by pouring it on an aluminum foil. The pressure profile presented in Figure 1 had no effect on the conversion and the molecular weight of the products. It was applied to reduce the loss of lactic acid.

Hydrolysis of Oligomers

The samples were hydrolyzed with a sodium hydroxide solution to lactic acid. For this purpose, in a 500 mL round-bottom flask, 5.0 g (0.07 mol, 1 eq) of lactic acid oligomer, 5.6 g (0.14 mol, 2 eq) of sodium hydroxide and 150 mL of a 1:1 mixture of distilled water and EtOH were stirred under reflux at 85 °C for 2 h. The reaction progress was monitored using HPLC. After cooling, dilute hydrochloric acid was added to get pH 4. The solution was concentrated at 80 °C and 50 mbar up to precipitation of sodium chloride. The precipitate was filtered. The remaining clear and colorless liquid was dried over anhydrous magnesium sulfate for 1 h.

Results and Discussion

Molecular Weight of PLA

^{1}H-NMR spectroscopy determined the average degree of polymerization based on work of Espartero *et al.*[9]:

$$\langle DP_n \rangle = \frac{\text{Integral(C} - \text{H medium chain)}}{\text{Integral(C} - \text{H terminal chain)}} \tag{2}$$

According to equation (2), the monomer fraction does not need to be considered. A 7% average variation of molecular weights was found. In the case of an increasing number average molecular weight $\langle M_n \rangle$, that needs a higher sample concentration, signals of chain methylene and terminal methyl hydrogen atoms interfere with each other. As a consequence of a more difficult integral analysis, there is a steady increase in absolute error.

For these reasons, the molecular weights of oligomers were also determined by GPC (experimental error $\langle M_n \rangle \pm$ 10%). In the absence of PLA standards, the GPC results were based on polystyrene (PS) standards. Due to the fact that PLA and PS with nearly the same molecular weight differ in their chemical structure and in terms of their spherical size in solution, the equation (3) turns out to be a suitable conversion relation[10] with an output which is almost

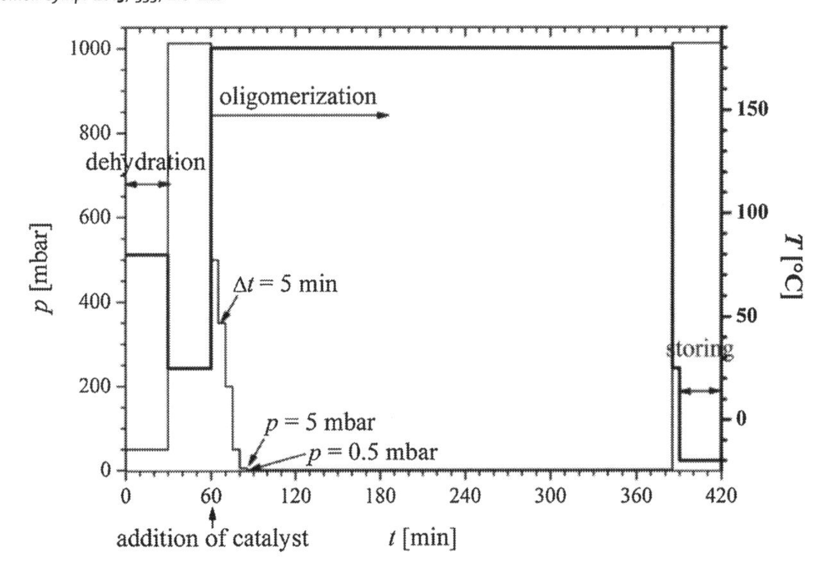

Figure 1.
Pressure and temperature profiles of dehydration and oligomerization of L-lactic acid.

synonymous with halving of GPC-PS results.

$$logM(\text{PLA}) = 0.95922 * logM(\text{PS}) - 0.1592 \quad (3)$$

It should be noted that the determined molecular weights are dependent on the choice of solvent and column material as well.[10] Therefore, as described in literature,[10] chloroform as eluent and a polystyrene/divinylbenzene matrix as stationary phase were chosen. Figure 2 shows the relation between the number average molecular weights determined by GPC

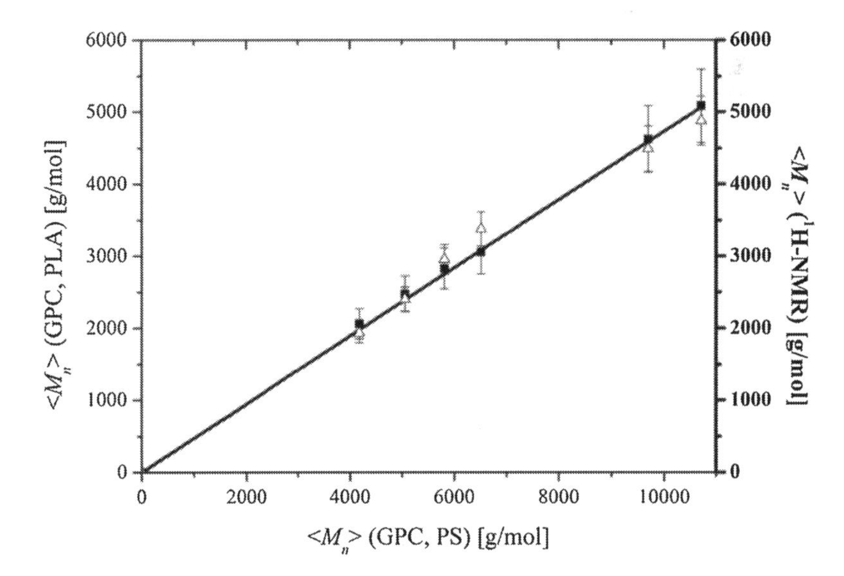

Figure 2.
Correlation between number average molecular weights determined by GPC (■) and ^{1}H-NMR spectroscopy (△) [PS: polystyrene standard; PLA: polylactic acid standard calculated from literature[10]].

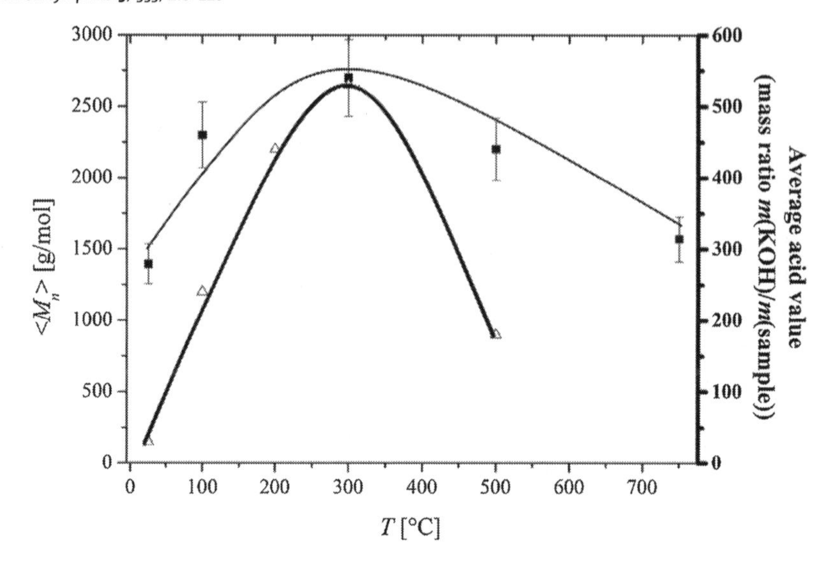

Figure 3.
Acidity (△) of zirconium sulfate tetrahydrate (ZST in methanol/acetone)[11] and number average molecular weights of oligomers (^{1}H-NMR ■) versus calcination temperature of ZST.

and ^{1}H-NMR spectroscopy for $\langle M_n \rangle \leq$ 5000 g/mol. A close correlation between ^{1}H-NMR spectroscopy and GPC data based on PLA standards was established. ^{1}H-NMR spectroscopy is appropriate to determine number average molecular weights of lactic acid oligomers.

Catalyst Calcination and Impregnation

Figure 3 presents the comparison between acid strength of ZST treated at different calcination temperatures and its impact on number average molecular weights of oligomers synthesized by melt condensation.

ZST will show the highest activity if it is calcined at 300 °C for 1 to 3 h before being added to the dehydrated lactic acid. This effect correlates with the maximum of the acid strength of calcined ZST which offers similar results for 1 and 3 h calcination times. 300 °C for 1 h is defined as standard calcination condition in the following section. For esterification, Huang *et al.*,[12] Arata *et al.*[13] and Chen *et al.*[14] found the maximum activity of ZST calcined at 750 °C due to the formation of sulfated zirconia. In contrast to these results, the activity of ZST treated at 750 °C is very poor (Figure 3).

30 wt% loaded silica SIPERNAT® 310 was produced by diffusion-controlled impregnation with ZST. In a SEM image of such prepared catalyst two distinguishable morphological structures can be seen (Figure 4. B). Free ZS crystal needles proven in EDX spectra are evident for a non-homogeneous loading of the support material.

For catalysts synthesized by the diffusion-controlled impregnation followed by filtration and washing, neither SEM nor TEM images of manifoldly analyzed sample areas were able to localize released ZS crystals, even though only a small area could be imaged. Moreover, the TEM images contained few regions where the electron beam completely penetrated through the material due to the relatively large particle thickness. Another problem of ZS was the instability against radiation. Depending on the selected magnification, the illuminated surface had to be adjusted. Therefore, the radiation dose necessarily increased. After two minutes of electron irradiation, the silica did not show any changes in characteristics, but almost immediately, blistering on ZS was observed. Therefore, in the investigation of impregnated silica, it could occur that ZS was indeed

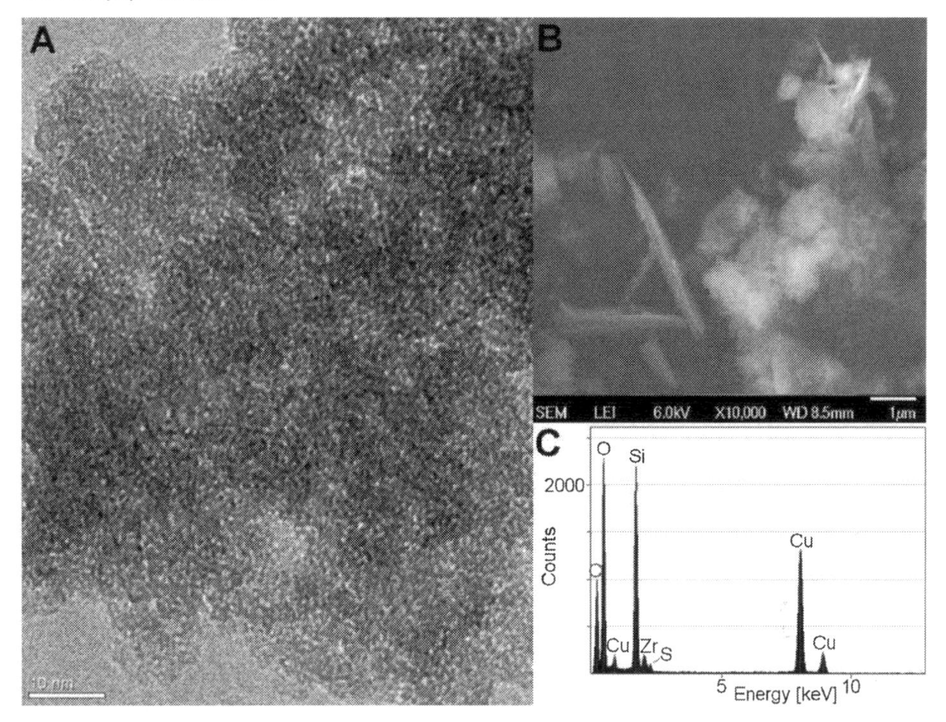

Figure 4.
A: TEM image of calcined zirconium sulfate tetrahydrate (ZST) supported on SIPERNAT® 310 (30 wt%) by diffusion-controlled impregnation followed by filtration and washing; B: SEM image of calcined ZST supported on SIPERNAT® 310 (30 wt%) by diffusion-controlled impregnation without filtration; C: EDX spectrum of catalyst shown in image A [TEM image by ZELMI, TU Berlin].

supported; however, it was evaporated by the electron radiation.

An EDX spectrum of the catalyst section shown in Figure 4. A detects zirconium and sulfur on the silica surface. This suggests that the black dots in the TEM image (Figure 4. A) which occur in places where the picture is lighter, i.e. the sample is thinner, represent ZS; hence, nanoscale ZS particles seem to be evenly distributed on the silica surface by diffusion-controlled impregnation followed by filtration/washing.

Impregnation Methods

By employing the incipient-wetness method, the experimental error of the oligomer molecular weights was substantially greater than by carrying out the diffusion-controlled impregnation (15% vs. 3%). This effect can be explained by the drop technique used. Lower molecular weights result from areas with very high loading (Figure 5). Therefore, only the diffusion-controlled impregnation followed by filtration and short-time washing was applied in the next experiments.

Effect of Pressure

In condensation reactions, the pressure plays a decisive role: the lower the reaction pressure the higher the molecular weight but the smaller the yield. A product synthesized within 5 h at 180 °C, 100 mbar and with 0.3 wt% ZS (30% loading) showed a molecular weight as if no catalysis took place. But for a five hour reaction with 0.3 wt% ZS there was effectively no difference between using 0.5 or 10 mbar because the lower the pressure the higher the dilactide formation (shifting of equilibrium). A pressure of 0.5 mbar showed a yield of 50% whereas 10 mbar allowed a yield of

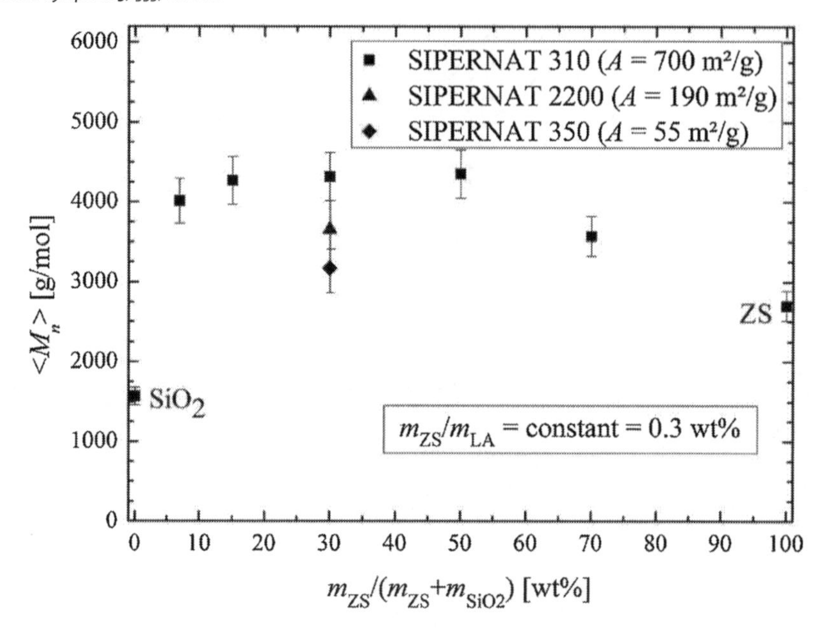

Figure 5.
Number average molecular weight (^{1}H-NMR) of oligomers as a function of support loading and different supports.

80%. The choice of 0.5 mbar is to enable an efficient removal of water out of the highly viscous reaction mixture for longer reaction times.

Effect of Support Loading

As shown in Figure 5, SIPERNAT® 310-supported ZS allowed an activity increase of 60% compared to the unsupported catalyst. As a result of supporting, products showed molecular weights multiplied by a factor of 3 relative to products synthesized by adding pure SIPERNAT® 310. As shown in Figure 5 a 14-fold increase of the specific support area led to a 30% increase on molecular weight although there was no evidence of non-homogeneous loading (Figure 4). Dissolving ZS-catalyzed lactic acid oligomers in chloroform, free sulfate ions formed insoluble barium sulfate after adding a barium nitrate solution. Hauser[15] described the formation of different zirconyl sulfuric acids when ZS gets in contact with water. Neglecting zirconyl sulfuric acids which form sulfuric acid in water,[15] ICP-OES results allowed to determine 0.1 wt% free sulfuric acid in oligomers synthesized by using 0.3 wt% ZS. Comparing molecular weights of oligomers catalyzed by 0.1 wt% sulfuric acid to catalysis with 0.3 wt% ZS, similar results were obtained. Using the same concentration of sulfuric acid, there was not any difference between concentrated or dilute sulfuric acid as catalysts. Lower concentration of released active sulfuric acid can be explained by the following reason. SIPERNAT® 310 is known to form agglomerates of which the exact size remains unknown. Regarding to particle diameters of 5 μm during impregnation, the silica particles were ten times bigger after impregnation and drying. After getting in contact with water, it is assumed that sulfuric acid is released from the heterogeneous material ZS. Due to the fact that sulfuric acid appears to be the active catalyst species, it is clear that homogeneous catalysis takes place. Calcination of ZST at 300 °C and supporting seem to prefer an improved dissolution of sulfuric acid. Using 0.1 wt% sulfuric acid, the crystallinity of the oligomers amounted to 2% in contrast to 40% (using 0.3 wt% ZS) under standard conditions. With the addition of pure SIPERNAT® 310 before condensation with sulfuric acid, the crystallinity did not change. It is possible that crystalline ZS acted as a nucleating agent.

With decreasing silica loading, the mass of support material increased. Lower activity of lightly loaded support material (7 wt%) in comparison to highly loaded supports (15–50 wt%) can be explained by incomplete wetting of the catalyst with the reaction mixture. After thermal treatment, ZS supported on macroporous silica (SIPERNAT® 350) did not show any appreciable change in the activity compared to the unsupported catalyst, especially due to the relatively small specific surface area.

XRD measurements confirm that nanoscale ZS particles were evenly distributed on the surface of the support material or/and that the ZS crystals were very small for loadings between 7 and 50 wt% (Figure 6). For this reason, there are no detectable lattice planes and the activity of the catalysts with these loadings is considered to be constant.

A support loading of 70 wt% decreased the catalyst activity considerably. Figure 6 presents an X-ray powder diffractogram of this support material showing lattice planes that are partially shifted relative to ZS. These were probably clusters on the silica surface. In comparison to literature[12], these clusters consist of ZS as well as monoclinic zirconia which is formed during the calcination at 300 °C.

Effect of Catalyst Concentration

Up to a concentration of 0.3 wt% ZS, a linear relation between the catalyst concentration and the number average molecular weight of lactic acid oligomers can be detected (Figure 7). The synthesis of lactic acid oligomers during melt condensation under given conditions is largely independent of mass transport. An oligomer, synthesized by using 0.3 wt% of 30 wt% silica loading at 180 °C for 5 h, shows a crystallinity of 40%; the conversion amounts to 98.4% calculated from Carothers equation[16] (4).

$$X = \frac{\langle PD_n \rangle - 1}{\langle PD_n \rangle} \tag{4}$$

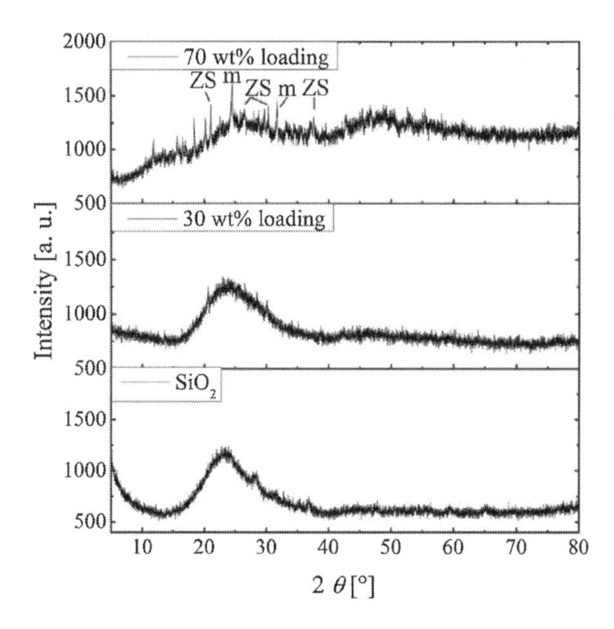

Figure 6.
Wide angle X-ray diffractograms of support SIPERNAT® 310 and supported catalysts with different loadings of zirconium sulfate (ZS) after calcination at 300 °C for 1 h [m: monoclinic zirconia[12]].

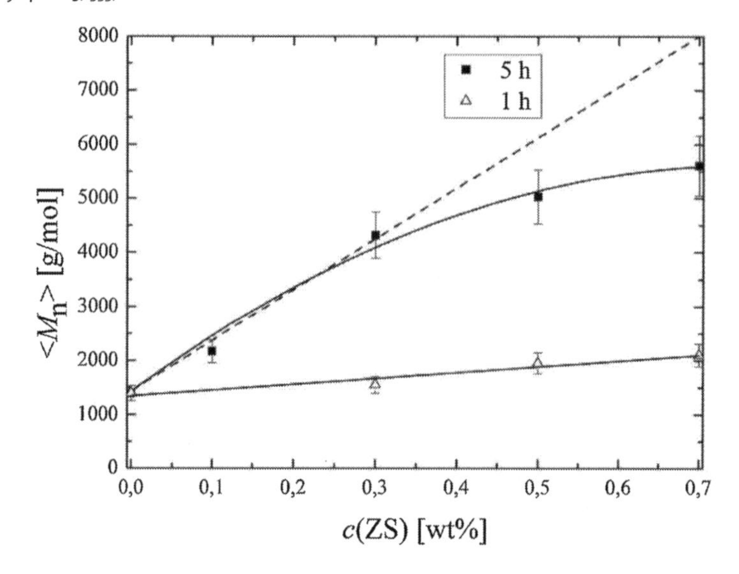

Figure 7.
Correlation between molecular weight (^{1}H-NMR) and concentration of zirconium sulfate (30% SIPERNAT® 310 loading) for different reaction times under standard conditions.

With increasing catalyst concentration, there is an increase in deviation from the linear trend. Reasons for this nonlinearity can be:

1. mass transport limitations regarding the removal of water
2. side reactions of condensation reaction (oligomer degradation)

Figure 7 shows a linear correlation between molecular weight and catalyst concentration for one hour reactions with supported ZS under standard conditions; hence, mass transport limitations regarding the removal of water, side reactions and an inefficient mixing related to a high support mass did not play any role in this case. The polydispersity index of oligomers formed after 5 hh h of reaction time changed from 1.8 (for 0.3 wt% ZS) to 2.2 (for 0.7 wt% ZS). In this case, mass transport and side reactions could play a certain role; however, both products attained a degree of crystallinity around 40%.

Analysis of the Stereoselectivity

Figure 8 presents the HPL chromatograms of the L- and D-lactic acid enantiomers. The L- lactic acid shows a maximum absorption after 2.7 and 3.9 min. The first peak is assigned to the monomer due to the highest intensity. In a 1:1 mixture of L- and D-lactic acid, the peaks of L- and D-lactic acid are reached within 3 to 8 min. The different retention time of L-lactic acid results from the interaction between the two enantiomers. The maximum absorption of the hydrolyzed sample appears after 2.8 and 4.3 min. The latter shoulder is attributable to an impurity of the L-lactic acid. D-lactic acid could not be detected even though the absorption of D-lactic acid will be reached within 7 to 10 min if D-lactic acid is detectable. A minimal interaction between the L- and D-lactic acid can be seen as a retention time shift. Using higher catalyst concentrations, the stereoselectivity of the catalyst did not visibly change.

Conclusion

By supporting ZST on acidic silica with a specific surface area of 700 m^2/g and calcination at 300 °C for 1 h, number average molecular weights of lactic acid oligomers increased by a factor of 3 in

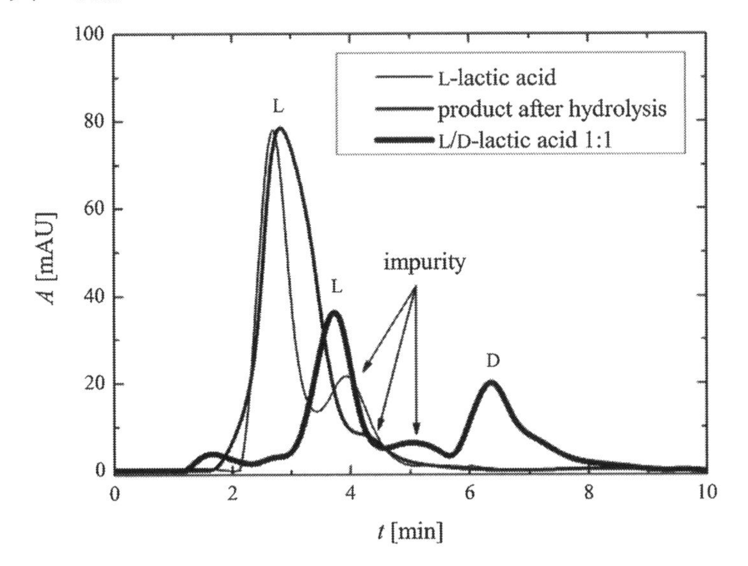

Figure 8.
HPL chromatograms of L-lactic acid, L/D-lactic acid (1:1) and a hydrolyzed lactic acid oligomer [standard synthesis with 30 wt% SIPERNAT® 310 loading].

relation to uncatalyzed reaction and by a factor of 2 related to unsupported ZS at corresponding conditions. To avoid free ZS crystals and to get an even distribution of ZS on the silica surface, the diffusion-controlled impregnation followed by filtration and short-time washing seems to be the best impregnation method. Choosing 0.3 wt % of a catalyst with 15–50 wt% silica loading and running the reaction at 180 °C under reduced pressure in a rotary evaporator, nearly enantiopure and high-yield lactic acid oligomers with a crystallinity of 40% were synthesized even at a high conversion. With increasing catalyst concentration, mass transport limitations and side reactions play a certain role in longer reaction times. ICP-OES results allowed to determine 0.1 wt% free sulfuric acid in oligomers synthesized by using 0.3 wt% ZS. In comparison with this case, oligomers catalyzed by 0.1 wt% sulfuric acid showed similar molecular weights and nearly no crystallinity. ZS may act as a nucleating agent. It is assumed that sulfuric acid is released from ZS during condensation. Lower concentration of released sulfuric acid can be explained by support agglomerates after impregnation. Due to the fact that sulfuric acid appears to be the active catalyst species, homogeneous catalysis takes place. Calcination and supporting of ZST seem to prefer an improved dissolution of sulfuric acid.

Acknowledgments: The authors thank Uhde Inventa-Fischer for GPC measurements. J. M. Raase is grateful to Evonik Foundation for financing her scholarship.
This paper was corrected after initial publication 29th November. Several mistakes have been introduced into the author's paper and we are very sorry for our oversight.

[1] H. Frey, F. K. Wolf, *Polymilchsäure – ein vielseitiges Material für Anwendungen in der Medizin*, in *Natur & Geist* **2010**, JGU Mainz, pp. 12–14.
[2] T. Walker, *Akt Dermatol* **2010**, 36(8/09), 300–304.
[3] K. Hiltunen, J. V. Seppälä, M. Härkönen, *Macromolecules* **1997**, 30(3), 373–379.
[4] G. Perego, G. D. Cella, C. Bastioli, *J. Appl. Polym. Sci.* **1997**, 59(1), 37–43.
[5] E. W. Fisher, H. J. Sterzel, G. Wegner, *Kolloid Z. Z. Polym.* **1973**, 251, 980–990.
[6] Astec CHIROBIOTIC Handbook **2004**, 5th ed., pp. 19–23.
[7] IUPAC *Quantities, Units and Symbols in Physical Chemistry* **1993**, 2nd ed., Blackwell Scientific Publications, Oxford pp. 41.

[8] G. Ertl, H. Knözinger, F. Schüth, J. Weitkamp, *Handbook of Heterogeneous Catalysis*, **1997**, 1st ed., Wiley-VCH, Weinheim pp. 2271.
[9] J. L. Espartero, I. Rashkov, S. M. Li, N. Manolova, M. Vert, *Macromolecules* **1996**, 29(10), 3535–3539.
[10] Deutsches Kunststoff-Institut Darmstadt *Polymilchsäure (PLA) für neue biobasierte Verpackungen*, Annual Report **2010**, pp. 41–44.
[11] A. L. Penagos, *Oligomerization of L-Lactic Acid in Melt Phase using Zirconium(IV) Sulphate Tetrahydrate as a Solid Acid Catalyst* **2011**, Master's thesis, TU Berlin
[12] Y. Huang, B. Zao, Y. Xie, *Appl. Catal. A: Gen.* **1998**, *173*, 27–35.
[13] K. Arata, M. Hino, N. Yamagata, *Bull Chem. Soc. Jpm.* **1990**, *63*, 244–246.
[14] C. Chen, T. Li, S. Cheng, H. Lin, C. J. Bhongale, C. Y. Mou, *Micropor. Mesopor. Mater.* **2001**, *50*, 201–208.
[15] O. Hauser, H. Herzfeld, *Über die basischen Zirkonsulfate und den molekularen Zustand des Zirkonsulfats in wäßriger Lösung* **1918**, Technologisches Institut der Universität Berlin pp. 1–8.
[16] W.H. Carothers, *Transaction of the Faraday Society* **1936**, 32, 39–49.

Polyolefin Composite Synthesis: From Small Scale to kg Material

Saskia Scheel,[1] André Rosehr,[1] Artur Poeppel,*[2] Gerrit A. Luinstra*[1]

Summary: Polyolefin composites were prepared by the in situ polymerization technique. Experiments were performed with mixtures of aluminum silicates and multi walled carbon nanotubes as fillers in 1 L reactors. The mixture of fillers was sonicated in toluene solution before polymerization, leading to a more stable dispersion in the solvent. Scale up to 10 L reactions gives access to kg material with well-dispersed filler in polypropylene. The thermal properties (Tm, Tc) of the composite are basically unchanged to polypropylene without filler, like the crystallinity is.

Keywords: clay; CNT; composites; in situ polymerization; poly(propylene)

Introduction

The property profile of polyolefins can be strongly enhanced by generating composites with appropriate fillers. A large change in mechanical properties can in particular be reached with nanoparticles with a high aspect-ratio, and already at a low filler content. A good and lasting dispersion of the filler and an intimate contact of matrix and filler is a prerequisite for an efficient and effective use of materials. The polarity of in particular inorganic filler and polyolefin may be very different and this imposes a challenge upon reaching and maintaining such a good dispersion and contact. One processing procedure to arrive at such composites, melt-compounding, is thus often hampered by agglomeration of the filler particles, especially when these are also agglomerated in the bulk starting material.[1,2] We use *in situ* polymerization according to the *polymerization filling technique* (PFT), reported by KAMINSKY.[3–5] The process of *in situ* polymerization, i.e. generation of the matrix in the presence of a filler, has shown to be a simple method to prepare well-dispersed polyolefin composites (Figure 1).[6]

The multi-step batch procedure comprises the steps of supporting a catalyst / cocatalyst system onto a filler (dispersed in a solvent) and addition of olefin, thus resulting in an *in situ* polymer formation in a slurry polymerization. The product is a ready to use composite or a master batch. The accessible dispersion level of filler in the matrix by the PFT is very high and basically independent of the filler content. In this study, mixtures of clays (μm scale) and multi walled carbon nanotubes (MWCNT) were used as a mixture of fillers. Screenings for PFT reaction conditions were performed in a 1 L glass reactor. Scale up of relevant reactions was successful in a 10 L steel reactor yielding up to 1.5 kg of fine, powdery composites. These can be processed further via melt-blending in a twin-screw extruder to tune the filler content.

[1] Institut für Technische und Makromolekulare Chemie, Universität Hamburg, Bundesstr. 45, 20146 Hamburg, Germany
Fax: +49 (0) 40 42838 6008;
E-mail: luinstra@chemie.uni-hamburg.de
[2] Honda R&D Europe (Deutschland) GmbH Carl-Legien-Straße 30, 63073 Offenbach, Germany
E-mail: artur_poeppel@de.hrdeu.com

Experimental Part

Materials

Argon (grade 5.0) and propylene (grade 3.5) were purchased from Linde. Propylene was purified over columns with molecular

Figure 1.
Cartoon for the preparation of polyolefin composites by *in situ* polymerization.

sieve (4 Å) and BASF-Catalyst R3–11. Toluene was distilled and dried by passing over columns similar to those used for the purification of propylene. The MAO (solution in toluene, 10 wt %, Crompton GmbH) was used as received. The clay was dried 3 h at 300 °C in vacuum before use (courtesy of Applied Minerals) as well as the MWCNT (Nanocyl NC7000). All reactions were carried out in an inert atmosphere and use of dry equipment.

Reactors

Screening experiments were performed in a 1 L glass Büchi-reactor with a horseshoe mixer charged with 300 mL dry toluene and a stirring rate of 270 rpm. A 10 L stainless steel reactor of Juchheim was used for scale-up reactions. Data management and parameter controlling was with Delphi® hard/ and software. The volume stream of propylene (max. 500 mL/min), temperature and pressure were recorded.

Preparation of Poly(propylene)

A typical experiment is described. A reactor of 1 L volume was charged with 300 mL of dry toluene, heated to a temperature of 30 °C. MAO solution (2 mL, 10 wt %) was added. The resulting mixture was subsequently saturated with propylene at a pressure of 0.2 MPa. The polyreaction was started by introducing a catalyst solution into the reactor by a syringe. After 30 minutes the polymerization was terminated by removal of propylene and quenched by addition of 5 mL of ethanol. The products were subsequently hydrolyzed with a solution of hydrochloric acid (7%), filtered, washed and dried in vacuum at 40 °C to constant weight, after which the powdery material was degassed in a twin screw extruder. The extrusion experiments were carried out on a Prism Eurolab 16 modular twin screw extruder by Thermo Scientific with an L/D ratio of 40:1. The screws were set up for typical PP composite preparation, i.e. with a residence time of about 2 minutes. The temperature of all zones was at 210 °C.

Preparation of Clay/MWCNT-Composites

A suspension was prepared by mixing an aluminum silicate and multi walled carbon nanotubes as solids and adding 300 mL of dry toluene. The suspension was sonicated and treated with MAO (10 wt% solution in toluene). The filler material was impregnated with the cocatalyst at a ratio of 4.5 mL of the MAO solution per gram clay/MWCNT mixture. The suspension was stirred for 24 hours.

Polymer Characterization

Testing bars: Flat specimens were used of type 5B, DIN EN ISO 527-2. The injection temperature was set at 240 °C and higher, the mold temperature to 50 °C and the pressure to 500 bar (20 s) with a post pressure of 300 bar (25 s). Molecular weight were determined on a PL GPC 220 of Varian in 1,2,4-trichloro benzene at 130 °C against polystyrene as reference. The signal results from an RI-detector. Melting temperatures (T_m), the crystallization temperature (T_c) and the crystallinity were determined by differential scanning calorimetry (DSC) using a Mettler-Toledo GmbH instrument (DSC 821). Generally, samples of 5–6 mg were used. T_m was taken from the heating curve of the second heating (rate of 10 °C/min, range from -20 to 200 °C). Crystallization temperatures T_c were determined from the cooling curve (cooling rate 10 °C/min) after melting was complete at 200 °C and were hold for 5 min. Thermogravimetric measurements for determining filler contents was on a Perkin Elmer instrument (in air, temperature program 30–800 °C, 10 K/min). SEM pictures (Gemini Leo 1525 field emission microscope) were taken from flat test specimen obtained from injection molding. These were cryo-cracked at −196 °C the clear breaking edge is depicted.

Results and Discussion

The synthesis of PP can be achieved with any standard metallocene catalyst with a high activity and yielding a product with high isotacticity.[7] The activity for an indenyl zirconocene was studied at various temperatures. It was found that the molecular weight is highly affected by temperature (Table 1). It is thus of high importance to have a good control over the reaction temperature profile in order to attain the molecular weight desired, and for reproducibility. Table 2

The polyreaction was monitored by mass flow and a temperature controller. The formation of polypropylene is strongly exothermic and control over the temperature can be easily lost (Figure 2). On the other hand, a broader molecular weight distribution can be reached by performing the polymerization at a temperature gradient. This usually gives the product superior flow properties. The same procedures were used to prepare composites of aluminum silicate clay and polypropylene. In that case, the clay was pretreated with the cocatalyst before addition of propylene.

Table 1.
Polypropylene products as function of the temperature of the polyreaction.

temperature [°C]	activity [t/h·bar·mol]	Mn [g/mol]	Mw [g/mol]
15	49.7	201 300	2 681 900
30	56.1	65 900	441 100
45	73.5	37 500	142 150

The synthesis procedures could be successfully scaled to a 10 L steel reactor. In this manner, up to 800 g of polypropylene and/or composite material is received in 60 min to 90 min of polymerization time. The calculated filler content from the propylene take up is within an error of 15% in accordance with the actual concentration. This allows to set the filler content within a certain range.

The major point in the study was to achieve a good distribution. It was found, that the clay particles are separated from each other and a good distribution at the breaking surface is seen in a SEM analysis. It was also found, that after fracturing the sample, a good contact between matrix and filler persistently exists (Figure 3).

Clay/MWCNT-Composites

After successful *in situ* polymerization of aluminum silicates with a high modulus into a fine dispersion in polypropylene, multi walled carbon nanotubes (MWCNT) were introduced as an additive filler compound. It has been shown in epoxy composites that combinations of fillers with several dimensions influence the mechanical properties more than just the additive of the two.[8,9] The polymerizations were performed in 10 L scale, no adjustment of the reaction parameter was found to be necessary. In the past, much effort was put into achieving a

Table 2.
Data on in situ preparation of polypropylene composites.

	yield [g]	clay [%]	clay calc. [%]	Tm [°C]	Tc [°C]	crystallinity [%]
Ref. iPP	685	0	0	158	113	44
1	705	5.65	5.1	160	119	42
2	518	10.5	12.0	163	117	48

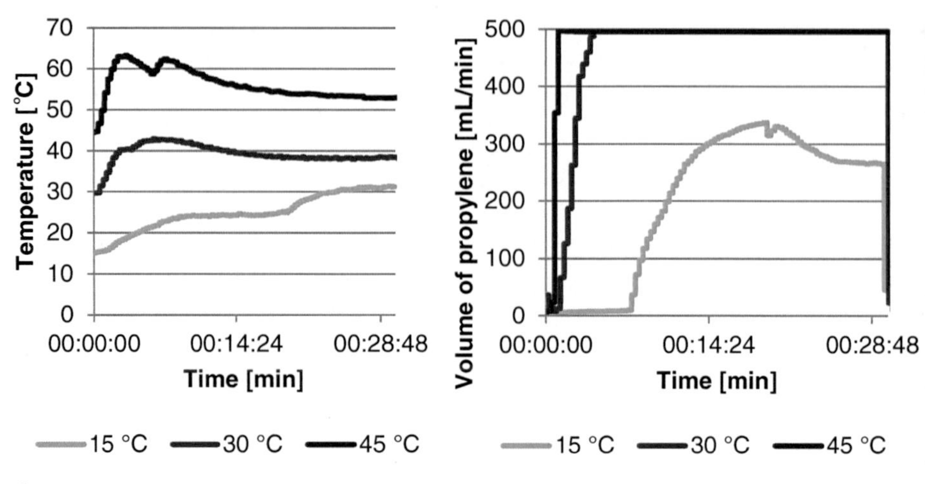

Figure 2.
Typical temperature profiles obtained from synthesis of pure poly(propylene) (maximum of 500 mL/min of propylene gas).

good dispersion of MWCNTs in polypropylene.[10,11] In particular, it was described how ultrasonic treatment or highly reactive reagents are useful to untie MWCNT from their bulky agglomerates. We find that indeed sonicating MWCNT in the presence of aluminum silicates in toluene suspension is a promising method to obtain clay covered MWCNT with little tendency to self-agglomerate. After mixing MWCNT with excess clay (clay/MWCNT ratios from 40:1 up to 100:1) and sonication in toluene suspension, the resulting fine powdery mixture was examined after drying in SEM. Good adhesion of MWCNT onto the silica was observed as well as entanglements between the both (Figure 4). Further improvement of the filler dispersion in toluene was observed after addition of MAO.

The thus prepared suspensions were transferred to a 10 L reactor and subsequently treated with propylene and metallocene catalyst. All of these *in situ* polymerizations yielded a fine powdery dark grey product. Reactor fouling was at the level of traces. The actual filler content of the composites was determined by TGA (Table 3). The dried composites were processed by extrusion and were pelletized yielding a completely black material. Specimens for SEM analysis were obtained by

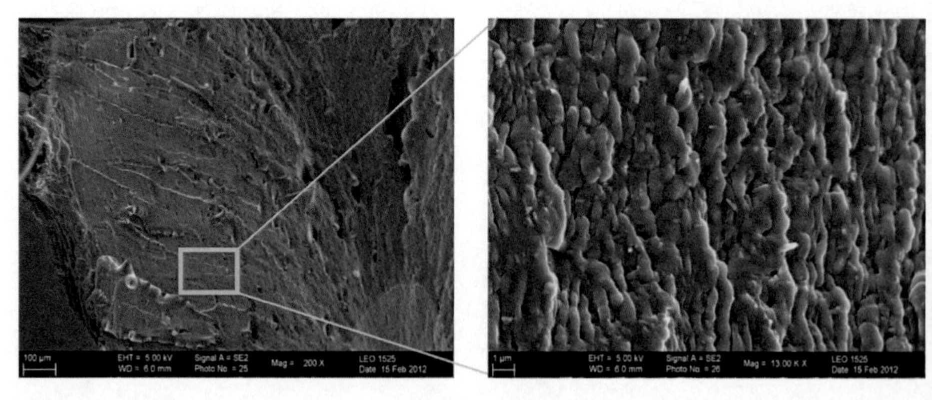

Figure 3.
SEM pictures of a stressed clay sample with 5.65% filler content.

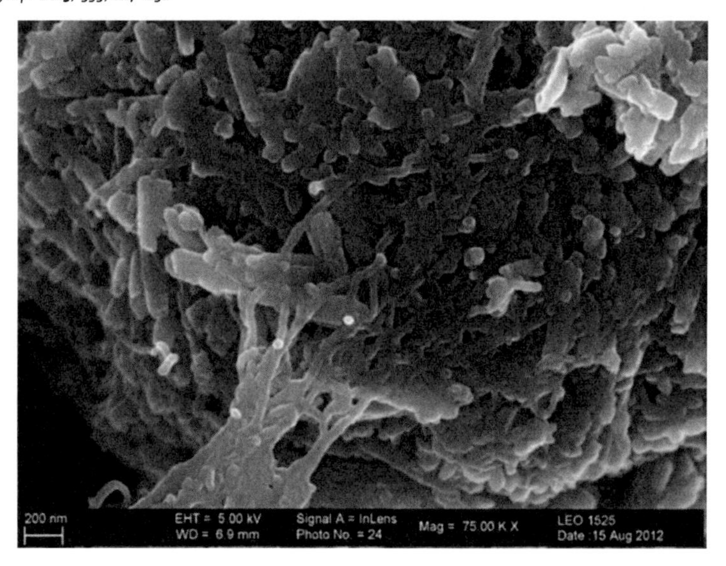

Figure 4.
Sonicated mixtures of aluminum silicates and MWCNT.

Table 3.
Results of the *in situ* polymerization of propylene and clay/MWCNT mixtures.

	yield [g]	clay [%]	clay calc. [%]	MWCNT calc. [%]	Tm [°C]	Tc [°C]	crystallinity [%]
clay/MWCNT	783	5.0	5.1	0.13	162	121	51
clay/MWCNT	620	6.1	6.1	0.12	157	118	47
clay/MWCNT	616	6.6	6.5	0.06	158	125	47
clay/MWCNT	612	6.1	6.5	0.16	161	122	48

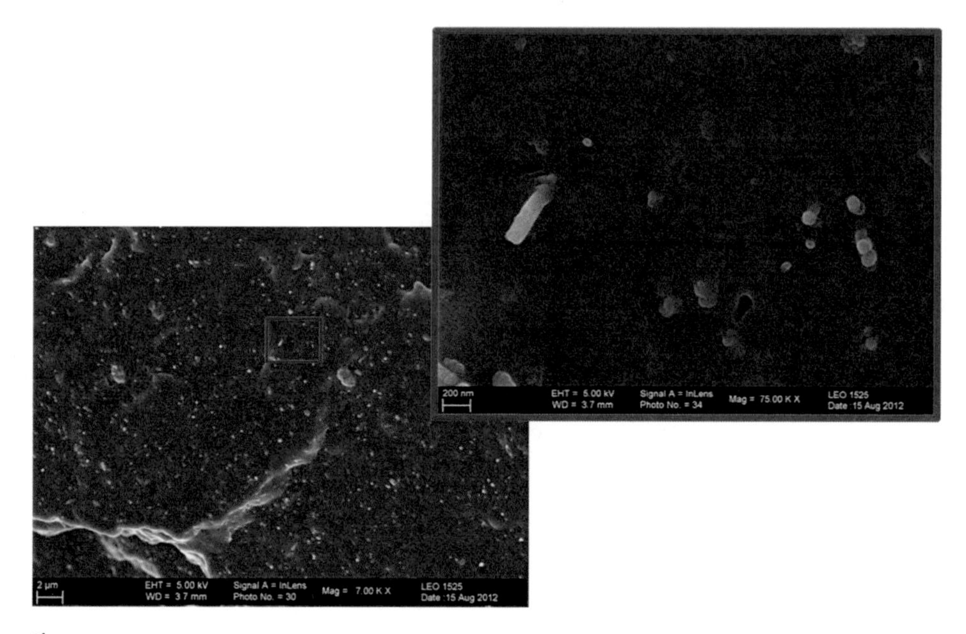

Figure 5.
SEM pictures of a clay/MWCNT cryo-fractured sample.

injection molding at temperatures not exceeding 220 °C. Isolated MWCNT and clay in the polymer matrix were observed by SEM of cryo fractures from specimen after injection molding (Figure 5).

Conclusion

A synthesis procedure was developed for the preparation of a multi-filler composite with polypropylene as matrix by an *in situ* polymerization with a clay/MWCNT mixtures as filler material. Up to 800 g of composite was synthesized in a multi-step one pot reaction in toluene. Reactor fouling was found to be at a low level typical for supported catalysts. The mixture of clay and MWCNT in toluene was found to be well-dispersed and had little tendency to agglomerate. MAO is absorbed onto the surface of the fillers within one day to give a support that is active in the formation of polypropylene after contacting with metallocene catalyst. The dispersion of the filler is excellent as judged from SEM analysis.

[1] M. B. Abu Bakar, Y. W. Leong, A. Ariffin, Z. A. Mohd. Ishak, *J. Appl. Polym. Sci.* **2007**, *104*, 434–441.

[2] Y. W. Leong, M. B. Abu Bakar, Z. A. Mohd. Ishak, A. Ariffin, B. Pukanszky, *J. Appl. F Polym. Sci.* **2004**, *91*, 3315–3326.

[3] W. Kaminsky, A. Funck, C. Klinke, *Top Catal.* **2008**, *48*, 84–90.

[4] K. Wiemann, W. Kaminsky, F. H. Gojny, K. Schulte, *Macromol. Chem. Phys.* **2005**, *206*, 1472–1478.

[5] W. Kaminsky, A. Funck, *Macromol, Symp.* **2007**, *260*, 1–8.

[6] N. Guo, S. A. DiBenedetto, P. Tewari, M. T. Lanagan, M. A. Ratner, T. J. Marks, *Chem, Mater.* **2010**, *22*, 1567–1578.

[7] J. Severn, R.L. Jones *Handbook of Transition Metal Polymerization Catalysts* R., Hoff, R.T., Mathers, Eds. J. Wiley,Sons,Inc. Hoboken, **2010**, Ch. *7*, 157–230.

[8] J. Sumfleth, M. Sriyai, L. Prado, M. H. G. Wichmann, K. Schulte, *Mater, Res. Soc. Symp. Proc.* **2008**, *1056*, MRS Fall Meeting 2007–Conference Proceeding, Boston. USA.

[9] M. H. G. Wichmann, K. Schulte, H. D. Wagner, *Compos, Sci. Technol.* **2008**, *68*, 329–331.

[10] A. A. Koval'chuk, A. N. Shchegolikhin, V. G. Shevchenko, P. M. Nedorezova, A. N. Klyamkina, A. M. Aladyshev, *Macromolecules.* **2008**, *41*, 3149–3156.

[11] W. Kaminsky, A. Funck, H. Hähnsen, *Dalton, Trans.* **2009**, 8803–8810

Macromol. Symp. 2013, 333, 233–241 DOI: 10.1002/masy.201300076

Specialised Tools for a Better Comprehension of Olefin Polymerisation Reactors

Timothy F.L. McKenna,[*1] Christophe Boisson,[1] Vincent Monteil,[1] Elena Ranieri,[1,2] Estevan Tioni*[1,2]

Summary: 2 laboratory-scale reactor systems suitable for gas phase, and for solution or slurry polymerisations are discussed. The underlying concept behind the design and use of these reactors is that they can be used to understand the impact of conditions specific to different time scales and/or length scales inherent to large reactors that are difficult to recreate at the laboratory scale. For instance the fixed bed gas phase reactor is used to study the influence of different relative gas/solid velocities on the evolution of the molecular weight distribution of the nascent polymer. It is shown that in certain conditions, changing the heat transfer characteristics does not change the observed yield, but will impact the polymer properties. In the case of the solution reactor, the concept is to design and use a reactor to study the activation of unsupported metallocenes. Here it is shown that different metallocenes have very different activation profiles to a point where a stopped flow reactor might not be the ideal tool for their study.

Keywords: kinetics (polym.); metallocene catalysts; polyolefins; stopped flow reactors

Introduction

Despite the fact that polyolefins – polyethylene and polypropylene – are the most widely produced polymeric materials in the world, some fundamental aspects of these catalytic polymerisation are still not fully understood. Issues include: how the morphology of the particles is set during the first instants; how the catalyst precursors are activated; and what kinds of activity profiles can we see during the initial phase of the reaction? In addition to these questions, it also appears that polymer properties (T_m, crystallinity, MWD) are not the same during the first few seconds of the reaction as they are at later stages.[1,2,3]

[1] C2P2 - LCPP Group, UMR 5265 CNRS, Université de Lyon, ESCPE Lyon, Bat 308F, 43 Bd du 11 novembre 1918, F-69616 Villeurbanne, France
E-mail: timothy.mckenna@univ-lyon1.fr
[2] Dutch Polymer Institute DPI, POBox 902, 5600 AX Eindhoven, The Netherlands

A better understanding of the transient phenomena occurring during the early seconds of reaction could lead to significant improvements in the production process. However, issues such as the moderately high pressures (several bars, or tens of bars of monomer are common in industry), the extreme sensitivity of the catalysts used to impurities in the reactor, the rapid changes in particle size and morphology, as well as the fact that the reactions can be carried out in different phases are among the factors that contribute to making it difficult to evaluate what occurs during olefin polymerisation on supported catalysts at short times.[4] Additionally, to use a well-coined phrase, polymers are “product-by-process”[5] in the sense that implementing the same chemistry in different processes leads to different products.

The concrete outcome of this is that if we are to cast any light on the issues mentioned above, we will need to develop reactor systems that allow one to understand how the process influences the chemistry and vice versa. As we will discuss below in more detail, it can be very difficult reproduce the hydrodynamics and residence time distribution of a full scale, heterogeneous reactor at the laboratory scale (not to mention prohibitively expensive when the processes are continuous). If one cannot perform a well-defined scale-down on an industrial process, the alternative is to develop reactors that are capable of mimicking parts of the process in clearly identifiable ways. This further implies that no single reactor system will be able to fulfill all of our needs in terms of studying the initial moments of the polymerisation. For this reason, our research group has constructed different specialised reactors over the course of the past several years for slurry, solution, and gas phase polymerisations. In the current paper we will illustrate that it is occasionally advantageous (but not always!) to use specialised reactors that allow us to isolate different aspects of the process in order to better understand the evolution of the chemistry and properties of the final product.

Specialised Reactors – Gas Phase Polymerisation

Heat transfer in a gas phase olefin polymerisation reactor is of primary concern, and in fact is the limiting factor in terms of the volumetric production rate with these systems. If we consider Figure 1, which shows a schema of a fluidised bed reactor (FBR), the work-horse of polyolefin production in the gas phase (and the only commercially viable way to make polyethylene in the gas phase), it can be seen that relative gas-particle velocities in different zones of the reactor will lead to different

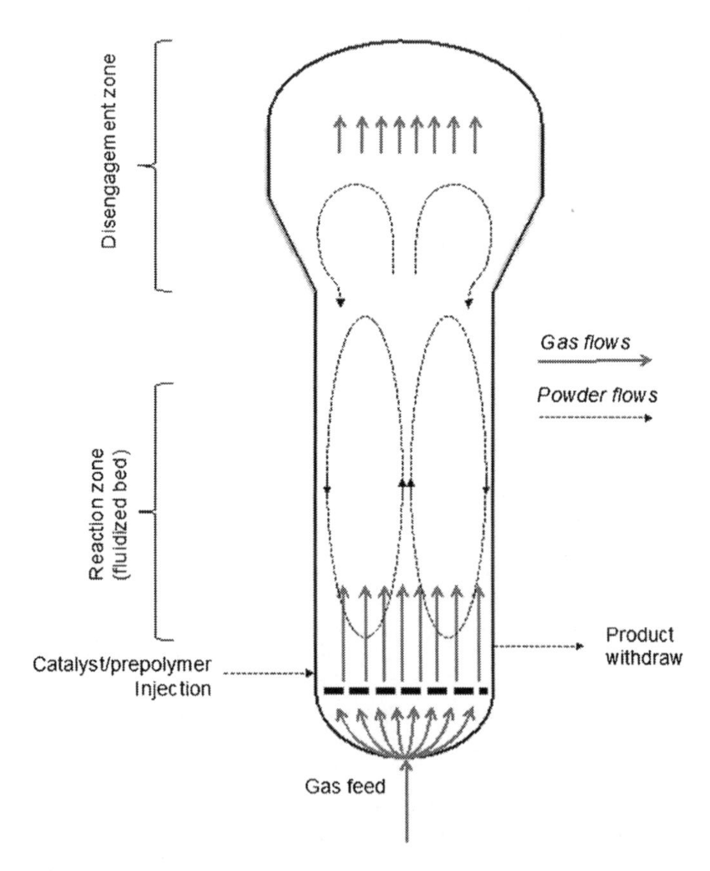

Figure 1.
Schema of a fluidised bed reactor (FBR) that shows different flow regimes inside the reactor.

conditions of heat transfer. In this type of reactor, the gas phase is assumed to move upwards in an approximation of plug flow, and the powder bed is thought to flow upward through the centre of the reactor, until it reaches the disengagement zone (where the enlarged reactor diameter leads to a drop in the superficial gas velocity), and then falls back down the outsides of the reactor. Typical recycle times for a fluidised bed reactor are on the order of 30-60 seconds, with average residence times in an FBR being on the order of several hours. This means that, as a first approximation, the powder phase has a residence time distribution similar to that of an ideal stirred tank reactor (of course the reality of this can be more complex[6,7,8]). Without going into too much detail this tells us that the relative gas-particle velocity will be different at different points in the bed, and that the particles can experience a range of hydrodynamic conditions over the time that they remain in the bed.

Of course, there are certain inherent challenges in building a miniature FBR. If one excludes the high cost associated with running experiments in a continuous laboratory-scale reactor, there remains the fact that it is difficult (not to say impossible) to reproduce the same hydrodynamics in a small reactor that one could obtain in a full-scale version.

An alternative approach is to develop reactors capable of operating at close to industrial conditions, with well-controlled conditions that can be varied in order to study a range of realistic reactor configurations. Stopped flow reactors have been demonstrated to provide such tools for slurry phase polymerisation, and earlier studies have shown that they can be operated in realistic conditions[2,9] in order to obtain information on the reaction rate, polymer particle morphology and molecular weight distribution. This reactor, which has been adapted for use in studying solution polymerisation systems, consists of two upstream reservoirs under pressure containing the catalyst slurry in one and the cocatalyst and monomer in solution in the other that are connected to a T-mixer. The fluids meet and the mixer, and flow through a tubular reactor and finally into a quench vessel that allows for recovery of the polymer particles. Given the nature of a slurry system, it is practical to cause the catalyst phase to flow continuously for periods of up to 4 seconds. However, this is significantly more challenging with a solid suspended in a gas, and in addition would require significant quantities of catalyst for experiments lasting upwards of a minute.

For this reason, we adapted a fixed-bed reactor configuration for use as a gas phase stopped flow reactor. The choice of a fixed bed reactor for this purpose might not be immediately evident, but since heat transfer is controlled (to a large extent) by the relative gas-particle velocity in an FBR, the fact that the particles are not moving in the fixed bed stopped flow reactor will not be a problem. In fact the advantage is that we can be sure that all of the particles in the bed experience similar hydrodynamic conditions, and we can therefore identifiy the reaction conditions with confidence – allowing us to separate chemical and physical effects more accurately that might be possible with other reactor configurations. In addition, the design of the fixed bed is such that reactions can be run at very short times (less than one second) with precision and in a repeatable manner – again something that is difficult even with standard stirred bed reactors.

The reactor in question, shown in Figure 2, consists of an inner sleeve (A) that is filled with a mixture of catalyst and salt. The cylindrical portion is in stainless steel, and the end is a porous metal frit with an average pore size of 7 microns. The reactor is filled, then placed in the outer sleeve (B) which is subsequently closed with an end cap (C). This end cap is of the same material as the reactor bottom. The assembly is screwed together as shown in the bottom portion of Figure 2. The reactor is connected to a coiled copper feed line, and the reactor and coiled portion of the feed line are placed in a water bath at the desired reaction temperature.

Figure 2.
Schema of a fixed bed reactor for use in studying the gas phase polymerisation of olefins on supported catalysts.

A metering valve is present on the outlet stream allowing the regulation of the gas flow rate inside the reactor. Thermocouples at the inlet and outlet of the reactor allow us to record the temperature rise of the gas phase as it flows through the bed. Pressure, temperature and flow rate of the feed are controlled. The reactor is equipped with three miniature solenoid valves (ASCO Joucomatic, France) controlled by a PLC equipped with software (Crouzet, Millennium II+, France): one for the feed, one for the quenching gas (CO_2) and one for degassing. The minimum time between subsequent actions of the solenoid is 0.1 s, and runs of up to 75 seconds are possible. The Reynolds number for the reaction conditions treated in this study varies from 0.1 to 20. From the Ergun equation the maximum calculated pressure drop with reaction conditions used in this study is of 30 mbar. This allows us to consider constant pressure along the bed. A typical polymerisation is conducted as follows: initially the reactor is filled with a mixture of seedbed and catalyst, then swept with argon and then the feed solenoid is opened to allow the feed mixture to flow through the catalytic bed. After a predetermined time the quenching and degassing solenoid valves are opened and the feed one is closed. This allows catalyst poisoning and complete removal of residual monomer even from the smaller pores of the catalyst thus stopping immediately the reaction. More details are available in a paper by Tioni et al.[10] The reaction conditions used throughout this work are (unless specified):

- Superficial gas velocity of 5-20 cm/s;
- 9 bar of total pressure with 3 bars of helium in the feed and 6 bars of ethylene;
- Use of fine NaCl as inert catalyst diluent;
- Catalyst mass between 30 and 80 mg according to reaction time and activity.
- A zirconene (*rac*-$EtInd_2ZrCl_2$) supported on Grace-Davison 948 silica prepared as described previously.[1]

Productivities are measured by weighing the reactor before and after the polymerisation step in a glove box (after a drying period to ensure that there is no residual water on the reactor assembly). The polymer is recovered from the fixed bed after 3 washing steps: one with demineralised water followed by one with a solution of 10% w/w of HCl in demineralised water and finally one with pure demineralised water again. The polymer recovered is the dried under vacuum at 80°C for at least 1 hour to eliminate the last traces of water. The molecular weight distributions (MWD) of polymer samples were characterized by SEC (Waters, Alliance GPCV 2000). The system was equipped with two detectors (a refractometer and a viscometer) and with three columns (PL gel Olexis 7*300 mm from Varian). Analyses were performed in trichlorobenzene (TCB; Biosolv, Valkenswaard, the Netherlands) at a flow rate of 1 mL/min. The molecular weight distributions were calculated by a calibration based on polyethylenes of different weight average molecular weights and only the RI signal was used in order to erase possible artefacts coming from experimental errors during the determination of the polymer mass.

This reactor can be used for many purposes, including a recent publication on the study of the fragmentation of silica-supported metallocene catalysts that showed a correlation between the advancement of the reaction in the particle and the crystallisation temperatures measured on the reactor powder. However, as mentioned above, the fundamental purpose of this reactor is to help elucidate the relationship between the process-chemistry-properties of olefin polymerisation. One of the ways that this can be done is to use the reactor and measurements of the productivity to calculate the actual temperature of the particles.

Tioni et al. showed that an estimate of the average solid temperature of the particles in the reactor for short times (assuming negligible temperature gradients in the particles) can be written:

$$T_{solid} = T_0 + \frac{Q_{gen} - Q_{transfer}}{m_{solid} C_{p,solid}} \quad (1)$$

Where $T_{\neg 0}$ is the initial solid temperature and $m_{\neg solid}$ is the mass of the bed (assumed to reach uniform temperature very quickly), $C_{p,solid}$ the heat capacity of the bed, $Q_{\neg gen}$ the total amount of heat generated at a given time (the mass of polymer produced times the enthalpy of polymerisation), and $Q_{transfer}$ is the total amount of energy removed by the flowing gas:

$$Q_{transfer} = \dot{q} \rho_g C_{p,g} \int_0^t (T_{g,out} - T_{g,in}) dt \quad (2)$$

where $\dot{q}$ is the volumetric flow rate of the gas phase, ρ_g its density, and $T_{g,out}$ and $T_{g,in}$ are the outlet and inlet gas temperatures. This is of course an upper bound on the average surface temperature.

The results of a series of typical experiments can be seen in Figure 3. Figure 3a shows the activity as a function of reaction time at two different gas velocities: 5.5 cm/s (giving poor heat transfer conditions) and 20 cm/s (leading to better heat transfer conditions). A look at this graph suggests that the reaction rates are not functions of the gas flow rates (both experiments gave very similar yields as well). Figure 3b also shows that the measured outlet temperature profiles are similar (an equipment problem meant that data for the measured outlet temperature at 20 cm/s was not available after the interval shown), again suggesting little difference in the kinetics of the 2 runs. However, when one looks at the surface temperatures calculated using equation (1), it can be seen that the particles experiencing the lowering superficial gas velocity are significantly hotter (almost 40 K) after a few seconds. The impact of the temperature difference can be seen in Figures Figure 3c and Figure 3d. These 2 graphs show the molecular weight distributions at 5.5 and 20 cm/s respectively for the different reaction times. For the very first value, measured on powder recovered after a fraction of a second, and where the particles are at similar temperatures, the MWD is narrow, and similar for both velocities. However as the reaction times get longer, a low molecular weight tail forms in the case of the 5.5 cm/s runs. The

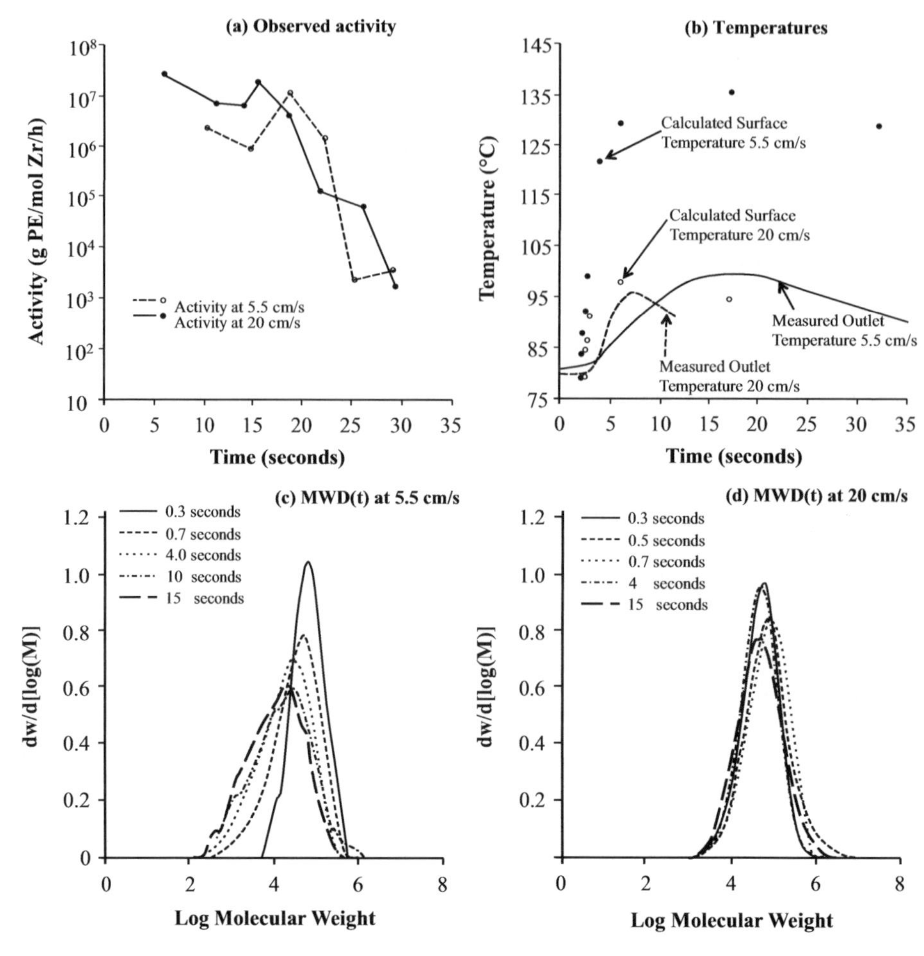

Figure 3.
The results of a series of homopolymerisation runs with an inlet temperature of 80 °C, 6 bars of ethylene, 3 bars of helium, for different reaction times and 2 superficial gas velocities.

difference in terms of product quality between the 2 runs is quite remarkable. It should be pointed out that it would be extremely challenging to isolate this type of effect with a standard lab scale reactor system.

Other uses for this type of reactor system include the exploration of the evolution of catalyst size and particle morphology,[11] explorations of the role of different process gases,[1] as well as detailed studies of the evolution of the molecular weight distribution, catalyst activity, comonomer incorporation for both ethylene and propylene polymerisations. As was stated earlier, these studies can also be done for well-defined hydrodynamic conditions (at precise time scales) that mimic the conditions found in different parts of a much larger commercial reactor.

Specialised Reactors – Polymerisation with soluble catalysts

In a different vein, polyethylene can be made via a solution polymerisation process, where the catalyst, monomer, polymer and a chemically inert solvent such as cyclohexane form one single phase in the reactor. While it is less necessary to separate "physical" from "chemical" effects since the active centres of the catalyst are not supported in particulate form (and therefore we should experience no resistances to heat and mass transfer), it is important to understand how the catalysts are activated, how the activity varies as a function of time, and what fraction of the metal atoms in the reactor actually participate in the reaction itself. Different methods are available to look at the number of active sites and the propagation rates,[12] but one of the more interesting is that proposed by Giulio Natta in 1959.[13] The method based on number of macromolecules is relied on relationship between the degree of polymerisation, P_n, and the concentration of active sites. P_n is defined as (moles of molecules of monomer which react at time t)/(number of polymer chain formed at time t) ratio which can be expressed by equation (2):

$$P_n = \frac{\int_0^t R_p dt}{M^* + \int_0^t \sum_{r=1}^{n} R_r dt} \tag{3}$$

where R_p is rate of propagation of growing chain and $\sum_{r=1}^{n} R_r$ is the termination rate of growing chains. Since P_n increases with time during the very early stages of the polymerisation before the first transfer reactions occur, it is possible to determine k_p and M^* (this implicitly supposes all the sites are active at the same time). If chain transfer reactions are negligible, then the number of macromolecules present during the living regime of a polymerisation is equal to the number of active sites in the reactor. Furthermore, if only a low fraction of chains has undergone a chain transfer reaction which corresponds to the initial controlled regime then equation (3) can be rewritten as:

$$\frac{1}{P_n} = \frac{1}{k_p[Mon]t} + \frac{\sum_{i=1}^{n} k_{tr}[X^i]t}{k_p[Mon]t} \tag{4}$$

If it can be assumed that R_p and R_t are time independent and that chain-terminating reactions are the only kind of chain transfer process, then a graph of (1/ P_n) as function of reciprocal of time (1/t) will be a straight line with a slope of $1/k_p$[Mon] and an intercept that gives us the value of frequency of chain transfer. Furthermore, if k_p and [Mon] are independent, then an integral of the rate of polymerisation as a function of time gives us the yield Y, which can be written:

$$Y = k_p[Mon] \int_0^t M^* dt \tag{5}$$

At low conversions and at short times, this last expression allows us to find the value of M^* by plotting the polymer yield as a function of time, and introducing the value of k_p found from equation (4).

In order to extract meaningful data using this approach several requirements must be satisfied. First and foremost, in order to obtain a reliable value of M^* it is necessary to work in a "controlled" regime where the chain transfer reactions are very limited. In this case, all active sites have a chain growing on them, and there are nearly no dead chains in the system. If we can adapt our experimental conditions in such a way that this happens, we can "capture" the chains before the transfer reactions become important, it should be a straight forward exercise to "count" the sites. It should be possible to do this if the reactor can be run for times going from fraction of seconds to at most one or two seconds. Of course this also implies that the activation of the chains is instantaneous, no significant transfer reactions occur. This approach clearly relies on a number of simplifying hypotheses and assumptions; however in order to discover whether or not they are applicable it is necessary to have a specially designed reactor system. In our case, we chose to work with a stopped flow (or quenched flow) reactor system.

Ranieri et al.[14] modified a stopped flow system originally designed by Di Martino[9] by separating the upstream part of the device into two distinct zones, with one gas storage tank per upstream reservoir (one for ethylene and one for argon; c.f. Figure 4). Different pressure regulators for gas distribution were introduced in order to minimize any delay in upstream gas flow, and a better overflow valve was installed in order to minimize the pressure

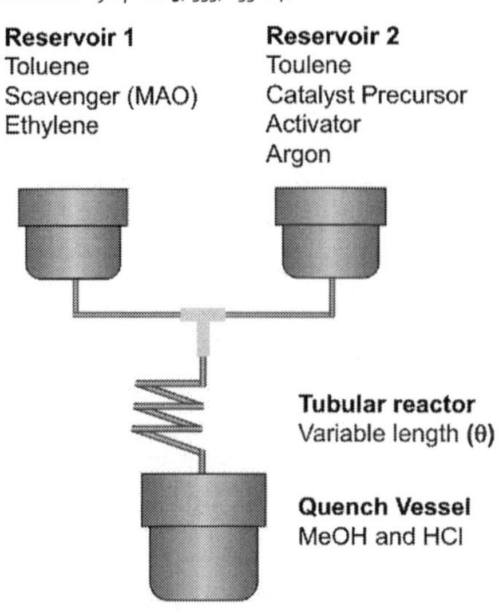

Figure 4.
Schematic diagram of a stopped flow reactor used for solution polymerisation. The two upstream reservoirs are connected to separate ballast tanks (as opposed to being joined on the same on in Di Martino's original design[9]) in order to speed up response times.

fluctuations. These modifications produced a reactor system (c.f. Figure 4) capable of running reproducible solution experiments with reactor residence times as short as 0.09 seconds. In a typical experiment Vessel 1 was filled with a solution of an alkyl-aluminium (MAO or iBu3Al) in approximately 250 mL of toluene, and pressurized with ethylene at the desired temperature and pressure. Vessel 2 was filled with a toluene solution (approximately 250 mL) containing the co-catalyst and the metallocene precursor. The mixture was stirred for 10 minutes at the desired temperature under argon (same pressure of ethylene as in vessel 1). The two solutions were then forced to flow simultaneously through tubes of equal length until they met at the T-mixer in Figure 4 where the polymerization started. The polymerization occurred in Teflon tubes of various lengths and the reaction was finally stopped in the quench vessel. The polymerizing mixture was quenched in 300 ml of MeOH/HCl (10% in weight) solution. The mixture was degassed and the polymer (on the order of 2 to 100 mg) was recovered by filtration, and then analyzed.

Ranieri et al.[14] reached several important conclusions in their full experimental study, notably that under these moderate conditions, a catalyst made from the precursor $Ph_2C(Cp)(9\text{-Fluorenyl})ZrCl_2$ showed an induction period that was well over one second long, regardless of the type of activator considered, and the pressure used (1.5 to 6 bars of ethylene). On the other extreme, with a precursor of *rac*-$Me_2Si(2\text{-Me-4-Ph-Ind})_2ZrCl_2$, the reaction was so rapid, that it was almost impossible to obtain controlled living conditions for reactions much longer than 100 ms, which made it very difficult to obtain amounts of polymer sufficient for SEC analysis. Nevertheless, if the reactions were carried out at low temperatures (25 °C) and pressures on the order of one atmosphere, it was possible to estimate k_p and M^* for catalysts based on $Cp^*{}_2ZrCl_2$. In this case, the authors once again observed that only 10-20% of the Zr atoms introduced into the reactor are active in ethylene polymerizations. Finally, a series of experiments using the precatalyst bis(phenoxy-immine) titanium dichloride revealed that this system does show controlled, or living behaviour at one bar of ethylene and a range of temperature from 25 – 75 °C. The authors claimed that, once again, approximately 25% of the Zr atoms introduced in the reactor were active in polymerisation.

They also showed that rapid stopped (quenched) flow methods are only appropriate for certain catalysts, and for certain reaction conditions. They cannot always be used to create the living conditions necessary to study kinetics using the chain counting method, which implies for metallocene catalysts that even at temperatures as low as 40 °C the time for full chain growth is less than 100 ms. This means that the stopped flow method is not adapted to studying highly active catalysts. While it might be possible to develop reactors operating on time scales capable of applying the chain counting approach to the study of k_p and the fraction of active metal atoms, it is unlikely that doing so would be economically justifiable, and in these cases it would be reasonable to say that the catalysts activate instantaneously.

On the other hand, some catalysts show induction periods that can be quite long (several seconds), making them difficult to study with standard stopped flow equipment. Stopped flow reactors similar to the ones developed here, but with much larger reservoirs (1 or 2 litres instead of 500 ml each) might be one solution to overcoming this problem.[15]

Conclusion

In this paper we show that the thoughtful design of different reactor systems is essential if we are to explore the chemistry, the process and their interactions during catalysed olefin polymerisations. With the gas phase fixed bed reactor we have shown that rather than attempting to reproduce fluidized bed reactor conditions at the laboratory scale, identifying the fact that heat transfer is a controlling issue and then designing a reactor that does not resemble a commercial scale allows us to learn a great deal about the process. In this paper we have shown that the ability to change flow rate, combined with a basic engineering model for heat transfer allows us see the impact of changing heat transfer conditions on catalyst behaviour. In this manner it is more straightforward to explain why the molecular weight distributions evolve differently, even if the kinetics appear to be similar for different reactor conditions. Perhaps more importantly, such reactor systems allow us to better isolate what can be attributed to the intrinsic chemistry of the process, and what is more a result of the process.

On the other hand, in instances such as the solution phase stopped flow reactor described here, even careful design of the reactor system is not enough to make the tool totally polyvalent. It can be used to differentiate between modes of activation of different metallocenes for instance; helping to identify those that are rapid or slow, or perhaps in certain cases (not discussed here) distinguishes between true single site catalysts, and those that, at least for a short period of type, have two sorts of active sites.

Continuous, iterative use and improvement of such reactors will lead to a new generation of reactor tools.

Acknowledgement: The authors are grateful to the Dutch Polymer Institute for financial and technical support in the context of DPI Projects 635 and 636.

[1] E. Tioni, J. P. Broyer, V. Monteil, T. F. L. McKenna, *Ind. Eng. Chem. Res.* **2012**, *51*, 14673.
[2] A. Di Martino, G. Weickert, T. F. L. McKenna, *Macromol. React. Eng.* **2007**, *1*, 165.
[3] J. B. P. Soares, T. F. L. McKenna, *Polyolefin Reaction Engineering*, Wiley - VCH, Mannheim, Germany **2012**.
[4] T. F. McKenna, J. B. P. Soares, *Chem. Eng. Sci.* **2001**, *56*, 3931.
[5] J. M. Asua, *Polymer Reaction Engineering*, Blackwell Publishing Ltd, Oxford UK **2007**.
[6] J. J. Zacca, J. A. Debling, W. H. Ray, *Chem. Eng. Sci.*, **1996**, *51*, 4859.
[7] J. Y. Kim, K. Y. Choi, *Chem Eng. Sci.*, **2001**, *56*, 4069.
[8] G. Dompazis, V. Kanellopoulos, V. Touloupides, C. Kiparissides, *Chem. Eng. Sci.*, **2008**, *63*, 4735.
[9] A. DiMartino, J. P. Broyer, R. Spitz, G. Weickert, T. F. L. McKenna, *Macromol. Rapid. Comm.*, **2005**, *26*, 215.
[10] E. Tioni, R. Spitz, J. P. Broyer, V. Monteil, T. F. L. McKenna, *AIChE J* **2012**, *58*, 256.
[11] E. Tioni, V. Monteil, T. F. L. McKenna, *Macromol.* **2013**, doi/10.1021/ma302150v
[12] T. F. L. McKenna, E. Tioni, M. M. Ranieri, A. Alizadeh, V. Monteil, C. Boisson, *Can. J. Chem. Eng.* **2013**, *91*, 669.
[13] G. Natta, *J. Polym. Sci.* **1959**, *34*, 21.
[14] M. M. Ranieri, J. P. Broyer, F. Cutillo, T. F. L. McKenna, C. Boisson, *Dalton Trans.* **2013**, DOI: 10.1039/c3dt33004d
[15] T. Taniike, S. Sano, M. Ikeya, V. Q. Thang, M. Terano, *Macromol. Reac. Eng.* **2012**, *6*, 275.

 DOI: 10.1002/masy.201300092

Condensed Mode Cooling in Ethylene Polymerisation: Droplet Evaporation

Arash Alizadeh,[1,2,3] *Timothy F.L. McKenna**[1]

Summary: A simplified model is used to explore the mechanisms for droplet vaporisation in fluidised bed reactors used for the production of polyethylene in the gas phase. Both homogeneous and heterogeneous vaporisation models are considered. It is shown that for a reasonable range of droplet sizes (less than 1 mm in diameter), droplet vaporisation will be essentially complete at bed heights on the order of 1 m.

Keywords: condensed mode cooling; fluidised bed reactor; polyethylene

Introduction

The rate of production in gas phase olefin polymerisation reactors, and in particular fluidised bed reactors (FBR) for ethylene polymerisation, is limited by the rate at which the heat of reaction can be removed. Condensed mode operation, where a partially liquefied recycle stream containing a compound such as i-pentane or n-hexane (inert condensing agents – ICA) is injected into the reactor with the feed. The liquid, in the form of droplets is thought to evaporate at hot spots in the bed, thereby using the associated latent heat of evaporation in the reactor to help cool the bed.

Therefore we would like to understand where/how this takes place inside the reactor. Obviously parameters like droplet size, size distribution, heat of vaporisation and properties of solid particle phase (like solid flux inside bed, size distribution, and heat capacity) as well as eventual contact between these two phases will control the overall vaporisation process of the liquid droplet in the presence of fluidising solid particles. Jiang et al. studied the heat exchanger unit of the condensed mode process in a series of papers.[1–3] Their analysis provides some interesting recommendations from process point of view. For example, it has been found that the lower operating pressure of heat exchanger inhibits the rate of heat transfer, or for optimizing the heat removal rate the makeup ethylene should be added after the heat exchanger while makeup hexene as condensable agent should be before the heat exchanger. While very useful, this global analysis does not explain what happens inside the reactor at the level of the droplets.

However, despite extensive experimental and modeling studies in FCC and the combustion literature, to the best of our knowledge, there is no single modeling approach which captures effect of all the parameters influencing the droplet vaporisation in the fluidised bed with a phenomenological description. The objective of the current paper is to attempt to quantitatively understand what influences the rate of evaporation of droplets of ICA injected into the bottom of a fluidised bed reactor for the production of polyethylene.[4]

[1] C2P2 - LCPP Group, UMR 5265 CNRS, Université de Lyon, ESCPE Lyon, Bat 308F, 43 Bd du 11 novembre 1918, F-69616, Villeurbanne, France
E-mail: timothy.mckenna@univ-lyon1.fr
[2] Dutch Polymer Institute DPI, POBox 902, 5600, AX, Eindhoven, the Netherlands
[3] Department of Chemical Engineering, Queen's University, Kingston, ON, Canada, K7L 3N6, Canada

Results

Droplet Vaporisation: How Long Do Droplets Exist?

Different scenarios for the liquid droplets introduced to a "hot" environment of fluidising particles can be found in the literature. If the droplet is small enough there is a chance that it will vaporise “homogenously” before making any contact with the solid particles fluidising in the bed. Otherwise, liquid-solid contact will be inevitable and we find “heterogeneous” evaporation.

We therefore need first to analyse the relative time scales for droplet heat up and vaporisation compared in case of homogenous vaporisation of the droplet. For heterogeneous vaporisation we consider heat transfer during (1) “sticky” collision between droplet and particles and (2) elastic collisions between droplet and particles. Elastic collisions will be important when the temperature of the particle approaches the *Leidenfrost* temperature. This value is approximately 150 °C above the saturated vapour temperature of the droplets.[5] Thus it is not expected that this will occur during ethylene polymerisation, the vaporisation time obtained in this case will be representative of upper limit of the vaporisation time of droplet in presence of solid particles, while vaporisation time assuming fast heat transfer between droplets and particles will be representative of lower limit of it.

Homogeneous Vaporisation

In the case of homogeneous vaporisation, the lifetime of the cold droplet introduced to a hot environment is split to two overlapping periods. First the droplet heats up rapidly, then the droplet vaporises at constant temperature i.e. the wet-bulb temperature in our case. It is assumed that droplet composition is pure isopentane and gas phase is of composition given in Table 1 at 80 °C and 20 bars, and the properties of gas phase at 80 °C and 20 bars and liquid phase at 80 °C are given in Table 2.

Table 1.
Representative gas phase composition for condensed mode.

Component	Mole fraction
Ethylene	0.5
Hydrogen	0.1
Nitrogen	0.25
Iso-pentane	0.15

It is essential to have reasonable estimation of relative velocity of liquid droplet and surrounding gas phase (the gas-liquid slip velocity) during the course of heating up and evaporation in our calculations. The gas-liquid slip velocity for the injection devices are reported ranging from 7-30 m/s in the literature.[15] By considering the fact that liquid droplets are introduced to a slow moving environment of fluidising particles where superficial gas velocity[4] is in the order of 1 m/s, the gas-liquid slip velocity of $u_{slip} = 10$ m/s has been used through our analysis in this section as a compromise. We could not find an indication about gas-liquid slip velocity in case of introduction of liquid droplets entrained in the gas phase in the literature nor patents, so will test the sensitivity of the calculations to this value.

The droplet size distribution depends on the type of injection devices with average size ranges being reported to be between 50-4000 μm[8]. In the case of the introduction of liquid as entrained droplets the holes of the distributor are of a maximum diameter of one half inch (12500 μm), so maximum droplet size of diameter 5000 μm seems a reasonable approximation. As a result our analysis will cover droplet size distribution of 50-5000 μm.

In order to calculate the time scale for heat up and vaporisation of droplet in the second step, it is essential to estimate the temperature of the droplet during the vaporisation. In our analysis, this will be approximately by the wet-bulb temperature. At this temperature the rate of heat transfer from the gas phase to the liquid will be equal to the rate of heat removal from droplet by evaporation:

$$h(T_g - T_w) = k_m(C_{eq} - C_b)\Delta H_v \quad (1)$$

where T_g and T_w are the gas phase and wet-bulb temperatures; C_{eq} and C_b are concentration of isopentane in gas phase in equilibrium with liquid isopentane and concentration of isopentane in bulk gas phase. C_{eq} can be obtained from isopentane vapour pressure. k_m, the mass transfer

Table 2.
Properties of gas phase of composition given in Table 1 at 80 °C and 20 bars and liquid isopentane and polymer particles at 80 °C.

Property	Symbol	Value	Units
Density - liquid[7]	ρ_l	553.4	kg/m^3
Density - gas[7]	ρ	21.8	kg/m^3
Density - particle	ρ_p	900 (estimate)	kg/m^3
Heat Capacity - liquid[7]	C_{pL}	2593.5	J/kg.K
Heat Capacity - gas[7]	C_p	1734.7	J/kg.K
Heat Capacity - particle[7]	C_{pp}	2000	J/kg.K
Kinematic viscosity – gas[6,7]	ν	6.2×10^{-7}	m^2/s
Thermal conductivity - gas[7]	k_f	3.1×10^{-2}	W/m.K
Latent heat of vaporisation[7]	ΔH_v	21077	kJ/kmol
Diffusivity isopentane[6*]	$D_{i\text{-}C5}$	5×10^{-7}	m^2/s
Vapor Pressure isopentane[7]	P_{vp}	4.6	bar

* Diffusivity of isopentane in gas phase of composition mentioned in Table 1.

coefficient, and h, the heat transfer coefficient, are obtained from the Ranz- Marshall correlations:

$$Sh = \frac{k_m d}{D} = 2 + 0.6Re^{1/2}Sc^{1/3} \qquad (2)$$

$$Nu = \frac{hd}{k_f} = 2 + 0.6Re^{1/2}Pr^{1/3} \qquad (3)$$

With a gas phase temperature of 80 °C, the wet-bulb temperature is estimated to be 61 °C.

It can be shown that the governing equation during the heat up period of an isolated droplet is given by

$$\tau \frac{dT}{dt} = T_g - T \qquad (4)$$

where

$$\tau = (\rho_1 C_{pl} d)/6h \qquad (5)$$

In this approach, a uniform temperature inside the droplet is assumed. The solution to equation (4) is:

$$t = -\tau \ln\left[\frac{T_g - T}{T_g - T_o}\right] \qquad (6)$$

Where T_g and T_0 are the gas and initial droplet temperature. For the droplet heat up step, the time needed for a droplet initially at 50 °C to reach a wet-bulb temperature of 61 °C is shown in Table 3 for different conditions.

It is found empirically that the heat transfer coefficient from gas phase to a droplet decreases when there is a substantial vaporisation from droplet.[15] As a result, the heat transfer coefficient for a vaporising droplet must be calculated differently than for a particle:[8]

$$Nu^* = \frac{h^* d}{k_f} = \frac{Nu}{(1+B)^n} \qquad (7)$$

where Nu is the Nusselt number obtained from Ranz-Marshall correlation, h^* and Nu^* are the effective heat transfer coefficient and Nusselt number for droplet in the presence of vaporisation. n is a constant equal to 0.7 and B is given by

$$B = \frac{c_p(T_g - T)}{\Delta H_v}\left[1 + \frac{Q_R}{Q_c}\right] \qquad (8)$$

c_p is heat capacity of gas phase, $\triangle H_v$ is the heat of vaporisation of liquid. Q_R/Q_C is the ratio of heat transferred to the droplet by radiation to convection, which can be assumed to be equal to zero in operational conditions of condensed mode polyolefin reactors.

Assuming that all the heat transferred to the droplet is consumed for evaporation of liquid from the droplet, it is possible to write the governing equation for evaporation as

$$\Delta H_v \rho_l \frac{dV}{dt} = -h^* A(T_g - T) \qquad (9)$$

where V and A are droplet volume and surface area. T is the evaporation temperature of the droplet which (wet-bulb

Table 3.
Estimation of time scales for isopentane droplet heat up and homogeneous vaporisation ($u_{slip} = 10$ m/s).

Droplet Size (μm)	50	100	300	1000	3000	5000
Heat up Time* (sec)	5.0×10^{-4}	1.5×10^{-3}	7.8×10^{-3}	4.9×10^{-2}	0.26	0.55
Vaporisation Time (sec)	1.3×10^{-2}	3.9×10^{-2}	2.1×10^{-1}	1.4	7.1	15.4

* Heat up time from 50 °C to the wet-bulb temperature of 61 °C.

temperature). Substituting for V and A allows us to rewrite equation (9) as:

$$\frac{d(d)}{dt} = \frac{-2h^{*}(T_g - T)}{\Delta H_v \rho_l} \qquad (10)$$

The solution to equation (10) results in following correlation for vaporisation

$$\tau_{vap} = -\frac{\Delta H_v \rho_l}{A'(T_g - T)} \left\{ \frac{2}{B'^4} \left[\frac{1}{3}(1 - v_0^3) - \frac{3}{2}(1 - v_0^2) + 3(1 - \upsilon_0) - \ln\left(\frac{1}{\upsilon_0}\right) \right] \right\} \qquad (11)$$

where A' and B' are clustered functions defined as

$$A' = \frac{k_f}{\left[1 + \frac{C_p(T_g - T)}{\Delta H_v}\right]^{0.7}} \qquad (12)$$

$$B' = 0.6 Pr^{1/3} \left(\frac{u}{2v}\right)^{0.5} \qquad (13)$$

and $v_0 = 1 + B' r_0^{0.5}$ in which r_o is the initial droplet radius.

The time needed for droplets of size between 50-5000 μm to vaporise homogeneously at slip gas-liquid velocity of $u_{slip} = 10$ m/s are given in Table 3. By comparing the time scale for droplet heat up and vaporisation, it can be concluded that the heat up time is negligible compared to vaporisation step. As a result, in our discussion about heterogeneous droplet vaporisation in the presence of solid particles we will focus on the vaporisation step assuming that droplet reaches steady state temperature of vaporisation instantaneously. It can also be seen that the time scale for vaporisation is relatively rapid as well, except for the largest droplets.

Heterogeneous Droplet Vaporisation

In order to estimate the time scale for the vaporisation of droplets in the presence of particles, Buchanan[15] considered two limit cases of heat transfer between droplets and particles. In the first limit, all of the heat possible from particles is transferred to droplet instantaneously. The formulation for this limit results in following correlation for estimation of droplet vaporisation

$$\frac{d(d)}{dt} = \frac{-\upsilon \rho_p (1 - \varepsilon) C_{pp}(T_p - T)}{2 \rho_l \Delta H_v} \qquad (14)$$

where ε is the bed porosity and we use the value of bed porosity at minimum fluidisation of 0.5 as an estimation.

It is assumed that the particle temperature (T_p) is the same as the temperature of the gas phase (T_g) and that the particle is cooled down immediately to the droplet vaporisation temperature when particles and droplets collide. The first assumption is not always valid, especially for the small active polymer particles. However, the correction for this will only result even shorter time scales for vaporisation of droplets. Consequently it is decided to keep this assumption despite its obvious imperfections. Similar to the case of homogeneous vaporisation, the vaporisation temperature of the droplet is estimated as wet-bulb temperature. In this event we are able to calculate the lower limit of vaporisation time as given in Table 4 for 2 different slip velocities.

In the second limiting case (upper limiting time, or lower limit of heat transfer), the collisions between droplet and particles are assumed to be elastic. Considering the observed trend of a decrease in heat transfer coefficient to

Table 4.
Heterogeneous vaporisation time – estimates of time scales for isopentane droplet heterogeneous vaporisation.

Droplet Size (μm)	50	100	300	1000	3000	5000
$u_{slip} = 10$ m/s						
Lower Limit (sec)	9.5×10^{-5}	1.9×10^{-4}	5.7×10^{-4}	1.9×10^{-3}	5.7×10^{-3}	9.5×10^{-3}
Upper Limit (sec)	2.4×10^{-3}	6.8×10^{-3}	3.6×10^{-2}	0.22	1.1	2.4
$u_{slip} = 1$ m/s						
Lower Limit (sec)	9.5×10^{-4}	1.9×10^{-3}	5.7×10^{-3}	1.9×10^{-2}	5.7×10^{-2}	9.5×10^{-2}
Upper Limit (sec)	9.7×10^{-3}	2.8×10^{-2}	1.5×10^{-1}	0.95	5.0	10.8

immersed objects by dilution of fluidised bed,[9–12] Buchanan proposed a correction for heat transfer coefficient for homogeneously vaporising droplet seen in equation (9). In this correction the gas phase density ρ in Re number is replaced by gas-solid density, $\rho_p(1\text{-}\varepsilon)$. Taking this correction into account will result in similar equation for estimation of droplet vaporisation as equation (11), except that the cluster parameter of B' in equation (13) is replaced by B":

$$B'' = B'\left(\frac{\rho_p(1-\varepsilon)}{p}\right)^{1/2} \tag{15}$$

This vaporisation time is intended to be representative of higher limit of droplet vaporisation time in the presence of solid particles. These results are also shown in Table 4. Both the upper and lower characteristic times are calculated at different slip velocities because of the uncertainties mentioned above.

Of course these time scales should be treated as a qualitative estimation due to the large number of assumptions used in their calculation. Also, this approach also does not take into account the local hydrodynamics of the bed, the heat capacity of solid phase, possibility of distribution of liquid between particles and droplet surface deformation. Nevertheless, it is not unreasonable to assume that the real vaporisation time to be between two limits of the heat transfer in the heterogeneous vaporisation. If this is the case, then it can be concluded that an isopentane droplet of 5000 μm will vaporise in a time of between 1 and 5 seconds. That limit is for the largest droplets in the range studied. It is much more likely that the injected droplets are really an order of magnitude smaller. This tells us that vaporisation is indeed almost instantaneous in terms the time the ICA will spend in the bed.

Conclusion

Based on these calculations expect that the major part of the liquid injected through the bottom of an FBR is vaporised at a height of between 1 and 2 m. As a result the heat of evaporation enhances the heat removal from the particles fluidising only in this fraction of bed. Given the impressive enhancement of productivity reported in the patent literature, it is not unreasonable to speculate that the presence of "inert" ICA might have other effects.

It is well-known that the nature of the gas phase in an FBR can have a significant impact on heat transfer if the standard inert component of nitrogen is replaced by a gas such as helium[13] or propane[14]. Since the evaporation process is quite rapid the gas phase will be quite rich in the heavier ICA. This will enhance the thermal conductivity and heat capacity of the gas phase.

It should also be noted that the alkanes used as ICA are more soluble in polyethylene (PE) than is ethylene, and since the feed will contain 10-30 wt% of such compounds, the polymer particles will contain considerable amount of them. The dissolution of heavier alkanes can have different effects, including a cosolubility effect that can increase the quantity of ethylene in the particles, as well as a plasticizing effect that might increase the diffusivity of the monomer through the polymer layer surrounding the active sites (these influences will be discussed in forthcoming papers from our group).

It can also be speculated that if the temperature of a polymer particle begins to increase, for instance due to the local defluidisation inside the reactor, the partial desorption of the sorbed ICA will also help to improve heat transfer from the particles. This would ultimately result in lower fluctuation of the temperature of growing polymer particles inside the reactor while lowering the risk of polymer fusion and local formation of polymer agglomerates. We will explore the significance of these different aspects in future papers.

Acknowledgement: The authors are grateful to the Dutch Polymer Institute for financial and technical support.

[1] Y. Jiang, K. B. McAuley, J. C. C. Hsu, *AIChE*, **1997**, 43, 13.
[2] Y. Jiang, K. B. McAuley, J. C. C. Hsu, *AIChE*, **1997**, 43, 2073.
[3] Y. Jiang, K. B. McAuley, J. C. C. Hsu, *Ind. Eng. Chem. Res.*, **1997**, 36, 1176.
[4] S. J. Rhee, L. L. Simpson, Fluidized Bed Polymerization Reactors, US Patent No. 4,933,149 (Jun. 12, **1990**).
[5] J. M. Le Corre, S. C. Yao, C. H. Amon, *Nuclear Engineering and Design*, **2010**, 240, 235.
[6] R. B. Bird, W. E. Stewart, E. N. Lightfoot, *Transport Phenomena* (Ed., John Wiley& Sons, Inc;, **2007**.
[7] R. C. Reid, J. M. Prausnitz, B. E. Poling, *The Properties of Gases and Liquids*, 4th Ed. McGraw-Hill Inc, **1987**.
[8] M. Renksizbulut, M. C. Yuen, *J. Heat Transfer-Transactions of the ASME*, **1983**, 105, 384.
[9] D. Kunii, O. Levenspiel, *Fluidization Engineering*, R. E. Krieger Publishing Co, **1977**.
[10] B. Leckner, *Heat Transfer in Circulating Fluid Bed Boilers. Circulating Fluidized Bed Technology III;*, P., Basu, M., Horio, M. Hasatani, Eds., Pergamon Press, **1991**.
[11] H. A. Vreedenberg, *Chem. Eng. Sci.*, **1958**, 9, 52.
[12] F. A. Zenz, D. F. Othmer, *Fluidization and Fluid-Particle Systems*, Reinhold, 235.
[13] Tioni, R. Spitz, J. P. Broyer, V. Monteil, T. F. L. McKenna, *AIChE J*, **2012**, 58, 256.
[14] M. Covezzi, *Macromol. Symp.*, **1995**, 89, 577.
[15] J. S. Buchanan, *Ind. Eng. Chem. Res.*, **1994**, 33, 3104.

 DOI: 10.1002/masy.201300094

Organoaluminum Initiators: Influence of Al Coordination on Polymerization and Product Properties

Ursula Tracht, Ricarda Leiberich, Hanns-Ingolf Paul*

Summary: Alkyl aluminum halides are well established initiators for cationic polymerizations. They enable homogeneous polymerization processes in non-chlorinated hydrocarbon processes. Activity and polymer properties can be significantly affected by the choice of the compound used for activating the organoaluminum halide. Three examples of organoaluminum chloride based initiator systems for cationic polymerizations are presented here illustrating different stages of a workflow established for the development and optimization of such initiators and corresponding polymerization processes. Initiator formation via addition of protic compounds to organoaluminum chloride precursors is studied by multiple spectroscopic techniques, in particular IR-spectroscopy. Differently prepared initiators are tested in lab scale batch polymerizations and continuous mini-plant polymerizations. The influence of aluminum coordination on polymer properties like molecular weights is analyzed.

Keywords: alkylaluminum halides; cationic polymerization; infrared spectroscopy; molecular weight distribution; monitoring

Introduction

Alkyl aluminum halides are well established initiators for cationic polymerizations. They enable homogeneous polymerization processes in non chlorinated hydrocarbon solvents. Activity and polymer properties can be significantly affected by the choice of the compound used for activating the organoaluminum halide. Even for a given initiator system in the same solvent, the ratio of its components and the preparation conditions influence polymerization performance. A better understanding of initiator formation and the dependence of polymerization kinetics on initiator choice therefore contribute to product development and process optimization.

For the system diethyl aluminum chloride (DEAC)/ethyl aluminum dichloride (EADC)/water an IR-spectroscopic monitoring technique for assessing initiator activity has been developed.[1] Results from olefin polymerizations initiated with this system are reported here. Reactions are performed with initiators of the same activity level but under different polymerization conditions, and also with differently prepared initiators under identical polymerization conditions. In the second part of this contribution, initiators activated with compounds other than water are presented for comparison.

Lanxess Deutschland GmbH, Leverkusen, Germany
E-mail: ursula.tracht@lanxess.com

Experimental Part

All organoaluminum compounds are handled in a glovebox under an atmosphere of dry nitrogen. Solutions of alkyl aluminum chlorides are used as received (diethyl aluminum chloride, DEAC, 1N solution in hexanes, i.e. 18.3 wt%, Acros Organics; ethyl aluminum dichloride, EADC, 1N solution in hexanes, i.e. 19.3 wt%, Sigma-Aldrich).

Hexane (Sigma Aldrich) is dried to <10 ppm water. Sample preparation and transfer to sealed tubes, vials or cells for analysis (e.g. NMR tubes, IR transmission cell) is performed in the drybox.

Precursor solutions are mixed in the desired ratio and diluted with hexane according to target concentration and reaction volumes. All compounds used for activating the initiator are added under vigorous mixing either as pure gas or liquid or using a separately prepared solution in hexane. Molar ratios of protic compound to aluminum between 0.5 to 1 and 3 to 1 are tested. A typical initiator composition contains equimolar amounts of proton source and aluminum.

Olefin polymerizations in hexane are performed either in a small scale batch reactor or in a continuous vessel cascade. The continuous set-up including initiator monitoring via infrared spectroscopy is depicted in Figure 1. In both experimental set-ups reactor temperature is controlled to enable isothermal polymerization experiments. Temperature fluctuates by less than 2.5K, even upon initiator addition to the batch reactor.

All infrared (IR) spectra are measured at room temperature on a Biorad FTS 40 PRO spectrometer with DTGS detector using a transmission cell with polyethylene windows. The cell spacing is adjusted to solution concentrations. A typical transmission pathlength is 1mm. Either a standard transmission cell is used and filled via syringe, or a flow cell is used for on-line measurements.

^{1}H-NMR spectroscopy is applied to the analysis of the nature and concentration of ethyl groups. Methylene signals of ethyl groups bound to aluminum are sufficiently removed from all solvent proton signals to be analyzed quantitatively. All NMR spectra are recorded on a Bruker 400 MHz spectrometer at room temperature.

Molecular weights are determined by gel permeation chromatography (GPC) using THF as eluent. Polystyrene standards are used for GPC calibration.

Polymerizations with Initiator System DEAC/EADC/Water

A typically initiator is prepared from precursor solutions of diethyl aluminium chloride, DEAC, and ethyl aluminium dichloride, EADC, in hexane, and is activated with water as described in.[1] The same initiator quality is used for the olefin polymerizations described in the following. Initiator quality is checked on-line by IR-spectroscopy monitoring and remains in the range of good activity corresponding to an activity indicator of 0.44 during all polymerization campaigns, compare Figure 2. The continuous polymerization process in combination with initiator monitoring is the method of choice for reliably determining product properties. Variability is significantly reduced compared to series of separate batch polymerizations because of the more consistent reaction conditions.

In order to determine the temperature dependence of product molecular weights multiple continuous polymerization campaigns are performed. Each polymerization runs over up to 2 days at variable temperatures. Molecular weight results from a typical run are shown in Figure 3. The reference molecular weight corresponds to

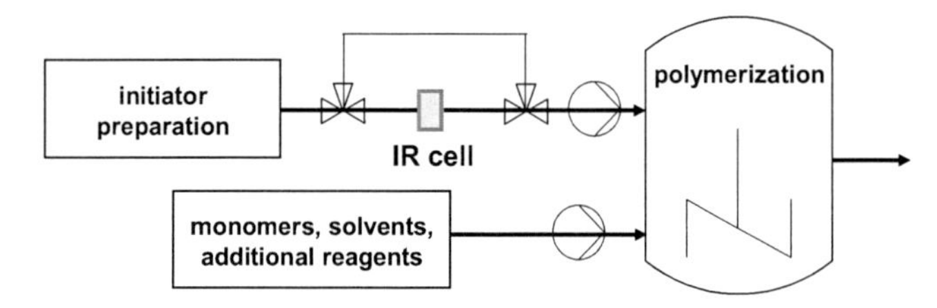

Figure 1.
Schematic of the continuous polymerization process with on-line IR monitoring of initiator quality.

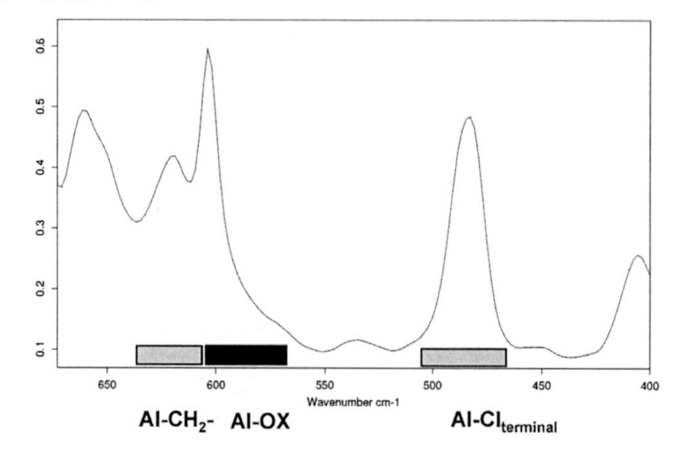

Figure 2.
Representative transmission IR spectrum of the initiator solution used in continuous olefin polymerizations. An activity parameter of 0.44 is determined from the spectrum according to.[1]

a degree of polymerization of ca. 2000. As expected, a linear dependence of the logarithm of M_n and M_w on inverse temperature is observed. Polydispersity also increases with decreasing temperature. Within the range of final conversions studied (10% to 25%) molecular weights do not depend on conversion.

Laboratory experiments with initiators differing in water to aluminum ratio, in water addition rate and mixing conditions upon water addition show that product properties may differ significantly even under identical polymerization conditions and for nominally the same initiator system. For a series of initiators with high degree of hydrolysis such differences are especially pronounced. The DEAC/EADC mixture is activated using always the same water to aluminum ratio of 2.5 but different water addition rates and different agitation periods before the clear solution is separated from the precipitate that forms in all cases. Differences in initiator preparation translate to differences in composition that are detectable in the IR spectra, see Figure 4. No attempt has been made to assign the observed changes in IR pattern to specific molecular structures. Differences are attributed to the formation of different fractions of insoluble aluminum compounds, to different numbers of oxygen, chlorine, and ethyl groups per aluminum, and to different degrees of association in the soluble fraction. Polyisoolefins produced with the differently

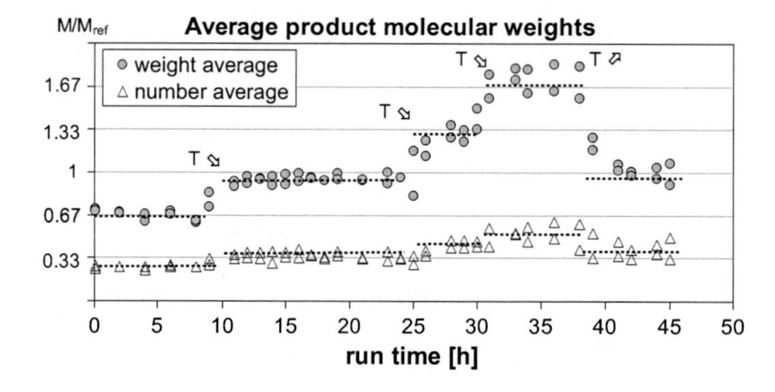

Figure 3.
Continuous isoolefin polymerization experiment at different temperature settings. Reactor temperature is decreased after 9h, 23h, and 29h run time. After 38 hours the temperature is raised again.

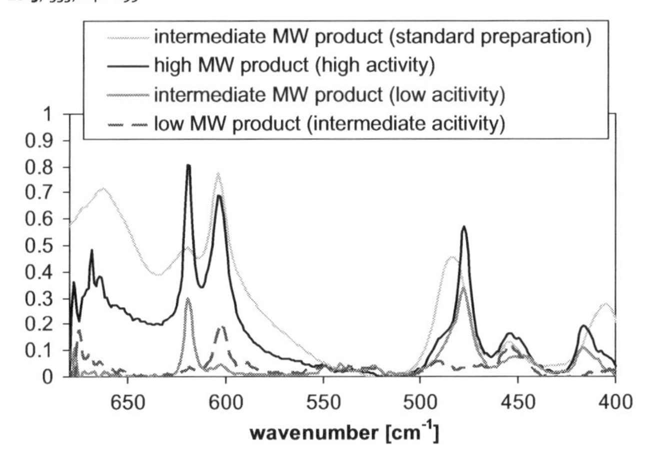

Figure 4.
IR spectra of initiator solutions prepared by water addition to DEAC/EADC in hexane. Concentrations are the same in all cases but the addition process is varied as described in the text. Molecular weights of polyisoolefins obtained from isothermal batch polymerizations using these initiators differ significantly. The lowest molecular weight is obtained in the experiment with highest conversion (highest amount of initiator added). Final conversion in the other two experiments is comparable. The temperature increase upon initiator addition is < 2.5 K in all three experiments.

prepared initiators have molecular weights between ca. 50% and 180% of the M_w obtained with an initiator prepared according to the standard procedure, see Figure 6. Of course small scale batch experiments are always prone to varying impurity levels and hence unknown levels of potential chain transfer agents. However, the range of molecular weights obtained here is well outside the range of scatter in M_w from multiple laboratory scale polymerizations using the same batch of initiator, compare duplicates in Figure 6. These results emphasize the importance but also the difficulty of controlling initiator preparation in the system DEAC/EADC/water in order to obtain reproducible product properties. Again, the IR monitoring technique proves to be an essential tool to provide well controlled reaction conditions in continuous cationic polymerization processes.

Comparison to DEAC/EADC/HCl Initiators

Even though water activated ethyl aluminium chlorides constitute reliable initiator systems for homogeneous polymerizations in halogen-free solvents, their disadvantage is that activation always goes along with a certain loss of active reagent because of the formation of inactive precipitate. Furthermore, subtle changes in the initiator preparation procedure may significantly change initiator performance and product properties. The local water concentration during activation is difficult to control. Therefore, an activation process that is easier to control and gives a fully homogeneous initiator system would be preferred. Protic compounds other than water in combination with DEAC/EADC are tested as alternative initiators. In a first step, the suitability of the established IR-spectroscopic method for initiator monitoring is checked. In a second step, polymerization performance and product properties are compared to the water activated system.

HCl in combination with ethylaluminum chlorides is known as efficient initiator system in cationic polymerizations.[2,3] HCl reacts with alkyl aluminium by replacing alkyl groups by chlorine and forming the corresponding alkane. In the case of DEAC or DEAC/EADC mixtures, the consecutive loss of ethyl groups with increasing HCl/Al addition can be followed by proton NMR as

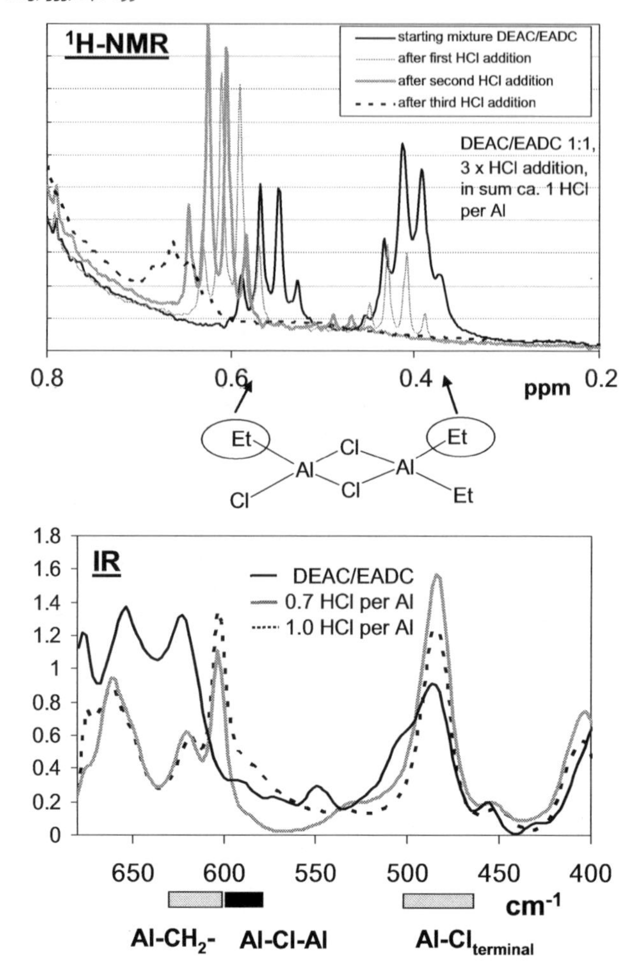

Figure 5.
^{1}H-NMR and IR spectra of HCl activated DEAC/EADC initiator solutions in hexane.

shown in Figure 7 for a 50:50 mixture of DEAC and EADC. HCl is introduced as pure gas into the DEAC/EADC solution. The amount of HCl added is determined via mass balance of the pressurized reservoir of HCl gas. With increasing HCl concentration, the proton NMR signal corresponding to the segment with two ethyl groups bound to the same Al atom disappears and the second signal corresponding to single ethyl groups is shifted further downfield and decreases. Residual ethyl groups are still detectable by proton NMR at the highest HCl concentration shown in Figure 7 corresponding to ca. 70% of the ethyl groups present in the starting solution. IR spectra also show characteristic changes in the case of HCl activation. The IR band attributed to Al-Et vibrations decreases whereas the band attributed to Al-$Cl_{terminal}$ vibrations increases and a new band attributed to vibrations involving bridging Al-Cl-Al structures is detected. When more HCl than ethyl groups present in the ethyl aluminium chloride starting mixture is added, $AlCl_3$ precipitates. At HCl concentrations corresponding to 70–80% of the ethyl group concentration, the initiator solution remains clear. These results indicate formation of oligomeric, chlorine bridged structures containing terminal ethyl groups.

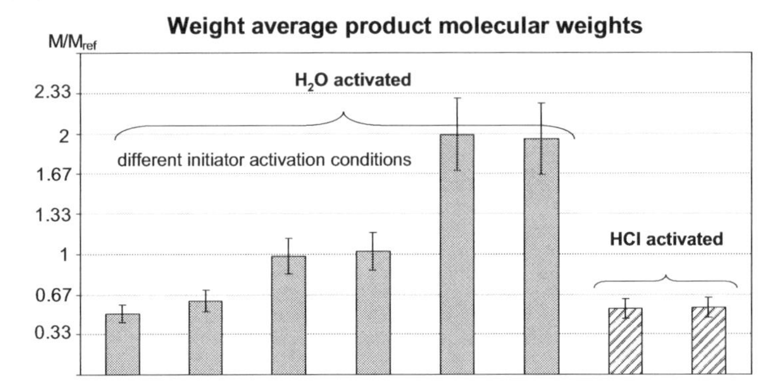

Figure 6.
Product molecular weights from isoolefin polymerizations initiated with differently activated initiators. Reactor temperature is the same in all experiments. For each initiator type, results from two experiments with initiator from the same batch or prepared according to the same activation procedure are given. The three water activated initiators correspond to the IR spectra shown in Figure 4.

The HCl activated initiator is tested in lab scale polymerizations to assess polymerization performance. In agreement with literature reports of cationic polymerizations with initiator systems with variable chlorine content[4,3], polymerization activity is higher for the HCl activated initiator in comparison to the water activated system by a factor of ca. 2. Molecular weights are significantly lower than in the water activated system under identical polymerization conditions. In order to obtain reliable molecular weight data, the HCl activated initiator system is tested in continuous polymerizations. Figure 6 presents results from such a continuous run in comparison to molecular weight levels for different water activated initiator systems tested at the same polymerization temperature. Weight average molecular weights are smaller by about a factor of 2 compared to the water activated system. Intermediate molecular weights are obtained when less HCl is used for activation. The increased chain transfer rate in the systems with higher chlorine content especially suppresses the formation of the higher molecular weight products (M_w significantly

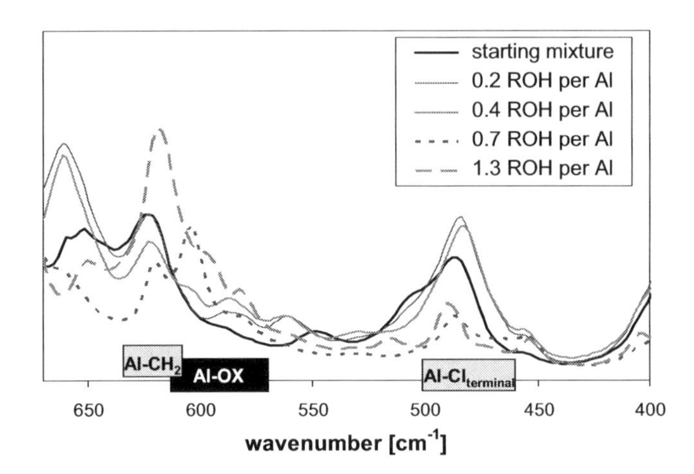

Figure 7.
IR spectra for the products from the reaction of DEAC/EADC mixtures with different amounts of substituted phenol (ROH). Spectra are scaled to identical Al-concentrations.

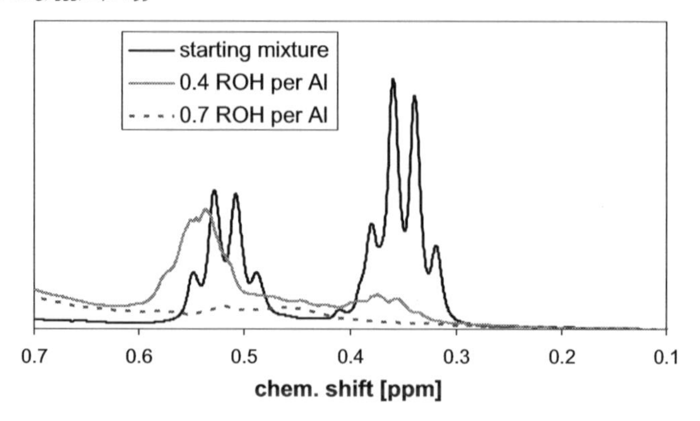

Figure 8.
^{1}H-NMR spectra (methylene proton range) for the products from the reaction of DEAC/EADC mixtures with different amounts of substituted phenol (ROH).

reduced) almost without reducing the chain length of the low molecular fraction of the distribution (little effect on M_n), i.e. the molecular weight distribution is broader for the higher molecular weight products obtained with the water activated initiator. The increased chain transfer rate may be attributed to differences in anion size and/or anion-cation association, i.e. smaller, or more strongly coordinating anions in the chlorine rich system. The temperature dependence of molecular weights obtained with the HCl activated initiator is somewhat weaker than for the water activated initiator.

In conclusion, the activation with HCl offers a more homogeneous, and highly active initiator system. Activation conditions are easier to control than for the water activated initiator.

Activation of DEAC/EADC by Alcohols

Alcohols in combination with ethyl aluminium chlorides also form active polymerization initiators. Another example of alternative initiators presented here uses a substituted phenol[5] for activating DEAC or DEAC/EADC mixtures.

As in the case of HCl and in contrast to the reaction with water, reaction with the phenol rapidly replaces ethyl groups from DEAC. The activation reaction is followed by IR and ^{1}H-NMR spectroscopy, see Figure 7 and Figure 8. Phenol preferably reacts with ethyl groups, and not with chlorine. The IR results furthermore indicate that bridging chlorine is converted to terminal chlorine. Phenoxide bridges are formed instead. According to[6] a well defined dimeric structure may be expected, but our NMR results indicate that the product formed here is not uniform. At low phenol concentrations some precipitate is formed but when the phenol concentration exceeds the ethyl group concentration, a clear solution is obtained. Ethyl groups are no longer detectable in the corresponding products by ^{1}H-NMR, but chlorine content is not significantly reduced compared to the starting mixture as concluded from the IR spectra and confirmed via elemental analysis. Phenol substituents in different coordination environments are detected by NMR.

The assessment of this initiator system in polymerization reactions is in progress.

Conclusion

The three presented examples of organoaluminum chloride based initiator systems for cationic polymerizations illustrate different stages of a workflow established for the development and optimization of such initiators and corresponding polymerization processes. Spectroscopic techniques, especially IR spectroscopy are applied to the characterization of initiator solutions. The IR method is also applicable to initiator monitoring in a continuous polymerization process for all presented initiator systems. Laboratory scale batch experiments are used for comparing differently prepared initiator systems and for assessing their performance in polymerizations of different monomers. For more detailed studies and in order to obtain reliable product properties, a larger scale continuous polymerization process is applied.

For the system DEAC/EADC/water a complex dependence of initiator performance and product properties on activation conditions has been observed. For identically prepared initiator, product molecular weights exhibit the expected dependence on temperature. When HCl or substituted phenols instead of water are reacted with ethyl aluminium chlorides, a better control of activation conditions is possible. The HCl activated as well as the phenol activated initiators provide higher activity than the water activated system.

Acknowledgements: Financial funding under BMBF projects #02PO2200 and #02PO2204 is gratefully acknowledged.

[1] U. Tracht, R. Leiberich, U. Wiesner, H.-I. Paul, *Macromol. Symp.* **2011**, *302*, 208.
[2] US3349065 (1964) Esso Research and Engineering Company, invs.: J.P. Kennedy, N.J. Clark.
[3] Ya. N. Prokofiev, V. P. Bugrov, N. V. Shcherakova, in "Synthetic Rubber" **1983**, p. 289.
[4] T. Cai, S. Liu, J. Qü, S. Wong, Z. Song, M. He, *Appl. Catal. A: General* **1993**, *97*, 113.
[5] The specific phenol was selected within the scope of BMBF projects #02PO2200 and #02PO2204.
[6] Z. Moravec, R. Sluka, M. Necas, V. Jancik, J. Pinkas, *Inorg. Chem.* **2009**, *48*, 8106.

 DOI: 10.1002/masy.201300059

Metallocene Catalyzed Ethylene Polymerization with Specially Designed Catalyst Supports and Reaction Systems

Sangyool Lee,[1] *Sung-Kyoung Kim,*[2] *Sangbok Lee,*[2] *Kyu Yong Choi**[2]

Summary: In heterogeneously catalyzed polymerization of α-olefins, the characteristics of a solid support material impact the catalyst activity, polymer particle morphology, and resulting polymer properties. Silica is the most widely used support for metallocene catalysts in α-olefin polymerization processes because of its large surface area and favorable surface properties for catalyst anchoring. Understanding the kinetics of heterogeneous olefin polymerization over a solid-supported catalyst is often quite complicated because of mass transfer effects and catalyst particle fragmentation during the polymerization. Incomplete or premature fragmentation of support material results in a large fraction of catalyst sites left unavailable for the polymerization, causing some inconsistencies in the performance of the catalyst. Silica-supported metallocene catalysts for α-olefin polymerization are known to follow the layer-by-layer fragmentation mechanism where the fracture of the silica/polymer layer begins from the surface region of a silica particle and it gradually continues into the center of the particle as fragmentation is complete. In this paper, we present new experimental results on ethylene polymerization with *rac*-Et(indenyl)$_2$ZrCl$_2$/MAO catalyst using different types of silica supports to quantitatively assess the effects of support geometry on intrinsic catalytic activity. Flat surface silica, nano-sized spherical silica, straight cylindrical pore silica, and conventional silica are used as supports. The presence or absence of intraparticle monomer diffusion resistance and particle fragmentation has been shown to have significant effects on the catalytic activity.

Keywords: kinetics (polym.); metallocene catalysts; polyethylene (PE); polymerization; silicas

Introduction

High porosity and high surface area silica particles are the most widely used support materials for high activity metallocene catalysts for α-olefin polymerization. One of the important requirements for an effective catalyst support is its ability to disintegrate with the growth of polymers in a controlled manner. Premature or incomplete fragmentation may cause severe diffusion resistance for monomer transport or blockage of catalyst sites for the monomer access.[1–3] It is not uncommon that only a small fraction of catalyst sites is actually used for the polymerization because of irregular or uncontrolled particle fragmentation, and the overall polymerization efficiency becomes quite low. It is also to be noted that the reproducibility of catalytic performance, polymer properties, and the resulting polymer particle morphology are dependent upon the particle fragmentation process.[3,4]

Unlike in Ziegler-Natta catalyzed polymerization processes where transition metal catalyst fragments formed at the beginning of polymerization become the

[1] Department of Chemical and Biomolecular Engineering, University of Maryland, College Park, MD 20742, USA
[2] Department of Chemistry and Biochemistry, University of Maryland, College Park, MD 20742, USA
E-mail: choi@umd.edu

nuclei for primary polymer particles and subsequent macroparticle growth, the fragmentation of silica-supported metallocene catalysts in olefin polymerization is described by the layer-by-layer fragmentation mechanism.[5–7]. According to this mechanism, the porous outer surface region of a silica particle is rapidly filled up with polymer at the beginning of polymerization, forming a dense polymer layer. This polymer layer causes a strong diffusion resistance for the monomer transport to the interior of a catalyst particle. Thus, the polymerization rate decreases as monomer experiences a strong intraparticle diffusion resistance. Eventually, as the polymer mass increases and the hydraulic force generated by the pore-filling polymer builds up, cracking of the silica/polymer layer occurs to allow the for the further diffusive penetration of monomer toward the particle core.[5,7] The sequence of events of polymer formation - pore filling - cracking of silica/polymer layer repeatedly occurs until the entire silica particle disintegrates. During the fracture or fragmentation, the active catalytic sites present in the silica pore surface are partially exposed and allow for the polymerization to continue.

Since the catalyst/polymer particle fragmentation is a complicated process of chemical reaction and physical transport events, quantifying the heterogeneous reaction kinetics of polymerization over silica-supported metallocene catalysts is generally a very difficult task. Furthermore, it is also of a practical interest to find a new way to improve the catalyst effectiveness by either having maximum active sites exposed to monomer early in the reaction process or minimizing intraparticle monomer diffusion limitations. In this paper, we report the kinetics of heterogeneous ethylene polymerization using *rac*-Et(indenyl)$_2$ZrCl$_2$/MAO catalyst with silica supports of different geometries. Figure 1 schematically illustrates these silica supports with growing polymer, and Table 1 summarizes their qualitative aspects of physical characteristics.

Here, the flat surface silica represents the most idealistic surface condition for the heterogeneous catalyst that all the active sites are equally exposed to monomer. The monomer diffusion resistance to these catalytic sites is minimal or even absent. Thus, it can be regarded as an ideal support surface for the assessment of catalytic performance. The second type of silica is a non-porous spherical silica nanoparticle of diameter c.a. 300 nm. Since it is non-porous, the active catalytic sites are present only at the exterior surface of the particle and hence, like in flat surface catalyst, maximum exposure to monomer is expected. The cylindrical tube of silica (silica nanotube) represents another well-defined support geometry. If both interior and exterior of a silica nanotube are deposited with the metallocene catalyst, it is expected that polymer will grow inside and outside of the tube. Finally, the porous micron-size silica represents a typical commercial grade support for a variety of high activity metallocene catalysts. It is characterized by non-uniform pore size distribution and potentially strong monomer diffusion resistance.

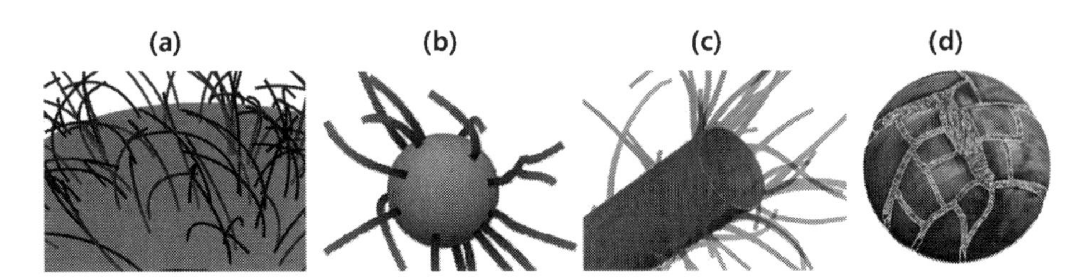

Figure 1.
Schematics of different support geometries for a metallocene catalyst in ethylene polymerization: (a) flat surface, (b) silica nanoparticle, (c) silica nanotube, (d) silica microparticle.

Table 1.
Characteristics of silica supports.

	Fragmentation by pore filling	Dynamics of fragmentation	Monomer diffusion resistance	Catalyst site usage
Flat surface	absent	n.a.	minimal	very high
Nonporous nano-size spherical particle	absent	n.a.	minimal	very high
Cylindrical nanotue	absent	n.a.	low	very high
Porous micron-size spherical particle (conventional silica)	present	time dependent (gradual)	strong	low (fragmentation dependent)

Experimental Part

Materials

Polymerization grade ethylene (Air products) was purified by passing through a stainless steel column packed with R3-11 Cu catalyst, 4Å molecular sieves, neutral alumina and activated carbon. We used toluene (Aldrich) as a diluents and it was purified over sodium and benzophenone in nitrogen atmosphere. *Rac*-Et(1-indenyl)$_2$ZrCl$_2$ (EBI) catalyst (Aldrich) and methylaluminoxane solution (MAO, Aldrich, 10 wt.% in toluene) were used without further purification.

Silica Supports

To make a flat surfaced silica-supported catalyst, we used a silicon wafer (University wafer, P-type). The wafer surface was treated with a mixed acid solution and calcined at 250 °C in a furnace. Nonporous solid nano-sized spherical silica particles were prepared by a sol-gel technique.[8,9] tetra-ethyl orthosilicate (TEOS)/ethanol solution was added into a mixture of ammonium hydroxide, water, and ethanol solution and reacted for 1 h. Then, HCl solution was added to neutralize the solution. Finally, the obtained nano particles were washed with excess amount of deionized water. Silica nanotubes (SNT) were prepared by surface sol-gel method.[10] A mixture of deionized water, hydrochloric acid, ethanol and TEOS was placed in an oven at 60 °C for 1 h. Then the solution of increased viscosity was dropped onto an anodized aluminum oxide (AAO) film several times and calcined at 120 °C for 12 h. Finally, the AAO template portion was removed by dissolving it in a solution of 20% phosphoric acid. Silica X is a conventional silica micro-particle that is commercially available (Its identity is not released in this paper due to confidentiality reason).

For each silica material, a supported metallocene catalyst was prepared as follows. A known amount of silica support was first calcined at 250 °C, and then it was soaked in a mixed solution of hydrogen peroxide (30%) and sulfuric acid (70%) for 30 min, washed with excess amount of deionized water, and treated with an MAO solution (10 wt.% in toluene) at ambient temperature for 24 h. After the MAO-treated silica support materials were washed with toluene and dried in vacuo overnight, they were immersed in the EBI catalyst/toluene solution for 24 hr, washed with toluene several times, and dried in vacuo overnight. The Zr loading amounts in the supported EBI catalysts were measured by inductively coupled plasma mass spectrometry (ICP-OES, ACTIVA, JY HORIVA), and the BET surface area was measured using Micrometrics ASAP2020 apparatus at 77K.

Table 2 shows the characteristic properties of the four silica-supported EBI catalysts. Notice that per support mass, a SNT has the largest Zr loading whereas flat silica has the smallest Zr loading. From the Zr loading for a flat silica shown in Table 2, the calculated value of the number of Zr molecules is 34 per nm^2, which does not seem to be feasible because the dimension

Table 2.
Properties of silica-supported EBI catalyst.

Catalyst system	Dimension	Specific surface area (m^2/g)	[Zr] loading (mol-Zr/g-support)	[Zr] loading (mol-Zr/nm^2)
Flat silica	n.a.	0.0015 [(a)]	8.56×10^{-8}	5.71×10^{-23}
Nano-silica	300 nm D	16.0 [(a)]	3.17×10^{-5}	1.98×10^{-24}
SNT	200 nm D × 10 μm L	75.9 [(a)]	4.00×10^{-4}	5.27×10^{-24}
Silica X	60 μm (avg.)	295.0 [(b)]	1.88×10^{-5}	6.36×10^{-26}

[(a)] Geometrical surface area without considering surface roughness after MAO deposition.
[(b)] Measured by BET method.

of EBI catalyst is about 0.3 nm. As noted in Table 2, the surface area per gram of the flat silica support was calculated from the wafer geometry. However, it was found from the scanning electron microscopic (SEM) images that MAO deposited on the flat surface formed layers and clusters, providing much larger surface area for the complexation with the EBI catalyst. Figure 2 illustrates the MAO clusters formed at the flat silica surface. Also, it should be noted that MAO is a mixture of oligomers with cage and cluster structures.[13] However, the commercial grade silica microparticle (Silica X) has about 0.04 Zr molecules/nm^2 or one zirconium molecule per 25 nm^2 of pore surface area. The data in Table 2 indicates that the amount of Zr loaded onto silica surface in commercial Silica X is only 0.11% of the Zr loading on the flat surface silica support. If the surface characteristics of the four silica support samples are assumed to be identical, the data in Table 2 suggests that catalyst immobilization onto micron-sized silica particles (Silica X) was the least effective. Or a large fraction of silica surface in the commercial silica microparticle might have not been easily reachable by the catalyst components and bulky MAO probably due to the inherently tortuous nature of the pores in the silica particles. In the synthesis of commercial silica particles, 10–50 nm-diameter spheroids or primary particles are first formed during the polymerization of silicic acid solution. These spheroids aggregate to 0.2–0.5 μm-diameter clusters and the channels between the primary particles and those

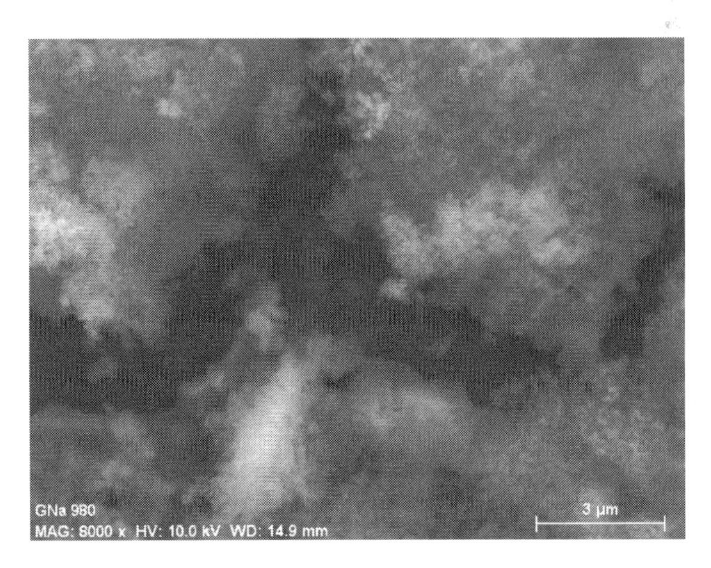

Figure 2.
MAO clusters on the flat surface silica.

between the clusters have pores with size ranging from 5 to 28 nm.[2]

Polymerization

The ethylene polymerization experiments were carried out in a 500 mL glass reactor charged with toluene and equipped with a mechanical agitator. The concentration of MAO in the reactor was 0.02 mol/L in all experiments. The reactor assembly was immersed in a constant temperature bath at 70 °C. The reactor pressure was raised to 2.07 bar by pure ethylene gas and the corresponding equilibrium ethylene concentration was 0.0157 mol/L in toluene (calculated by Henry-Gesetz equation). The reactor pressure (i.e., ethylene pressure) was kept constant by supplying ethylene on demand automatically with a pressure controller and hence, the ethylene flow rate to the reactor (mL/min) corresponds to the actual polymerization rate for a given mass of supported catalyst. It was monitored by an in-line mass flow meter. On-line data acquisition computer was used to record the ethylene mass flow rate, reactor temperature, and reactor pressure data. The recorded ethylene flow rate was numerically integrated with time to obtain the polymer yield data. The calculated and the actually measured polymer yield values agreed quite well within $\pm 7 \sim 9\%$. A loss of very fine polymer samples might have occurred during the sample collection process (separation and drying), contributing to the discrepancies in the calculated and measured yield values. After polymerization, the reaction mixture was filtered, washed with acidified methanol and dried in vacuo overnight. The morphology of polymer particles and support silica materials were analyzed by scanning electron microscopy (SEM, Hitachi S-4700). The sample particles were coated with carbon layers in a Denton DV-503 vacuum evaporator before taking SEM.

Results and Discussion

Figure 3 shows the polymerization yield data for the EBI catalyst supported on four different types of silica materials at 70 °C. The polymer yield data were obtained by numerically integrating the on-line logged ethylene flow rate (polymerization rate). The calculated yield and experimentally measured final yield values agreed very well within $\pm 7 \sim 9\%$ error. First of all, it is seen in Figure 3 that the conventional silica-supported catalyst exhibits the lowest polymer yield on millimolar Zr base whereas the flat surface-supported catalyst shows the highest yield. The polymerization rates for each silica-supported catalyst, derived from from ethylene flow rates are shown in Figure 4.

To characterize the catalytic reaction kinetics of ethylene polymerization over

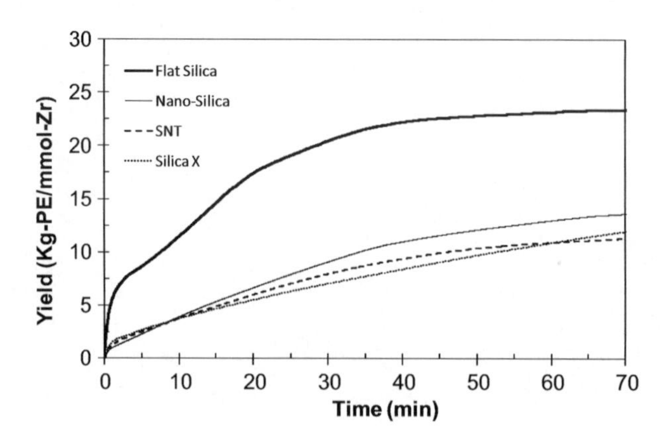

Figure 3.
Effect of support type on polymer yield.

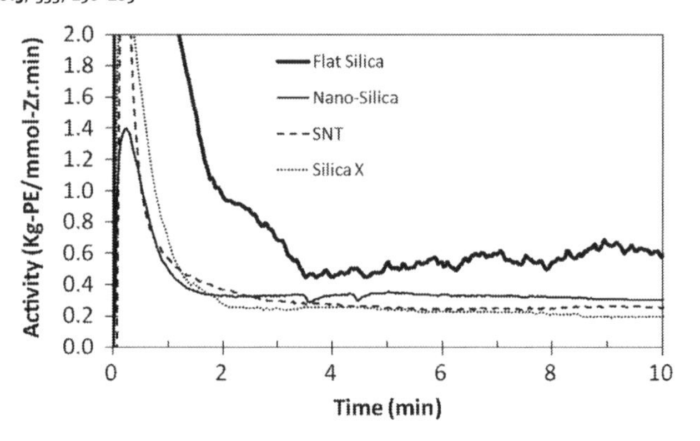

Figure 4.
Polymerization rate curves for silica-supported catalysts.

different support materials, we consider the following polymerization rate equations where $\tilde{R}_p$is in $g_{polymer}/g_{support}$.min:

$$\tilde{R}_p = k_p[M]_p f([Cat])w_m \quad (1)$$

k_p is the propagation rate constant (L/mol · min), $[M]_p$ is the monomer concentration (mol/L) at the catalytic site, w_m is the molecular weight of monomer (g/mol), $[Cat]$ is the active catalyst site concentration (mol-site, $[Zr]$/g-support). It should be noted that the concentration of active catalytic sites may not be equal to the initial catalytic site concentration or zirconium loading ($[Cat]_0$) because of imperfect complexation of Zr compounds with MAO, catalyst side deactivation, and incomplete support fragmentation during the polymerization. We also note that the catalyst site activity may also be dependent on the specific catalyst preparation procedure, silica surface properties, uniformity of complexation of zirconium site with MAO, silica morphology, and intraparticle monomer diffusion resistance. Thus, it is practically difficult to determine the exact concentration of active catalyst sites available for polymerization. Thus, in Equation (1), the catalyst activity function f ($[Cat]$) is expressed in a functional form. The monomer concentration in the solid phase, $[M]_p$, may not be same as in the bulk liquid phase and it can be approximated as $[M]_p = \eta[M]_b$ where $[M]_b$ is the bulk phase monomer concentration and η is the *apparent* or overall effectiveness factor ($\eta \leq 1.0$) that represents the intraparticle monomer diffusion effect. Here, we would like to point out that the parameter η is not the same as the effectiveness factor obtained by exact diffusion and reaction model for a solid catalyst particle because η can also reflect the effect of any external mass transfer resistance. We define another parameter,ψ, that represents the fraction of catalyst sites that are actually available for polymerization. Also, we define ϕ as the activity factor. Then, the catalytic activity function $f([Cat])$ is approximated by $f([Cat]) = (\psi[Cat]_0)\phi$ where $[Cat]_0$is the initial catalyst metal (zirconium) concentration in the silica support. For example, $\psi = 1.0$ indicates that all the initial zirconium sites are active and available for polymerization. In practice, the exact value of ψ will be difficult to measure independently but it will be reasonable to assume that ψ would be smaller than 1.0 due to, incomplete complexation/activation with MAO or incomplete fragmentation that causes some catalyst sites inaccessible for monomer during the reaction. ϕ value is set as 1.0 for the initial catalyst activity.

Then, Equation (1) can be rewritten as follows using the parameters we defined in the above:

$$\tilde{R}_p = k_p(\eta[M]_b)(\psi[Cat]_0)\phi w_m \quad (2)$$

In our ethylene polymerization system, we assume the first order deactivation kinetics for the catalyst, i.e., $\phi = \phi_0 e^{-k_d t} = e^{-k_d t}$ (note: $\phi_0 = 1$) where k_d is the deactivation rate constant. At constant reaction temperature, Equation (2) can be written as

$$\tilde{R}_p = k_p(\eta[M]_b)(\psi[Cat]_0)e^{-k_d t}w_m \qquad (3)$$

The initial polymerization rate is $\tilde{R}_{p0} = k_p \eta_0 [M]_b \psi_0 [Cat]_0 w_m$ where $\eta_0$is the initial effectiveness factor (i.e., $\eta_0 = \lim_{t \to 0} \eta$) and ψ_0 is the initial fraction of Zr sites that are catalytically available for reaction.

The polymerization rate can also be conveniently expressed in g/mol-Zr·min (i.e., $R_p = \tilde{R}_p/[Cat]_0$). Then, the initial polymerization rate (i.e., at t = 0) in g/mol-Zr·min can be written as

$$R_{p0} \equiv \frac{\tilde{R}_{p0}}{[Cat]_0} = \eta_0 \psi_0 k_p [M]_b w_m \qquad (4)$$

For the experimental conditions used in this study, the monomer concentration in the liquid phase ($[M]_b$) was kept constant by controlling the ethylene partial pressure constantly during the polymerization. The polymerization rate (g/mol-Zr·min) normalized by the initial polymerization rate is expressed as

$$\frac{R_p}{R_{p0}} = \frac{\eta\psi}{\eta_0\psi_0}\exp(-k_d t) \equiv \frac{\eta'}{\eta_0'}\exp(-k_d t) \qquad (5)$$

Equation (5) is rearranged to

$$-\ln\frac{R_p}{R_{p0}} = -\ln\frac{\eta'}{\eta_0'} + k_d t \qquad (6)$$

The plot -ln(R_p/R_{p0}) vs. t provides the estimates of $\frac{\eta'}{\eta_0'}$ and k_d. And if a straight line is obtained, the above model can be considered as a reasonably good representation of the kinetic behavior of the catalytic polymerization.[11,12] In what follows, we shall test Equation (6) for the ethylene polymerization with different silica support materials used in our experiments.

Figure 5 shows the test of Equation (6) for the polymerization data with four different silica-supported catalysts, and Table 3 summarizes the parameter values in Equation (6).

Here, we would like to note that measuring the initial polymerization rate (R_{p0}) value from the experimental data of ethylene supply rate was in fact quite difficult. It would be fair to say that unambiguous measurements of perfectly accurate initial polymerization rates will be very difficult, if not impossible. It is because when ethylene supply to a reactor charged with catalyst particles is started, ethylene dissolves in the liquid phase while ethylene is polymerized at the same time. Thus, in the short initial reaction period, both absorption of ethylene and consumption of ethylene occur simultaneously until the equilibrium ethylene concentration in the

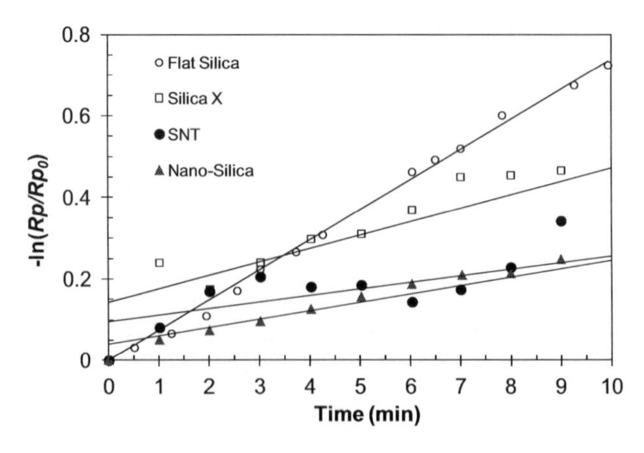

Figure 5.
Plot of Equation (6).

Table 3.
Catalyst kinetic parameters.

Catalyst system	Flat silica	Nano-silica	SNT	Silica X
η'/η_0'	1.0	0.96	0.91	0.85
k_d (min^{-1})	0.0613	0.0205	0.0160	0.0330

liquid phase is reached. Also, since ethylene flow rate is measured by a mass flow meter connected to a solenoid valve, the ethylene flow rate signal is often quite noisy. In our analysis, we used polymer yield vs. time data to construct polymerization rate vs. time plots because the yield data is a kind of filtered data that give rise to less noise sensitive estimates of polymerization rates.

For the conventional silica-supported catalyst, the rate data are well fitted by Equation (6) and from the intercept ($t \rightarrow 0$) we obtain $\eta'/\eta_0' = 0.85$. This value indicates that the mass transfer resistance and/or unavailability of all the catalytic sites might have affected the polymerization right after the polymerization has started. On the other hand, the flat silica and silica nano-particle supported catalysts show that the intercept is zero (i.e., $\eta' \approx \eta_0'$), which suggests that the effects of particle fragmentation and physical transport resistance on the reaction kinetics were minimal. The silica nanotube-supported catalyst shows the effectiveness parameter ratio $\eta'/\eta_0' = 0.91$ which is between the Silica X and silica nano-particle supported catalysts. As the interior of the silica nanotubes are filled up with polymers, monomer diffusion resistance limits the access of monomer to the active sites. Although the effectiveness factor can be changed with reaction time because of increasing polymer particle size and decreasing active site concentration, we assumed in the above analysis that the effectiveness factor is constant during the polymerization. One interesting observation we can make in Table 3 is that the catalyst deactivation parameter k_d for the flat silica support catalyst is much larger than the other three supported catalysts. The relatively longer catalyst life for Silica X-supported catalysts might have been caused by the gradual fragmentation of the support materials during the polymerization. In other words, the fragmentation has an effect of exposing the fresh catalytic sites to monomer and thereby extending the catalyst life. Thus, k_d value for the flat silica-supported catalyst can be regarded as near intrinsic first order deactivation rate constant for this heterogeneous metallocene catalyst.

Polymer Particle Morphology

The scanning electron microscopic (SEM) images of the polymers are shown in Figure 6 for four different types of silica supports. Before we discuss these images, it should be pointed out that these SEM images do not show the actual polymer morphology during the polymerization. It is because as the polymers on silica supports were dried after the reaction, they collapse onto the solid surface and form layers. Figure 6(a) for a flat surface catalyst shows that indeed polyethylene nanofibrils of diameter about 20–30 nm grow directly from the support surface and Figure 6(b) shows similar polymer nanofibril growth at the exterior surfaces of nano-silica particles. When silica nanotubes are used, polymers grow from both inside and outside the tubes and Figure 6(c) clearly confirms the growth behavior. It is interesting to note that the silica nanotubes are also ruptured during the polymerization as the tube ends are covered with polymers, which are extruded out from the tube inside or are accumulated by external surface-grown polymers while polymers continue to grow inside the tubes. The dimension of the nanofibrils seen in Figure 6(c) is also the same as observed in Figure 6(a) and (b). Figure 6(d) shows a disintegrating silica micro-particle. It is clearly seen that the outer region of the silica particle has fragmented and we can observe the stress fibrils between the

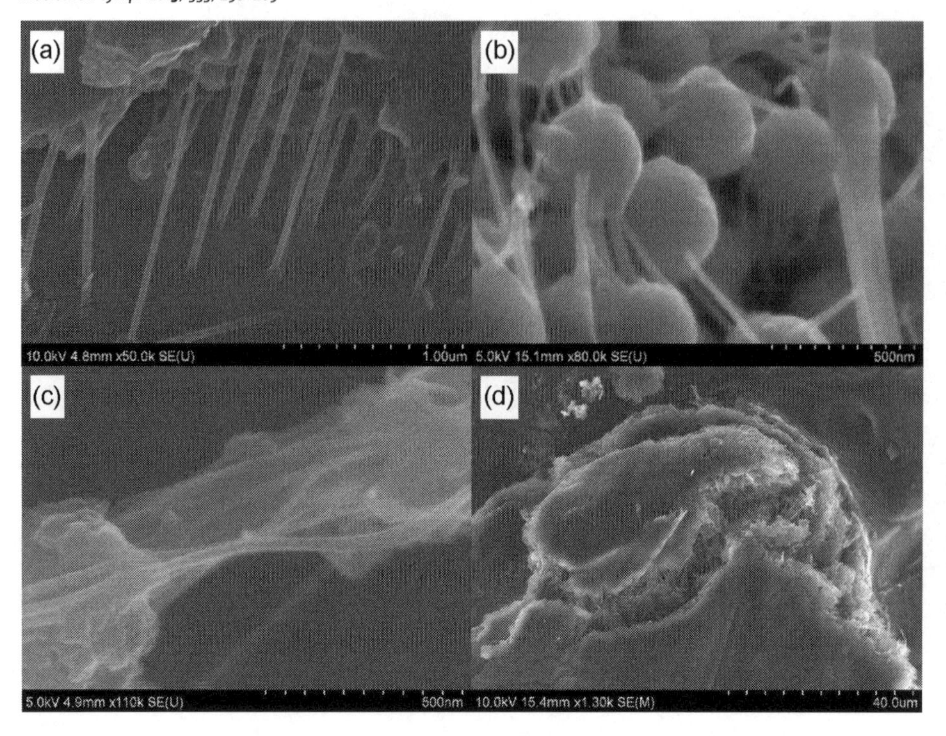

Figure 6.
Polymer morphologies: (a) flat silica, (b) silica nano-particle, (c) silica nanotube, (d) silica micro-particle (Silica X).

cracks. Here, it is to be noted that Figure 6 (d) shows the most representative image of the surface-cracked polymer particles. The silica particle size distribution as well as actual polymer particle size distribution was quite broad.

Conclusion

This paper presents new experimental results on the effects of silica support geometry on the kinetics of ethylene polymerization over *rac*-Et(indenyl)$_2$ZrCl$_2$/MAO catalyst. With four silica materials of different geometries, the experimental data indicate that the nature or characteristics of catalyst support material affects the catalyst activity as well as the polymer morphology. In particular, we have shown that the fragmentation of conventional silica micro-particles and silica nanotubes impart delays in catalytic activities whereas non-disintegrating flat surface silica and silica nano-particle supported catalysts exhibit minimal physical transport effects and thus provide near-intrinsic heterogeneous kinetics. The usage of catalyst metal loaded per unit surface area of the support was the lowest for conventional silica micro-particles because of imperfect fragmentation and complex pore size and structures.

[1] S. Floyd, K. Y. Choi, T. W. Taylor, W. H. Ray, *J. Appl. Polym. Sci.* **1986**, *32*, 2935.
[2] W. D. Niegisch, S. T. Crisafulli, T. S. Nagel, B. E. Wagner, *Macromolecules* **1992**, *25*, 3910.
[3] M. Hassan Nejad, P. Ferrari, G. Pennini, G. Cecchin, *J. Appl. Polym. Sci.* **2008**, *108*, 3388.
[4] X. Zheng, J. Loos, *Macromol. Symp.* **2006**, *236*, 249.
[5] F. Bonini, V. Fraaije, G. Fink, *J. Polym. Sci., Part A: Polym. Chem.* **1995**, *33*, 2393.
[6] G. Fink, B. Steinmetz, J. Zechlin, C. Przybyla, B. Tesche, *Chem. Rev.* **2000**, *100*, 1377.
[7] A. Alexiadis, C. Andes, D. Ferrari, F. Korber, K. Hauschild, M. Bochmann, G. Fink, *Macromol. Mater. Eng.* **2004**, *289*, 457.

[8] J. D. Wells, L. K. Koopal, A. de Keizer, *Colloids Surf., A* **2000**, *166*, 171.
[9] B. Zhao, C. Tian, Y. Zhang, T. Tang, F. Wang, *Particuology* **2011**, *9*, 314.
[10] D. Zhao, Q. Huo, J. Feng, B. F. Chmelka, G. D. Stucky, *J. Am. Chem. Soc.* **1998**, *120*, 6024.
[11] S. Y. Lee, S.-K. Kim, T. M. Nguyen, J. S. Chung, S. B. Lee, K. Y. Choi, *Macromolecules* **2011**, *44*, 1385.
[12] S. Y. Lee, K. Y. Choi, *Ind. Eng. Chem. Res.* **2012**, *51*, 9742.
[13] E. Zurek, T. Ziegler, *Prog. Polym. Sci.* **2004**, *29*, 107.

 DOI: 10.1002/masy.201300061

Micro-Cellular Polystyrene Foam Preparation Using High Pressure CO_2: The Influence of Solvent Residua

Andra Nistor,[1,2] *Adam Rygl,*[1] *Marek Bobak,*[1] *Marie Sajfrtova,*[3] *Juraj Kosek**[1,2]

Summary: Polystyrene (PS) foams are commonly used as heat insulators. Their heat insulation properties can be greatly improved by decreasing the cell size into the range of micrometres, resulting in the so-called micro-cellular foams. We prepared micro-cellular PS foams and studied the influence of toluene residua on the foam structure. First, PS films were prepared by the dip-coating method using toluene as the solvent. Then, the PS films were foamed by the pressure induced foaming method using high pressure CO_2. The foam structures were examined by Scanning Electron Microscopy and X-ray micro-tomography. For the industrial PS sample containing nucleation agents, the influence of temperature and depressurization rate on cell size was observed in a way consistent with the literature. However, the obtained cell sizes were much larger. Therefore, the effect of toluene residua on the foaming process was systematically studied. A PS sample without nucleation agents was used to exclude their effect on the foaming process. Toluene acts as a co-solvent and enhances the CO_2 solubility in PS, which should lead to a large amount of small cells. However, at toluene concentrations below 8 %wt, the plasticizing effect of toluene dominates, resulting in enhanced coalescence and cell growth. Foam heat insulation properties improve with increasing porosity. The toluene residua increased the porosity by increasing the cell sizes and lowering the thickness of the compact skin at the film surface. We conclude that toluene residua greatly influence the foam structure and this finding can be potentially used to control the foaming process.

Keywords: high pressure CO_2 foaming; micro-cellular foams; micro-tomogra; morphology; phypolystyrene

Introduction

Polymer foams are used as heat or acoustic insulators, in separation processes as membranes, and in filtration and absorption devices. PS foams are mainly used as heat insulators.[1] The heat insulation properties of PS foams can be improved, for instance, by reducing the cell size to the range of micrometres or less resulting in the so-called micro-cellular foams. According to mathematical modelling, the heat insulation properties of micro-cellular foams are improved by Knudsen effects[2] (for cells smaller than several micrometres) and reduced heat radiation[3] (for cells smaller than approx. 100 μm) while keeping the porosity around 95%.[4]

Heat insulation properties of micro-cellular foams are determined by the overall structure, such as the cell size distribution, cell shape and orientation, thickness of the walls between the cells and porosity.[4] Most authors in the literature were focused mainly on the cell size reduction.[5] However, we are focused on

[1] Institute of Chemical Technology Prague, Dept. of Chemical Engineering, Technicka 5, 166 28 Prague 6, Czech Republic
E-mail: Juraj.Kosek@vscht.cz
[2] New Technologies – Research Centre, University of West Bohemia, Teslova 9, 306 14 Pilsen, Czech Republic
[3] Institute of Chemical Process Fundamentals, Czech Academy of Sciences, Rozvojova 2/135, 165 02 Prague 6, Czech Republic

the overall structure of the foams in order to improve the heat insulation properties.

Micro-cellular foams can be prepared using environmentally friendly and sustainable blowing agents such as CO_2. The batch CO_2-based foaming process can be temperature induced (TIF) or pressure induced (PIF). A rapid increase in temperature or decrease in pressure in the TIF or PIF method, respectively, leads to an oversaturation of the system polymer-CO_2. Phase nucleation and separation then occur followed by cell growth, Oswald ripening and coalescence of cells, until the polymer vitrificates and stabilizes its structure. The formation of nuclei followed by the cell growth is a complex process, which is still not fully understood.[5]

The PIF method for the PS-CO_2 system with the objective to control the final foam structure has been studied only by few researchers. The cell shape and orientation of micro-cellular foams can be influenced by the shape of the used pressure vessel.[6] Foams with the cell size in the range of micro- or even nanometres and with uniform cell size distribution can be prepared, for instance, by the incorporation of nanosilica particles[7] or nanostructured CO_2-philic block copolymers[8] into PS. A bimodal cell size distribution can be obtained when the pressure is reduced in two steps.[6] The PS molecular weight and polydispersity do not significantly affect the resulting foam structure.[9] However, small amounts of low molecular weight components (solvents) have a pronounced effect on the foam structure,[9] which could open a new way of controlling the foaming process.

In this work, we systematically investigated the influence of toluene residua on the structure of PS foams prepared by the PIF method from thin PS films impregnated by CO_2. The PS films were prepared by the dip-coating method from a solution of PS in toluene. The structure of the produced micro-cellular foams was studied by Scanning Electron Microscopy (SEM) and X-ray micro-tomography (micro-CT).

Foaming Experiments

Two PS samples were used for the foaming experiments: PS_A (with nucleation agents, glass transition temperature $T_g = 105\,°C$) and PS_B (without nucleation agents, $T_g = 100\,°C$ and molecular weight $M_W = 280\,000$).

We prepared thin PS films with the film thickness in the range from 50 to 100 μm by a custom-built dip-coating apparatus using toluene as the solvent. The prepared films were dried at room conditions for several days in order to obtain samples with various residual toluene concentrations. The toluene concentration was retrospectively determined gravimetrically. For this purpose, we prepared PS films of various thicknesses and dried them at room conditions. Each day, we measured the mass loss of the sample caused by the toluene desorption. Then, the samples were melted and we measured the mass of the pure PS, that is, without any toluene residua. The time evolution of toluene concentration in the samples was then calculated. The foaming of PS in the form of a film enables to have a well-defined system suitable for studying the fundamental aspects of the foaming process such as nucleation and cell growth. However, industrial semi-continuous processes of microcellular sheet production were also proposed.[10]

The prepared and dried PS film sample was placed in a high pressure vessel. The vessel was heated to the desired temperature and pressurized with CO_2. After reaching the sorption equilibrium (the estimation of the required absorption time is described below), the vessel was depressurized in 5 or 45s. In the literature, this pressure-induced foaming method is usually called the pressure-quench method.

The produced PS foam samples were analysed first by SEM and then by micro-CT. SEM images were used to obtain first insight on the PS foam structure. When the cell sizes were larger than 10 μm, we examined the foams also by micro-CT. The used micro-CT (MicroXCT-400 from XRadia) has a practical resolution of about 1 μm for the PS foam samples. Therefore, only cells with sizes of at least 10 μm can be accurately analysed, as they comprise enough pixels to determine the structure. The micro-CT images were analysed in the software Lucia from Laboratory Imaging Company.

During the CO_2 sorption, both temperature and pressure have to be set so that the system is above its T_g once it reaches the sorption equilibrium. When CO_2 is desorbed below the T_g of the mixture, the PS sample will not foam. The dependency of T_g depression on CO_2 pressure (equilibrium CO_2 concentration in PS) reported by various researchers differs strongly. It was reported that the T_g of the PS-CO_2 mixture is well below the room temperature at CO_2 pressures above 200 bar.[6]

The absorption period needed for establishing the sorption equilibria has to be estimated. Arora et al.[6] determined experimentally the diffusion coefficient of CO_2 in PS at 80 °C and 243 bar to be $D = 2.4 \times 10^{-10}\,m^2/s$. Therefore, we can assume D to be of the order of 10^{-11} to $10^{-10}\,m^2/s$. To be on the safe side, we chose a lower diffusivity of $1 \times 10^{-11}\,m^2/s$ and estimated the absorption time τ for all films as

$$\tau = l^2/D,$$

where l is the half-thickness of the film. However, we enabled the sample to absorb CO_2 for at least 5τ (usually about two hours) to assure near equilibrium conditions at the start of the foaming.

As indicated above, the PIF method enables to control the following parameters: initial impregnation temperature and pressure, and depressurization time. These parameters affect the T_g,[11] polymer viscosity, CO_2 solubility[12] and diffusivity of CO_2 in PS. In the following paragraphs, we will discuss these parameters and their influence on the foam structure.

Toluene Residua: The Influence of Temperature and Depressurization Time on Average Cell Size

The effect of temperature on the average cell size of the PS foam is complex. As the temperature increases the CO_2 solubility decreases and fewer nuclei will be formed.[6,13] An increase in temperature will also decrease the viscosity of the PS-CO_2 mixture, increase the diffusivity of CO_2, and increase the time necessary for vitrification. All of these factors lead to cell growth and thus increase the average cell size (Figure 1).

In the PIF method, foaming is initiated by creating an oversaturated system. The oversaturation dynamics can be controlled by the depressurization time t. The faster we depressurize the system the higher is the temporal degree of oversaturation, which will lead to a higher nucleation density and thus smaller cells (Figure 1).

In Figure 1, the above mentioned effects of impregnation temperature and depressurization time can be clearly seen despite the fact that the individual PS films contained various amounts of toluene residua. However, pure PS foamed at the same conditions had the average cell sizes much lower.[6,8] In addition, the difference of the average cell size between the experiments with shorter ($t = 5$ s) and longer ($t = 45$ s) depressurization time is smaller than reported in the literature.

Stafford et al.[9] observed that even a small amount (< 4%wt) of solvent residua greatly increased the cell size. To verify this assumption, we prepared a PS film by the dip-coating method. The film was

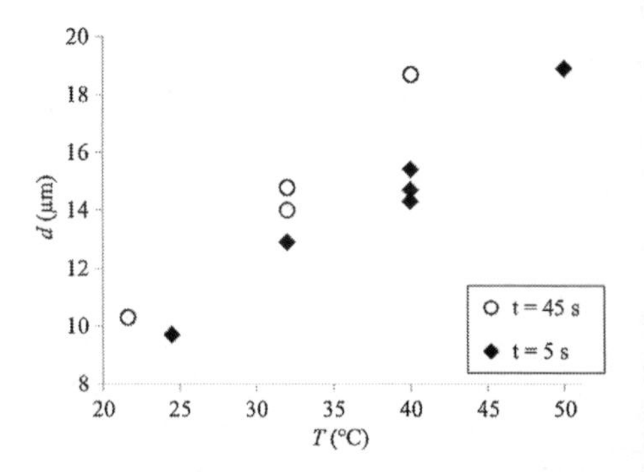

Figure 1.
The influence of impregnation temperature T and depressurization time t on the average cell size d of foamed PS_A samples with various residual toluene concentrations. The CO_2 saturation pressure was 285 bar.

dried at room conditions and subsequently degassed under vacuum at 40°C for 4 days. This sample was foamed and compared with a sample impregnated at the same conditions, but dried just at room conditions. We observed that the sample degassed under vacuum had a smaller average cell size ($d = 3\,\mu m$) and a thicker compact skin at the surface than the room-temperature degassed sample ($d = 9.7\,\mu m$) (Figure 2).

The formation of the skin at the surface was observed in almost all relevant articles.[5] Upon depressurization, the diffusion of CO_2 is so fast that nuclei are not formed or disappear quickly close to the sample surface. However, the surface was cracked at several places by the release of CO_2. In addition, CO_2 expansion during depressurization causes cooling,[8] which shortens the time needed for vitrification.

Moreover, the toluene residua seem to influence the foam structure. Therefore, we focused on foaming processes which are affected by the solvent residua. To exclude the influence of nucleation agents (present in sample PS_A) on the foaming, we used the sample without nucleation agents PS_B for all following experiments.

The effect of toluene in the PS film on the temperature dependence of the average cell size is shown in Figure 3. When the CO_2 was in the liquid state (i.e., at 25 °C), we observed a great increase in the average cell size with the increasing toluene concentration. When the CO_2 was in the supercritical state (i.e., at temperature equal to or higher than 32 °C) an approximately constant average cell size was observed. However, the cell size was much larger than reported in the literature as mentioned above. Thus, it seems that the presence of toluene enhanced the coalescence of cells and Ostwald ripening, which resulted in the formation of larger cells.

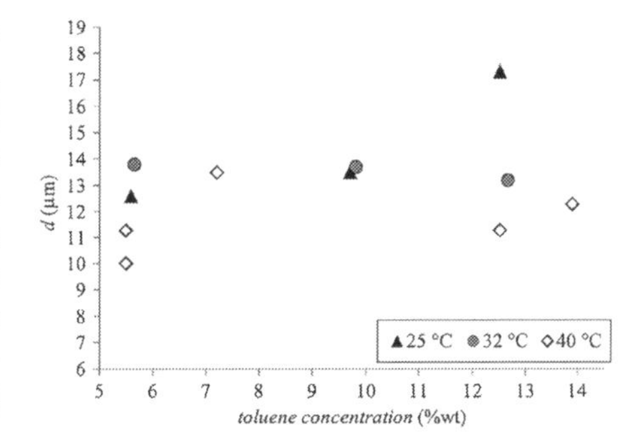

Figure 3.
The influence of temperature on the average cell size d at different concentrations of toluene. The PS_B sample was impregnated at 285 bar and depressurized in 5s.

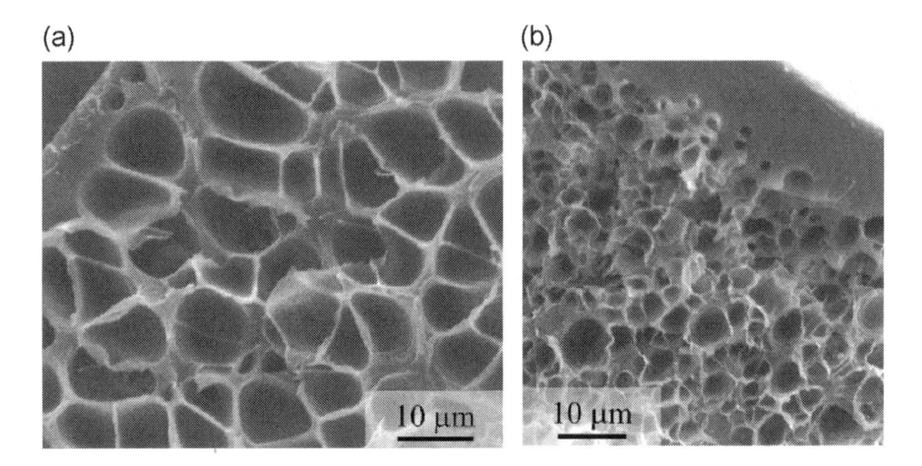

Figure 2.
SEM images of PS_A foams prepared from samples a) dried at room conditions for 36 days ($d = 9.7\,\mu m$) and b) dried at room conditions for 34 days and then degassed under vacuum at 40 °C for 4 days ($d = 3.0\,\mu m$). Both experiments were carried out at 25 °C, 285 bar of CO_2 and depressurized in 5s.

In addition, the presence of the toluene reduces the T_g of the PS film. Neat PS films did not foam at conditions at which PS films with toluene residua foamed, for example, at 25 °C, 180 bar and depressurization period of 5s. Thus, the enhancement of the cell growth could be also affected by T_g reduction, which extends the time for vitrification.

However, we observed that for sample PS_B without nucleation agents the temperature influence on the average cell size is reverse, in other words, the higher is the impregnation temperature the smaller are the cells (Figure 3). This indicates that the temperature effect on foaming is overcome by other effects. The toluene should mainly affect the CO_2 solubility in the PS sample. Therefore, in the following paragraph we will discuss the influence of toluene on the solubility of CO_2 in PS.

Toluene as Co-Solvent: The Influence on CO_2 Solubility and Nucleation

Toluene acts as a co-solvent and should enhance the solubility of CO_2 according to the theory of solubilities in ternary systems.[14] Therefore, the presence of residual toluene in the PS samples should lead to foams with a higher amount of smaller cells. However, the cells are larger (Figure 4), as mentioned above. It seems that at toluene concentrations below 8%wt, the plasticizing effect overcomes the enhanced solubility of CO_2 in PS.

The CO_2 solubility in PS is mainly affected by temperature and pressure. According to theory, the average cell size should decrease as the CO_2 saturation (impregnation) pressure increases. This is because the CO_2 solubility in PS increases with increasing pressure leading to the formation of more nuclei and thus smaller cells.[5,6,13] However, we observed almost constant cell size independent on the impregnation pressure (Figure 4). We would suggest that this is also due to the plasticizing effect of toluene, which dominates over the effect of increased pressure (CO_2 solubility).

The CO_2 solubility in PS mainly affects the number cell density. We observed that (i) the cell number density increased only slightly as the saturation pressure (i.e., the CO_2 solubility) increased, and (ii) the number cell densities of the samples with different contents of toluene were almost the same (Figure 5). The number cell densities are proportional to the number of nuclei. From this we can conclude that although toluene enhanced the CO_2 solubility, it mainly affected the cell growth and coalescence by the plasticizing effect at toluene concentrations below 8%wt.

As the CO_2 impregnation temperature increases, the solubility of CO_2 in PS

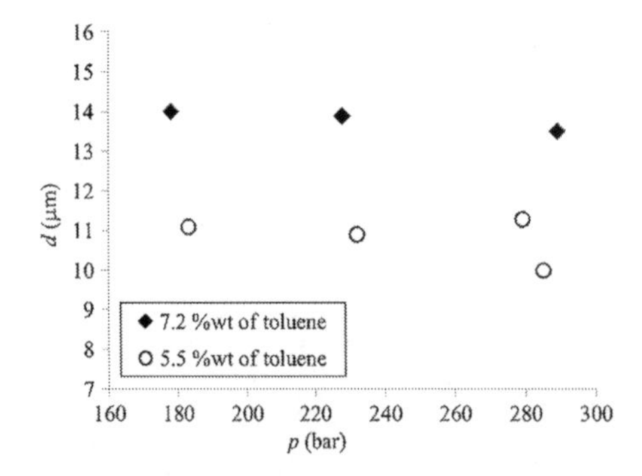

Figure 4.
The influence of CO_2 saturation pressure on the average cell size at different concentrations of toluene in the PS_B samples. All experiments were carried out at 40 °C with a depressurization period of 5s.

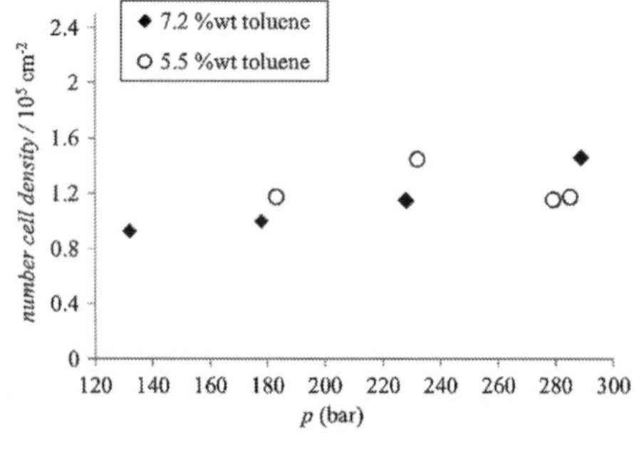

Figure 5.
The influence of the initial CO_2 saturation pressure on the number cell density at different contents of toluene in PS_B. All experiments were carried out at impregnation temperature $T = 40$ °C and depressurization period of 5s.

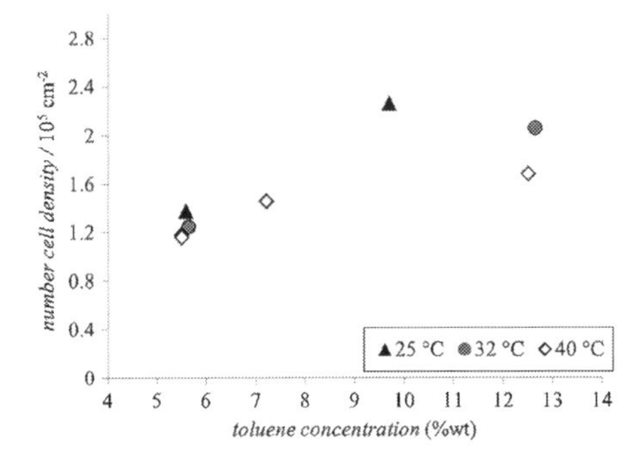

Figure 6.
The influence of impregnation temperature on the number cell density at different contents of toluene residua in PS_B impregnated at 285 bar and depressurized in 5s.

decreases, which results in a lower number cell density (Figure 6). Again, at toluene concentrations below 8%wt the number cell densities are similar. However, at higher toluene concentrations, we observed an increase of the number cell density that was probably caused by the co-solvent effect.

Influence of Toluene Residua on Porosity

Since the overall foam porosity greatly affects the heat insulation properties, it is an important foam parameter. The overall porosity is the volume percentage of cells in the foam and it thus includes the area of the compact skin at film the surface.

In Figure 7, we can see that lower impregnation temperatures and higher

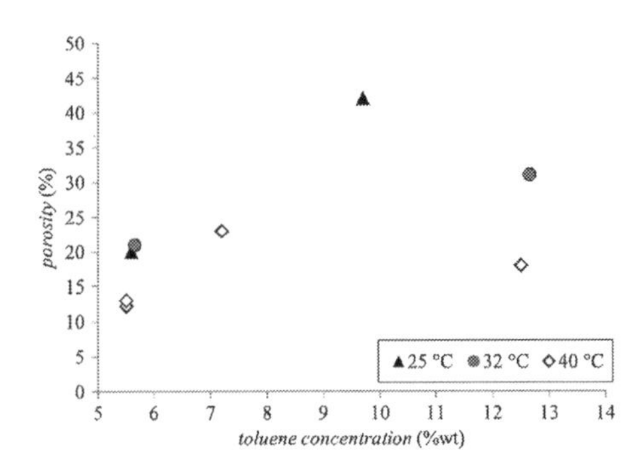

Figure 7.
The influence of toluene residua in PS_B on the overall foam porosity at different impregnation temperatures. Samples were impregnated at 285 bar and depressurized in 5s.

toluene concentrations lead to higher porosities. The overall porosity increased also due to the reduction of the compact skin at the foam surface. The compact skin at the surface lowers the porosity in some cases by more than 10%. However, the lower is the initial PS film thickness the faster is the CO_2 desorption to the surrounding. Thus, some nuclei close to the film surface will not grow at all.

Conclusion

We studied micro-cellular PS foams prepared by the pressure induced foaming method using high pressure CO_2. The PS films used for the foaming experiments were prepared by the dip-coating method using toluene as the solvent. The structure of the prepared PS foams was visualized by SEM and micro-CT.

For the PS sample containing nucleation agents and toluene residua, we observed that the average cell size increased with temperature. This is explained by the decrease of CO_2 solubility and PS viscosity with increasing temperature, and the longer time necessary for vitrification. Furthermore, when we shortened the depressurization, we observed smaller cells. The faster oversaturation of the system thus resulted in enhanced formation of nuclei and lead to smaller cells.

The cell sizes observed in our foaming experiments were larger than those reported in the literature. This difference was caused by the toluene residua present in the PS film samples. Therefore, we studied the influence of toluene residua on the foaming process. For this purpose, we used a PS sample without the nucleation agents to exclude their influence on foaming.

Toluene acts as a co-solvent and enhances the CO_2 solubility in PS, which should lead to the formation of a larger amount of smaller cells. However, we observed that at toluene concentrations below 8%wt the cell sizes increased due to enhanced cell growth and coalescence. This was mainly due to the plasticizing effect of toluene, which reduced the glass transition temperature. The influence of impregnation pressure on cell size was almost negligible. At higher toluene concentrations, the enhanced CO_2 solubility in PS resulted in the increase of the number cell density, which affects the porosity. Thus, the presence of toluene increased the porosity mainly due to enhanced cell growth, increased number cell density and the reduction of the compact skin at the film surface. The compact skin at the surface lowered the foam porosity in some case by more than 10%.

Our results suggest that toluene could be used to control the foaming process. As toluene lowers the T_g of the polymer, the size of the compact skin at the foam surface and increases the CO_2 solubility in the polymer, the foaming process could be carried out at lower temperatures and saturation pressures. In addition, by carefully optimizing the process, we could still prepare foams with high porosity and the average cell size below 10 μm.

Acknowledgements: Financial support from the Czech Grant Agency (project 106/12/P673) and specific university research (MSMT No 20/2013) is acknowledged. The result was developed with instruments available in the CENTEM project, reg. no. CZ.1.05/2.1.00/03.0088, co-funded by the ERDF as part of the Ministry of Education, Youth and Sports' OP RDI program.

[1] H. Weber, I. De Grave, E. Röhrl, **2000**, *Foamed Plastics. Ullmann's Encyclopedia of Industrial Chemistry.*

[2] R. Baetens, B. P. Jelle, J. V. Thue, M. J. Tenpierik, S. Grynning, S. Uvslokk, A. Gustavsen, *Energy and Buildings* **2010**, *42*, 147–172.

[3] A. Kaemmerlen, C. Vo, F. Asllanaj, G. Jeandel, *Journal of quantitative spectroscopy & radiative transfer* **2010**, *111*, 864–877.

[4] P. Ferkl, R. Pokorný, M. Bobák, J. Kosek, *Chemical Engineering Science* **2013**, *97*, 50–58.

[5] L. J. M. Jacobs, M. R. Kemmere, J. T. F. Keurentjes, *Green Chemistry* **2008**, 731–738.

[6] K. A. Arora, A. J. Lesser, T. J. McCarthy, *Macromolecules* **1998**, *31*, 4614–4620.

[7] H. Jannani, M. H. N. Famili, *Polymer Engineering & Science* **2010**, *50*, 1558–1570.

[8] J. A. R. Ruiz, M. Pedros, J.-M. Tallon, M. Dumon, *The Journal of Supercritical Fluids* **2011**, *58*, 168–176.

[9] C. M. Stafford, T. P. Russel, T. J. McCarthy, *Macromolecules* **1999**, *32*, 7610–7616.

[10] D. Eaves, Ed. *Handbook of Polymer Foams*, 1st ed. Rapra Technology Limited: UK, **2004**.

[11] J. E. Mark, (Ed., *Physical Properties of Polymers Handbook* (Ed., Springer, New York **2007**.

[12] M. Panatoula, C. Panayiotou, *Journal of Supercritical Fluids* **2006**, *37*, 254–262.

[13] I. Tsivintzelis, A. G. Angelopoulou, C. Panayitou, *Polymer* **2007**, *48*, 5928–5939.

[14] J. Chmelar, T. Gregor, H. Hajova, A. Nistor, J. Kosek, *Polymer* **2011**, *52*, 3082–3091.

Dynamics of Network Formation in Aqueous Suspension RAFT Styrene/Divinylbenzene Copolymerization

Miguel A. D. Gonçalves,[1] *Virgínia D. Pinto,*[1] *Rolando C. S. Dias,**[1]
Julio C. Hernándes-Ortiz,[2] *Mário Rui P. F. N. Costa*[2]

Summary: Experimental studies concerning the RAFT copolymerization of styrene and commercial divinylbenzene were performed in order to assess the use of controlled radical polymerization for the production of non-conventional polymer networks. Aqueous suspension polymerizations were carried out in stirred batch reactor and changes of a few operation parameters were tried in order to assess their effect on key product properties: reaction temperature, initial ratio monomer/RAFT agent, monomer dilution in the organic phase and crosslinker content in the monomer mixture were changed along the experimental program. Polymerizations were extended beyond the gel point in order to synthesize gel beads. The effect of different commercially available RAFT agents (e.g. S-thiobenzoyl thioglycolic acid and two trithiocarbonates) on network process formation was also studied. Dynamics of gel formation was measured using size exclusion chromatography with refractive index and light scattering detection. Iodine chloride titration yielded pendant double bonds concentration. *In-line* and *off-line* FTIR-ATR measurements were also performed in order to obtain information concerning the kinetics of formation and final structure of the gels.

Keywords: crosslinking; FTIR-ATR; networks; RAFT

Introduction

Among Controlled Radical Polymerization (CRP) techniques, Reversible Addition-Fragmentation Chain-Transfer (RAFT) polymerization is especially versatile because can be used with different classes of monomers (e.g. with organic and water compatible monomers) and considering a wide range of operation conditions (e.g. using a broad range of polymerization temperatures).[1,2] Due to these features, RAFT presents promising potentialities for industrial purposes and its application with dispersed systems was recently assessed in many important works.[3,4] It was also showed that, when CRP techniques in emulsion or mini-emulsion are considered, the combination between RAFT and mini-emulsion is advisable because the number of radicals per particle is not changed and monomer droplets become the main loci of particle nucleation when oil-soluble initiators are used.[5,6] Distinctive features of RAFT polymerization have been explored to synthesise linear polymers with well-defined topologies and also advanced materials based on non-linear polymerization, such as networks,[7,8] amphiphilic netwoks,[9,10] nanoparticles/nanocapsules,[11,12] and to control branching/hyperbranching considering organic[13] and aqueous polymerization systems.[14,15]

RAFT production of styrene/divinylbenzene networks was also reported in few recent works considering bulk polymerization in sealed ampoules[16] and in supercritical carbon dioxide.[17] The ultimate goal of these researches (see[16,17] and references therein) is the assessment of the use of CRP

[1] LSRE-Instituto Politécnico de Bragança, Quinta de Santa Apolónia, 5300, Bragança, Portugal
E-mail: rdias@ipb.pt
[2] LSRE-Faculdade de Engenharia da Universidade do Porto, Rua Roberto Frias s/n, 4200-465 Porto, Portugal

techniques aiming the production of networks with improved structural homogeneity. Results here presented are a contribution to this research line considering the aqueous suspension RAFT polymerization of S/DVB as case study. Operation with aqueous suspension is explored to perform single-pot reactions, eventually with gel formation, maintaining good stirring and heat dissipation conditions, as recently showed with NMRP for the same chemical system.[18]

Dynamics of RAFT crosslinking is studied through the analysis by SEC/RI/MALLS and *off-line* FTIR of reaction samples collected at different polymerization times. Chemical analysis is also used to measure the dynamics of the pendant double bond concentration (PDB) and the usefulness of *in-line* FTIR-ATR to obtain information concerning the RAFT building-up of the networks is assessed. Influence of some operation parameters (e.g. polymerization temperature, initial composition, RAFT agent used) on the kinetics of polymerization and crosslinking was studied by changing these parameters along the experimental program.

Materials

Styrene (S) of 99% purity stabilized with 0.005% w/w 4-tert-butylcatechol, commercial grade of divinylbenzene (DVB) of 80% purity stabilized with 0.1% w/w 4-tert-butylcatechol, AIBN of 98% purity, xylene of 98.5% purity and toluene of 99.7% purity were purchased from Sigma Aldrich and used as received. Commercial DVB is a mixture of isomers, 56.2% *m*-divinylbenzene and 24.2% *p*-divinylbenzene, plus 19.6% of ethylvinylbenzene. Xylene solvent is also a mixture of xylenes plus ethylbenzene (25% in the latter component). The following three commercially available RAFT agents were also purchased from Sigma Aldrich and used as received: 2-(dodecylthiocarbonothioylthio)- 2-methyl-propionic acid (DDMAT) of 98% purity, S-(thiobenzoyl)thioglycolic acid (TBTGA) of 99% purity and cyanomethyl dodecyl trithiocarbonate (CDT) of 98% purity.

Polymerization Runs

A set of polymerization runs performed in this research is described in Table 1. The following main parameters were changed along the experimental program:

- Polymerization temperature (T).
- Initial mole fraction of crosslinker in the monomer mixture (Y_{CL}).
- Volumetric fraction of the monomer in the organic phase (Y_M).
- Initial mole ratio between initiator (AIBN) and monomer (Y_I).
- Initial mole ratio between RAFT agent and initiator (Y_I^{RAFT}). Correspondent initial mole ratio between monomer and RAFT agent (Y_{RAFT}^M) is also presented in Table 1.
- Three different RAFT agents (DDMAT, CDT and TBTGA) were alternatively used in the experimental program.

An atmospheric reactor with maximum capacity of 2.5 L was used to perform the polymerizations at low temperature. A Parr 5100 pressurized glass reactor with 1 L

Table 1.
A set of polymerization runs performed in the suspension RAFT polymerization of styrene (S) and Divinylbenzene (DVB).

Run	T (°C)	Y_{CL} (%)	Y_M (%)	Y_I (%)	Y_I^{RAFT}	Y_{RAFT}^M	RAFT Agent	w_g
1	70	0	50	0.216	1.869	247.8	DDMAT	–
2	70	0	50	0.213	1.884	249.5	CDT	–
3	70	0	50	0.208	1.841	261.0	TBTGA	–
4	70	100	100	0.107	1.855	503.8	DDMAT	1.0
5	70	100	50	0.107	1.876	498.2	DDMAT	1.0
6	130	0	50	0.214	1.882	248.3	DDMAT	–
7	130	5.15	35	0.205	2.006	261.0	DDMAT	0.81

maximum capacity was used to perform aqueous suspension experiments at higher temperatures. Technical description of such apparatus can be found elsewhere.[18–20] Typically, these aqueous suspension RAFT polymerization experiments were performed with 10% volume of dispersed phase comparatively to the global volume (continuous + dispersed phase). Polyvinyl alcohol (PVA) was chosen as the suspending agent with a typical concentration in the aqueous phase of 0.1% (w/w). Influence of these parameters and stirring on reactor sampling were discussed elsewhere.[18]

Product Analysis by SEC/RI/MALLS

A similar procedure to that before described[18] was here considered: reaction samples were collected into previously cooled 20 mL glass vials and immediately poured in the refrigerator at $-14\,°C$ to stop the reaction. At least 24 hr later, samples were allowed to reach room temperature and aqueous/organic phases were separated. A small amount of the latter was diluted in THF and filtered to be injected in the SEC/RI/MALLS system. The SEC/RI/MALLS apparatus is composed of a Polymer Laboratories PL-GPC-50 integrated SEC system with differential refractometer working at 950 ± 30 nm, attached to a Wyatt Technology DAWN8$^+$ HELEOS 658 nm Multi Angle Laser Light Scattering (MALLS) detector. The polymer samples were fractionated by molecular size using a train of 3 GPC columns PL gel (300×7.5 mm) with nominal particle size 10 μm and pore type MIXEDB-LS, maintained at constant temperature of 30 °C and using THF as the eluent at a flow rate of 1 mL/min. Molecular weight averages, size distribution, z-average radius of gyration of the soluble polymer phase and monomer conversion were thus estimated.

In-Line and *Off-Line* FTIR-ATR Measurements

In-line monitoring of some polymerization runs was performed using an Attenuated Total Reflection (ATR) immersion probe coupled to a Fourier Transform Infra-Red (FTIR) instrument. Technical details of this apparatus were presented elsewhere.[19] Isolated polymer samples, after drying, were also *off-line* characterized by FTIR using the powdered products mixed with KBr and pressed into pellets.

PDB Concentration Measurement

Quantification of pendant double bonds (PDB) concentration in the RAFT S/DVB polymer networks was performed using chemical analysis (ICl titration), as recently reported for NMRP S/DVB networks.[18] Dynamics of PDB concentration was thus estimated, which is an important issue in the crosslinking process. This chemical method is an alternative/complementary approach to the FTIR analysis above described.

Results and Discussion

In Table 2 are presented some vibrational assignments correspondent to styrene and divinylbenzene monomers.[21–23] These features can also be observed in Figure 1(a) where the *off-line* IR spectra collected for S, DVB, RAFT polystyrene and RAFT S/DVB (95/5) are compared. Specially important for the study of the kinetics of crosslinking are the assignments correspondent to C=C bonds that can be identified at around 992, 1019, 1410, 1452 and 1630 cm^{-1}. For this chemical group, strong absorptions and well defined peaks (minimizing the interference with other structures) are observed at 992 and 1630 cm^{-1}. These characteristic frequencies are therefore good candidates to obtain information concerning the monomers carbon-carbon double bonds consumptions and also concerning the presence of pendant double bonds in the polymer/network. In fact, these bands are not present in isolated RAFT polystyrene and very low responses are observed in these regions for isolated RAFT S/DVB 95/5 networks due to the relative small amount of DVB used (see Figure 1(a)). Potentialities of *off-line* FTIR analysis of isolated products are further enhanced in Figure 1(b)-(d) where RAFT

Table 2.
Some IR vibrational assignments for styrene and divinylbenzene.

Wave number (cm^{-1})	Peak assignment
776	puckering ring CH out-of-plane bending
795	*m*-disubstituted ring bending
841	ring CH out-of-plane bending
909	ring CH out-of-plane bending and $=CH_2$ wag
992	C=C torsion and $= CH_2$ wag
1019	$= CH_2$ rock and ring CC stretching
1083	ring CC stretching
1202	CC stretching
1410	*p*-disubstituted phenyl ring stretching and $=CH_2$ scissoring
1452	$=CH_2$ scissoring and and ring C-H in-plane-bending
1494	ring C-H in-plane-bending
1510	*p*-disubstituted phenyl ring stretching
1600	CC stretching
1630	C=C stretching

materials with high pendant double bonds content (e.g. resulting from DVB homopolymerization) are considered. Clear qualitative information about the presence of the PDBs in the networks can be obtained and the comparison of products with different reaction times and/or resulting from different initial compositions can also be performed.

Attempts to use FTIR-ATR for *in-line* monitoring of aqueous suspension RAFT crosslinking polymerization are presented in Figures 2 and 3. In such cases (runs 4 and 5), polymerization conditions were designed in order to try maximize the C=C responses in the IR spectra (only DVB in run 4 and DVB/toluene 50/50 in run 5). Nevertheless, the strong influence of water cannot be eliminated (measurements were performed using water background) and only the lower frequencies regions of the spectra can be considered (see Figures 2–3). Despite these issues, good IR responses are observed in some useful bands (specially for run 4), namely with the 992 cm^{-1} peak. These data, collected by *in-line* FTIR-ATR, were used to estimate the double bonds conversion (p) using the 992 cm^{-1} for C=C quantification and, alternatively, the peaks at 909, 841 or 795 cm^{-1} as internal references. Similar results were obtained for run 4 considering different references but, for run 5, calculations were only possible with 909 cm^{-1} peak as internal reference. Important discrepancies between the estimated double bond conversion using *in-line* FTIR-ATR and gravimetric or SEC measurements are observed. Monomer conversion (p_M) measured by SEC and gravimetry was used to estimate the double bond conversion in the framework of the equal reactivity model: $p_M = p(2\text{-}p)$. Coating of the FTIR-ATR probe, catastrophic coagulation (see e.g.[24] and references therein) are some issues with possible negative impact on the *in-line* FTIR-ATR measurements. On other hand, the choice of a reference peak with constant intensity along the polymerization it is not a straightforward issue because shifts in peaks intensities are possible, as before reported with Raman spectroscopy. Further developments on the *in-line* FTIR-ATR measurements here reported are therefore needed in order to improve the reliability of the results.

Off-line FTIR analysis of isolated products provides important qualitative information concerning the presence of PDBs in the networks, as above described. In principle, quantification of PDBs in these materials is also possible using *off-line* FTIR if an appropriated internal reference in the spectra is found. This possibility was explored in the present work considering the main C=C absorptions at 992 and 1630 cm^{-1} and trying different internal references, namely those correspondent to 841, 909, 1494 and 1600 cm^{-1}. Some

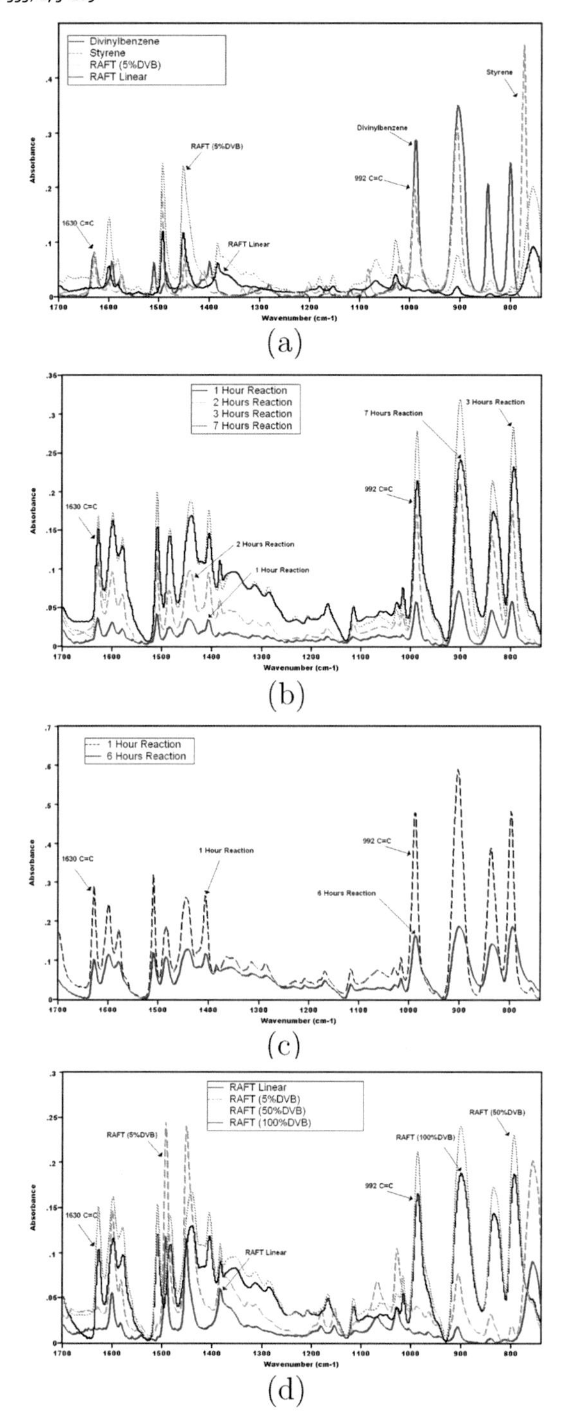

Figure 1.
Observed *off-line* FTIR spectra for styrene and divinylbenzene monomers and for isolated RAFT synthesized S/DVB polymers. (a) S, DVB, polystyrene (run 6) and poly(S/DVB) with 5% DVB (run 7). (b) DVB networks correspondent to different polymerization times in run 5. (c) DVB networks correspondent to different polymerization times in run 4. (d) Final samples correspondent to runs 4, 5, 6 and 7.

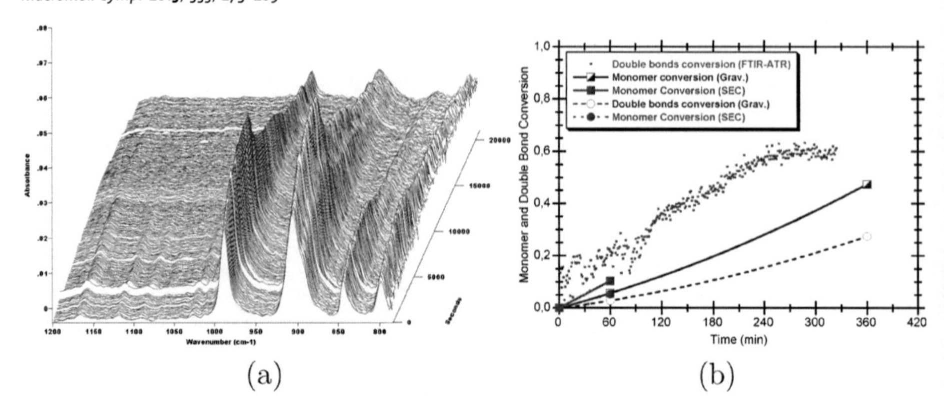

Figure 2.
(a) FTIR-ATR spectra observed during the *in-line* monitoring of the aqueous suspension RAFT polymerization of divinylbenzene at 70 °C with 100% DVB in the organic phase (run 4). (b) Estimated pendant double bond conversion using *in-line* FTIR-ATR monitoring and estimated monomer conversion using SEC (run 4).

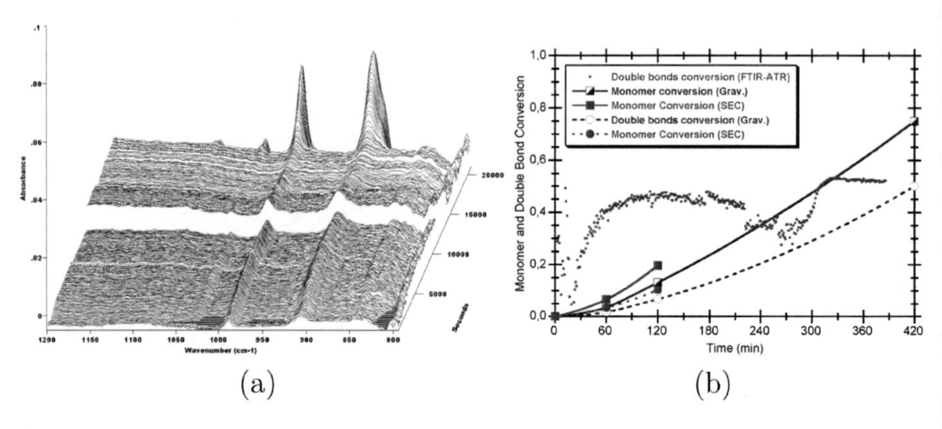

Figure 3.
(a) FTIR-ATR spectra observed during the *in-line* monitoring of the aqueous suspension RAFT polymerization of divinylbenzene at 70 °C with 50% DVB/50% Toluene in the organic phase (run 5). (b) Estimated pendant double bond conversion using *in-line* FTIR-ATR monitoring and estimated monomer conversion using SEC (run 5).

results of this analysis are presented in Figure 4 considering samples collected at different polymerization times during runs 4 and 5. Consistent profiles for the change of PDBs with reaction time where obtained when the 1630 cm^{-1} was considered to monitor this chemical group and the 1494 cm^{-1} was used as internal reference. In fact, for long enough polymerization time, the decrease of PDBs concentration with reaction time was before experimentally observed considering alternative measurement techniques[18,23] (see also Figure 7 (b) in this paper) and also predicted using kinetic modelling[25,26] with or without cyclization inclusion. With this analysis was also observed that the 1630 cm^{-1} peak is a better choice to monitor PDBs than the 992 cm^{-1} alternative peak. Better results would be probably obtained if the 1630 cm^{-1} was also considered in the *in-line* FTIR-ATR analysis above described but the presence of water interferes with measurements at this region.

One possibility to perform a quantitative analysis of the *off-line* FTIR data is through the comparison of normalized peak intensities for the isolated polymer/network and the constitutive monomer, as described by Equation (1):

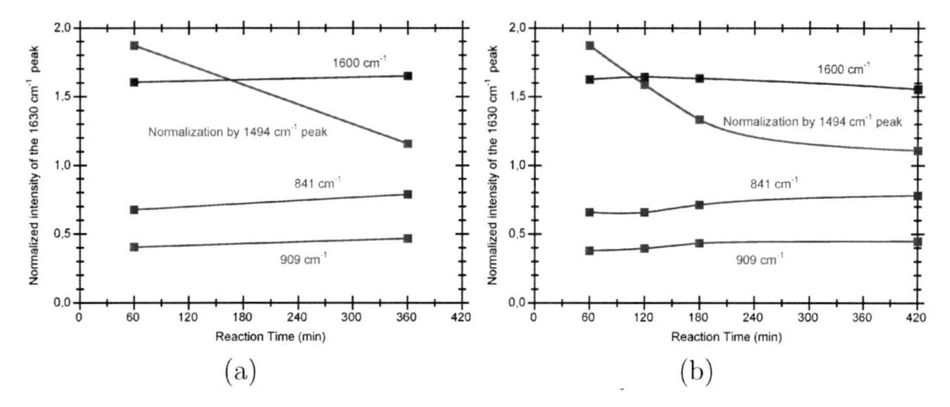

Figure 4.
(a) Normalized intensity of the FTIR 1630 cm^{-1} peak correspondent to C=C bonds considering different internal references (841, 909, 1494 and 1600 cm^{-1}). The values presented are correspondent to *off-line* FTIR analysis of samples collected at different reaction times in run 4. (b) Similar analysis described in (a) for run 5.

$$IPDB = \frac{\left(\frac{I_{1630}}{I_{1494}}\right)_{Network}}{\left(\frac{I_{1630}}{I_{1494}}\right)_{DVB}} \quad (1)$$

This *index of pendant double bonds* (*IPDB*) should represent the number of PDBs per aromatic ring in the network, comparatively to the DVB monomer, if the proportion between the IR responses of the chemical groups in the monomer and network is the same. Results for this index with the RAFT networks synthesized in this work are presented in Figure 5(a). However, as presented in that Figure, too high values are obtained for this index (approaching the unrealistic value *IPDB* $=1$ in the early stages of polymerization) which means that the proportion of the IR responses of these chemical groups is not the same in the polymer and in the monomer. Partially, these inconsistencies can be a consequence of the DVB commercial grade used in this work that includes two isomers (*m*-divinylbenzene and *p*-

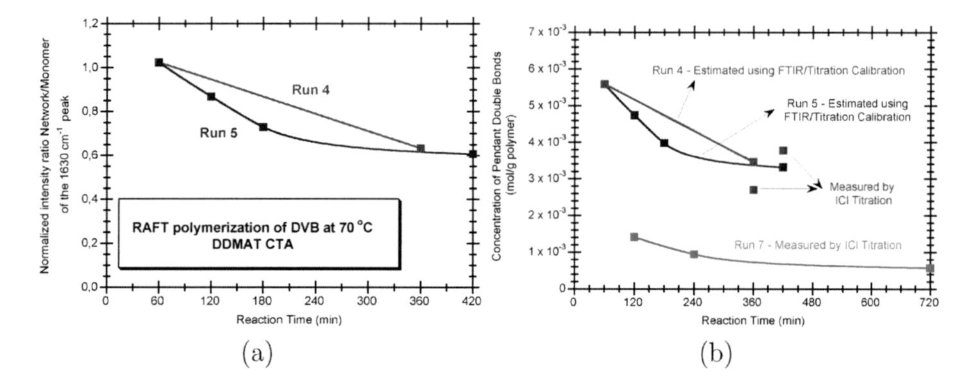

Figure 5.
(a) Observed ratios between network and DVB considering the normalized FTIR 1630 cm^{-1} peak (using the 1494 cm^{-1} peak as internal reference). Measurements are correspondent to samples with different polymerization time in runs 4 and 5 (see Equation 1). (b) Dynamics of the PDB concentration (mol/g polymer) measured by ICl titration (run 7) and considering also the calibration between the normalized FTIR 1630 cm^{-1} peak intensity (I_{1630}/I_{1494}) and PDB concentration. This calibration was obtained considering the final samples of runs 4, 5 and 7 which measurements using both methods (FTIR and ICl titration) were performed.

divinylbenzene) and also ethylvinylbenzene. The use of purified isomers is preferable to perform fundamental studies on this polymerization system.[23] Moreover, the consideration of model molecules (e.g. based on 4-isopropyl styrene) seems to be the better choice to obtain reliable results with IR-spectroscopy of PDBs.[23] This approach was not tested in the present work and the practical usefulness of the measurements performed was explored through the calibration between the normalized FTIR 1630 cm^{-1} peak intensity (I_{1630}/I_{1494}) and PDBs concentrations measured by ICl titration. To obtain this calibration, the final samples correspondent to runs 4, 5 and 7 were considered. Measurements using both methods (FTIR and ICl titration) were performed for these networks. The following relation between the normalized FTIR 1630 cm^{-1} peak intensity (I_{1630}/I_{1494}) and the concentration of PDBs in the network (expressed in mol/g polymer) was thus estimated:

$$\frac{I_{1630}}{I_{1494}} = 335 \times [PDB] \quad (2)$$

This calibration was used to estimate the PDB concentration for networks with a single (and more simple) FTIR measurement. These results are presented in Figure 5(b). Dynamics of PDBs concentration for runs 4 and 5 were obtained and are compared with the reaction time evolution of the same variable for run 7 which was fully measured using ICl titration. Note that a much smaller amount of DVB was used in run 7 and, under these circumstances, chemical titration in preferable to FTIR spectroscopy due to the low response of PDBs observed (when compared with that correspondent to other groups). Results presented in Figure 5(b) show the (small) effect of intramolecular cyclization that is probably caused when monomer is diluted from bulk to a 50% monomer/solvent solution (see comparison between runs 4 and 5). Strategy here presented should be extended to more diluted polymerization systems in order to assess the possible effect of cyclization on network formation, even with RAFT polymerization. Comparison of the incidence of these intramolecular mechanisms in FRP, NMRP[18,25,26] and RAFT of S/DVB is an expected result of this research line. Results here presented (and expected extensions) can be used to develop kinetic modelling studies including intramolecular cyclizations with RAFT S/DVB polymerization, as recently

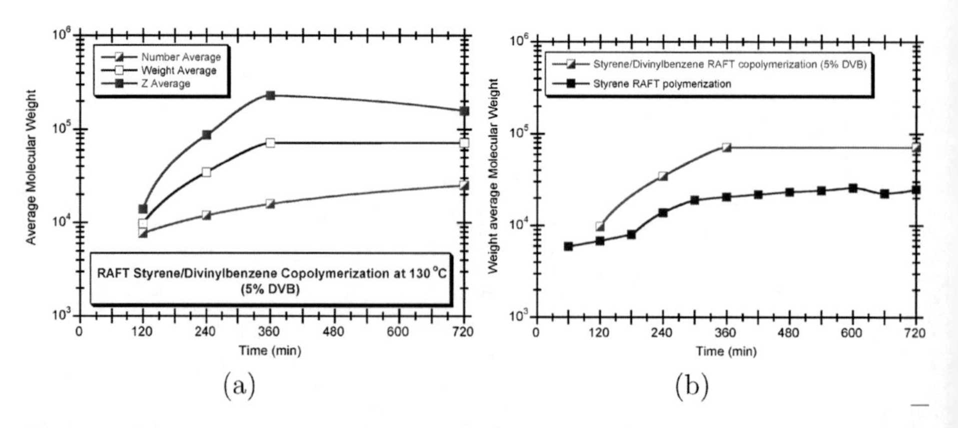

Figure 6.
(a) Measured dynamics of number-, weight- and z-average molecular weight ($\overline{M}_n$, $\overline{M}_w$ and $\overline{M}_z$) in aqueous suspension RAFT copolymerization of styrene/divinylbenzene (run 7). (b) Comparison of the observed dynamics for the weight-average molecular weight ($\overline{M}_w$) in aqueous suspension RAFT polymerization of styrene (run 6) and RAFT copolymerization of styrene/divinylbenzene (run 7). Both polymerizations at 130 °C and using DDMAT CTA.

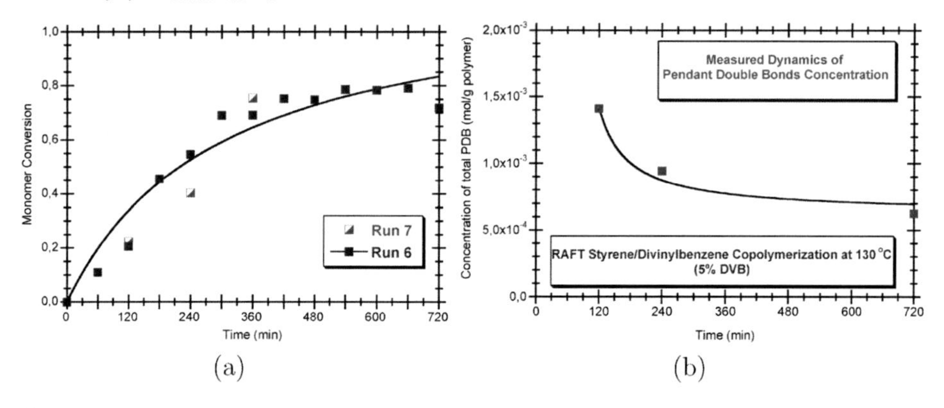

Figure 7.

(a) Measured dynamics of monomer conversion in aqueous suspension RAFT polymerization of styrene (run 6) and RAFT copolymerization of styrene/divinylbenzene (run 7). Both polymerizations at 130 °C and using DDMAT CTA. (b) Measured dynamics of pendant double bonds (PDB) concentration in RAFT copolymerization of styrene/divinylbenzene (run 7).

performed with NMRP of the same chemical system.[25,26]

SEC/RI/MALLS proved to be a valuable technique to obtain insights on the RAFT crosslinking polymerization. Figure 6 (a) shows the measured dynamics of $\overline{M}_n$, $\overline{M}_w$ and $\overline{M}_z$ for S/DVB (95/5) RAFT copolymerization at 130 °C (run 7). Gelation at around 360 min is identified with these conditions. Dissimilitudes between linear (S) and non-linear (S/DVB) RAFT polymerization are highlighted in Figure 6 (b) where the measured dynamics of $\overline{M}_w$ for runs 6 and 7 are compared. SEC estimated monomer conversion for these same two runs is showed in 7 (a) and similar kinetics for the linear and non-linear cases is observed under these circumstances. Measured time-evolution of PDBs concentration for S/DVB (95/5) RAFT copolymerization (run 7) is showed in 7 (b). These concentrations were measured through the ICl titration method above described.[18]

Impact of some operation conditions on the dynamics of S/DVB RAFT products formation was studied through the change of particular parameters, such as the kind of RAFT CTA considered. In Figure 8 (a) is

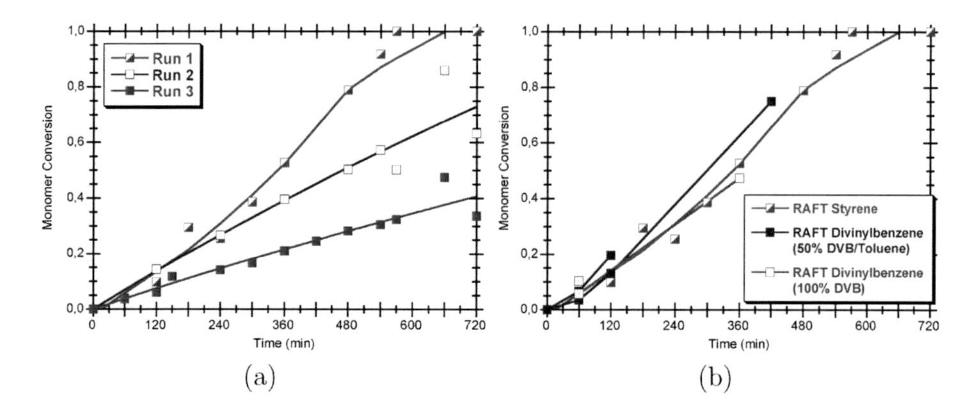

Figure 8.

(a) Measured dynamics of monomer conversion in aqueous suspension RAFT polymerization of styrene at 70 °C. Different RAFT CTAs were used in these polymerization runs (runs 1, 2 and 3). (b) Comparison of the measured dynamics of monomer conversion in aqueous suspension RAFT polymerization of styrene and DVB at 70 °C (runs 1, 4 and 5).

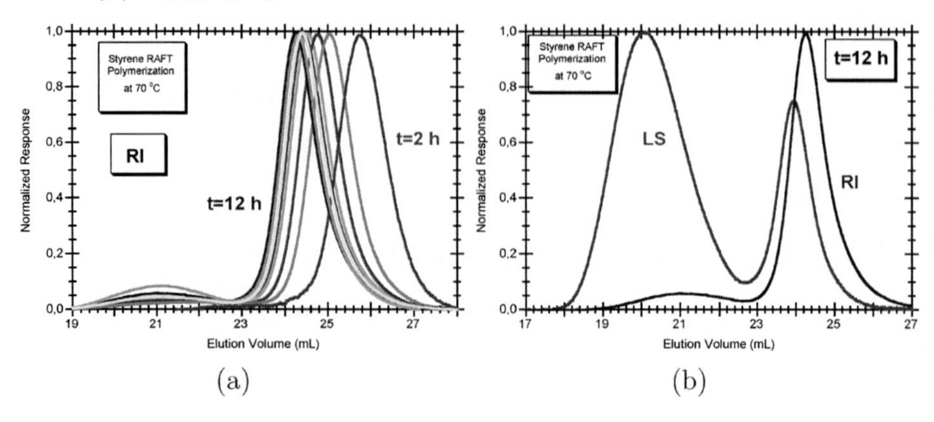

Figure 9.
(a) Normalized RI signal of polystyrene samples with different polymerization time. (b) Comparison of the normalized RI and MALLS signals of a polystyrene sample with polymerization time $t = 12$ h. In both cases the synthesis is correspondent to aqueous suspension RAFT polymerization of styrene at 70 °C using CDT (run 2).

presented the dynamics of monomer conversion for S RAFT polymerization at 70 °C but considering three different RAFT CTAs, namely DDMAT, CDT and TBTGA. Different conversion profiles are observed and the retardation effect associated to dithiobenzoates (TBTGA) becomes clear when compared with trithiocarbonates (DDMAT and CDT). Some scattering observed for the measured values of monomer conversion is possibly due to the non-ideal sampling of the reactor along polymerization time due to the spacial heterogeneity of the suspension (adhesion of organic phase to reactor walls was evident with some polymerization runs). This issue is an important shortcoming of the polymerization conditions here explored. The effect of the initial organic phase composition (e.g. S, DVB and diluent amounts) on the dynamics of global monomer conversion was briefly assessed through comparison of runs 1, 4 and 5 (see Figure 8(b)) where the same reaction temperature and RAFT CTA were considered (70 °C and DDMAT). Similar dynamics of monomer conversion where measured within the ascribed RAFT polymerization conditions.

Key features of the RAFT crosslinking mechanism become evident when the SEC traces of products with different polymerization times are compared. Observation of these SEC traces for linear (S) and non-linear (S/DVB) runs highlights central issues of network formation. In Figure 9 (a) are showed the SEC RI signals of polystyrene samples correspondent to different polymerization times. Besides the growth with reaction time of the main polymer population, a secondary sub-population with higher size and low concentration can be identified in these chromatograms. This issue is enhanced in Figure 9(b) where the LS and RI signals of the final sample correspondent to the same run (run 2) are compared. The huge size of the secondary population becomes evident through the respective LS signal. This means that, even in the linear case, non-ideal mechanisms can be involved in RAFT polymerization leading to the increase of the products dispersity. Slow fragmentation mechanisms in RAFT leading to bimodal distributions formation are a possible justification for these observations.[27] Differences/similarities between S and DVB RAFT polymerizations are illustrated in Figure 10(a) where the observed time evolution of $\overline{M}_w$ for runs 1, 4 and 5 are compared. For the DVB RAFT polymerization runs (4 and 5), the values presented are correspondent to the soluble phase which, in contrast to S polymerization, vanishes after some polymerization time. Coexistence of sol and gel is observed when

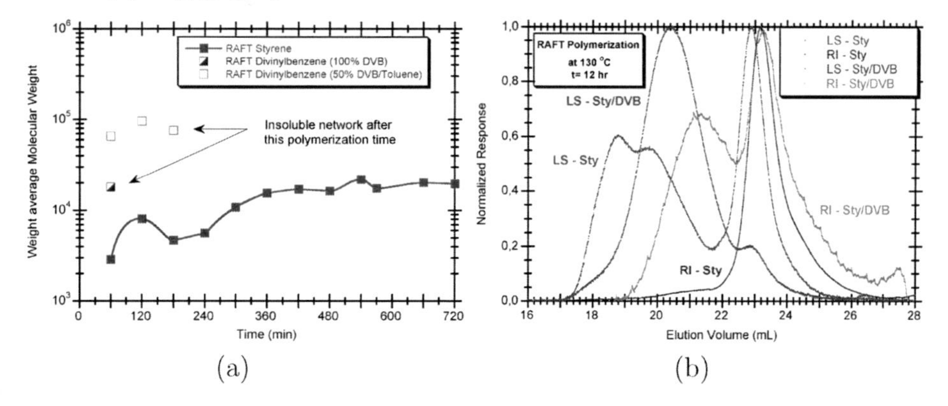

Figure 10.
(a) Measured dynamics of $\overline{M}_w$ for RAFT styrene polymerization (run 1) and RAFT DVB polymerization (runs 4 and 5). Polymerisations at 70 °C. (b) Normalized RI and LS signals observed for polystyrene (run 6) and soluble poly(S/DVB) network (run 7) synthesized by RAFT polymerization at 130 °C (both samples correspondent to $t = 12$ hr reaction time).

S/DVB RAFT polymerization is promoted, even with low amount of DVB (e.g. run 7). In spite of these important differences, some similarities in the SEC traces of RAFT polystyrene and soluble RAFT poly(S/DVB) can be identified, as show in Figure 10(b). RI signal shows a bimodal population for poly(S/DVB) and the very high molecular size of the secondary (crosslinked) set of chains is highlighted by the correspondent LS signal. Molecular size of the secondary population developed with S RAFT polymerization (as above discussed) is located in a region close to that observed with non-linear RAFT polymerization but the correspondent concentration is significantly lower comparatively to the latter.

Comparisons for the molecular architecture of the different RAFT products synthesized in this work are also illustrated in Figure 11. SEC observed structural dissimilitudes of RAFT poly(S) and soluble poly(DVB)s are showed in Figure 11(a). Even analysing only the soluble phase, the highly crosslinked nature of the latter systems show a clear contrast with the linear case (see also Figure 10(b) for comparision with poly(S/DVB) with low

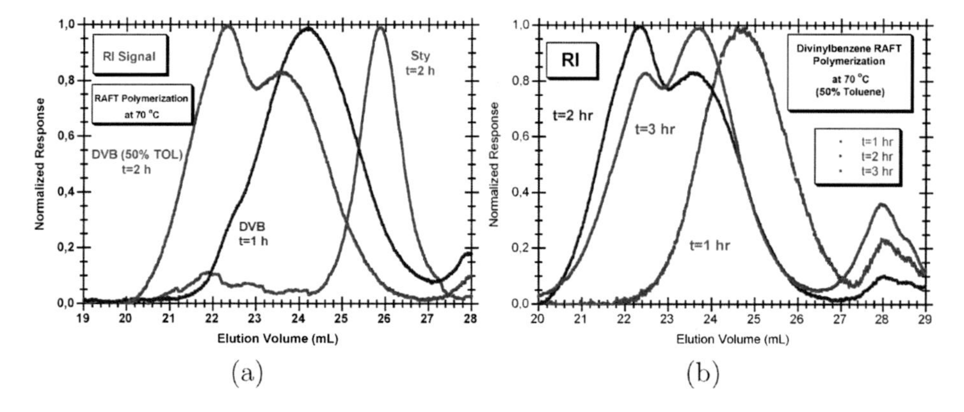

Figure 11.
(a) Comparison of the observed RI signal for RAFT synthesized polystyrene (run 1) and soluble poly(DVB) (runs 4 and 5). (b) Observed RI signal for RAFT synthesized soluble poly(DVB) samples. Different stages of the crosslinking process (reaction times 1, 2 and 3 hours) are compared.

amount of DVB). Figure 11(b) compares the SEC traces of poly(DVB) samples, prepared with 50% toluene, and collected at different reaction times. Dynamics of two different sub-populations can be observed and differences between samples correspondent to 2 and 3 hours of reaction time are a consequence of gelation. Results present in Figure 11 also show structural differences between RAFT poly(DVB) products synthesized using different dilutions (bulk and 50% toluene), evidencing the possible effect of cyclization due to intramolecular propagation. Additional developments on this important issue are expected within this research line, namely extending recent results on mathematical modelling of cyclization in NMRP of S/DVB[25,26] to RAFT polymerization of the same chemical system.

Conclusion

RAFT styrene/divinylbenzene copolymerization was experimentally studied in stirred batch reactor, running with aqueous suspension. Influence of a few operation parameters on the dynamics of polymerization was assessed: reaction temperature (70 to 130 °C), amount of DVB in the initial monomer mixture (0 to 100%), dilution in the organic phase and the kind of RAFT agent. Three different RAFT agents, commercially available, were used in these studies (two trithiocarbonates and one dithiobenzoate). Some polymerization runs were designed in order to promote gel formation and the dynamics of network production was thus assessed in a single pot reaction.

Isolated products were *off-line* characterized using FTIR and information concerning the presence of double bonds in the networks was obtained (important issue when materials functionalization is sought). PDBs in the networks were also quantified through chemical titration with ICl. This latter method is a complement to FTIR analysis when the concentration of these functional groups is low (e.g. due to a small amount of DVB in the initial mixture). *In-line* FTIR-ATR was also tried in order to obtain real-time information concerning these RAFT polymerizations. Nevertheless, these measurements are probably affected by probe coating/catastrophic coagulation[24] and further developments are needed in order to increase the reliability of this approach.

SEC products analysis with simultaneous RI and MALLS detection proved to be a very useful technique to obtain insights on the RAFT dynamics of crosslinking. Non-ideal RAFT mechanisms, leading to the formation of a secondary population with very high molecular size where identified, even with the polymerization of styrene. These effects are more pronounced when trithiocarbonates (e.g. DDMAT or CDT) are used and are probably related with slow fragmentation mechanisms leading to bimodal distributions.[27] Inspection of the molecular architecture of the soluble part provided by SEC shows that the RAFT crosslinking process, in a single pot reaction, also has a random nature and (as also showed with NMRP[18]) only a limited control on the network structure is possible replacing FRP by RAFT. Effect of dilution on gelation was also detected, which provides an evidence for the ocurrence of intramolecular reactions (cyclizations) that cannot be totally suppressed using RAFT. Production, without gelation, of high functionalized materials based of vinyl/multivinyl copolymerization is possibly the major advantage of the CRP techniques (comparatively to FRP) in this context. Combination of CRP techniques with a multi-step polymerization (e.g. synthesis of functionalized primary polymer chains followed by crosslinking) seems to be a more effective strategy to obtain networks with controlled molecular architecture.

Acknowledgments: Miguel Gonçalves acknowledges the financial support by FCT and FSE (Programa Operacional Potencial Humano/POPH) through the PhD scholarship with reference SFRH/BD/76587/2011. Virgínia Pinto also acknowledges the financial support by FCT through researcher scholarship in the framework of the project PTDC/EQU-EQU/098150/2008 (Ministry of Science and Technology of Portugal/Program COMPETE - QCA III/ and European Community/FEDER).

This research has also been supported through the Marie Curie Initial Training Network "Nanopoly" (Project: ITN-GA-2009-238700) and by the program SAESCTN - PIIC&DT/1/ 2011, Programa Operacional Regional do Norte (ON.2), contract NORTE-07-0124-FEDER-000014 (RL2_P3 Polymer Reaction Engineering).

[1] G. Moad, E. Rizzardo, S. H. Thang, *Aust. J. Chem.* **2012**, *65*, 985.
[2] W. A. Braunecker, K. Matyjaszewski, *Prog. Polym. Sci.* **2007**, *32*, 93.
[3] M. Cunningham, *Prog. Polym. Sci.* **2008**, *33*, 365.
[4] P. B. Zetterlund, Y. Kagawa, M. Okubo, *Chem. Rev.* **2008**, *108*, 3747.
[5] A. Butté, G. Storti, M. Morbidelli, *Macromolecules* **2000**, *33*, 3485.
[6] A. Butté, G. Storti, M. Morbidelli, *Macromolecules* **2001**, *34*, 5885.
[7] Q. Yu, Y. Zhu, Y. Ding, S. Zhu, *Macromol. Chem. Phys.* **2008**, *209*, 551.
[8] Q. Yu, S. Xu, H. Zhang, Y. Ding, S. Zhu, *Polymer* **2009**, *50*, 3488.
[9] T. C. Krasia, C. S. Patrickios, *Macromolecules* **2006**, *39*, 2467.
[10] M. Achilleos, T. Krasia-Christoforou, C. S. Patrickios, *Macromolecules* **2007**, *40*, 5575.
[11] F. Lu, Y. Luo, B. Li, *Ind. Eng. Chem. Res.* **2010**, *49*, 2206.
[12] Y. Liu, W. Tu, D. Cao, *Ind. Eng. Chem. Res.* **2010**, *49*, 2707.
[13] M. A. Pinto, R. Li, C. D. Immanuel, P. A. Lovell, J. F. Schork, *Ind. Eng. Chem. Res.* **2008**, *47*, 509.
[14] D. Wang, X. Li, W.-J. Wang, X. Gong, B.-G. Li, S. Zhu, *Macromolecules* **2012**, *45*, 28.
[15] D. Wang, W.-J. Wang, B.-G. Li, S. Zhu, *AIChE J.* **2012**, *58*, DOI: 10.1002/aic.13890
[16] M. Roa-Luna, G. Jaramillo-Soto, P. V. Castañeda-Flores, E. Vivaldo-Lima, *Chem. Eng. Technol.* **2010**, *33*, 1893.
[17] G. Jaramillo-Soto, E. Vivaldo-Lima, *Aust. J. Chem.* **2012**, *65*, 1177.
[18] M. A. D. Gonçalves, V. D. Pinto, R. C. S. Dias, M. R. P. F. N. Costa, L. G. Aguiar, R. Giudici, *Macromol. React. Eng.* **2013**, *7*, 155.
[19] M. A. D. Gonçalves, V. D. Pinto, R. C. S. Dias, M. R. P. F. N. Costa, *Macromol. Symp.* **2010**, *296*, 210.
[20] M. A. D. Gonçalves, R. C. S. Dias, M. R. P. F. N. Costa, *Macromol. Symp.* **2007**, *259*, 124.
[21] N. B. Colthup, L. H. Daly, S. E. Wiberley, "*Introduction to Infrared and Raman Spectroscopy*", Third Edition, Academic Press, Boston **1990**.
[22] C. H. Choi, M. Kertesz, *J. Phys. Chem. A* **1997**, *101*, 3823.
[23] M. Hecker, "*Experimentelle Untersuchungen und Monte-Carlo-Simulation netzwerkbildender Copolymerisationen.* " Fortschritte der Polymerisationstechnik II. (H. U. Moritz Ed.) Wissenschaft & Technik Verlag Berlin, ISBN 3-89685-353-8, **2000**.
[24] S. Salehpour, M. A. Dubé, *Macromol. React. Eng.* **2012**, *6*, 85.
[25] L. G. Aguiar, *Synthesis of styrene-divinylbenzene copolymers using free radical and nitroxide mediated polymerization: experiments and mathematical modelling.* PhD thesis (in Portuguese), University of São Paulo, Brasil **2013**.
[26] L. G. Aguiar, M. A. D. Gonçalves, V. D. Pinto, R. C. S. Dias, M. R. P. F. N. Costa, R. Giudici, *Macromol. React. Eng.* **2013**, DOI: 10.1002/mren.201300105.
[27] I. Zapata-González, E. Saldívar-Guerra, J. Ortiz-Cisneros, *Macromol. Theory Simul.* **2011**, *20*, 370.

DOI: 10.1002/masy.201300056

Models in the Polymer Industry: What Present? What Future?

H. Vale, A. Daiss, O. Naeem, L. Šeda, K. Becker, K.-D. Hungenberg*

Summary: An industrial perspective on the present and future role of kinetic and process models in the polymer industry is presented. A number of selected examples illustrate the current value of model-based approaches in supporting and accelerating tasks like process optimization, product development and design of polymerization reactors. In addition, future trends for the application of models in the polymer industry are discussed and areas where further improvement is thought to be needed are carefully underlined.

Keywords: modeling; polymerization; processing

From the Present to the Future

Over the years, model-based approaches have conquered a solid position among the set of tools available to support and promote the development of the polymer industry. This is obviously true for the polymerization step itself – where the methods of polymer reaction engineering (PRE) fit like a glove –, but also for downstream unit operations, such as devolatilization, compounding, extrusion, and so forth.

This state of affairs cannot be disconnected from the fact that the polymer industry has achieved a certain degree of maturity. Most current processes could indeed be termed 'classical', in the sense that they are known and/or operated for quite some time now. And fortunately, for many of such classical processes, the dominant mechanisms are sufficiently well established (often with the help of models!). For this subset of processes, the success of model-based approaches is thus highly probable, as the various examples given in Section 2 demonstrate.

Polymer Processing & Engineering, BASF SE, Carl-Bosch-Strasse, 67056 Ludwigshafen, Germany
E-mail: hugo.vale@basf.com

However, the polymer industry is changing at a fast pace. In order to escape the commoditization trend, more and more companies are focusing their business on new polymeric materials of high added-value. For such companies, the main goal is to imagine and bring such innovative products to the market in record time. As part of this strategy, new processes and production concepts that enable cost-efficient, flexible and decentralized production at small capacities must also be engineered. Meanwhile, for economic and practical reasons, 'classical' polymers will remain dominant in terms of world demand. Numerous manufacturers will therefore operate in these traditional markets. There, however, the pressure for cost reduction is high, and companies that wish to last must ensure that their production processes are run at maximum efficiency.

This view is perhaps simplistic, but it has the merit of emphasizing the increasing importance that these three fields (cf. Figure 1) will have in the near future. In order to align with this trend, modeling methods must evolve as well, and in diverse ways. For sure, we will need to deliver even *better* models of the kind that we are used to develop, but probably also models of a *different nature*. How these will actually look like and how to build them it is hard to

	Optimization of 'classical' polymer processes	Development of new polymeric materials	Development of new processes (for specialties)
Expected role of models	Increase productivity	Reduce time to market	Accelerate development, reduce costs and technical risks
Challenges	Room for improvement can be small or hard/expensive to implement	Develop basic knowledge and tools in a short period. Establish link between structure and properties.	Non-standard reactor concepts: strong coupling between reaction and flow/transport processes.
Product complexity	Low to moderate (in comparison to new products)	High: sophisticated chemical composition, architecture, morphology, etc.	Moderate to high
Background knowledge	Often significant knowledge from literature or own experience	More often than not insufficient quantitative information	More often than not insufficient quantitative information
Expected modeling detail/accuracy	High	As possible!	Normal (focus on physical coherence)

Figure 1.
The three major fields where model-based approaches are expected to support the evolution of the polymer industry. Product complexity and modeling detail/accuracy should be understood in relative terms; for instance, certain 'classical' products can actually be quite complex and/or hard to model, but probably newer products of the same family would be even more complex.

say, but some ideas are outlined in Section 3.

Some Examples from the Present

The examples described here are taken from the authors' own experience and illustrate the value of model-based approaches in understanding and improving some fairly intricate polymerization processes. These examples clearly do not cover the whole spectrum of processes/problems, but we do believe that they give a nice flavor of what the possibilities are.

Making Recipe Scale up Less of a Problem

After successful recipe development at the lab or pilot scale, the next key step is to transfer the recipe to industrial scale. However, when scaling up polymerizations carried out in stirred-tank reactors, heat transfer limitations can arise and lead to significant differences between the desired (recipe) and real temperature profiles.

The occurrence of such problems in advanced phases of the project can be avoided if the recipe is developed hand in hand with a kinetic and heat transfer model. This is exactly what we did during the development of a new recipe for a polymer produced batchwise. Process limitations (maximum and minimum service side temperature) were considered from the very beginning of the project in order to avoid the temperature in the reactor jacket crossing the physical and/or safety limits (cf. Figure 2). In addition, this kind of approach was also used to compensate for the particular dynamics of the industrial temperature control loop. The final result was a scale up without surprises or variations in final product properties.

Avoiding Demixing with a Little Help from Kinetics

Two things are well-known to a polymer chemist:

- Polymers with different composition are usually immiscible and give a turbid two-phase system.
- When synthesizing a copolymer from various co-monomers in a batch process, these monomers usually do not react equally, but show a shift in copolymer composition.

To avoid such a shift, a common approach is to feed the comonomers at a constant rate over a certain time, assuming

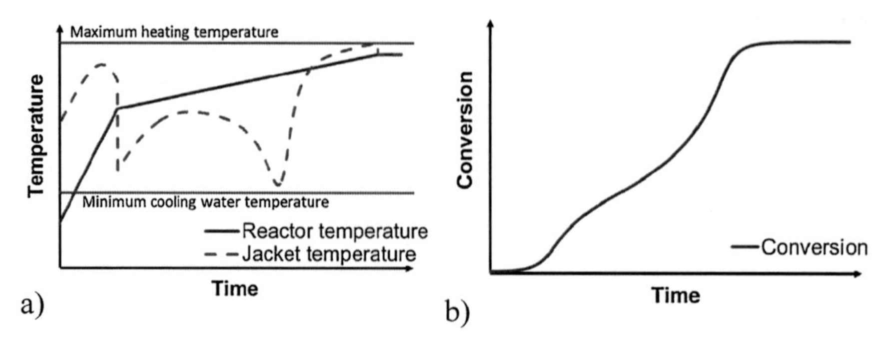

Figure 2.
Computed (a) jacket and reactor temperature trajectories and (b) monomer conversion for a new polymer recipe developed to take into account the specific process constraints of the industrial plant.

that they will be incorporated according to the feed composition. However, if the reactivity of the monomers is too different, this approach will not help.

This second example is precisely about a multi-comonomer product that was synthesized in such a manner and showed turbidity and demixing effects. A careful model-based analysis of the copolymerization kinetics of this system showed that with the classical feed strategy – constant feed rates for all monomers – monomer X was consumed rather fast during the last part of the reaction, after the end of monomer dosing. As a result, a copolymer without X units was formed at the end; obviously this polymer fraction was immiscible with the polymer formed during the dosing period.

By small changes in the dosing profile (cf. Figure 3) – less monomer X in the beginning and an equivalent extra amount at the end – a more homogenous incorporation of monomer X was achieved, and demixing no longer occurred.

Getting the Most Out of Polymerization Processes: Online Optimization

Online optimization methods such as model-based predictive control (MPC) use a process model to compute and track optimal trajectories efficiently in real time. MPC (cf. Figure 4) exploits process variations (seasonal, educts quality, fouling, etc.) and forces operation close to the 'real' process constraints by coordinating control moves of manipulated variables (feed profiles and jacket inlet temperature and flowrate) resulting in batch time reduction.[1,2]

This strategy was recently applied to a polymerization process in order to meet the following two objectives: improve

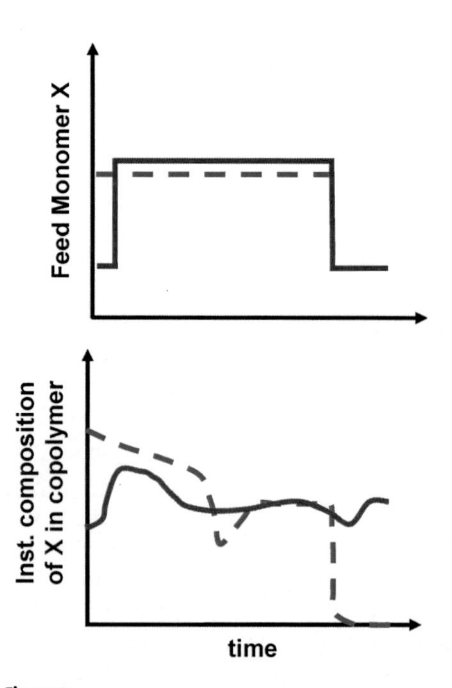

Figure 3.
A small modification of the dosing profile of monomer X led to a more homogeneous incorporation of this comonomer and demixing problems were solved. The dashed and continuous lines denote, respectively, the old and new conditions.

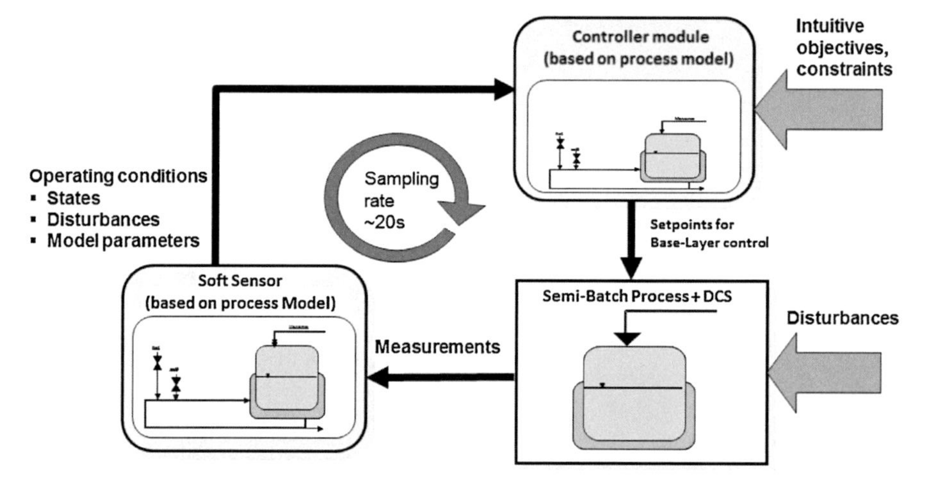

Figure 4.
Model predictive control of a semi-batch polymerization process.

temperature control and reduce batch time. Reaction temperature control is an implicit product quality measure. Furthermore, tighter temperature control at set-point allows operating the process close to the 'real' constraints rather than having conservative constraints and set-points with large bounds.

The process model consisted of a rigorous kinetic scheme coupled with a reactor periphery model. The model was reduced (structure, stiffness) to ensure real time computation before its implementation for online optimization and control. Figure 5a shows the reactor temperature profiles for a number of batches carried out with conventional PID control. As shown, the reactor temperature remains within ± 2 °C for the initial 30% of batch time, while afterwards the reactor temperature varies between ± 4 °C.

Figure 5b shows the reactor temperature profiles for batches performed with MPC. It is obvious that the reactor temperature control improved significantly. The temperature was maintained within a narrow bound of ± 1 °C around the set-point for the first 25% of batch time, and bounded within ± 0.5 °C for the rest of the batch time. Figure 6 shows the dosing profiles of different batches carried out with MPC compared to the standard dosing profile obtained with PID control. As noticeable, MPC varied the dosing profiles based on the actual conditions of the batch (for every batch), to achieve the

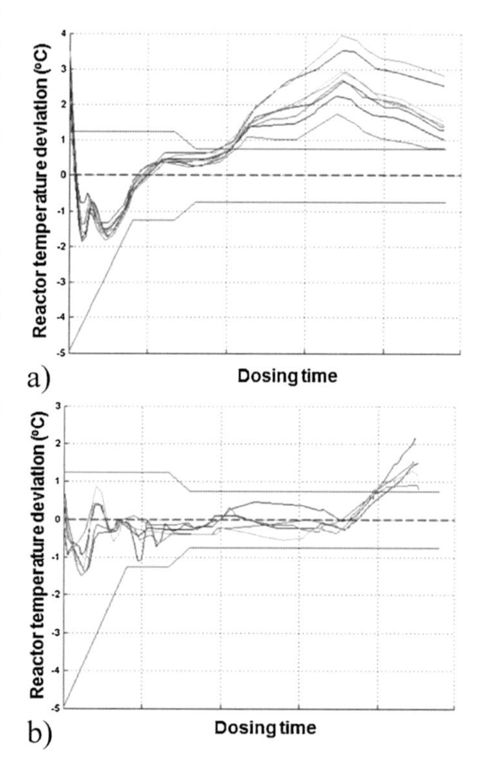

Figure 5.
Comparison of reactor temperature profiles obtained with (a) conventional control and (b) MPC.

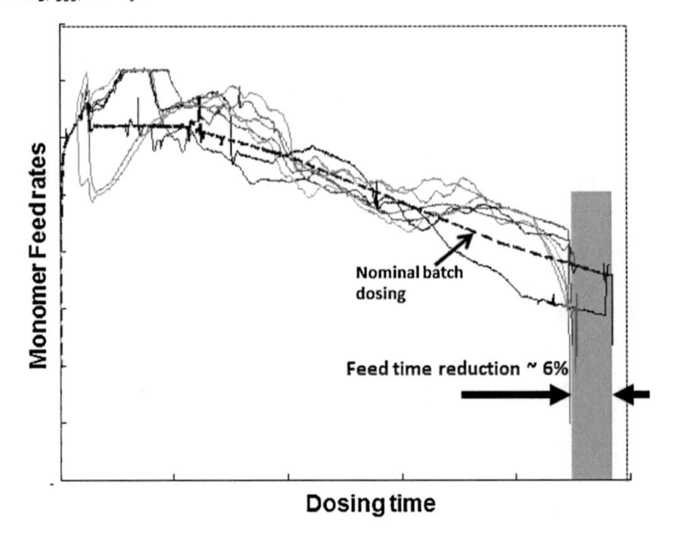

Figure 6.
Comparison of monomer dosing profiles obtained with conventional control and with MPC.

optimization targets. This model-based method allowed a dosing time reduction of approximately 6% for an already offline optimized process.

Reactor Analysis and Design with CFD

Computational fluid dynamics (CFD) is a powerful tool to characterize the fluid dynamics, heat and mass transfer in chemical reactors. In order to determine the influence of reactor geometry and operating conditions on polymerization processes, two approaches are possible:

1) CFD-based determination of correlations for a simplified analysis of the polymerization reaction.
2) Full coupling of fluid dynamics, heat and mass transfer and polymerization chemistry within CFD.

The second approach is computationally far more expensive. In the simplest case of a homogenous polymerization process, it requires the calculation of the population balances for the molecular weight distributions (MWD) of the polymers in each control volume of the computational grid. To make this task tractable, typically only the first $n+1$ statistical moments of the MWD are determined. This means that for each polymeric component $n+1$ additional scalar equations need to be calculated. Normally, it is sufficient to take $n=2$.

We use both these approaches on a regular basis to analyze and design new polymerization reactors and improve existing equipment. As an example, Figure 7 shows the temperature distribution, conversion and polydispersity index for a fast free-radical solution polymerization process carried out in a pilot plant static mixer heat exchanger reactor. The calculation was done by approach 2. As can be seen, the reactor shows mostly a one-dimensional variation of the three quantities. However, locally, in stagnant zones behind the baffles of the mixer elements, deviations from the ideal 1D scenario can be observed.

Figure 8 shows the conversion and the deviation of the average reactor temperature from the wall temperature as a function of the normalized residence time calculated with both approaches (the normalized residence is calculated based on the axial position in the reactor and the cross-sectionally averaged axial flow velocity profile). For the full coupling, a slightly

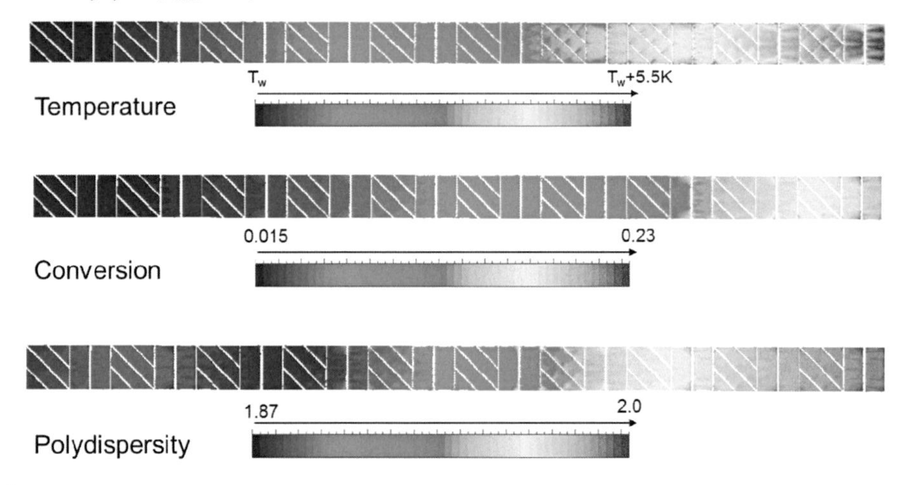

Figure 7.
Temperature, conversion and polydispersity index in an axial cross section through the first part of a pilot plant static mixer heat exchanger reactor (CSE/XR) for a fast free-radical solution homopolymerization process.

lower conversion is observed and hence also the temperature in the reactor is somewhat lower. The explanation is that in the middle of the flow the temperature is the highest but the residence time is significantly smaller than at the wall, while at the wall the residence is long but the temperature is low. In addition to a high accuracy, these methods allow to capture non-idealities like fouling and to describe systems with strongly varying material properties.

Preparing for the Future

Whether the goal is optimizing classical process or creating new polymeric materials and associated processes, model development remains an iterative, modular,

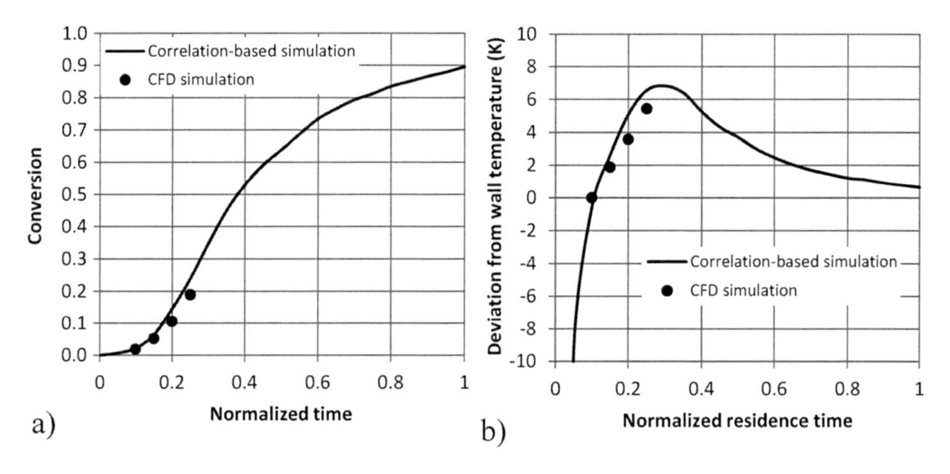

Figure 8.
Conversion to polymer (a) and temperature deviation from the wall temperature (b) in a static mixer heat exchanger reactor as a function of the normalized residence time. The calculations are performed with (i) a 1D reactor model using heat and mass transfer correlations calculated by CFD (solid lines) and (ii) a full coupling of CFD and polymerization kinetics (symbols).

multi-scale activity. The following general principle should therefore apply:

$$\left\{\begin{matrix}\text{modeling}\\\text{success}\end{matrix}\right\} \propto \left[\left\{\begin{matrix}\text{fundamental}\\\text{knowledge}\end{matrix}\right\}^{-1} + \left\{\begin{matrix}\text{phenomenologic}\\\text{knowledge}\end{matrix}\right\}^{-1} + \left\{\begin{matrix}\text{numerical methods}\\\text{and software}\end{matrix}\right\}^{-1} + \left\{\begin{matrix}\text{reliable}\\\text{data}\end{matrix}\right\}^{-1} + \left\{\begin{matrix}\text{problem pertinence}\\\text{and perseverance}\end{matrix}\right\}^{-1}\right]^{-1}$$

Just like in a resistances-in-series circuit, the success of a model is essentially limited by the value of the highest resistances. The relative values of these five terms are system (and time) dependent, which makes it quite hard to do a truly universal analysis. Despite that, we do think that some fairly general patterns emerge; especially if we take into account the priorities mentioned in the introductory section (cf. Figure 1). These key areas are pointed out and discussed below. It should be clear to the reader that the list is intentionally restricted; other important topics have been considered in a previous publication.[3]

Fundamental Knowledge

Strengthening our fundamental (microscopic and/or molecular) knowledge is, in our opinion, the most important task. The reason for this is that insufficient basic knowledge is already a stumbling block for the modeling of many polymerization systems. And, of course, this problem can only become more prominent for the polymeric materials of tomorrow, given the natural tendency toward more elaborate chemical composition, molecular architecture, morphology, etc.

Simply and pragmatically put, the problem is that reaction and process models contain too many (intercorrelated) parameters which can only be estimated by fitting because we ignore how they are linked to the actual properties of the system. In the best case, this lack of knowledge just limits the range of applicability of the model; in the worst case, it can make a model useless (no predictive power).

Some good examples of this difficulty can be found in the classical field of free-radical kinetics. The most well-known is perhaps the dependence of k_t on reaction conditions; although a significant number of theories and equations are available to correlate observed behavior,[4,5] microscopic understanding and predictability are still limited.[6,7] The evaluation of reactivity ratios in aqueous-soluble systems is another source of difficulty, as these can strongly depend on concentration, pH, ionic strength, etc. in ways that are not yet clear.[8,9] Another good example has to do with redox initiation,[10] a method commonly employed in industry; despite some interesting studies,[11–14] the exact initiation mechanism and kinetics of redox systems of practical interest remain essentially a mystery.

As one could expect, the situation is even worse in heterogeneous processes. For instance, in the context of polyolefin reactor modeling, Soares et al.[15] wrote: "until a more quantitative understanding of polymerization kinetics steps, particle fragmentation and growth (...) is achieved, the integration of all these phenomena in a single reactor model may be viewed as a stimulating academic exercise but will hardly lead to mathematical models with improved predictive capabilities." In emulsion (co)polymerization too the challenges are significant, especially in the fields of particle nucleation (primary and secondary), particle aggregation and radical entry/exit.[16–21] Of course, the list could go on.

One must however recognize that there are good reasons for things being the way they are – the complexity of the phenomena at hand is certainly a respectable argument. Finding the right answers will therefore not happen overnight, which is why these activities need to be strongly encouraged and continuously supported. Such long term dedication undeniably pays back, as demonstrated by the significant advances made over the last decade on topics like: k_p dependence of aqueous-soluble and ionisable monomers,[22–25] chain-length-dependence of termination rate coefficients[26,27] and effect of stabilizers in radical entry/exit rates[28,29] in emulsion latexes.

Finally, we should also add that most, if not all, these unsolved issues lie at the (lower-end) border of classical PRE competencies. Close collaboration with researchers from more fundamental disciplines such as Physical Chemistry, Colloid Chemistry, Interfacial Chemistry, Computational Chemistry, etc. is a therefore a must. Several examples of this collaborative work are already in place, hopefully others will follow.

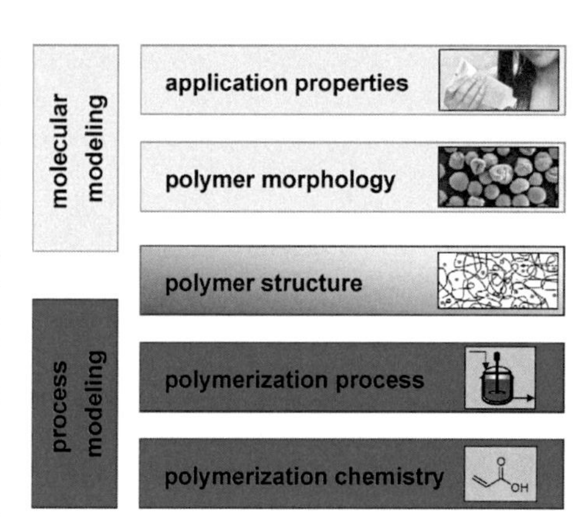

Figure 9.
The different links of the chain that connects polymer chemistry to application properties.

Methods and Software: Structure-Properties Relationship

Properties play a crucial role in the field of polymeric materials. On one hand, *application* properties are the reason why synthetic polymers are produced in the first place. On the other hand, properties like solubility, viscoelasticity or adhesion strongly influence the production process of polymeric substances.

Establishing solid relationships between the structure and composition of polymeric materials and their macroscopic physicochemical properties, and ultimately their application properties (cf. Figure 9), is therefore crucial for efficient product/process design. The interest spans over all fields of application (e.g, water treatment, energy storage, cosmetics, coatings, pharma, etc.) and types of physicochemical properties (e.g., electrical, optical, interfacial, rheological, transport, etc.).

Thanks to molecular modeling methods, the progress in this area has been considerable; many material properties can now be estimated when the models are adequately parameterized.[30] However, the challenges are still numerous. For instance, predicting the 'simple' viscoelastic properties of a pure linear homopolymer is now within reach,[31] but remains a formidable task given the wide time and length scales involved. As one could expect, the situation is much less favorable in the area of nanocomposite materials, given the complex interactions between the polymer matrix and the particle surface.[32]

Methods and Software: Product Development

First-principles models offer unique possibilities for product/process development and optimization. However, as discussed above, these models rely on knowledge and kinetic data that is not always available, especially for new polymeric systems. In addition, mechanistic models only deliver molecular properties, which may be hard to relate to the desired application properties (cf. Section 3.2). What should one do in those cases?

Should one start a parallel research project for model development? Maybe, but only if time and expected benefits are worth the investment. In certain cases, depending on product complexity and previous know-how, we might get to the conclusion that, by reasoning and trial and error, a satisfactory product can be made long before we can finish our nice first-principles model…

What else then? Well, for these special but important cases, perhaps fully mechanistic models are not the best alternative. One possibility could be to use some sort of *adjustable* 'hybrid' (or gray box) model. This designation refers to models that combine a first principles part and an empirical part (e.g., artificial neural network); the latter usually represents a portion of the process that is incompletely elucidated or that is difficult to describe. These methods date from the mid-90s[33] and there are by now several reports of successful application to polymerization processes.[34,35] So far, they have been employed mostly for process modeling and control, but we think that there is an opportunity for product development as well. The problem is that such techniques are currently not user-friendly enough for daily usage in industry. For that, one would require a specific software tool offering the possibility to combine, at will, mechanistic (e.g., Predici, Aspen Polymers) and empirical/statistical (e.g., ProSensus) strategies. Of course, such approach would not put an end to all modeling problems, but the idea seems attractive.

Methods and Software: Reactive CFD

Particularly in the area of polymer specialties, there is a growing demand for processes that are both cost efficient and flexible at small production capacities. From a technical point of view, this usually translates into some form of 'process intensification'. The result is often a technology with a strong integration of process functions (or unit operations) that would otherwise be physically separated. As a simple example, we can cite twin-screw extruders, by means of which one can in principle carry out such diverse operations like mixing, heating/cooling, reaction, devolatilization, and so forth. The 'by-product' of these integrated approaches is, of course, a strong coupling between reaction and flow/transport processes. Because of this interdependence, the design of such processes is heavily dependent on reactive CFD simulations (cf. Section 2.4, approach 2). Fortunately, the application of CFD methods to polymerization systems is advancing rapidly, making it possible to tackle rather complex problems.[36–39] Nonetheless, there are still several areas where improvements would be more than welcome. Specific examples are heterogeneous polymerization systems (e.g., emulsion polymerization), two-phase systems (e.g., suspension polymerization, emulsification processes, physical deodorization and drop polymerization) and flows in complex geometries involving multiple scales (e.g., resin transfer molding). A particular problem of the modeling of polymerization processes by means of CFD is the occurrence of an extremely thin concentration boundary layer due to the usually high Schmidt numbers (up to several millions).

Problem Pertinence and Perseverance

This last topic is as much about the future as it is about the present; it is actually more of a heads up. Hopefully, it is not a secret to anyone: there are many industrially relevant problems for which best solution is *not* a model (not even a data-driven one). Model-based approaches are just tools (not ends per se) and, as such, must always be evaluated in comparison to other tools, not the least of which is pure empiricism (cf. Section 3.3). Forgetting this simple principle, can lead to some frustration and ultimately to general reluctance in applying models.

Unfortunately, it is not always easy to identify which problems are suited for model-based approaches. One important and necessary criterion, though, is the expected value and usage of the acquired knowledge, both in the present and in the future. Can a model really teach us something that we would not be able to learn otherwise (e.g., just by looking at the data needed anyway to build the model)? Is this knowledge usable, directly or indirectly, in future projects? If the answers to both questions are yes, then developing a model might indeed be the right choice.

Finally, perseverance should not be forgotten. This is clearly not an exclusive characteristic of model developers, but given the time frames usually involved in going from the original idea to its achievement, it is certainly a necessary condition!

Conclusion

The rapid evolution of the polymer industry is creating or reinforcing needs in three main areas: development of innovative polymeric materials, design of new processes for specialties, and optimization of 'classical' processes. Model-based approaches are ideally placed to support and accelerate such activities. However, for this to happen in practice, our models need to progress as well. In particular, their predictive character must increase considerably and, at the same time, the time required for their development must decrease. Achieving such opposing goals is not simple, but can certainly be made easier if the efforts of industry and academia are combined in a fruitful and sustainable way. The opportunities for improvement are numerous, be it in the fields of fundamental knowledge, methods and software or measurement techniques.

Acknowledgements: The authors are grateful to the numerous colleagues that worked with us on the projects mentioned here. The authors thank BASF SE for permission and support to publish this work.

[1] H. Seki, M. Ogawa, S. Ooyama, K. Akamatsu, M. Ohshima, W. Yang, *Control Engineering Practice* **2001**, *9*(8), 819.
[2] N. Hvala, F. Aller, T. Miteva, D. Kukanja, *Computers & Chemical Engineering* **2011**, 35(10), 2066.
[3] K. D. Hungenberg, in: Papers of the 8th International Workshop on Polymer Reaction Engineering, Hamburg. **2004**.
[4] E. Vivaldolima, A. E. Hamielec, P. E. Wood, *Polym. Reac. Eng.* **1994**, 2(1-2), 17.
[5] D. S. Achilias, *Macromol. Theory Simul.* **2007**, *16*(4), 319.
[6] G. A. Oneil, M. B. Wisnudel, J. M. Torkelson, *Macromolecules* **1996**, *29*(23), 7477.
[7] M. Buback, M. Egorov, R. G. Gilbert, V. Kaminsky, O. F. Olaj, G. T. Russell, P. Vana, G. Zifferer, *Macromol. Chem. Phys.* **2002**, 203.
[8] I. Rintoul, C. Wandrey, *Polymer* **2005**, *46*(13), 4525.
[9] A. Paril, A. Giz, H. Catalgil-Giz, *J. Appl. Polym. Sci.* **2013**, *127*(5), 3530.
[10] A. S. Sarac, *Prog. Polym. Sci.* **1999**, *24*(8), 1149.
[11] M. Schneider, C. Graillat, S. Boutti, T. F. McKenna, *Polymer Bulletin* **2001**, *47*(3-4), 269.
[12] S. Boutti, R. D. Zafra, C. Graillat, T. F. McKenna, *Macromol. Chem. Phys.* **2005**, *206*(14), 1355.
[13] N. Kohut-Svelko, R. Pirri, J. M. Asua, J. R. Leiza, *Journal of Polymer Science Part A-Polymer Chemistry* **2009**, *47*(11), 2917.
[14] Z. Kechagia, O. Kammona, P. Pladis, A. H. Alexopoulos, C. Kiparissides, *Macromol. React. Eng.* **2011**, 5(9-10), 479.
[15] J. B. P. Soares, T. F. McKenna, C. P. Cheng, in: *Polymer reaction engineering*, J. M. Asua, (Ed.), Blackwell Publishing, Oxford **2007**, pp. 29–117.
[16] A. M. van Herk, A. L. German, *Macromol. Theory Simul.* **1998**, *7*(6), 557.
[17] H. M. Vale, T. F. McKenna, *Prog. Polym. Sci.* **2005**, *30*.
[18] C. S. Chern, *Prog. Polym. Sci.* **2006**, *31*(5), 443.
[19] A. M. van Herk, *Macromol. Symp.* **2009**, *275ΓÇô276* (1), 120.
[20] K. Tauer, H. Hernandez, S. Kozempel, O. Lazareva, P. Nazaran, *Colloid Polym. Sci.* **2008**, *286*(5), 499.
[21] H. F. Hernandez, K. Tauer, *Macromol. React. Eng.* **2009**, *3*(7), 375.
[22] M. Stach, I. Lacik, P. Kasak, D. Chorvat, A. J. Saunders, S. Santanakrishnan, R. A. Hutchinson, *Macromol. Chem. Phys.* **2010**, *211*(5), 580.
[23] S. Santanakrishnan, M. Stach, I. Lacik, R. A. Hutchinson, *Macromol. Chem. Phys.* **2012**, *213*(13), 1330.
[24] I. Lacik, S. Beuermann, M. Buback, *Macromol. Chem. Phys.* **2004**, *205*(8), 1080.
[25] I. Lacik, L. Ucnova, S. Kukuckova, M. Buback, P. Hesse, S. Beuermann, *Macromolecules* **2009**, *42*(20), 7753.
[26] C. Barner-Kowollik, G. T. Russell, *Prog. Polym. Sci.* **2009**, *34*(11), 1211.
[27] J. Barth, M. Buback, G. T. Russell, S. Smolne, *Macromol. Chem. Phys.* **2011**, *212*(13), 1366.
[28] S. Caballero, J. C. de la Cal, J. M. Asua, *Macromolecules* **2009**, *42*(6), 1913.
[29] S. C. Thickett, R. G. Gilbert, *Polymer* **2007**, *48*(24), 6965.
[30] H. Weiss, P. Deglmann, *Macromol. Symp.* **2011**, *302*(1), 6.
[31] Y. Li, S. Tang, B. C. Abberton, M. Kroger, C. Burkhart, B. Jiang, G. J. Papakonstantopoulos, M. Poldneff, W. K. Liu, *Polymer* **2012**, *53*(25), 5935.

[32] J. Jancar, J. F. Douglas, F. W. Starr, S. K. Kumar, P. Cassagnau, A. J. Lesser, S. S. Sternstein, M. J. Buehler, *Polymer* **2010**, *51*(15), 3321.
[33] M. L. Thompson, M. A. Kramer, *AIChE J.* **1994**, *40*(8), 1328.
[34] S. Curteanu, F. Leon, *Polymer-Plastics Technology and Engineering* **2006**, *45*(9), 1013.
[35] R. A. M. Noor, Z. Ahmad, M. M. Don, M. H. Uzir, *Canadian Journal of Chemical Engineering* **2010**, *88*(6), 1065.
[36] P. Cassagnau, V. Bounor-Legare, F. Fenouillot, *International Polymer Processing* **2007**, *22*(3), 218.
[37] C. Serra, G. Schlatter, N. Sary, F. Schonfeld, G. Hadziioannou, *Microfluidics and Nanofluidics* **2007**, *3*(4), 451.
[38] W. C. Yan, J. Li, Z. H. Luo, *Powder Technology* **2012**, *231*, 77.
[39] J. Pohn, M. Heniche, L. Fradette, M. Cunningham, T. McKenna, *Chemical Engineering & Technology* **2010**, *33*(11), 1917.

Modelling of Spray Polymerisation Processes

Winfried Säckel, Ulrich Nieken*

Summary: Spray polymerisation is a technique to carry out polymerisations in a highly efficient manner. As experimental optimisation of process parameters is tedious, simulations of the processes inside a single droplet are desirable. Today's models used in spray polymerisation assume ideal mixing and thus do not resolve the particle state along the radius. In this contribution we derive a single particle model for the case of free radical polymerisation, which spatially resolves the physical and chemical processes. The model allows for predicting inhomogeneity of the polymeric product as a function of drying conditions. The state of the polymer is calculated using the method of moments.

Keywords: drying; modelling; polymerisation; simulations; spray

Introduction

Spray drying processes are very widely used in industries. Reactive drying processes such as spray polymerisation are a further development of spray drying, with additional chemical reactions taking place within the spray droplets when they fall down inside the dryer. Several process steps may be integrated into one apparatus, which makes spray polymerisation an interesting option with respect to process intensification. However, an appropriate choice of process conditions is rather difficult, as a great number of physical and chemical processes takes place simultaneously. Moreover, literature on spray polymerisation is for the most part based on experiments, so that hardly any theoretical models have been published (compare Krüger's extensive literature and patent review).[1] A model based on the assumption of an ideally mixed droplet (like a batch reactor) was developed by Walag for the case of emulsion polymerisation.[2] Still, a model for polymerisation in solution, accounting for spatial inhomogeneities, is missing.

Institute of Chemical Process Engineering, University of Stuttgart, Böblinger Str. 78, 70199 Stuttgart, Germany
E-mail: winfried.saeckel@icvt.uni-stuttgart.de

Therefore, a new model for spray polymerisation will be derived, which resolves the processes of free radical polymerisation inside the droplets one-dimensionally in radial direction. This paper will focus on the simulation of single droplets, as polymer properties rely strongly on the processes taking place inside the droplets. Spray generation or coalescence in droplet ensembles will not be discussed.

Simulating Spray Polymerisation Based on Single Droplet Drying Models

In the last decades, a great number of drying models for single droplets have been proposed. An overview can be found in Mezhericher et al.[3] Typically, these models assume a spherically symmetric droplet shape and, consequently, spatial distributions of concentrations and temperature in radial direction. Very often, only a binary mixture is considered, so that solely the mass fraction of one component needs to be computed. Inside the droplet diffusive transport is taken into account. The boundary to the drying gas is modelled using appropriate correlations for the dimensionless Nusselt and Sherwood numbers. For the simulation of the second drying period, a porous crust around the liquid core is often taken into account as a quasi-homogeneous medium with transport properties fitted to experiments.

Regarding spray polymerisation, additional physical and chemical processes need to be taken into account. Contrary to spray drying a multi-component mixture needs to be considered. Chemical reactions change the mass fractions of the polymer and monomer components and, additionally, induce convective fluxes when the densities of educts and products are different. Moreover, the complex thermodynamics of polymeric mixtures has to be accounted for, when diffusion is modelled by the Maxwell-Stefan equations. An evaluation of the chain length distribution or its statistical averages is desirable, as the "chemical homogeneity" of the product is an important feature.

In the following the model equations are set up in radial direction r as indicated by Figure 1. Transport is calculated at inner points of the droplet and heat and mass transfer is considered at the droplet surface Γ at $r = R$. In principle, a formulation of model equations using concentrations would be straightforward for both the calculation of chemical reactions as well as the evaluation of the moments of the polymer. However, the transport of the individual components is coupled and, due to the drying process, solvent on the one hand and monomer and polymer on the other hand will diffuse to opposite directions. With the concentrations of solvent and polymer being different in several orders of magnitude a preliminary implementation of transport with a formulation based on concentrations showed large numerical errors. Hence, the transport equations for a species j is written using the partial densities ρ_j

$$\frac{\partial \rho_j}{\partial t} = -\frac{1}{r^2}\frac{\partial}{\partial r}\left(r^2 \rho_j v + r^2 j_j\right) + r_j^F\, MW_j, \tag{1}$$

with the subscript j being solvent (S), monomer (M) or polymer (P) and r the coordinate in radial direction. Diffusive fluxes j_j need to be calculated with an appropriate approach, such as a generalised Fick's law or the Maxwell-Stefan equations. MW_j is the molar weight and r_j^F the rate of formation of the species j. The convective velocity v is nonzero in case of varying density. Treating the liquid phase as an incompressible, ideal mixture the partial volumes V_j of the single components can be expressed by their partial densities and the densities of the pure substances ρ_j^0. With the sum of all volume fractions being unity one gets

$$\sum \frac{V_j}{V} = \sum \left(\frac{V \rho_j}{\rho_j^0}\right) \frac{1}{V} = \sum \frac{\rho_j}{\rho_j^0} = 1 \tag{2}$$

and

$$\frac{\partial}{\partial t} \sum \frac{\rho_j}{\rho_j^0} = \sum \frac{1}{\rho_j^0} \frac{\partial \rho_j}{\partial t} = 0. \tag{3}$$

This expression in combination with equation 1 can be implemented as an algebraic equation for the iteration of the velocity with the boundary condition at the droplet centre

$$v(r = 0) = 0. \tag{4}$$

The partial differential equations of the species is of second order, thus requiring two boundary conditions at the droplet

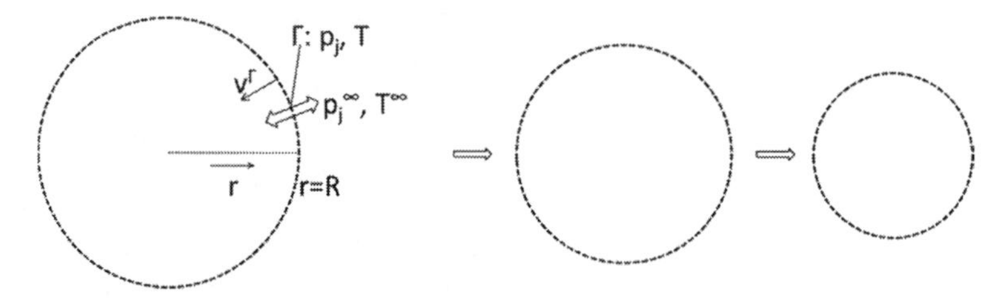

Figure 1.
Schematic drawing of the droplet evolution and nomenclature.

centre and the outer boundary Γ to the surrounding gas at the droplet surface $r = R$. The fluxes to the boundary Γ add up to zero

$$\left(v - v^{\Gamma}\right)\rho_j + j_j - \Omega_j = 0, \tag{5}$$

where v^{Γ} is the velocity of the droplet surface and Ω_j the evaporative flux of the component j. For non-volatile components Ω_j equals zero. Equations 2 and 5 provide $m+1$ conditions for the evaluation of m components and the velocity v^{Γ} of the receding droplet surface. Because of the droplet symmetry the boundary condition at the droplet centre is

$$\left.\frac{\partial \rho_j}{\partial r}\right|_{r=0} = 0. \tag{6}$$

The statistical averages of the chain length distribution can be calculated with the well-known method of moments. The moments λ^P, λ^D for the distributions of the living (index P) and dead (index D) chains are always formulated in concentrations. Therefore, the moments' transport needs to be coupled to the transport of the polymer component in an appropriate way. According to Bird, Stewart and Lightfoot[4] in a molar point of view diffusive fluxes J_j^* and the molar average velocity v^* are different from their counterpart formulated with respect to mass/densities. Therefore, the transport equation for the moments in a molar system is

$$\frac{\partial \lambda_k}{\partial t} = -\frac{1}{r^2}\frac{\partial}{\partial r}\left(r^2 \lambda_k v^* + r^2 J^*_{\lambda_k}\right) + r^F_{\lambda_k}. \tag{7}$$

The component velocity v_j is the sum of diffusion and convection and independent from the underlying reference system:

$$v_j = \frac{j_j}{\rho_j} + v = \frac{J^*_{j_1}}{c_j} + v^*. \tag{8}$$

The first moment λ_1 represents the total concentration of polymerised monomer. The relation to the partial density of the polymer reads

$$\lambda_1 = \lambda_1^P + \lambda_1^D = \frac{\rho_P}{MW_M}, \tag{9}$$

so that the component velocity of the polymer

$$v_P = \frac{j_P}{\rho_P} + v = \frac{J^*_{\lambda_1}}{\lambda_1} + v^*. \tag{10}$$

can be used to reformulate equation 7 for the first moment

$$\frac{\partial \lambda_1}{\partial t} = -\frac{1}{r^2}\frac{\partial}{\partial r}\left(r^2 \lambda_1 v + r^2 \frac{\lambda_1}{\rho_P} j_P\right) + r^F_{\lambda_1}. \tag{11}$$

Whereas the diffusive flux of the first moment can be obtained from equation 10, the fluxes of the other moments are still unknown. Despite the fact that diffusion of polymer chains depends on the chain length, implementing this effect into the present model would be rather complicated. For the diffusive fluxes of the moments it is therefore assumed that they are scaled according to their relation to the first moment

$$J^*_{\lambda_k^{P/D}} = \frac{\lambda_k^{P/D}}{\lambda_1} J^*_{\lambda_1}, \tag{12}$$

so that the transport equation of the moments can finally be formulated as

$$\frac{\partial \lambda_k^{P/D}}{\partial t} = -\frac{1}{r^2}\frac{\partial}{\partial r}\left(r^2 \lambda_k^{P/D} v + r^2 \frac{\lambda_k^{P/D}}{\rho_P} j_P\right) + r^F_{\lambda_k}. \tag{13}$$

The boundary condition at the droplet's centre is again

$$\left.\frac{\partial \lambda_k^{P/D}}{\partial r}\right|_{r=0} = 0. \tag{14}$$

Applying a boundary condition to the moments analogous to equation 5 is difficult. If equation 10 and 12 are used for the coupling to polymer transport, all moment boundary conditions will be linear combinations of equation 5 for the polymer. Thus, for $r = R$ instead of an algebraic boundary condition the moment transport is computed according to equation 13.

The change of droplet radius R can simply be computed via

$$\frac{\partial R}{\partial t} = v^{\Gamma}. \tag{15}$$

Inside the droplet heat conduction and the heat of reactions are considered, whereas the effect of convective heat transport is negligibly small:

$$\frac{\partial T}{\partial t} = -\frac{1}{\rho c_p}\frac{1}{r^2}\frac{\partial}{\partial r}\left(r^2 k \frac{\partial T}{\partial r}\right) - \sum_i r_i \Delta h_{R,i}. \tag{16}$$

T is the temperature, ρ the density, c_p the heat capacity and k the heat conduction of the mixture. In the reaction term r_i and $\Delta h_{R,i}$ are the rate of reaction i and the corresponding heat of reaction, respectively. Heat and mass fluxes at the droplet surface to the surrounding drying gas are calculated using linear driving forces with the heat and mass transfer coefficients α and β. These coefficients can be computed via appropriate correlations for the dimensionless Nusselt and Sherwood numbers Nu and Sh, such as the Ranz-Marshall equations.[5] At the droplet surface the boundary condition is

$$k \left.\frac{\partial T}{\partial r}\right|_{r=R} + \alpha(T(r=R) - T^{\infty}) + \sum_j \Omega_j \Delta h_{V,j} = 0 \tag{17}$$

with the superscript $^{\infty}$ denoting the surrounding drying gas and $\Delta h_{V,j}$ the heat of vaporisation of species j. For reasons of symmetry the following condition is used at the droplet centre:

$$\left.\frac{\partial T}{\partial r}\right|_{r=0} = 0. \tag{18}$$

The mass flux over the interface can be expressed via

$$\Omega_j = \frac{\beta MW_j}{\Re \cdot 0.5(T(r=R) + T^{\infty})}\left(p_j(r=R) - p_j^{\infty}\right). \tag{19}$$

$\Re$ is the universal gas constant. In the denominator the average temperature of the boundary layer is applied. The partial pressures p_j at the interface can be calculated applying Raoult's law on the saturation pressure and the Flory-Huggins theory for calculating the components' activities.

The Flory-Huggins theory can be used as well for calculating diffusive fluxes via the Maxwell-Stefan equations. Still, all binary and self-diffusion coefficients need to be known for a wide range of concentrations, so that it can be rather difficult to obtain a reliable set of parameters for a specific application. Ternary or multi-component Fickian approaches use a matrix of (often) concentration-dependent diffusion coefficients based on experiments, which are not identical with the data of binary mixtures.[6] Whereas the Maxwell-Stefan approach is the only correct formulation according to thermodynamics, for reasons of simplicity in the following simulations the (pseudo-) binary law of Fickian diffusion will be used

$$j_j = -\rho D \frac{\partial w_j}{\partial r}. \tag{20}$$

The diffusive flux of a component j is therefore depending on the gradient of the corresponding mass fraction w_j, the mixture density and the diffusion coefficient D. This law can only be applied to a multi-component mixture if diffusional transport is identical for all species. For an analysis of the model this simplification has the advantage that diffusive behaviour is characterised by one parameter. Still, equation 1 is formulated in a universal way, so that other approaches for diffusion can be chosen as appropriate.

Numerical Results

In the following the equations derived above will be applied to the polymerisation of N-vinylpyrrolidone in water to polyvinylpyrrolidone. Admittedly, PVP polymerisation is a rather slow process and therefore an unusual choice for spray processes, where the reaction time is very limited. On the other hand a longer time scale is preferable for experiments in ultrasound levitators, which will be undertaken in collaboration with other groups in the future in order to validate the model experimentally at well-defined process conditions. Moreover, the main focus of this contribution is to facilitate a theoretical framework, which allows for a spatially resolved calculation of the processes inside the droplet. In this respect the polymerisation of PVP is a legitimate choice. For the same reason, the simplifications stated below, which needed to be made due to a lack of data, are still acceptable. Kinetics of aqueous PVP polymerisation have recently been measured by Santanakrishnan et al.[7] The Flory-Huggins parameters for water/PVP interactions have been extrapolated from measurements of Karimi et al.[8] As thermodynamic data for NVP was not fully available, the Flory-Huggins parameterisation for NVP was chosen the same as for the water component. For the same reason monomer evaporation was not considered in the calculations, despite the fact that NVP is a volatile component. However, experiments of a NVP/water/PVP mixture in an ultrasound levitator indicate, that almost only water is evaporated as long as it is present in the droplet.[9] Furthermore, kinetic data was measured at a maximum monomer concentration of 25 w% NVP. Therefore the reaction mechanism is extrapolated at higher monomer contents, which may be prone to errors. Spray polymerisation is a very attractive concept, when high solvent contents are required in batch processes due to high polymer viscosity and cooling necessities, as these effects are of minor importance in a spray droplet. Still, in most cases available kinetic data is often limited to small monomer/polymer and high solvent concentrations.

The "natural" choice of process conditions for a spray polymerisation process lies in rather harsh drying conditions, so that the solvent evaporates quickly at the beginning of the process. Due to the low wet-bulb temperature during this drying period virtually no chemical reactions take place inside the droplet until the solvent is mostly evaporated. This drying period is then followed by a bulk polymerisation in absence of a solvent. In this case spatial inhomogeneities may result from temperature gradients inside the droplets and different monomer contents due to monomer evaporation. If, for any reason, polymerisation shall be operated in presence of the solvent, mild drying conditions with a well-chosen solvent partial pressure in the drying gas need to be applied. Furthermore, an inhibition of the drying effect might occur if polymer is already present in the droplet, so that a film of polymer with low diffusivity can occur at the droplet surface. These kinds of process operation are considerably more difficult to apply and shall be used for example calculations with the new polymerisation model.

Simulations have been carried out with a Matlab implementation of the model using the method of lines[10] in order to discretise the droplet in radial direction and Matlab's ODE suite for time integration. The results will be compared with a 0D batch reactor model of the droplet which assumes ideal mixing in the droplet and thus neglects radial gradients.

Figure 2 shows results for a droplet of 50 μm initial size, with an initial content of 25 wt% of monomer and 0.01 wt% of initiator V-50 at drying gas conditions of 80 °C and 0% relative humidity. The diffusion coefficient was set to 10^{-9} m^2/s. The calculation shows an almost uniform distribution of values in radial direction. This is not surprising with respect to the long process time of 10^3 s, where diffusion is fast enough to smooth all radial gradients. However, the one dimensional model meets the results of the 0D calculation. While the solvent content decreases very fast, the monomer concentration increases and virtually no chemical reactions take place due to the low droplet temperature. After the evaporation period the polymerisation reactions start. Because of transfer and branching reactions the length average of the polymer decreases while the mass average increases, which is in consistency with the kinetic measurements of Santanakrishnan et al.[7] The droplet radius and temperature are plotted in Figure 3 as a function of time for the first

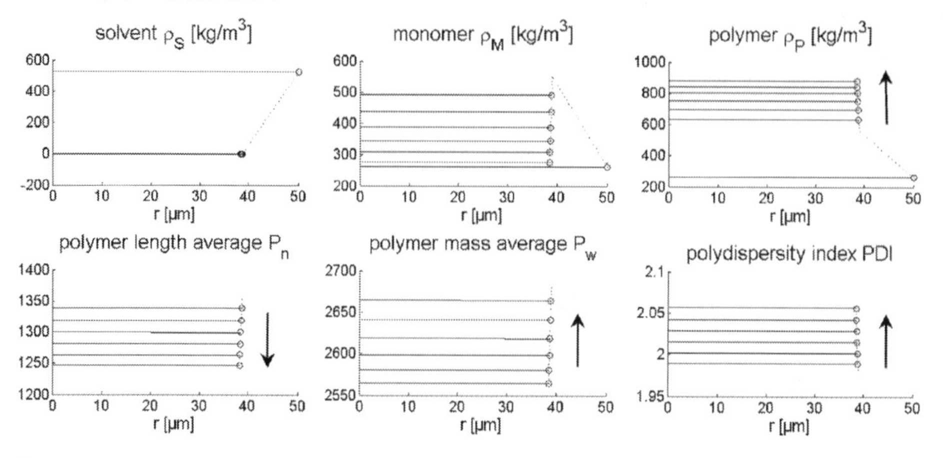

Figure 2.
Radial profiles of free radical polymerisation of a 50 µm NVP/water droplet at 80 °C and 0% rel. humidity. The diffusion coefficient was set to 10^{-9} m^2/s. Results were calculated at times 0, 170, 330, 500, 670, 830 and 1000s (time 0 s not printed for the polymerisation results). Dashed lines with circles mark the corresponding results assuming an ideally mixed droplet.

10 s of the process. Both models provide virtually the same results. The initial drying period is short, as the droplet is rather small. It can also be seen, that the temperature control of the spray polymerisation process is excellent. The droplet temperature stays constantly at the drying gas temperature as soon as the drying period is finished. One major advantage of reactive spray processes consists in the large ratio of surface (cooling) area to reaction volume. Therefore, the heat of reaction can be removed by the surrounding gas very efficiently.

If the diffusion coefficient is considerably smaller, the drying process will slow down due to a decrease of solvent content at the droplet surface. Figure 4 shows results where the diffusion coefficient was set to 10^{-11} m^2/s. In order to limit the process time to realistic values, the reaction rate of chain propagation was multiplied by a factor of 100. As the process is now taking place within about 10 s, it can clearly be seen that large gradients of all species occur in radial direction, whereas the assumption of ideal mixing leads to considerable deviations. The drying process is inhibited, leaving a large amount of solvent in the droplet core region. For obvious reasons a large fraction of remaining solvent inside the droplet can be rather problematic in

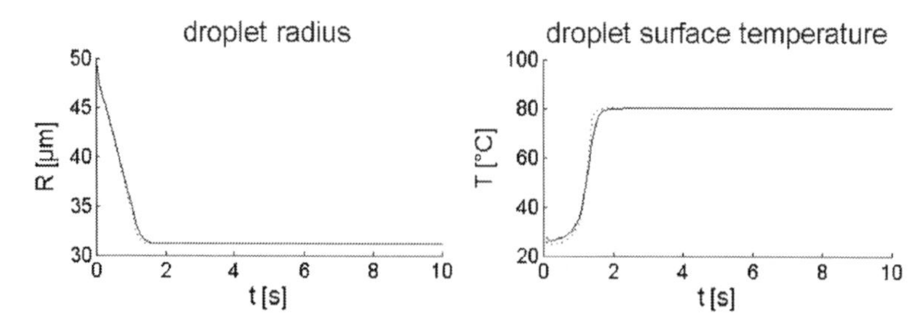

Figure 3.
Droplet radius and surface temperature as a function of time, conditions the same as in Figure 2. Dashed lines indicate results assuming an ideally mixed droplet.

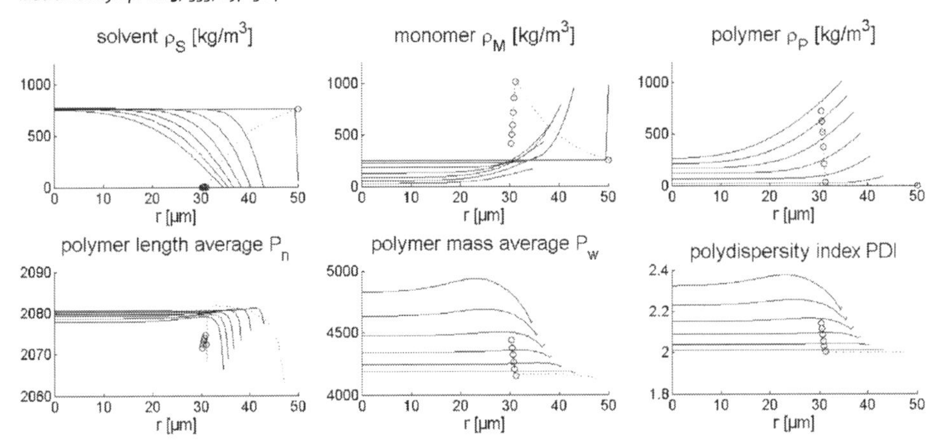

Figure 4.
Radial profiles of free radical polymerisation of a 50 μm NVP/water droplet. In comparison to the results of Figure 2 propagation reactions were faster by a factor of 100 and the diffusion coefficient was set to 10^{-11} m^2/s. Results were calculated at times 0, 1.7, 3.3, 5, 6.7, 8.3 and 10 s (time 0 s not printed for the polymerisation results). Dashed lines with circles mark the corresponding results assuming an ideally mixed droplet.

practice. The resulting polymer shows significant radial inhomogeneity regarding average molecular mass and polydispersity. Moreover, the results of the 0D model deviate strongly from the values in the inner part of the droplet. The deficiencies of the 0D approximation can also be observed at the values of droplet radius and temperature drawn in Figure 5.

Conclusion

The presented single droplet polymerisation model allows for a spatially resolved simulation of the physical and chemical processes, which take place inside the droplet simultaneously. Radial inhomogeneity can be studied for the species in the mixture as well as for the polymer formation. Model equations have been set up in a general way, so that different approaches for diffusion can be used. A formulation of model equations with respect to mass is straightforward from the viewpoint of drying and was the best choice in first numerical studies. However, errors in the calculation of moments may arise as a result of the coupling of the (molar) moment transport to the polymer component, the assumptions, which were made in order to obtain diffusive fluxes for all

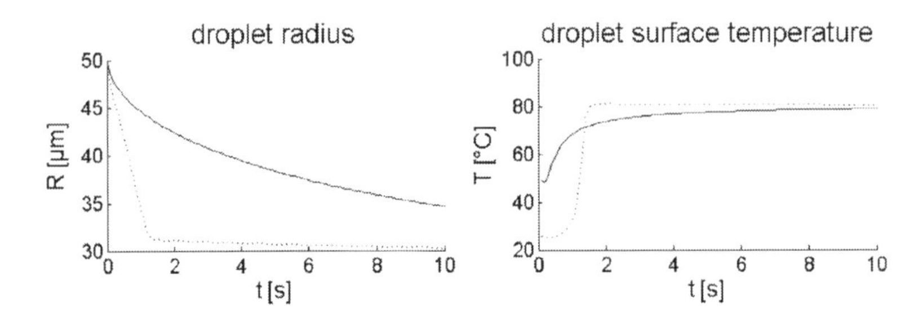

Figure 5.
Droplet radius and surface temperature as a function of time, conditions the same as in Figure 4. Dashed lines indicate results assuming an ideally mixed droplet.

moments, and the lack of an appropriate boundary condition for the calculation of moments at the droplet surface. If a formulation of the model equations in concentrations could be further developed, these shortcomings might be overcome. Still, the present model provides a significant advancement with respect to 0D models and can be used to derive sensible process parameters for spray polymerisation, when spatial inhomogeneities need to be considered.

A well-chosen parameterisation of the kinetics as well as of the thermodynamics is mandatory and may require future investigations regarding desired polymerisation reactions. Online experimental observation of a spray polymerisation process is very difficult, so that a validation of the presented model has not been accomplished yet. In the future, single droplet experiments in an ultrasound levitator shall be used in order to validate the proposed model.

[1] M. Krüger, "*Sprühpolymerisation: Aufbau und Untersuchung von Modellverfahren zur kontinuierlichen Gleichstrom-Sprühpolymerisation*", Wiss.-und Technik-Verl., Berlin **2004**, p. 23ff.
[2] K. Walag, "*Sprühpolymerisation: Modellierung der Mikropolymerisation im Turmreaktor*", Wiss.-und-Technik-Verl., Berlin **2011**.
[3] M. Mezhericher, A. Levy, I. Borde, *Drying Technol.* **2007**, *25*, 1025.
[4] R. B. Bird, W. E. Stewart, E. N. Lightfoot, "*Transport phenomena*", J. Wiley, New York, **2007**.
[5] W. E. Ranz, W. R. Marshall, *Chem. Eng. Prog.* **1952**, *48*, 141–146 & 173–180.
[6] R. Taylor, *Ind. Eng. Chem. Fundam.* **1982**, *21*, 407.
[7] S. Santanakrishnan, L. Tang, R. A. Hutchinson, M. Stach, I. Lacík, J. Schrooten, P. Hesse, M. Buback, *Macromol. React. Eng.* **2010**, *4*, 499.
[8] M. Karimi, W. Albrecht, M. Heuchel, T. Weigel, A. Lendlein, *Polymer* **2008**, *49*, 2587.
[9] J. Laackmann, S. Ahmed, R. Sedelmayer, M. Klaiber, W. Pauer, S. Simon, H.-U. Moritz, in: "*Proc. 12th Triennial International Conference on Liquid Atomization and Spray Systems*", E., Gutheil, C. Tropea, Eds., Heidelberg, **2012**, 1308b.
[10] W. E. Schiesser, "*The numerical method of lines: Integration of partial differential equations*", Academic Press, San Diego **1991**.

Numerical Simulation of Reactive Extrusion for Polycondensation of Poly(*p*-phenylene terephthalamide) (PPTA)

*Hao Tang, Yuan Zong, Zhenhao Xi, Ling Zhao**

Summary: Poly(*p*-phenylene terephthalamide) (PPTA), as a typical high performance fiber with excellent strength and modulus property, is usually synthesized by low temperature solution polycondensation with p-phenylenediamine(PPD) and terephthaloyl chloride(TPC) in twin screw extruders. Since the polycondensation during the process is a multicomponent reaction with high reactivity and wide viscosity range, the molecular weight and its distribution of PPTA are strongly influenced by mixing behavior of the extruder. In this paper, the polycondensation process in fully-filled conveying elements of twin screw extruder were analyzed by three-dimensional numerical simulation using the finite element method. The effects of conveying element geometry, rotating speed on mixing state were studied. In order to understand the mechanism behind the reactive extrusion process of PPTA, reaction extent and instantaneous efficiency of stretch in extruder were calculated simultaneously. The results indicate that the reaction process and mixing behavior of the PPTA flow are strongly interrelated. The increase of rotating speed promotes the reaction extent by enhancing the stretch effect, while larger screw pitch accelerates the reaction by rearranges the proportion of energy for stretch.

Keywords: mixing; numerical simulation; polycondensation; reaction extent; reactive extrusion

Introduction

Reactive extrusion (REX) is a continuous process for polymerization. This is an efficient method that combines polymer reaction engineering and polymer extrusion into a screw extruder. The main reason for suitability of twin screw extruders for these applications is the complexity of the flow in the intermeshing region, which provides them good mixing and compounding characteristics. However, the complexities of the flow make it difficult to predict the performance of twin-screw extruders.[1] *Rauwendaal*[2] proposed that theoretical analysis of twin screw extruders should include mixing characteristics and flow behavior of the intermeshing region. This statement was widely accepted and intensive 3D numerical simulations have been employed for analysis.[3] *Avalosse*[4] analyzed the pressure profiles and temperature fields in twin screw extruders and compared different extruders from a mixing point. *Barrera*[5] simulated the velocity and pressure fields in co-rotating twin screw extruder with Newtonian fluid and non-Newtonian fluid. For reactive extrusion process, numerical simulation has also been extensively adopted. *Zhu*[6] analyzed the effects of screw rotating speed, screw pitch and initial conversion on radical polymerization of ε-caprolactone. Both mixing mechanisms and competition of energy sources were investigated. *Rodríguez*[7] numerically investigated the effect of initial peroxide concentration and mass throughput on the final M_W and PDI of the degraded resin of

[1] East China University of Science and Technology, Shanghai, China
E-mail: zhaoling@ecust.edu.cn

polypropylene. Due to the diffusion limitation of multicomponent reaction, distributive mixing in polycondensation is indispensable because the reaction only takes place at the interface between the components.[8] However, distributive mixing has not been intensively discussed in the research of reactive extrusion.

In order to understand the relationship between reaction and mixing, 3D numerical simulation was employed in this paper to illustrate the interaction between the flow field of the extruder and the polycondensation reaction of PPTA. And the effect of screw pitch and rotating speed on mixing state were analyzed. The results could offer an optimization direction for the reactive extrusion for PPTA.

Numerical Simulation

Reaction Kinetics

PPTA is synthesized by low-temperature solution polycondensation of p-phenylene diamine(PPD) and terephthaloyl chloride-(TPC). The reaction of PPTA is presented as below:

$$n\,Cl-C(=O)-C_6H_4-C(=O)-Cl + n\,NH_2-C_6H_4-NH_2 \longrightarrow \left[-C(=O)-C_6H_4-C(=O)-NH-C_6H_4-NH-\right]_n + 2n\,HCl$$

The activation energy and enthalpy of reaction is 6.3 kJ/mol and −199 kJ/mol, respectively. According to kinetics studies by Kuz'min et al,[9] the polycondensation of PPTA does not obey Flory principle (equal reactivity of the functional groups). Instead, this process can be concluded into four stages by an overall scheme:

$$\begin{aligned}
A:\ & R_1 - Ar - NH_2 + ClOC - Ar \\
& - R_1 \xrightarrow{k_1} \ldots\Phi - R_2 + HCl \\
B:\ & R_1 - Ar - NH_2 + ClOC - Ar \\
& - R_2 \xrightarrow{k_2} \ldots\Phi - R_2 + HCl \\
C:\ & R_2 - Ar - NH_2 + ClOC - Ar \\
& - R_1 \xrightarrow{k_3} \ldots\Phi - R_2 + HCl \\
D:\ & R_2 - Ar - NH_2 + ClOC - Ar \\
& - R_2 \xrightarrow{k_4} \ldots\Phi\ldots + HCl
\end{aligned}$$

where $k_1, k_2, k_3$and k_4 are the rate constants of the corresponding stages; R_1 is $-NH_2$ or $-COCl$; R_2 is $-NH - CO - Ar - R_1$; Φ is a fragment of polymer chain.

Based on kinetics theory, stage (D) has the lowest pre-exponential factor and it determines the whole reaction process. Therefore, only stage (D) has been taken into consideration. In extrusion, heat removal by liquid nitrogen[10] or n-pentane evaporation[11] is used to keep a low reaction temperature and the flow process is treated as an isothermal one.

In order to quantify the reaction progress, reaction extent p is defined as follow (assuming $C_{NH_2} = C_{COCl} = C_A$):

$$p = \frac{C_{A0} - C_A}{C_{A0}} \tag{1}$$

where, C_{A0} is the concentration of reactant present initially and C_A is the concentration of reactant after a time of t of polycondensation.

The polymerization can be described by:

$$\begin{aligned}
\frac{dC_{[\mathrm{NH_2}]}}{dt} = \frac{dC_{[COCl]}}{dt} &= kC_{[\mathrm{NH_2}]}C_{[COCl]} \\
&= kC^2_{[NH_2]} = kC^2_{[COCl]}
\end{aligned} \tag{2}$$

Then the molecular weight can be estimated:

$$M_n = M_0 \times x_n + M_{pp} \tag{3}$$

where, $x_n = \frac{1}{1-p}$, $M_0$is the molecular weight of the repeat unit, which is almost equal to 238, and M_{pp} is the initial molecular weight after perpolycondensation.

Rheology Model

The viscosity of the reaction system was measured by ARES (Advanced Rheology Expanded System, TA company, USA). Figure 1. shows a typical result of steady rate sweep test.

The viscosity of the reaction system can be modeled with the Yasuda-Carreau equation:

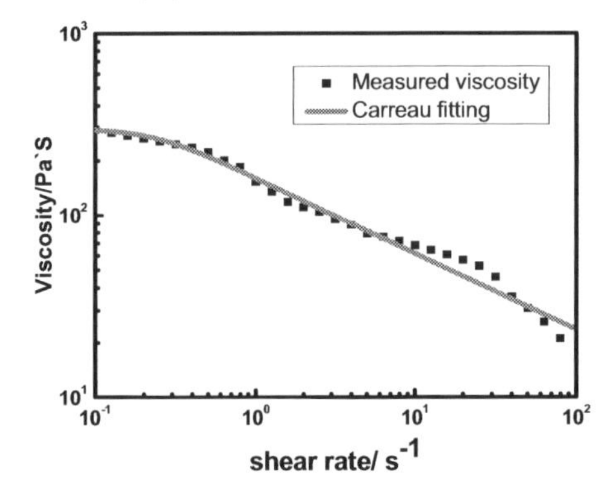

Figure 1.
The Carreau fitting of measured viscosity against shear strain rate. ($\overline{M_n} = 16104$, measured by GPC).

$$\eta(\dot{\gamma}) = \frac{\eta_0}{[1 + (\tau\dot{\gamma})^a]^{(1-n)/a}} \quad (4)$$

where n is power index, $\tau$is natural time and a represents the width of the transition region between η_0 and the power-law region. From the figure, it is clear that the application of Yasuda-Carreau equation could obtain good fitting results and n, a are equal to 0.58, 2.65, respectively.

In the simulation, only low reaction degree was considered, which represents the mass weighted molecular weight was lower than the critical enganlement molecular weight($M_c = 20000g/mol$) for PPTA system. Hence, Eq(5) was adopted to correlate the relationship between viscosity and the reaction:

$$\eta_0 = K_1 M_n \quad (M_n \leq M_c) \quad (5)$$

The above relationships were realized by user defined functions coupling with other procedures in POLYFLOW.

Simulation Details

The geometric specifications of conveying elements are listed in Table 1. 3D finite element mesh (Figure 2) was constructed covering the whole space. POLYFLOW, a commercial computational fluid dynamics (CFD) software, was adopted to solve the problem in extruder.

The conservation forms of governing equations for the simulations are as follows:

Continuity equation:

$$\frac{\partial \rho}{\partial t} + \nabla \cdot (\rho U) = 0 \quad (6)$$

Momentum conservation equation:

$$\frac{\partial(\rho u)}{\partial t} + \nabla \cdot (\rho u U) = -\frac{\partial p}{\partial x} + \mathrm{div}(\eta\, \mathbf{grad} u) + S_u + \rho f_x \quad (7)$$

Table 1.
Geometry of the extruder.

Barrel diameter	42.0 mm	Screw root diameter	27.4 mm
Screw tip diameter	40.0 mm	Screw pitch	40, 50 or 60 mm
Centerline distance	33.9 mm	Flow rate	3000 mm³/s

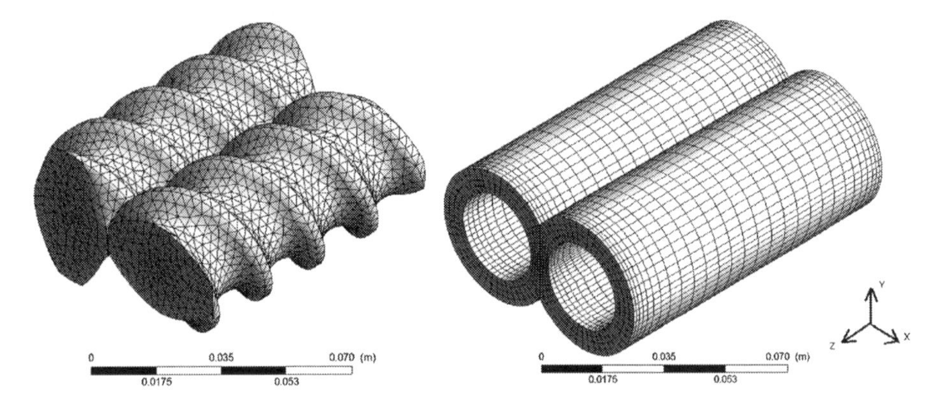

Figure 2.
Geometry sketch of flow domain.

$$\frac{\partial(\rho v)}{\partial t} + \nabla \cdot (\rho v U) = -\frac{\partial p}{\partial y} + \mathrm{div}(\eta \mathbf{grad} v) + S_v + \rho f_y \quad (8)$$

$$\frac{\partial(\rho w)}{\partial t} + \nabla \cdot (\rho w U) = -\frac{\partial p}{\partial z} + \mathrm{div}(\eta \mathbf{grad} u) + S_w + \rho f_z \quad (9)$$

where f_x, f_y, f_z are the body force per unit mass of x, y, z components. In the simulation, they refer to gravitational forces, which are negligible because of high viscosity fluid,[12] i.e., $f_x = f_z = 0, f_y = -g \approx 0$; S_u, S_v, S_w are the generalized source terms of momentum conservation equation, which can be expressed as:

$$S_u = \frac{\partial}{\partial x}(\eta \frac{\partial u}{\partial x}) + \frac{\partial}{\partial y}\left(\eta \frac{\partial v}{\partial x}\right) + \frac{\partial}{\partial z}\left(\eta \frac{\partial w}{\partial x}\right) + \frac{\partial}{\partial x}(\lambda \mathrm{div} U) \quad (10)$$

$$S_v = \frac{\partial}{\partial x}\left(\eta \frac{\partial u}{\partial y}\right) + \frac{\partial}{\partial y}\left(\eta \frac{\partial v}{\partial y}\right) + \frac{\partial}{\partial z}\left(\eta \frac{\partial w}{\partial y}\right) + \frac{\partial}{\partial y}(\lambda \mathrm{div} U) \quad (11)$$

$$S_w = \frac{\partial}{\partial x}\left(\eta \frac{\partial u}{\partial z}\right) + \frac{\partial}{\partial y}\left(\eta \frac{\partial v}{\partial z}\right) + \frac{\partial}{\partial z}\left(\eta \frac{\partial w}{\partial z}\right) + \frac{\partial}{\partial z}(\lambda \mathrm{div} U) \quad (12)$$

where, η is the molecular viscosity coefficient and λ is the second viscosity coefficient.

In reactive extrusion, the viscosity depends not only on the shear rate, but also on the molecular weight, seen in Eq(4)~(5), the molecular weight can be estimated by Eq(1)~(3). The concentrations of species are crucial factors, which can be calculated by laminar diffusion equations:

$$u_x \frac{\partial C_A}{\partial x} + u_y \frac{\partial C_A}{\partial y} + u_z \frac{\partial C_A}{\partial z} = D_{AB}\left(\frac{\partial^2 C_A}{\partial x^2} + \frac{\partial^2 C_A}{\partial y^2} + \frac{\partial^2 C_A}{\partial z^2}\right) + R_A \quad (13)$$

for reactants, $R_A = -(-\frac{dC_A}{dt}) = -kC_{[COCl]}C_{[NH_2]} = -kC_A^2$, while for productions, $R_A = -\frac{dC_A}{dt} = kC_{[COCl]}C_{[NH_2]} = kC_A^2$. Therefore, there exist 3 species transport equations and 1 closure equation for –NH_2, –COCl, –NH–CO– and HCl, respectively. Including the reaction equation, 5 kinetics equations are incorporated with flow equation in POLYFLOW.

Figure 3 shows the simulation route of reactive extrusion for polycondensation of PPTA. Firstly, the kinetics was reformed into a laminar diffusion equation, i.e., Eq. (13) and rheological model was established, including the effect of molecular weight and shear rate, i.e., Eq(4)~(5); then rotating speed and screw pitch were imported as independent variables and simulation began. Governing equations, geometrical and operating parameters determined the flow behavior in extruder, with laminar diffusion equation, the mixing state was revealed and species distributions were obtained. The viscosity of reaction system, which was an important parameter

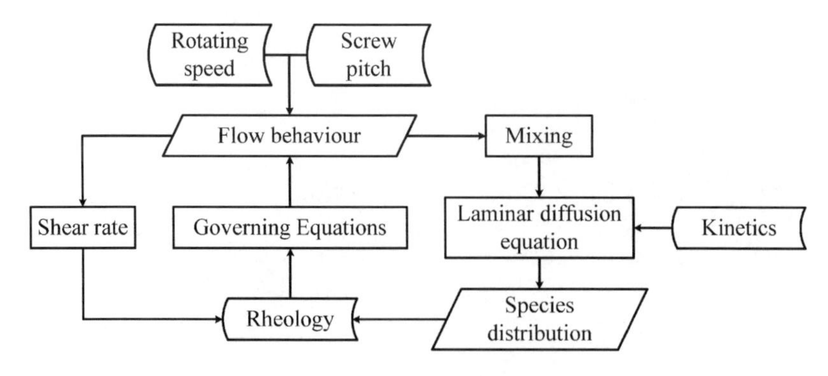

Figure 3.
The interaction of reactive extrusion for polycondensation of PPTA.

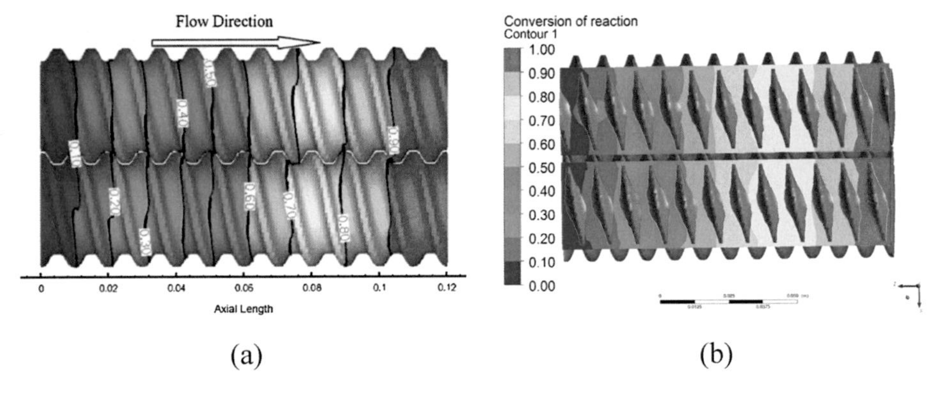

Figure 4.
Conversion profiles at z-plane. (a) simulated by FLUENT; (b) simulated by POLYFLOW.

for the governing equations, could be estimated according to Eq(1)~(5). The iterations for all parameters were continued until the residuals were less than 0.001.

In the simulation, solid walls were set as no-slip condition. At the inlet, volumetric flow rate was implemented and zero normal and tangential forces were specified at the outlet. For reacting species, concentrations were defined as a fixed value at the entrance of the flow domain. Rotating speed of 30, 60, 90 rpm were investigated since the rotating speeds in polycondensation of PPTA always lower than general extrusion. The pre-exponential factor, k, was iterated with EVOL feature and the viscosity, η, was iterated with PMAT feature in order to achieve convergence of the equations.

Validation

A comparison of simulation between FLUENT[6] and POLYFLOW was processed under the same condition. Figure 4 and 5 compare the conversion profiles and area weighted average conversions for both cases. Obviously, the results from both cases are consistent, which confirmed the route we used is feasible and correct.

Parameters Definition

Improving spatial homogeneity of the reactants is crucial for multicomponent reactions because they are diffusion limited. To evaluate the reaction extent and distributive mixing state precisely, average reaction extent $\overline{p}$, variance of reaction extent $\bar{\delta}_p2$ and instantaneous efficiency of stretching e_λ were defined:

$$\overline{p} = \frac{1}{n}\sum_{i=1}^{n} p \tag{14}$$

$$\bar{\delta}_p^2 = \frac{1}{n}\sum_{i=1}^{n} \frac{(p-\bar{p})^2}{\bar{p}^2} \tag{15}$$

$$e_\lambda = \frac{\dot{\lambda}/\lambda}{D} = \frac{rate\_of\_stretching}{rate\_of\_dissipation} \tag{16}$$

Results and Discussions

Effect of Rotating Speed

Figure 6 shows the profiles of reaction extent along flow direction with different

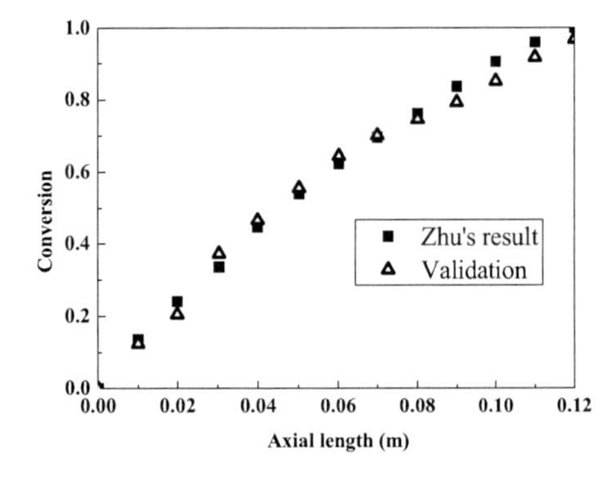

Figure 5.
Comparison of area weighted average conversion along the axial length of screw elements.

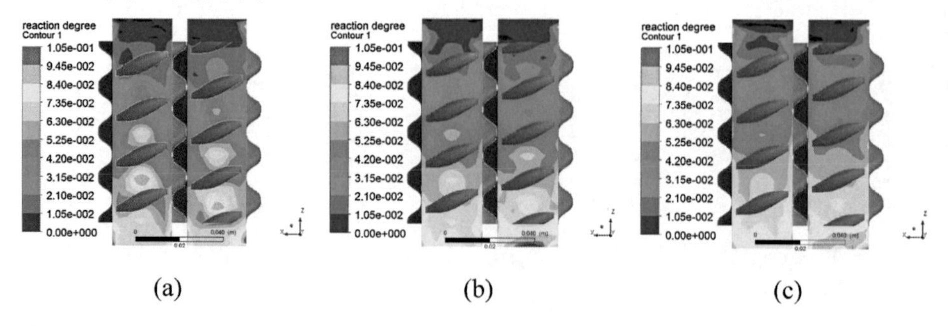

Figure 6.
Distribution of reaction extent at plane y = 16 mm with various rotating speeds. (a): 30 rpm; (b): 60 rpm; (c): 90 rpm.

rotating speed. Because intermeshing twin screw extruder is crosswise closed, there is no material transfer between adjacent channels and the distribution of reaction extent is separated by screw flight. Since the velocities between the screw flights are lower than those near screw flights, local residence time is longer at the center of channel. Therefore, reaction extent at the center of channel is higher than that in the remaining regions.

To evaluate the reaction extent precisely, reaction extent and its variance near the outlet(at surface Z = 10 mm) are shown in Table 2. It can be found that increasing rotating speed accelerates the reaction while the variance of reaction extent decreases dramatically. These results may due to the increase of interfacial area and backflows in screw channels generated by the improved distributive mixing at higher rotating speed.

To verify the assumption above, distributive mixing was evaluated by two parameters, i.e., logarithm of stretching and instantaneous efficiency of stretch. In Figure 7, the probability function of logarithm of stretching (ln eta) is observed. It could be seen that with increasing ln eta, the probability initially does not show significant variation. After ln eta = 0, a sharp increase is observed for the probability which reaches a saturation value after ln eta reaches a special value. Besides, the probability curves shift to a higher level with increasing rotating speed, indicating that the stretching effect is enhanced.

Additionally, the proportion of energy for stretch can be evaluated by the instantaneous efficiency of stretching. When $e_\lambda > 1$, the energy is mainly utilized for stretching, which facilitates reactants mixing; $e_\lambda < 1$ referred to compress energy. The three curves show little difference in Figure 8, indicate that rotating speed does not have significant effect on the instantaneous efficiency of stretching.

Table 2.
Average reaction extent and its variance at plane Z = 10 mm.

Rotating speed	30 rpm	60 rpm	90 rpm
$\bar{p}/10^{-1}$	0.7293	0.7568	0.7706
$\delta_p^2/10^{-3}$	51.367	6.928	2.503

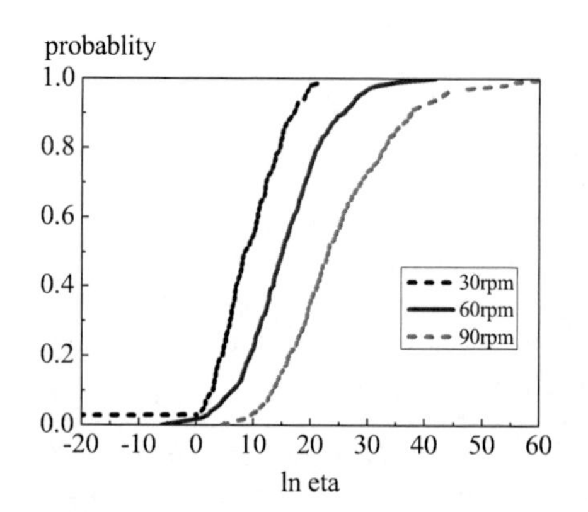

Figure 7.
Probability function of logarithm of stretching at plane Z = 10 mm.

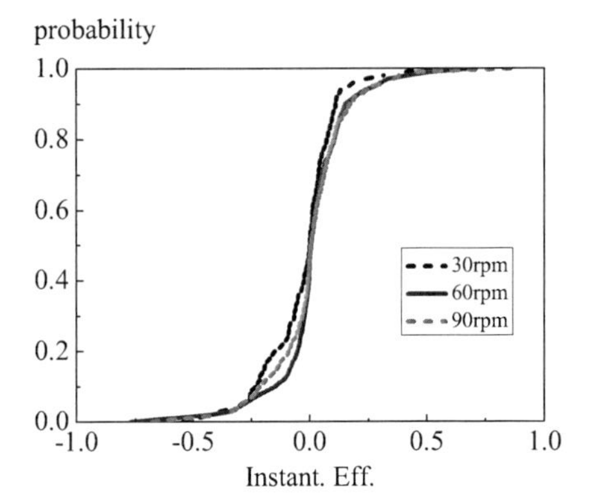

Figure 8.
Probability function of instantaneous efficiency of stretch at plane Z = 10 mm.

Table 3.
Average reaction extent and its variance at plane Z = 10 mm.

Screw pitch	40 mm	50 mm	60 mm
$\bar{p}/10^{-1}$	0.7293	0.7984	0.8230
$\bar{\delta}_p^2/10^{-3}$	51.367	18.472	8.691

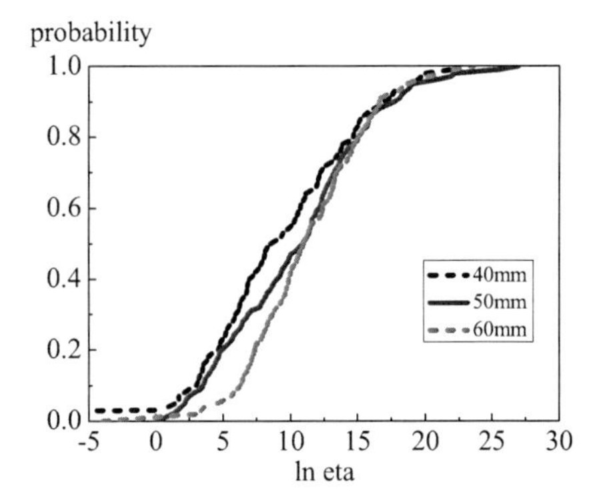

Figure 10.
Probability function of logarithm of stretching at plane Z = 10 mm.

Effect of Screw Pitch

The reaction extent distributions with different screw pitches are shown in Figure 9. As seen from the figure, higher reaction extent can be obtained with larger screw pitch. The difference is believed to derive from the increasing volume with longer local residence time.

Average reaction extent and its variance are shown in Table 3. Larger screw pitch improves backflows and materials transportation in extruder by offering broader screw channel. Therefore, higher reaction extent and narrower reaction extent distribution are obtained.

The curves in Figure 10 show the same trends as that in Figure 7. Besides, the probability curves shift to a higher level with increasing screw pitch, indicating that the stretching effect is enhanced. Consequently, it can be concluded that large screw pitch could improve stretching significantly. Further, instantaneous efficiency of stretch is shown in Figure 11. The curves also shift to higher efficiency with increasing screw pitch, that means the increase proportion of energy for stretch. Both figures prove that a better distributive mixing is obtained.

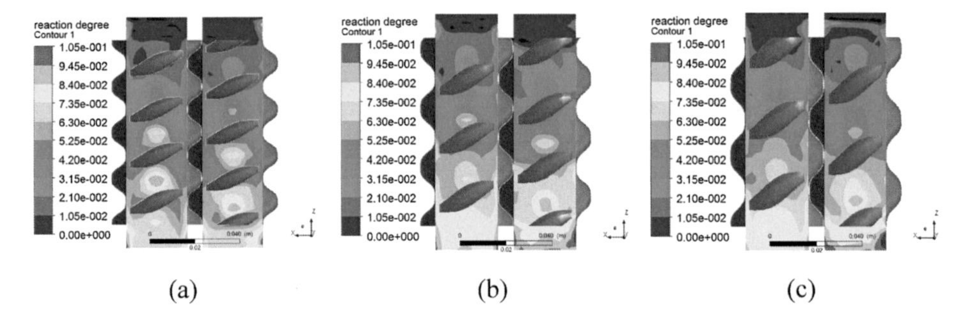

Figure 9.
Distribution of reaction extent at plane y = 16 mm with various screw pitches. (a): 40 mm; (b): 50 mm; (c): 60 mm.

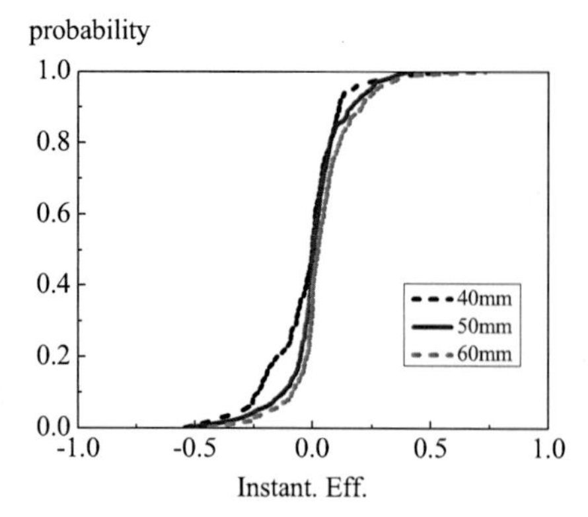

Figure 11.
Probability function of instantaneous efficiency of stretch at plane Z = 10 mm.

Conclusions

In order to investigate the relationship between mixing behavior of the flow field in the extruder and the polycondensation reaction of PPTA, numerical simulation was conducted to analyze the effect of rotating speed and screw pitch. The results show that increase of rotating speed promotes the reaction extent by enhancing the stretching flow, which offers a larger interface between reactants. Moreover, larger screw pitch is found to accelerates the reaction by improving instantaneous efficiency of stretching, which rearranges more proportion of energy for stretching. Therefore, it can be concluded that reaction and distributive mixing have a close relationship, and the reaction process in the twin screw extruder could be controlled by geometrical and operational parameters optimization.

This article was published online in on 29 November, 2013. Several errors in the first publication were identified and corrected for the online version of the manuscript. The Editorial office apologizes for any inconvenience caused.

[1] A. Shah, M. Gupta, *ANTEC. Conf. Process.* (Conference paper). **2004**, 443.
[2] C. Rauwendaal, *Adv. Polym. Tech.* **1996**, *15*, 127.
[3] V. L. Bravo, A. N. Hrymak, J. D. Wright, *Polym. Eng. Sci.* **2000**, *40*, 525.
[4] T. Avalosse, Y. Rubin, L. Fondin, *J. Reinf. Plast. Comp.* **2002**, *21*, 419.
[5] M. Barrera, J. Vega, J. Martinezsalazar, *J. Mater. Process. Tech.* **2008**, *197*, 221.
[6] L. Zhu, K. A. Narh, K. S. Hyun, *Int. J Heat Mass Tran.* **2005**, *48*, 3411.
[7] E. O. Rodríguez, "*Numerical Simulations of Reactive Extrusion in Twin Screw Extruders*", (Thesis). University of Waterloo, **2009**.
[8] K. J. Ganzeveld, L. P. B. M. Janssen, *Polym. Eng. Sci.* **1992**, *32*, 457.
[9] N. Kuz'min, B. Zhizdyuk, A. Chegolya, *Fibre Chem.* **1986**, *18*, 13.
[10] T. Zhang, G. H. Luo, F. Wei, W. Z. Qian, X. L. Tuo, *2011 Int. Symp. Chem. Eng. Mater. Proper.* (Conference paper). **2011**, 56.
[11] T. Zhang, G. H. Luo, F. Wei, Y. Y. Lu, W. Z. Qian, X. L. Tuo, *Chinese Chem. Lett.* **2011**, *22*, 1379.
[12] T. Kajiwara, Y. Nagashima, Y. Nakano, K. Funatsu, *Polym. Eng. Sci.* **1996**, *36*, 2142.